Edited by
Igor Agranovski

Aerosols – Science and Technology

Related Titles

Hirscher, M. (ed.)

Handbook of Hydrogen Storage

New Materials for Future Energy Storage

2010

ISBN: 978-3-527-32273-2

Salthammer, T., Uhde, E. (eds.)

Organic Indoor Air Pollutants

Occurrence, Measurement, Evaluation

Second, Completely Revised Edition

2009

ISBN: 978-3-527-31267-2

Wicks, G., Simon, J. (eds.)

Materials Innovations in an Emerging Hydrogen Economy

Ceramic Transactions, Volume 202

2009

ISBN: 978-0-470-40836-0

Press, R. J., Santhanam, K. S. V., Miri, M.J., Bailey, A. V., Takacs, G. A.

Introduction to Hydrogen Technology

2009

ISBN: 978-0-471-77985-8

Züttel, A., Borgschulte, A., Schlapbach, L. (eds.)

Hydrogen as a Future Energy Carrier

2008

ISBN: 978-3-527-30817-0

Vincent, J. H.

Aerosol Sampling

Science, Standards, Instrumentation and Applications

2007

ISBN: 978-0-470-02725-7

Klöpffer, W., Wagner, B. O.

Atmospheric Degradation of Organic Substances

Data for Persistence and Long-range Transport Potential

2007

ISBN: 978-3-527-31606-9

Seinfeld, J. H., Pandis, S. N.

Atmospheric Chemistry and Physics

From Air Pollution to Climate Change

2006

ISBN: 978-0-471-72018-8

Edited by Igor Agranovski

Aerosols – Science and Technology

WILEY-VCH Verlag GmbH & Co. KGaA

The Editor

Prof. Dr. Igor Agranovski
Griffith University
Griffith School of Engineering
170, Kessels Road, Nathan Cam.
Brisbane, Queensland 4111
Australia

Library of Congress Card No.: applied for

British Library Cataloguing-in-Publication Data
A catalogue record for this book is available from the British Library.

Bibliographic information published by the Deutsche Nationalbibliothek
The Deutsche Nationalbibliothek lists this publication in the Deutsche Nationalbibliografie; detailed bibliographic data are available on the Internet at <http://dnb.d-nb.de>.

Composition Laserwords Private Ltd., Chennai, India
Printing and Bookbinding betz-druck GmbH, Darmstadt
Cover Design Formgeber, Eppelheim

Printed in the Federal Republic of Germany
Printed on acid-free paper

ISBN: 978-3-527-32660-0

Contents

Aerosols – Science and Technology. Edited by Igor Agranovski

ISBN: 978-3-527-32660-0

List of Contributors

Igor E. Agranovski
Griffith University
School of Engineering
170 Kessels Road
Nathan Campus
Brisbane
Queensland 4111
Australia

Irina S. Andreeva
Federal Service for Surveillance in
Consumer Rights Protection and
Human Well-Being
State Research Center of Virology
and Biotechnology "Vector"
Koltsovo, 630559
Novosibirsk
Russia

Boris D. Belan
Siberian Branch of the Russian
Academy of Sciences
V.E. Zuev Institute for
Atmospheric Optics
Akademicheskii Avenue 1
634055 Tomsk
Russia

Galina A. Buryak
Federal Service for Surveillance in
Consumer Rights Protection and
Human Well-Being
State Research Center of Virology
and Biotechnology "Vector"
Koltsovo, 630559
Novosibirsk
Russia

Aladár Czitrovszky
Research Institute for Solid
State Physics and Optics
Department of Laser Application
P.O. Box 49
1525 Budapest
Hungary

Sergey A. Grinshpun
University of Cincinnati
Department of
Environmental Health
3223 Eden Avenue
107 Kettering Building
Cincinnati, Ohio
OH 45267
USA

Aerosols – Science and Technology. Edited by Igor Agranovski

ISBN: 978-3-527-32660-0

Lev S. Ivlev
Saint Petersburg
State University
7–9 Universitetskaya
Naberezhnaya
Saint Petersburg 1999034
Russia

Alexander A. Kirsch
Russian Research Center
"Kurchatov Institute"
Kurchatov Square, 1
123182 Moscow
Russia

Vasily A. Kirsch
Russian Academy of Sciences
Frumkin Institute of Physical
Chemistry and Electrochemistry
Leninski Prospect, 31
119991 Moscow
Russia

Alexey A. Lushnikov
Karpov Institute of
Physical Chemistry
10, ul Vorontsovo Pole
103062 Moscow
Russia

and

University of Helsinki
Faculty of Mathematics and
Natural Sciences
Physics, Atmospheric Sciences
and Geophysics Department
Gustav Hällströmin katu 2
00014 Helsingen Yliopisto
Finland

Arkadi Maisels
Evonik Degussa GmbH
Industriepark Wolfgang
Rodenbacher Chaussee 4
63457 Hanau
Germany

Albert G. Nasibulin
NanoMaterials Group
Department of Applied Physics
and Center for New Materials
Aalto University
Puumiehenkuja 2
00076 Espoo
Finland

Boris I. Ogorodnikov
Karpov Institute of
Physical Chemistry
10, ul Vorontsovo pole
105064 Moscow
Russia

Sergei E. Olkin
Federal Service for Surveillance in
Consumer Rights Protection and
Human Well-Being
State Research Center of Virology
and Biotechnology "Vector"
Koltsovo, 630559
Novosibirsk
Russia

Mikhail V. Panchenko
Siberian Branch of the Russian
Academy of Sciences
V.E. Zuev Institute for
Atmospheric Optics
Akademicheskii Avenue 1
634055 Tomsk
Russia

Olga B. Popovicheva
Skobeltsyn Institute of
Nuclear Physics
Division of Microelectronics
Moscow State University
1(2) Leninskie gory
119991 Moscow
Russia

Irina K. Reznikova
Federal Service for Surveillance in
Consumer Rights Protection and
Human Well-Being
State Research Center of Virology
and Biotechnology "Vector"
Koltsovo, 630559
Novosibirsk
Russia

Bogdan F. Sadovsky
Karpov Institute of
Physical Chemistry
10, ul Vorontsovo Pole
103062 Moscow
Russia

Aleksandr S. Safatov
Federal Service for Surveillance in
Consumer Rights Protection and
Human Well-Being
State Research Center of Virology
and Biotechnology "Vector"
Koltsovo, 630559
Novosibirsk
Russia

Alexander N. Sergeev
Federal Service for Surveillance in
Consumer Rights Protection and
Human Well-Being
State Research Center of Virology
and Biotechnology "Vector"
Koltsovo, 630559
Novosibirsk
Russia

Sergey D. Shandakov
Laboratory of Carbon
NanoMaterials
Department of Physics
Kemerovo State University
Krasnaya 6
Kemerovo, 650043
Russia

Valery A. Zagaynov
Karpov Institute of
Physical Chemistry
10, ul Vorontsovo pole
105064 Moscow
Russia

List of Symbols

a	amount of vapor adsorbed (Chapter 5)
a	fiber radius (Chapter 10)
a	particle radius (Chapter 1)
a_0	radius of molecule of condensable substance
a_g	radius of g-mer
a_m	molecular radius
a_m	monolayer coverage
a_s	characteristic particle radius, for normalization of particle size
a_v	equilibrium concentration of vapor
A	acceleration (Chapter 7)
A	Hamaker constant (Chapter 11)
$A(t)$, $B(t)$	algebraic functions of time
B	ion mobility (Chapter 1)
B	particle mobility (Chapter 6)
c	filter packing density
c^*	critical vapor concentration level
$c_0(Z_p)$	concentration of particles at inlet
c/c_c	supersaturation
c_e	equivalent filter packing density
$c_g(t)$	g-mer concentration
c_M	concentration of M-mer
$c_{out}(Z_p, r, t)$	concentration of particles at outlet
c_p	filter packing density
$c(r, t)$	particle concentration at point r at time t
C	condition number (Chapter 8)
C	Cunningham correction coefficient (Chapter 11)
C	monomer number concentration (Chapter 4)
C	vapor concentration (Chapter 4)
$C_0(t)$	concentration at time t

Aerosols – Science and Technology. Edited by Igor Agranovski

ISBN: 978-3-527-32660-0

C_a	aerosol concentration at filter inlet
$C(a)$	correction factor
C_c	Millikan correction factor
$C_c(Kn)$	slip correction factor
C_D	drag coefficient
$C(r)$	density–density correlation function
C_S	slip correction factor
Cu	Coulomb number
d	diameter of adsorbate molecule (Chapter 5)
$\mathbf{d}$	dipole moment
d	particle diameter (Chapter 2)
d_1	spherule diameter
d_1'	diameter of monomer
d_{50}	particle diameter at which 50% of particles are collected
d_A	radius of the equivalent projected sphere
d_b	diameter of bubble
d_f	fiber diameter
d_k	diameter of particle in size class k
d_m	transition mobility diameter
d_{max}	maximal size of a fractal aggregate
d_{mc}	mobility diameter of fractal aggregate in continuum regime
d_{mk}	mobility diameter of fractal aggregate in kinetic regime
dN	number of particles within size range from x to $x + dx$
d_{opt}	optical diameter
d_p	particle diameter
$d\mathbf{S}$	element of particle surface
d_V	volume equivalent diameter
$d\sigma_e/d\Omega$	differential elastic cross-section
D	active factor dose (Chapter 14)
D	average coefficient of diffusion (Chapter 10)
D	diffusivity (Chapter 1)
D	ion diffusivity (Chapter 1)
D	molecular diffusivity (Chapter 1)
D	tube diameter (Chapter 3)
$\langle D \rangle$	average diffusion coefficient
D_d	diameter of drop
D_f	fractal dimension
D_{gA}	diffusivity of reactant molecule A in gas phase
D_i	particle diffusion coefficient for spherical particle of diameter d_i
D_{ion}	diffusion coefficient for ions
D_S	diffusion coefficient
$\underline{D}_{st}$	geomagnetic disturbance storm time index
D_X (X = A,B)	diffusivity of reactant molecules inside particle

e	coefficient of restitution (plastic and elastic deformation) (Chapter 11)
e	elementary charge (Chapter 4)
e/m	ion's charge-to-mass ratio
e_{pl}	coefficient of restitution (plastic deformation only)
e_{pl}	microscopic yield pressure
E	filter efficiency (Chapter 10)
E	kinetic energy for single vapor molecule (Chapter 4)
$\mathbf{E}$	electric field strength
E_a	activation energy
E_A	activation energy
E_f	filter efficiency
E_g	bandgap
$E(r, t)$	distribution of electric field
$E_r(r, z)$	electrical intensity along radial coordinate
$E_z(r, z)$	electrical intensity along longitudinal coordinate
f^+	velocity distribution function of molecules flying toward particle surface
f^-	velocity distribution function of molecules flying outward from particle
$f(a)$	particle size distribution
f_A	distribution function of A molecules over coordinates and velocities
$f_G(a)$	generalized gamma distribution
f_L	total fiber length in filter sample
$f_L(a)$	log-normal distribution
$f(x)$	particle size distribution
F	drag coefficient
$\mathbf{F}$	electric force
F^*	drag force acting on unit length of fiber
F_{drag}	drag force acting on particle
g	gravity (Chapter 11)
g	number of spherules comprising fractal aggregate (Chapter 1)
g	particle mass (Chapter 1)
G	cutoff particle mass
G_g	gas flow rate
G_y	total liquid supply at filter cross-section at height y
h	half distance between neighboring fibers (Chapter 10)
h	Planck constant (Chapter 2)
H	classical Hamiltonian (Chapter 4)
H	dimensionless Henry's constant (Chapter 4)
H	filter thickness (Chapter 10)

H_C	Henry's constant for reaction product C
H_S	Henry's constant as defined by Seinfeld and Pandis
$I(t)$	particle productivity (number of particles produced per unit volume per unit time)
$\boldsymbol{j}$	density of total flux of particles
j_A	total flux of A molecules trapped by particle
J_m	Bessel functions
j_r	normal component of density of overall flux of particles
$j(r)$	steady-state density of ion flux
$j(x)$	dimensionless nucleation rate
J	flux of evaporated atoms (Chapter 1)
J	total flux of condensable vapor (Chapter 4)
J_0	nucleation rate
$J(a)$	steady-state ion flux
$J(a)$	steady-state molecular flux
$J(t)$	nucleation rate
$\mathcal{J} = AC^{G*}$	nucleation rate for fluctuation-controlled nucleation
$\mathcal{J}$	steady-state rate of new particle production
$\mathcal{J}_2(c_1)$	rate of dimerization
k	Boltzmann constant (Chapter 1)
k	hydrodynamic factor (Chapter 10)
k^*	number of condensable monomers in critical size nucleus
k_B	Boltzmann constant
k_D	fractal prefactor
K	particle breakthrough
$K_0(z)$	modified Bessel function
K_d	dissolution coefficient
$K(g, l)$	coagulation kernel
Kn	Knudsen number
$\mathrm{Kn}_{\mathrm{ion}}$	Knudsen number for ions
K_X	enrichment coefficient of element X
$K(x, y)$	coagulation kernel
l	distance deflected from original trajectory (Chapter 7)
l	mean free path of carrier gas molecules (Chapter 1)
l	mean free path of condensing molecule in carrier gas (Chapter 4)
l_C	Coulomb length
l_m	height of mid-section
L	characteristic length of the flow (Chapter 1)
L	fiber length per unit surface area of filter (Chapter 4)
L	fiber length per unit volume of filter (Chapter 10)
L_c	total length of fibers in cell

m	mass of foreign molecule (Chapter 1)
m	mass of particle (Chapter 2)
$\langle m \rangle$	mean particle mass
m_0	mass of monomer
m_1	mass of monomer
m_{el}	mass of electron
m_{g}	mass of g-mer
$m(r)$	density at point r
M	total mass of fractal aggregates
M	particle mass (Chapter 1)
Mie	Mie number
n	dimensionless particle concentration (Chapter 4)
n	refractive index of particle (Chapter 7)
n_0	inlet particle concentration
$\overline{n}^{(1,2)}$	first and second moments of fractal aggregate size distribution function
n_∞	ion density far away from particle
n_a	number concentration of vapor molecules at particle surface
n_{A}^*	concentration of reactant in liquid phase immediately beneath surface
n_{A}^+	concentration of particles flying outward (Chapter 4)
n_{A}^+	concentration of reactant immediately above particle surface (Chapter 4)
$n_{\mathrm{A}\infty}$	concentration of A far away from particle
n_{Ae}	equilibrium concentration of A molecules
$n_{\mathrm{exact}}(r)$	exact ion/vapor concentration profile (Chapter 1)
$n_{\mathrm{fm}}(r)$	ion/vapor concentration profile in free-molecule zone (Chapter 4)
n_{g}	concentration of clusters of mass g
$\overline{n}_{\mathrm{g}}(t)$	average occupation number
n_{ion}^-	concentration of negative ions
$n^{(J)}(r)$	steady-state ion concentration profile corresponding to total ion flux J
$n^{(J)}(r)$	steady-state vapor concentration profile corresponding to flux $J(a)$
$n_{\mathrm{p},i}$	number of primary particles of fractal aggregate i
n_R	ion/vapor concentration at distance R from particle center
n_{s}	equilibrium concentration of vapor molecules over planar surface of liquid
$n_{\mathrm{X}}(r)$	concentration profile
$n(\gamma,\tau)$	particle mass spectrum
n, m	number of screens in diffusion battery
(n, m)	chiral indices
N	molecular number concentration (Chapter 4)
N	number of spores (Chapter 13)
N	total particle concentration (Chapter 8)

N_0	particle number concentration/total number of particles
N_1	fraction of condensed-matter particles of smallest size
$N_1'(t)$	number concentration of condensing monomers
N_A, N_B	total number of molecules of reactants
N_C	number of molecules of reaction product
$N_{E_i}(Y_i)$	distribution function of aerosol particles with respect to Y_i
N_i	density of ions
N_i^q	aerosol fraction with particle diameter d_i and charge q
$N_k(t)$	fraction of particles containing k monomers at time t
N_p	number of primary particles
$N(t)$	total number concentration of coagulating particles
$N(x)$	number of particles with size less than x
p	pressure (Chapter 5)
p	probability of causing reaction in organism (Chapter 14)
p_s	saturation vapor pressure
Pe	Peclet number
P_f	perimeter of fibers
P_i	penetration through battery with n_i screens
P_{int}	internal pressure at embryo surface
P_m^l	associated Legendre polynomial
$P(n)$	penetration function
$P(n, D)$	penetration of particles with diffusion coefficient D through diffusion battery with n screens
$P(x)$	reading of instrument measuring property x
q	electrical charge
Q	volumetric flow rate
Q_a	flow rate of aerosol gas carrier
Q_{sh}	flow rate of buffer gas or filtered air
$\boldsymbol{r}$	position of particle
r	radial coordinate of particle
r_0	radius of spherule
r^2	correlation coefficient
r_2	radius of outer cylinder surface
r_E	equivalent film radius
r_f	fiber radius
r_i	position of the ith spherule
$\langle r_i \rangle$	average particle size of fraction i
r_p	nanoparticle radius
(r,θ)	dimensionless polar coordinates
R	channel radius (Chapter 8)
R	distance (Chapter 1)
R	gas constant (Chapter 5)

R	gyration radius of fractal aggregate (Chapter 1)
R	radius of limiting/constraining sphere
Re	Reynolds number
$R(x, a)$	linear response function of instrument
s	particle surface area
s_1	monomer surface area
s_{SC}	surface area of the completely sintered particle (volume-equivalent sphere)
S	ratio of the jet-to-plate distance (Chapter 13)
S	measured specific surface area (Chapter 5)
S	total particle area (Chapter 2)
$S_1(\Theta)$	normalized amplitude of flux polarized normal to the scattering plane scattered through angle Θ
$S_2(\Theta)$	normalized amplitude of flux polarized parallel to the scattering plane scattered through angle Θ
S_c	critical supersaturation
S_e	equivalent surface area of filter
S_{H_2O}	surface area covered by water
Stk	Stokes number
t	number of years/time
t^*	time at which spontaneous nucleation process starts
t^{**}	time at which spontaneous nucleation process stops
t_c	critical time
T	absolute temperature
T	fluid temperature (Chapter 2)
T	thickness of filter (Chapter 11)
T_0	bulk melting temperature (1535 °C)
T_0	spot temperature (Chapter 1)
$T_{1/2}$	half-life
T_f	front temperature
T_m	melting temperature for given particle
u	constant uniform velocity of incoming flow
$\mathbf{u}$	flow velocity vector
u_0	average flow velocity
$\mathbf{u}(\mathbf{r})$	flow field at time t
$u_r(r, z)$	particle velocity along cylinder radius
u_t	tangential component of velocity
$u_z(r, z)$	particle velocity along cylinder axis
u_ξ	normal component of velocity
U	potential difference between plates
$U(r)$	ion–particle interaction potential

$U_z(r)$	velocity distribution of flow across cylinder radius
U_τ	velocity of circulating gas at surface of bubble
v	macroscopic flow velocity speed of carrier gas
v_1	molecular volume
v_a	volume per added molecule of A
$v_{a,b,c}$	molecular volume of reactants A, B, and C
$v_{i,j}$	relative thermal velocity between particles i and j
$\boldsymbol{v}_k$	molecular velocities
v_T	thermal velocity of condensable gas molecules
V	filter face velocity of aerosol carrier (Chapter 11)
V	mole volume (Chapter 5)
V	volume of metal molecule (Chapter 3)
V_0	initial particle volume (Chapter 4)
V_0	potential difference (Chapter 8)
$V(a)$	average volume of a void of size a
V_b	velocity of rise of bubble
V_c	critical velocity
V_{fiber}	fiber volume
V_R	volume of constraining sphere
V_T	average speed of ion's thermal movement
W	binding energy of surface film (Chapter 5)
W	impactor's nozzle size (Chapter 13)
W	width of filter (Chapter 11)
W_{DF}	dry filter weight
$W_{i,j}^{p,q}$	stability function
W_L	weight of liquid remaining on filter after drainage
$W(n_g,t)$	probability for realization of given set at time t
$W(N,t)$	probability to find exactly N particles at time t
x	distance of separation between center of mass of particle and surface (Chapter 11)
x	particle geometry (Chapter 8)
x, y	masses of colliding particles (Chapter 1)
Y_i	scattered light intensity
z	longitudinal coordinate of particle
Z	partition function for single vapor molecule (Chapter 4)
Z	total particle charge in units of e (Chapter 1)
Z_g	partition function of g molecules inside sphere
Z_i	charge on ion in units of e
Z_p	charge on particle in units of e

α	particle polarizability (Chapter 1)
α	filter packing density (Chapter 10)
α_1	rate of dimer formation
$\alpha(a)$	charging efficiency as function of a (Chapter 1)
$\alpha(a)$	condensation efficiency (Chapter 4)
$\alpha(a, R)$	charging efficiency as function of a at distance R
α_{coll}	collision parameter
$\alpha_{\mathrm{fm}}(a)$	condensation efficiency in free-molecule regime
$\alpha_{\mathrm{fm}}(a, R)$	free-molecule form of $\alpha(a, R)$
α_{g}	condensation coefficient
$\alpha(g)$	condensation efficiency
β	coagulation kernel (coefficient) of two colliding particles
β	sticking probability
β'	collision frequency of particles and monomers
β_{C}	sticking probability of molecules C
$\beta_{i,j}$	projected surface area between particles i and j
β_i^q	ion attachment coefficient
β_{M}	scattering coefficient from Mie scattering theory
β_{p}	particle scattering coefficient
$\beta^{q \to q-1}$	ion attachment coefficient
β_{R}	scattering coefficient from Rayleigh scattering theory
γ	shape factor
Γ	velocity gradient
$\Gamma(x)$	Euler gamma function (Chapter 1)
$\Gamma(\gamma)$	Euler's gamma function (Chapter 8)
δ	Kronecker delta (Chapter 2)
δ_{D}	thickness of diffusion boundary layer
δ_{E}	equilibrium film thickness
δ_{max}	maximum thickness of the film
$\delta(x)$	Dirac delta function
$\delta(y)$	film thickness on fibers at filter vertical elevation y
Δ	three-dimensional Laplace operator (Chapter 10)
ΔH_{fus}	latent heat of fusion
Δt	time between pulses
Δv	change in velocity
Δx	width of thin slot
$[\Delta p]$	standard resistance of material
ϵ	dielectric permeability (Chapter 1)
ε	fraction of water-soluble compounds
ϵ	rate of dissipation of kinetic energy of the turbulent flow

η	dynamic gas viscosity (Chapter 2)
η	fiber collection efficiency (Chapter 10)
η	trapping efficiency (Chapter 8)
η_D	efficiency of diffusion deposition
η_i	efficiency of inertial deposition
θ	adsorption coverage (in monolayers) (Chapter 5)
θ	latitude angle measured from zero at direction of rise (Chapter 11)
θ/Θ	scattering angle (Chapter 1)
$\theta(\epsilon)$	Heaviside step function
ϑ_i^q	combination coefficient
$\Theta(x)$	Heaviside step function
κ	binary reaction rate constant
λ	homogeneity exponent (Chapter 1)
λ	mean free path of carrier gas molecules
λ_g	mean free path of gas molecules
λ_u	average length of ion's mean free path
Λ	thermal conductivity of carrier gas
μ	dynamic viscosity
μ	liquid viscosity (Chapter 11)
μ	smallness parameter (Chapter 4)
$\mu_\pm$	ion mobility
μ_g	dynamic viscosity of gas
ν	kinematic viscosity of carrier gas
$\overline{v}_{ion}$	mean ion thermal velocity
ξ_m, ψ_m	Riccati–Bessel functions
ρ	density
ρ_0	density of spherule
ρ_f	front density
ρ_{FM}	filter material density
ρ_g	carrier gas density
ρ_p	density of particle/particulate material
ρ_L	liquid density
σ	average distribution width (Chapter 8)
σ	scattering coefficient (Chapter 12)
σ	surface tension
σ_{abs}	absorption cross-section
σ_i	average distribution width of fraction i
σ_{sca}	elastic scattering cross-section
σ_{sl}	surface tension between liquid and solid
Σ	dimensionless surface tension parameter

τ	lifetime
τ_A	characteristic time for chemical reaction of A molecules in liquid phase
$\tau_{changes}$	characteristic time of substantial chemical changes inside particle
τ_{chem}	characteristic reaction time for diffusion-controlled reaction
τ_g	characteristic time of non-stationarity in gas phase
τ_l	characteristic time in liquid phase
τ_S	characteristic sintering time
φ	potential function
φ_2	second moment
$\varphi(D)$	diffusion coefficient distribution
$\varphi(l, q)$	interaction potential between ion and q-charged particle
$\Phi^{q \to q+1}$	work function
ψ	stream function
$\psi(x)$	universality function
$\Psi(z, t)$	generating function for probability
$\Omega(Z_p)$	transfer function for the differential mobility analyser
∇	gradient operator (Chapter 11)
$\sum \beta$	activity concentration of mixture of beta-emitting nuclides

Introduction

Dear Reader

For more than a decade I have had the idea of producing this book. This is why I accepted a corresponding offer from the Publisher with great pleasure. My frequent teaching and research related trips to various countries allowed me to meet many colleagues who, similarly to me, have originated from Eastern European countries and inherited glorious traditions of the Russian school of aerosol science established by Prof Nikolai A. Fuchs and Academician Igor V. Petryanov in the first half of the last century. Some of my colleagues still work in their countries of origin, whilst others, due to various reasons, have moved to other places and currently work at leading research, industrial and educational organisations around the world. I am very grateful to all contributors who shared my idea about this book and accepted my invitation to participate in this project, which presents a collection of fourteen invited Chapters produced by scientists representing various institutions of six countries – Australia, Finland, Germany, Hungary, Russia, and the USA.

This book was not planned to be an encyclopaedia type project comprehensively covering all aspects of aerosol science and technology. In contrast, I requested all contributors to focus on aspects not commonly discussed in classic aerosol books. Of course, this issue does not exclude some coverage of traditional concepts and theories widely used in the field. In addition, a significant amount of information provided in this book has never been published in English before and is not known by Western readers.

The book consists of 14 Chapters divided into four sections; Aerosol Formation, Aerosol Measurements and Characterisation, Aerosol Removal, and Atmospheric and Biological Aerosols.

Chapter 1 (Aerosol Fundamentals) is written by Prof Alexey A. Lushnikov (Karpov Institute of Physical Chemistry, Moscow, Russia – University of Helsinki, Finland) who inherited Headship of the Laboratory of Physics of Aerodisperse Systems at Karpov Institute of Physical Chemistry, Moscow, Russia directly from Prof Fuchs.

Aerosols – Science and Technology. Edited by Igor Agranovski

ISBN: 978-3-527-32660-0

He is laureate of prestigious Fuchs Memorial Award (2002) and Christian Junge Award (2007). The chapter summarizes all important theoretical and practical issues widely used by aerosol scientists and engineers. It contains information and formulas describing aerosol behaviour in gas carriers and theoretical methods for particle analysis.

Chapter 2 (High Temperature Aerosol Systems) is produced by Dr. Arkadi Maisels (Evonik Degussa GmbH, Hanau, Germany). Amongst different sources of aerosol particles, high temperature processes are very common in both nature and industry. Therefore, understanding of aerosol formation and dynamics in high temperature processes is of immense environmental and industrial importance. In this chapter, an overview is provided of different high temperature aerosol reactors. The properties of resulting particles are considered with respect to reactor design. Besides the engineering of aerosol particles, main dynamic formation processes are described.

Chapter 3 (Aerosol Synthesis and Properties of Carbon Nanotubes) is written by Dr. Albert G. Nasibulin (Academy Research Fellow of Finnish Academy of Science and a Docent at Helsinki University of Technology, Finland). The Chapter briefly reviews the research in the field of Carbon Nanotubes (CNTs): discovery, properties and applications. Special attention is devoted to the development of the synthesis methods. Advantages of aerosol methods in the controlled production of CNTs for both laboratory and industrial purposes are thoroughly reviewed.

Chapter 4 (Aerosol Nucleation, Evaporation and Condensation) is written by Prof Alexey A. Lushnikov. Aerosol nucleation, evaporation and condensation processes are of primary importance for the fate of any aerodisperse system. Starting with the Boltzmann equation the equations for the rates of birth-growth-death processes have been derived. The approximate solution of the kinetic equation describing the time-spatial behaviour of the species moving toward the particle is matched with the solution to the diffusion equation describing the concentration profile far away from the particle. The matching distance is then found from the condition of the absence of jumps of the first spatial derivatives of the concentration profile. This approach allows one to find the efficiencies of the mass-charge transfer from (to) the particle.

Chapter 5 (Combustion-Derived Carbonaceous Aerosols (Soot) in the Atmosphere: Water Interaction and Climate Effects) is written by Dr Olga B. Popovicheva (Moscow State University, Russia). This Chapter presents a comprehensive analysis of water interaction with various transport engine-generated and laboratory-made combustion particles at atmospheric conditions. Gravimetrical measurements of water uptake coupled with chemical composition and porosity analysis clarifies the mechanism of water interaction with aircraft engine soot, ship exhaust residuals, and different fuel burning particles for wide range of relative humidites up to the condensation regime. Systematic analysis demonstrates two mechanisms of water/soot interaction, namely the bulk dissolution into soot water soluble coverage (absorption mechanism) and the water molecule adsorption on surface active sites (adsorption mechanism).

Chapter 6 (Radioactive Aerosols – Chernobyl Nuclear Power Plant Case Study) is written by Prof Boris I. Ogorodnikov (Principle Research Fellow at Karpov Institute of Physical Chemistry, Moscow, Russia). Since year 1985, Prof Ogorodnikov has spent a significant amount of time studying the Chernobyl disaster and following aerosol related contamination of the environment. The concentration dynamics and size distribution of radioactive aerosols over the 23 year period after the disaster are presented. Sampling methods and instruments are discussed.

Chapter 7 (Optical Properties of Aerosols) is written by Dr Aladár Czitrovszky (Research Institute for Solid State Physics and Optics, Budapest, Hungary). In this Chapter, the following issues are described: optical properties of aerosols; light scattering, absorption and extinction of aerosols; methods of measurement of the optical parameters; application of the new measurement methods for determination of the complex refractive index, concentration, size distribution, etc.; new instruments for study of atmospheric pollution by aerosols.

Chapter 8 (Inverse Problem and Aerosol Measurements) is presented by Dr Valery A. Zagaynov (Deputy Head of the Laboratory of Physics of Aerodisperse Systems, Karpov Institute of Physical Chemistry, Moscow, Russia. Dr Zagaynov was the last PhD student supervised by Prof Fuchs). One of the main tasks of aerosol science and technology is representative determination of particle size distribution. At the same time, this problem is very acute and ambiguous to solve. There are two obstacles in resolving this problem. First of all, any monitoring equipment has defined sensitivity, which could leave substantial particle quantity not registrant. The concentration of such particles may be even greater, than the concentration of counted aerosols. Secondly, some uncertainty is related to an inverse problem. In this Chapter, the instrumentation along with theoretical approach to attack the problem is discussed.

Chapter 9 (History of Development and Present State of Polymeric Fine-Fiber Unwoven Petryanov Filter Materials for Aerosol Entrapment) is written by Prof Bogdan F. Sadovsky (Karpov Institute of Physical Chemistry, Moscow, Russia). This Chapter provides some historical and modern aspects of the development of filter materials, traded as "Petryanov's Filters", by electrospinning process over the last few decades in the Soviet Union and Russian Federation. The main parameters of these materials along with their applications are discussed in the Chapter.

Chapter 10 (Deposition of Aerosol Nanoparticles in Model Fibrous Filters) is written by Dr Vasily Kirsh (Frumkin Institute of Physical Chemistry and Electrochemistry, Moscow, Russia) and Prof Alexander Kirsh (Russian Research Center "Kurchatov Institute", Moscow, Russia). Prof Kirsh was one of the main co-workers of Prof Fuchs. The Chapter discusses mechanisms of deposition of aerosol nanoparticles in model fibrous filters at low Reynolds numbers and a wide range of Peclet numbers. The deposition of nanoparticles in model filters with ultra-fine fibers, with fibers with elliptical, strip-like, porous and composite fibers, and in model filters with non-regular arrangement of fibers is considered. The deposition of nanoparticles on the square screens from the three-dimensional flow is calculated. The validity of the formulas used for the estimation of the coefficient of diffusion of nanoparticles from the measured penetration of nanoparticles through screen diffusion batteries is discussed.

Chapter 11 (Filtration of Liquid and Solid Aerosols on Liquid-Coated Filters) is written by Prof Agranovski (Griffith University, Brisbane, Australia). The Chapter discusses theoretical and practical aspects of the removal of particles of mineral and biological nature on liquid irrigated filters. Special attention is given to consideration of the filters' regeneration mechanisms for industrial applications. In addition, utilization of irrigated filters for bioaerosol monitoring and control applications is discussed.

Chapter 12 (Atmospheric Aerosols) is written by Prof Lev S. Ivlev (St. Petersburg State University, Russia). This chapter is devoted to atmospheric aerosols. It provides detail overview of formation of soil, marine and volcanic aerosols and their global transportation at different elevations in atmosphere. A significant amount of the results presented in this Chapter were obtained by monitoring stations located across Former Soviet Union Republics and Eastern European Countries.

Chapter 13 (Biological Aerosols) is presented by Prof Sergey A. Grinshpun (Professor of Environmental Health and Director of the Center for Health-Related Aerosol Studies, University of Cincinnati, USA). The chapter is devoted to the particles of biological origin, including viruses, bacteria, fungi, pollen as well as their products, fragments and aggregates. Physical and biological characteristics of these particles are discussed, as well as the aerosolization, sampling, analysis and filtration of bioaerosol particles. Additionally, respiratory protection and air purification techniques related to the bioaerosol exposure reduction are reviewed.

Chapter 14 (Atmospheric Bioaerosols) is written by Dr Alex S. Safatov (Aerobiology Laboratory, FSRI SRC VB "Vector", Koltsovo, Novosibirsk region, Russia) and his colleagues. The chapter is devoted to the results of ten-year study of the biogenic components of tropospheric aerosol at the altitudes of up to 7000 meters and comparison of the results obtained at various regions of the planet. The most important bioaerosol components are the total protein as an indicator of all substances of biological origin and culturable microorganisms as a component, which is the most harmful to humans and animals. The ten-year dynamics of the above concentrations, their variations and altitude profiles will be presented. The properties of the observed bioaerosol, its possible sources and potential influence on human health are discussed.

I hope that the aerosol community will find the information presented in this book to be both useful and interesting. Happy reading!

April 2010

Igor E. Agranovski, Ph.D., Professor
Griffith School of Engineering
Griffith University, Brisbane
Australia

1
Introduction to Aerosols

Alexey A. Lushnikov

1.1
Introduction

Aerosol science studies the properties of particles suspended in air or other gases, or even in vacuum, and the behavior of collections of such particles. A collection of aerosol particles is referred to as an aerosol, although the particles may be suspended in some other gaseous medium, not just air. The term *cosmosol* is used for a collection of particles suspended in vacuum. Although attempts to give a strict definition of aerosol have appeared from time to time, to date no commonly acceptable and concise definition of an aerosol exists. In my opinion, it is better not to make any attempts in this direction, especially because intuitively it is clear what an aerosol is. For example, it is clear that birds or airplanes are not aerosol particles. On the other hand, smoke from cigarettes, fumes from chimneys, dust raised by the wind, and so on, *are* aerosols. Hence, there are some essential features that allow us to distinguish between aerosols and other objects suspended in the gas phase. There are at least two such features: (i) aerosol particles can exist beyond the aerosol for a sufficiently long time; and (ii) an aerosol can be described in terms of the concentration of aerosol particles, or, better, the *concentration field.* From this point of view, it is clear why birds are not aerosols. Interestingly, clouds are also not aerosols! Of course, we can introduce the concentration of cloud droplets. But if we isolate a cloud particle, it will immediately evaporate. The cloud creates a specially designed environment inside it – the humidity and the temperature fields – the conditions in which a water droplet does not evaporate during a long time.

Aerosols are divided into two classes, namely *primary aerosols* and *secondary aerosols*, according to the mechanisms of their origination. Primary aerosol particles result, for example, from fragmentation processes or combustion, and appear in the carrier gas as already well-shaped objects. Of course, their shape can change because of a number of physico-chemical processes such as humidification, gas–particle reactions, coagulation, and so on. Secondary aerosol particles appear in the carrier gas from "nothing" as a result of gas-to-particle conversion. For example, such aerosols regularly form in the Earth's atmosphere and play a key role in a number

Aerosols – Science and Technology. Edited by Igor Agranovski

ISBN: 978-3-527-32660-0

of global processes such as the formation of clouds. They serve as the centers for heterogeneous nucleation of water vapor. No aerosols – no clouds! One can imagine how our planet would look without secondary aerosol particles.

Primary and secondary aerosols are characterized by the size, shape, and chemical content of the aerosol particles. As for the shape, one normally assumes that the particles are spheres. Of course, this assumption is an idealization necessary for simplification of the mathematical problems related to the behavior of aerosol particles. There are very many aerosols comprising irregularly shaped particles. The non-sphericity of particles creates many problems. There exist also agglomerates of particles, which in some cases reveal fractal properties. We shall return to the methods for their description later on.

There are a number of classifications of particles with respect to their size. For example, if the particles are much smaller than the molecular mean free path, they are referred to as "fine" particles. This size range stretches from 1 to 10 nm under normal conditions. But from the point of view of aerosol optics, these particles are not small if the wavelength of the incident light is comparable with their size. This is the reason why such very convenient and commonly accepted classifications cannot compete with natural classifications based on the comparison of the particle size with a characteristic size that comes up each time when one solves a concrete physical problem.

1.2 Aerosol Phenomenology

1.2.1 Basic Dimensionless Criteria

It is convenient to characterize aerosols by dimensionless criteria. The most commonly used in the area of aerosol science are listed below. Each of these criteria contains the particle size a. In what follows we consider spherical particles of radius a.

1.2.1.1 Reynolds Number

The Reynolds number Re is introduced as follows:

$$\mathrm{Re} = \frac{ua}{\nu} \tag{1.1}$$

Here ν is the *kinematic viscosity* of the carrier gas and u is the particle velocity with respect to the carrier gas. Small and large Re correspond to laminar or turbulent motion of the particle, respectively.

1.2.1.2 Stokes Number

The Stokes number Stk characterizes the role of inertial effects:

$$\mathrm{Stk} = \frac{2a^2 u}{9\nu L} \tag{1.2}$$

Here L is the characteristic length of the flow. The Stokes number Stk is seen to increase on increasing the particle size.

1.2.1.3 **Knudsen Number**

The Knudsen number Kn characterizes the discreteness of the carrier gas:

$$\mathrm{Kn} = \frac{l}{a} \tag{1.3}$$

Here l is the mean free path of the molecules of the carrier gas,

$$l = \frac{1}{\sqrt{2}\sigma^2 N} \tag{1.4}$$

σ is the size of a carrier gas molecule, and N is the molecular number concentration. If a foreign molecule moves toward the aerosol particle, then Kn can be expressed in terms of the *molecular diffusivity D*,

$$\mathrm{Kn} = \frac{2D}{v_T a} \tag{1.5}$$

where

$$v_T = \sqrt{\frac{8kT}{\pi m}} \tag{1.6}$$

is the molecular thermal velocity, m is the mass of the foreign molecule, k is the Boltzmann constant, and T is the absolute temperature (K).

1.2.1.4 **Peclet Number**

The Peclet number Pe defines the regimes of energy transfer from particles to the carrier gas. It is introduced similarly to Kn (Eq. (1.7)):

$$\mathrm{Pe} = \frac{2\Lambda}{v_T a} \tag{1.7}$$

Here Λ is the thermal conductivity of the carrier gas.

1.2.1.5 **Mie Number**

The *Mie number* given by the dimensionless group

$$\mathrm{Mie} = \frac{2\pi\lambda}{a}$$

defines the optical properties of the particle. Here λ is the wavelength of the incident light.

1.2.1.6 **Coulomb Number**

The *Coulomb number* Cu given by

$$\mathrm{Cu} = \frac{l_C}{a} = \frac{Ze^2}{akT} \tag{1.8}$$

is important in the processes of particle charging. Here e is the elementary charge, Z is the total particle charge in units of e, and

$$l_C = \frac{Ze^2}{kT} \tag{1.9}$$

is the Coulomb length. This is the distance at which the influence of the Coulomb forces cannot be ignored.

1.2.2 Particle Size Distributions

Particle size distributions play a central role in the physics and chemistry of aerosols, although direct observation of the distributions are possible only in principle. Practically, what we really measure is just the response of an instrument to a given particle size distribution,

$$P(x) = \int R(x, a) f(a)\, da \tag{1.10}$$

Here $f(a)$ is the particle size distribution (normally a is the particle radius), $P(x)$ is the reading of the instrument measuring the property of aerosol x, and $R(x, a)$ is referred to as the linear response function of the instrument. For example, $P(x)$ can be the optical signal from an aerosol particle in the sensitive volume of an optical particle counter, the penetration of the aerosol through the diffusion battery (in this case x is the length of the battery), or something else. The function $f(a)$ cannot depend on the dimensional variable a alone. The particle size is measured in some natural units a_s. In this case the distribution is a function of a/a_s and depends on some other dimensionless parameters or groups. The particle size distribution is normalized as follows:

$$\int_0^\infty f(a/a_s)\, \frac{da}{a_s} = 1 \tag{1.11}$$

The length a_s is a parameter of the distribution. Although the aerosol particle size distribution is such an elusive characteristic of the aerosol, it is still convenient to introduce it because it unifies all the properties of aerosols.

In many cases the distribution function can be found theoretically by solving dynamic equations governing the time evolution of the particle size distribution, but the methods for analyzing these equations are not yet reliable, not to mention the information on the coefficients entering them. This is the reason why the phenomenological distributions are so widely spread.

There is a commonly accepted collection of particle size distributions, which includes those outlined in the following subsections.

1.2.2.1 The Log-Normal Distribution

The *log-normal distribution* is given by

$$f_L(a) = \frac{1}{\sqrt{2\pi}\,(a/a_s)\ln\sigma} \exp\left[-\frac{1}{2\ln^2\sigma}\ln^2\left(\frac{a}{a_s}\right)\right] \tag{1.12}$$

Here a is the particle radius. This distribution depends on two parameters, a_s and σ, where a_s is the characteristic particle radius and σ ($\sigma > 1$) is the width of the distribution. Equation (1.12) is known as the log-normal distribution. It is important to emphasize that it is not derived from theoretical considerations.

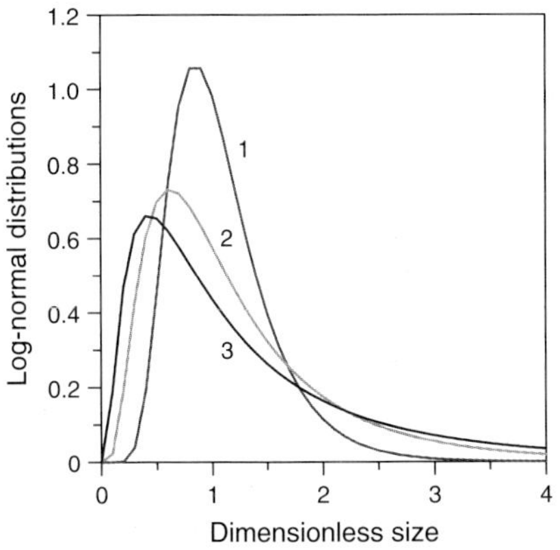

Figure 1.1 The log-normal distributions with $\sigma = 1.5$ (curve 1), 2.0 (curve 2), and 2.5 (curve 3). The parameter σ defines the width of the distribution. The dimensionless size is defined as a/a_s.

Rather, it was introduced by hand. The function $f_L(a)$ is shown in Figure 1.1 for different σ.

1.2.2.2 **Generalized Gamma Distribution**

The generalized *gamma distribution* is given by

$$f_G(a) = \left(\frac{a}{a_s}\right)^k \frac{j}{\Gamma((k+1)/j)} \exp\left[-\left(\frac{a}{a_s}\right)^j\right] \tag{1.13}$$

Here $\Gamma(x)$ is the Euler gamma function. The distribution f_G depends on three parameters, a_s, k, and j. Figure 1.2 displays the generalized gamma distribution for three sets of its parameters.

Once the particle size distributions are known, it is easy to derive the distribution over the values depending only on the particle size:

$$f(\psi_0) = \int \delta(\psi_0 - \psi(a)) f(a) \frac{da}{a_s} \tag{1.14}$$

Here $\delta(x)$ is the Dirac delta function. For example, if we wish to derive the distribution over the particle masses, then $\psi(a) = (4\pi a^3/3)\rho$, where ρ is the

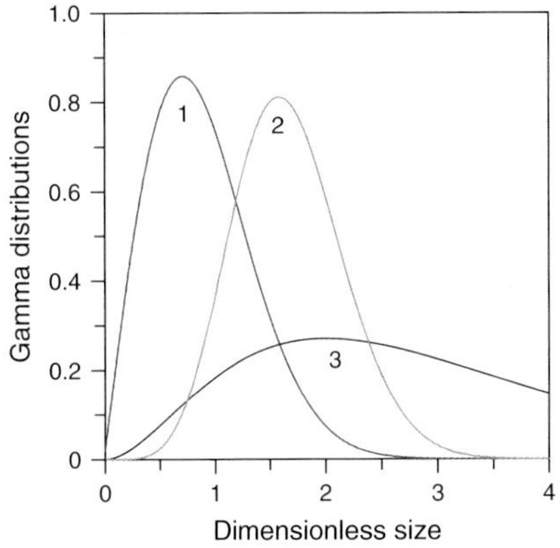

Figure 1.2 The gamma distributions with three sets of parameters: $k = 1$, $j = 2$ (curve 1); $k = 2$, $j = 1$ (curve 2); and $k = 5$, $j = 2$ (curve 3). These parameters define the shape of the distribution. Again, the dimensionless size is defined as a/a_s.

density of the particle material. Of course, the properties of aerosols do not depend solely on their size distributions. The shape of aerosol particles and their composition are important factors.

The log-normal distribution often applies in approximate calculations of condensation and coagulation. Two useful identities containing the integrals of a product of log-normal distributions can be found in, for example, [1, 2]. A regular theory of the log-normal distribution is expounded in the book [3].

1.3 Drag Force and Diffusivity

If the carrier gas moving with speed v flows past a spherical particle of radius a, the drag force acting on it is

$$F_{\text{drag}} = \tfrac{1}{2} C_{\text{D}} \pi a^2 \rho v^2 \tag{1.15}$$

where C_{D} is the drag coefficient and ρ is the density of the carrier gas. The latter depends on Re as follows:

$$C_{\text{D}} = \frac{12}{\text{Re}} \qquad \text{Re} < 0.1$$

$$C_{\text{D}} = \frac{12}{\text{Re}}\left(1 + \frac{3}{8}\text{Re} + \frac{9}{40}\ln \text{Re}\right) \qquad 0.1 < \text{Re} < 2$$

$$C_{\text{D}} = \frac{12}{\text{Re}}(1 + 0.15\,\text{Re}^{0.687}) \qquad 2 < \text{Re} < 500$$

$$C_{\text{D}} = 0.44 \qquad 500 < \text{Re} < 2 \times 10^5$$

The particle mobility B is introduced as

$$\boldsymbol{v} = B\boldsymbol{F} \tag{1.16}$$

When a particle of radius a moves in the carrier gas, the latter resists particle motion. The force acting on the particle is proportional to a in the limit of small flow velocity $\text{Re} \ll 1$ and Kn (continuum regime),

$$\boldsymbol{F} = 6\pi \rho \nu a \boldsymbol{v} \tag{1.17}$$

where ρ is the gas density and ν is the kinematic viscosity. Equation (1.17) is the Stokes equation.

In the transition regime, Eq. (1.17) should be corrected to

$$F = \frac{6\pi \nu v a}{C_{\text{c}}} \tag{1.18}$$

with C_{c} being the *Millikan correction factor*,

$$C_{\text{c}} = 1 + \text{Kn}\left[1.257 + 0.4\exp\left(-\frac{1.1}{\text{Kn}}\right)\right] \tag{1.19}$$

The diffusivity D is connected with the mobility B by the *Einstein–Smoluchowski formula*

$$D = kTB \tag{1.20}$$

The diffusivity is then

$$D = \frac{kT}{6\pi a\nu\rho}C(a) \tag{1.21}$$

where $C(a)$ is the correction factor. We can use $C(a) = C_c$ or $C(a)$ found [4] theoretically,

$$C(a) = \frac{15 + 12c_1\mathrm{Kn} + 9(c_1^2 + 1)\mathrm{Kn}^2 + 18c_2(c_1^2 + 2)\mathrm{Kn}^3}{15 - 3c_1\mathrm{Kn} + c_2(8 + \pi\sigma)(c_1^2 + 2)\mathrm{Kn}^2} \tag{1.22}$$

with

$$c_1 = \frac{2-\sigma}{\sigma}, \qquad c_2 = \frac{1}{2-\sigma}$$

and $\sigma < 1$ being a factor entering the slip boundary conditions. The Knudsen number is $\mathrm{Kn} = \lambda/a$, with λ being the mean free path of the carrier gas molecules ($\lambda = 65$ nm for air at ambient conditions). The parameter σ changes within the range 0.79–1.0. Equation (1.22) describes the transition correction for *all* Knudsen numbers and gives the correct limiting values (continuum and free-molecule ones). In what follows we put $\sigma = 1$. The correction factors of Eqs. (1.19) and (1.22) are plotted as functions of Kn in Figure 1.3.

All the above formulas are more thoroughly discussed in aerosol textbooks, except Eq. (1.22). This formula was derived from a 13-moment approximate solution of the Boltzmann equation by Phillips in [4]. It is remarkable that the results of Millikan and Phillips almost coincide.

1.4 Diffusion Charging of Aerosol Particles

At first sight the process of particle charging looks similar to particle condensation: an ion moving in the carrier gas approaches the particle and sticks to it. However, the difference between these two processes (condensation and charging) is quite significant. Even in the case when the ion interacts with a neutral particle, one cannot ignore the influence of the image forces. As was explained at the very beginning of this chapter, the motion of the ion is defined by two parameters: $\mathrm{Kn} = 2D/v_T a$ (the Knudsen number) and $\mathrm{Cu} = Ze^2/akT$ (the Coulomb number). Next, in most practical cases $\mathrm{Cu} > \mathrm{Kn}$. For example, at ambient conditions and $Z = 1$, the Coulomb length $l_C = e^2/kT = 0.06$ µm. This value is comparable with the mean free path of molecules in air ($l = 0.065$ µm), which means that the free-molecule regime of particle charging demands some special conditions and can be realized, for example, in the ionosphere.

1.4.1
Flux Matching Exactly

The steady-state ion flux $J(a)$ onto the particle of radius a can be written as

$$J(a) = \alpha(a) n_\infty \tag{1.23}$$

that is, the flux is proportional to the ion density n_∞ far away from the particle. The proportionality coefficient $\alpha(a)$ is known as the *charging efficiency*. The problem is to find $\alpha(a)$.

Once again, a dimensional consideration shows that $\alpha(a)$ is a function of two dimensionless groups, $\mathrm{Kn} = l/a$ and $\mathrm{Cu} = Ze^2/akT$,

$$\alpha(a) = \pi a^2 v_T F(l/a, Ze^2/akT) \tag{1.24}$$

We can generalize Eq. (1.23) as follows:

$$J(a, R, n_R) = \alpha(a, R) n_R \tag{1.25}$$

where n_R is the ion concentration at a distance R from the particle center. It is important to emphasize that n_R is (still) an *arbitrary* value introduced as a boundary condition at the distance R (also arbitrary) to a kinetic equation that is necessary to solve for defining $\alpha(a, R)$.

The flux defined by Eq. (1.23) is thus

$$J(a) = J(a, \infty, n_\infty) \qquad \text{and} \qquad \alpha(a) = \alpha(a, \infty) \tag{1.26}$$

The value of $\alpha(a, R)$ does not depend on n_R because of the linearity of the problem.

Let us assume that we know the exact ion concentration profile $n_{\text{exact}}(r)$ corresponding to the flux $J(a)$ from infinity (see Eq. (1.23)). Then, using Eq. (1.25) we can express $J(a)$ in terms of n_{exact} as follows:

$$J(a) = J(a, R, n_{\text{exact}}(R)) = \alpha(a, R) n_{\text{exact}}(R) \tag{1.27}$$

Now let us choose R sufficiently large for the diffusion approximation to reproduce the exact ion concentration profile,

$$n_{\text{exact}}(R) = n^{(J(a))}(R) \tag{1.28}$$

with $n^{(J)}(r)$ being the steady-state ion concentration profile corresponding to a given total ion flux J. The steady-state density of the ion flux $j(r)$ is the sum of two terms,

$$j(r) = -D\frac{\mathrm{d}n^{(J)}(r)}{\mathrm{d}r} - B\frac{\mathrm{d}U(r)}{\mathrm{d}r} n^{(J)}(r) \tag{1.29}$$

where D is the ion diffusivity, $U(r)$ is a potential (here the *ion–particle interaction*), and B is the ion mobility. According to the *Einstein relation*, $kTB = D$. On the other hand, the ion flux density is expressed in terms of the total ion flux as follows: $j(r) = -J/4\pi r^2$, with $J > 0$. Equation (1.29) can be now rewritten as

$$\mathrm{e}^{-\beta U(r)} \frac{\mathrm{d}}{\mathrm{d}r}[n^{(J)}(r)\, \mathrm{e}^{\beta U(r)}] = \frac{J}{4\pi D r^2}$$

where $\beta = 1/kT$. The solution to this equation is

$$n^{(J)}(r) = \mathrm{e}^{-\beta U(r)}\left(n_\infty - \frac{J}{4\pi D}\int_r^\infty \mathrm{e}^{\beta U(r')}\frac{\mathrm{d}r'}{r'^2}\right) \tag{1.30}$$

On substituting Eqs. (1.28) and (1.30) into Eq. (1.27), one obtains the equation $J(a) = \alpha(a, R)n^{(J)}(R)$ or

$$J(a) = \alpha(a, R)\mathrm{e}^{-\beta U(R)}\left(n_\infty - \frac{J(a)}{4\pi D}\int_R^\infty \mathrm{e}^{\beta U(r')}\frac{\mathrm{d}r'}{r'^2}\right) \tag{1.31}$$

We can solve this equation with respect to $J(a)$ and find $\alpha(a)$:

$$\alpha(a) = \frac{\alpha(a, R)\mathrm{e}^{-\beta U(R)}}{1 + [\alpha(a, R)\mathrm{e}^{-\beta U(R)}/4\pi D]\displaystyle\int_R^\infty \mathrm{e}^{\beta U(r')}\mathrm{d}r'/r'^2} \tag{1.32}$$

Equation (1.32) is exact if $R \gg l$. We, however, know neither $\alpha(a, R)$ nor R.

1.4.2 Flux Matching Approximately

Current knowledge does not allow us to find $\alpha(a, R)$ exactly. We thus call upon two approximations:

1) We approximate $\alpha(a, R)$ by its free-molecule expression,

$$\alpha(a, R) \approx \alpha_{\mathrm{fm}}(a, R) \tag{1.33}$$

2) We define R from the condition

$$\mathrm{d}_r n_{\mathrm{fm}}(r)|_{r=R} = \mathrm{d}_r n^{(J(a))}(r)|_{r=R} \tag{1.34}$$

where $n_{\mathrm{fm}}(r)$ is the ion concentration profile found in the free-molecule regime for $a < r < R$. The distance R separates the zones of the free-molecule and the continuum regimes.

All currently used approximations for α can be derived from Eq. (1.32).

1.4.3 Charging of a Neutral Particle

In this case the ion–particle interaction is described by the potential of image forces,

$$U(r) = -\frac{e^2}{2a}\frac{a^4}{r^2(r^2 - a^2)} \tag{1.35}$$

This expression for $U(r)$ is valid for metallic particles. The case of dielectric spheres is much more complicated, and we do not analyze it – however, see [5]. As is seen from Eq. (1.35) the image forces are singular at the particle surface. Nevertheless, it

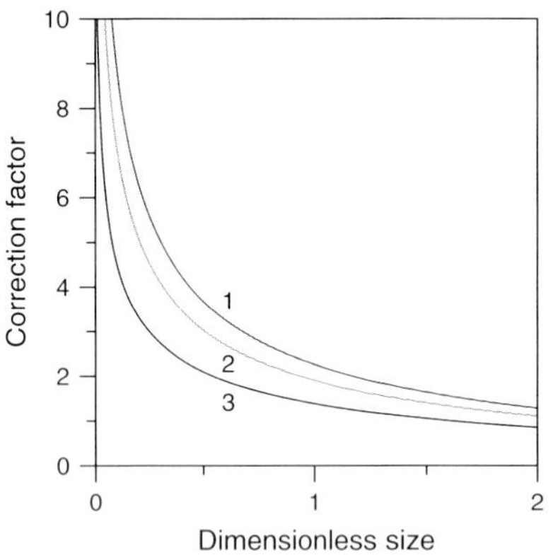

Figure 1.3 When an ion approaches a neutral particle, the image forces strongly enhance the efficiency of ion capture. The correction factors for the free-molecule efficiency versus dimensionless particle size av_T/D is shown here. It is seen that at large sizes the correction factor approaches unity. Curves 1–3 correspond to Coulomb numbers: Cu = 1, 3, and 5, respectively.

is possible to find the expression for the charging efficiency following the method of [6]. The final result has the form

$$\alpha(a) = \frac{2\pi a^2 v_T z(a)}{1 + \sqrt{1 + [a v_T z(a)/2D\zeta^2]^2}} \tag{1.36}$$

where

$$z(a) = 1 + \sqrt{\frac{\pi e^2}{2akT}} \tag{1.37}$$

and

$$\zeta^2 = 1 + \sqrt{\frac{2e^2}{\pi akT}} \tag{1.38}$$

Figure 1.3 shows the influence of the Coulomb number (see Eq. (1.8)) on the particle charging efficiency.

1.4.4 Recombination

Let us consider the situation when an ion carrying Z_i elementary charges approaches a particle of radius a carrying Z_p charges of opposite polarity. In this case Eq. (1.32) allows one to find the expression for the *recombination efficiency* in the continuum limit. We restrict our analysis to the case of non-singular Coulomb forces. Then we can approximate $R \approx a$, ignore unity in the denominator of Eq. (1.32), and come to the well-known *Langevin formula*,

$$\alpha(a) = \frac{4\pi D l_C}{1 - \exp(-l_C/a)} \tag{1.39}$$

where $l_C = Z_i Z_p e^2/kT$. In the limit of very small particles, the recombination efficiency is independent of particle size.

There are some difficulties in the case of smaller particles and image potential.

This section is based on the work by Lushnikov and Kulmala [6]. There exists an extensive literature on particle charging. Many authors addressed their efforts to deriving expressions for the charging efficiencies of an aerosol particle by ions. There are no problems in resolving this problem for the continuum limit, where the ion transport is described by the diffusion equation [7–9].

In the free-molecule regime the charging efficiency can be easily found only when the ion–particle interaction is described by the Coulomb potential alone. Attempts to take into account the image forces make the analysis much more difficult. Especially, this concerns the dielectric particles, in which case the ion–particle interaction is described by an infinite and slowly convergent series [10].

The first successful attempt to apply the free-molecule approximation for calculating the charging efficiencies of small aerosol particles was undertaken by Natanson [11, 12]. Since then, this problem has been considered by many authors [13–19]. None of these works could avoid the difficulty related to the very inconvenient expression for the ion–dielectric particle potential. The latter has been replaced by the ion–metal particle potential modified by the multiplier $(\epsilon - 1)/(\epsilon + 1)$, with ϵ being the *dielectric permeability* of the particle material.

Attempts to consider the transition regime using as the zero approximation the solution of the collisionless kinetic equation have been made [18–20] and very recently by us [6, 21]. The analysis of these authors clearly demonstrated the significance of the ion–carrier gas interaction in calculating the *ion–particle recombination efficiency*. The point is that the ion can be captured by the charged particle from bound states with negative energies. This effect has been considered in [20] by taking into account a single ion–molecule collision in the Coulomb field created by the charged particle. A new version of flux matching theory [11, 12, 22] has been applied by us [6] to take this effect into account explicitly. Results of experiments on particle charging can be found in [23–28].

1.5
Fractal Aggregates

It is now well established that fractal aggregates (FAs) appear in numerous natural and anthropogenic processes. Their role in the atmosphere may be immense, for FAs possess anomalous physico-chemical, mechanical, and optical properties, making them extremely effective atmospheric agents.

The main goal of this section is to overview the mechanisms of FA formation and their properties, and to discuss the sources and sinks of atmospheric FAs and their possible contribution to intra-atmospheric processes.

1.5.1
Introduction

The presence of aggregated structures in the atmosphere was detected very long ago: forest fires and volcanic eruptions are well known to produce tremendous amounts of ash and other aggregated particles. Many authors have attempted to estimate the role of the latter in the formation of the Earth's climate. Transport and industrial aerosol exhausts also often contain a considerable amount of aggregated particulate matter, not to mention such intense anthropogenic sources like oil and gas fires. Specialists on the "Nuclear Winter" did not push this problem to one side either.

Irregularly shaped particles have been studied for many years, but until fairly recently there was no unique and effective key idea for their characterization that would reflect the common origin of irregular aggregates or would allow the explanation of their physico-chemical behavior from a unique position.

Therefore, the fractal ideas introduced into physics (and other natural sciences) by Mandelbrot [29] immediately attracted the attention of aerosol scientists, who applied them for the characterization of atmospheric and laboratory-made aggregated aerosol particles.

So the fractal concept quickly found its way into the study of atmospheric aerosols. The success in its application to aerosols gave rise to a splash of fractal activity at the end of the 1980s and the beginning of the 1990s. The main efforts were directed at recording FAs in the atmosphere, attempting to define their fractal dimension, and returning the physics of FAs to the realm of the former and habitual ideas such as aerodynamic diameter, mobility, *coagulation efficiency*, and so on. Although the successes along this route were doubtless – even the optical properties of Titan's hazes were explained by assuming that they consist of FAs – the slight coolness that came later resulted, perhaps, from the impression that there is almost nothing to investigate any further. Of course, this is not so: the newly discovered physical and chemical properties of aggregated particles are pertinent to bear in mind in considering aerosol processes.

This section focuses on the properties of self-similar or, better, *scaling-invariant aggregates* – so-called fractals or fractal aggregates – whose structure is repeated within a considerable range of spatial scales (from tens of nanometers up to fractions of a centimeter or even more). This very kind of order stipulates many unusual properties of FAs.

The books edited by Avnir [30] and by Pietronero and Tosatti [31] contain sufficiently full information on the directions of the development of fractal physics and chemistry. The interested reader can find a regular account of fractal ideas in the book of Feder [32]. Colbeck *et al.* [33] reviewed the fractal concept and its application to environmental aerosols. The fairly recent textbook by Friedlander [34] also contains a chapter on fractals.

1.5.2 Phenomenology of Fractals

A typical FA consists of small spherules with diameters of several tens of nanometers united in an aggregate of size on the order of micrometers. It is important to stress that the sizes of the spherules are much less than the characteristic parameters in the atmosphere, such as the mean free path of molecules or the characteristic wavelength of solar radiation, whereas the total aggregate sizes are either comparable with these parameters or even exceed them. It is also not surprising that the main attention in studying the atmospheric FAs has been on soot aggregates.

In this section the main concepts characterizing FA are introduced.

1) **Mass of FA.** Any FA is characterized by its total mass M, which can also be measured in units of a spherule mass or, better, by the number g of spherules comprising the FA.
2) **Size of FA.** It is natural to introduce the *gyration radius* of an FA as

$$R^2 = \frac{1}{g(g-1)} \sum_{i \neq j} (\boldsymbol{r}_i - \boldsymbol{r}_j)^2 \tag{1.40}$$

where $\boldsymbol{r}_i$ is the position of the ith spherule. The maximal size of an FA can also be of use:

$$d_{\max} = \max|\boldsymbol{r}_i - \boldsymbol{r}_j| \tag{1.41}$$

1.5.2.1 Fractal Dimension

Not every irregular aggregate is a fractal. The main point of the definition of an FA is the self-similarity at every scale, which eventually leads to rather odd ramified structures of FAs whose local mass distribution cannot be so easily measured.

The most straightforward way to measure D is to follow its definition. Let the FA (or other fractal object) be covered with boxes whose size ϵ goes to zero. If the number N of boxes filled with the elements of the FA grows as $N \longrightarrow \epsilon^{-D}$, then D is identified with the fractal dimension of the FA.

The simplest and yet still non-trivial example of the application of the fractal concept to real objects is the measurement of the length of a diffusion trajectory. The diffusion displacement is given by $\Delta = \sqrt{2\mathcal{D}t_\Delta}$. If we represent Δ as the sum of smaller and smaller diffusion displacements $\epsilon = \sqrt{2\mathcal{D}t_\epsilon}$, then we find that the number $N(\epsilon)$ of the ϵ displacements necessary to cover the diffusion route is $N(\epsilon) = t_\Delta/t_\epsilon \propto \epsilon^{-2}$. The fractal dimension of the diffusion trajectory is thus $D = 2$.

FAs are very loose objects. Their fractal dimension D characterizes the part of space occupied by FA matter. This means that the mass of an FA grows with its gyration radius R more slowly than R^3: $M \propto R^D$, where $D < 3$. Such a dependence assumes that the FA density $\rho(r)$ changes with distance r from its center as

$$\rho(r) \propto \rho_0 \left(\frac{r_0}{r}\right)^{3-D} \tag{1.42}$$

at $r < R$ and as $\rho(r) = 0$ otherwise. Here ρ_0 is the density of the spherule and r_0 is its radius. Equation (1.42) provides the R^D dependence of the FA mass to hold:

$$g = k_D \left(\frac{R}{r_0} \right)^D \tag{1.43}$$

where k_D is the fractal prefactor

1.5.2.2 Correlation Function

The density–density *correlation function* also drops as a power of distance r:

$$C(r) = \left\langle \sum_i m(\boldsymbol{r}_i) m(\boldsymbol{r}_i + \boldsymbol{r}) \right\rangle \propto r^{-(3-D)} \tag{1.44}$$

where $m(\boldsymbol{r})$ is the density at the point $\boldsymbol{r}$, the sum goes over all centers of spherules, and the angle brackets stand for averaging over all possible spatial configurations of the spherules.

1.5.2.3 Distribution of Voids

FAs thus mainly consist of "empty space" distributed among voids whose size spectrum is of great importance for the characterization of FAs. This spectrum normalized to unity has the form

$$n(a) = \frac{3 - D}{R^{3-D}} a^{2-D} \tag{1.45}$$

One immediately sees that the total volume occupied by the voids is exactly $4\pi R^3/3$ once the shape factor γ defining the dependence of the average volume $V(a)$ of a void on its characteristic size a is given as $\gamma = 4\pi(6 - D)/3(3 - D)$ $(V = \gamma a^3)$.

1.5.2.4 Phenomenology of Atmospheric FA

Measurements of D of atmospheric FAs have shown the following:

1) Atmospheric FAs (mainly soot aggregates) are not well-developed fractal structures whose fractal dimensionality varies within the range 1.3–1.9, indicating that these FAs are of coagulation origin.
2) Such low fractal dimensions are explained by non-isotropy of observed FAs, which are mostly aligned in one direction. This anisotropy probably arises due to Coulomb or *dipole–dipole interaction* of FAs.
3) The fractal prefactor (Eq. (1.43)) for soot particles is $k_D \approx 27.46$ at $D = 1.75$.
4) The structure of atmospheric (soot) fractals may change by condensation–evaporation cycles: the loose FAs become more dense (D grows by 10–15%).

Katrinak *et al.* [35] analyzed urban aggregates within the size range 0.21–2.61 μm and found that D varies from 1.35 to 1.38. The maximal value of D found in [36] for diesel exhausts was $D = 1.2$, that is, their particles were strongly aligned. Considerable attention has been given by others [37–41] to the process of the transformation of FAs in a humid atmosphere. The chemical methods were applied by Eltekova *et al.* [42] for determining the D of soot FAs. The value of the fractal prefactor was discussed by Nyeki and Colbeck [43], who showed that k_f is close to 1.

1.5.3 Possible Sources of Fractal Particles

The sources of FAs are subdivided into two groups: natural and anthropogenic ones.

1.5.3.1 Natural Sources

Volcanos Volcanic eruptions produce a lot of volcanic ash, consisting of aggregated oxide particles of the size from fractions of a micrometer up to millimeters. In addition, extreme volcanic conditions produce a lot of smaller aggregates.

Forest fires These produce a huge amount of ash flakes whose sizes vary from fractions of a micrometer up to centimeters. Smaller aggregated particles accompany the combustion process (aggregated carbon plus hydrocarbon particles or, better, soot). The chemical content of the ash flakes is known: they consist of the mineral residue of the combustion process, resins, hydrocarbons, and the products of their chemical interaction with atmospheric air.

Thunderstorms High-energy lightning processes are able to release carbon from carbon-containing molecules and thus to produce small (nanometer-sized) charged carbon particles (maybe in the fullerene form), which then aggregate, forming FAs and even aerogels.

Intra-atmospheric chemical processes Intra-atmospheric chemical and photochemical processes are able to produce substances of low volatility that may then solidify into nanoparticles. On colliding, these objects form fractal structures.

1.5.3.2 Anthropogenic Sources

Industrial exhausts These produce a lot of smoke particles, FAs among them. The chemical content of these aggregates corresponds to the average content of the smoke. Unfortunately, what share of these particles is aggregated is not yet established.

Transport exhausts Transport produces aggregated aerosol particles consisting of nanometric soot particles. The sizes of these aggregates rarely exceed a micrometer.

Gas–oil fires Such fires produce aggregated soot particles (black smokes) consisting of nanometric units that reach sizes on the order of fractions of a centimeter.

There are many other less substantial sources of fractal aggregates.

1.5.4
Formation of Fractal Aggregates

One of the most important branches of fractal science is the study of the growth kinetics of fractal objects. There exist two commonly accepted approaches to this problem.

1) The first is the direct modeling of the growth process. The elements of fractal construction (spherules or fractal fragments) are assumed to move on a lattice, collide, stick together, and finally form a fractal structure. The whole process is modeled by computer from the very beginning up to the end. This approach allows one to investigate the structure of a single fractal aggregate, and to define its fractal dimension and other individual characteristics. In particular, it was shown that the fractal dimension D is totally stipulated by the type of growth process: namely, coagulation leads to the most loose structure, with $D \approx 1.8$; diffusion-controlled condensation gives more dense particles, with $D \approx 2$; and the collision-limited condensation process (low-efficiency collisions do not permit the spherule to join to the aggregate immediately after the first collision) produces the most dense FAs, with $D \approx 2.4$.
2) The growth process is considered within a kinetic scheme describing the time evolution of fractal mass spectra irrespective of the details of the motion of the fractal fragments, the latter being included via *kinetic coefficients* whose mass dependence alone defines the characteristic features of the mass spectra. In contrast to direct modeling, this approach accounts for the collective characteristics, first and foremost the mass distribution of growing fractal aggregates.

Below, the second (kinetic) approach – more traditional for aerosol physics – is used for studying the time evolution of the mass distribution of a collection of FAs growing by condensation and coagulation. The collective is assumed to consist of aggregates whose fractal dimension D does not change during the growth process. The initial stage of FA formation assumes the formation of monomers (spherules). We do not discuss this process, since it does not contain anything specific to fractal physics.

1.5.4.1 Growth by Condensation

The latter includes the joining of monomeric units (spherules or monomers) of unit mass by one, with the *condensation coefficients* α_g being known functions of the fractal aggregate mass g.

It is not very difficult to see that the condensation coefficients α_g should be proportional to the total number of spherules in the FA in the free-molecule limit, and to the FA size in the continuum regime. Indeed, an FA has a loose structure and the incident spherule readily reaches any point inside the FA where it can be captured, unless the collisions with the molecules of the carrier gas make the incident spherule trajectory very long and "knotty." In this latter case the FA becomes a "black absorber," that is, the incident spherule randomly walks inside the FA for long enough to be absorbed even if the absorption efficiency is not very high, and the average density of matter inside the FA is negligibly low. Hence

$$\alpha_g = \alpha_0 g \tag{1.46}$$

in the free-molecule regime, and

$$\alpha_g = A\mathcal{D}nr_0 g^{1/D} \tag{1.47}$$

in the opposite limit (the continuum regime).

The physical meaning of the constants entering Eqs. (1.46) and (1.47) is apparent: α_0 (Eq. (1.46)) is the rate of capture of an incident spherule by a vacancy incorporated into the FA. The right-hand side of Eq. (1.47) repeats the expression for the rate of condensational growth of a sphere in the continuum regime, except that the constant A is replaced by the usual coefficient 4π specific for spherical geometry. Equation (1.47) thus describes the diffusion growth of an FA whose radius is proportional to $g^{1/D}$. The values of α_0 and A cannot be found from theoretical considerations and should be thus considered as fitting parameters.

1.5.4.2 **Growth by Coagulation**

Coagulation seems to be the most effective mechanism of FA growth. Sufficiently large fractal aggregates grown by coagulation have rather low fractal dimensionality $D \approx 1.8$. The rate of the coagulation process depends on the form of the coagulation kernel – the efficiency for two colliding particles to produce a new one whose mass is equal to the sum of the masses of the particles. The coagulation kernel is the collision cross-section multiplied by the relative velocity of the colliding fragments. The easiest way to estimate the coagulation kernels is just to extend well-known expressions for the coagulation kernels for the free-molecule, continuum or transition regimes by substituting $R \propto g^{1/D}$ instead of $R \propto g^{1/3}$. So, one may expect that the following collection of coagulation kernels governs the time evolution of mass spectra of coagulating FAs:

- free-molecule regime

$$K(x, y) \propto (x^{1/D} + y^{1/D})^2 \sqrt{x^{-1} + y^{-1}} \tag{1.48}$$

- continuum regime

$$K(x, y) \propto (x^{1/D} + y^{1/D})(x^{-1/D} + y^{-1/D}) \tag{1.49}$$

- turbulent regime

$$K(x, y) \propto (x^{1/D} + y^{1/D})^3 \tag{1.50}$$

- coagulation of magnetic or electric dipoles

$$K(x, y) \propto x^{1/D} y^{1/D} \tag{1.51}$$

- coagulating FA form linear chains

$$K(x, y) \propto xy \tag{1.52}$$

Here x and y stand for the masses of the colliding particles. All the above kernels are homogeneous functions of the variables x and y: $K(ax, ay) = a^{\lambda} K(x, y)$. The homogeneity exponent $\lambda < 1$ for the first two kernels, and may exceed unity otherwise. The latter fact means that the *aerosol–aerogel transition* is possible in the last three cases.

1.5.4.3 **Aerosol–Aerogel Transition**

This remarkable phenomenon consists of the formation of a macroscopic object (or objects) from initially microscopic aerosol particles. Everyone has seen the web-like structures or lengthy filaments suspended in the air or attached to the walls of cleaning devices. Sometimes such objects spontaneously arise in the carrier gas as a consequence of the coagulation process in cases when the coagulation kernel grows sufficiently fast with the colliding particle masses ($\lambda > 1$). Aerosols consisting of fractal aggregates are the most probable candidates to form aerogels by coagulation.

Atmospheric aerogel objects may play a crucial role in the formation of ball lightning. According to the model developed by Smirnov [44], ball lightning is a plasma ball spanned on an aerogel framework. This aerogel framework may form after a linear lightning strike, which is able to produce fractal aggregates by ablation or directly from carbon-containing molecules in the air. Although the dynamics of this process is not yet fully understood, the aerogel model was shown to be a useful perspective for explanation of many properties of ball lightning.

A huge literature is devoted to computer modeling of FA formation. It is summarized in the review article by Meakin [45]. The mass spectrum of a growing FA meets the set of kinetic equations describing FA condensational growth. These equations were analyzed and solved by Lushnikov and Kulmala [46].

Coagulation of fractals in the free-molecule regime was theoretically investigated by Wu and Friedlander [47, 48], who found considerable broadening of the particle mass spectra on decreasing the fractal dimensionality. Similar results were reported by Vemury and Prastinis [49]. Wu *et al.* [50] proposed a method for definition of D from the kinetics of coagulation. The interested reader will find a rather simple introduction to fractal physics in the review by Smirnov [51], where considerable attention is given to the kinetics of FA formation.

The coagulation in the system with the kernel $K = xy$ was analyzed by Lushnikov [52–58], who showed that a gel should form from coagulating sol after a finite interval of time. Experimentally, this process was observed by Lushnikov *et al.* [59, 60], who supposed that the dipole–dipole interaction of FAs is responsible for this phenomenon.

1.5.5
Optics of Fractals

Atmospheric fractals reveal very specific optical properties interacting intensely with sunlight. This fact is linked closely with their structure: the geometrical size of the atmospheric FAs lies in the micrometer range, that is, the particle sizes are comparable with the wavelength of visible light and infrared radiation. On the other hand, FAs are composed of tiny nanometer-sized units whose electrodynamic properties often differ from those of macroscopic objects. This felicitous combination of micro- and macro-properties together with a kind of spatial order (scaling invariance) stipulate specific optical properties of fractal aggregates.

Strong spatial correlations of the nanospherules (Eq. (1.44)) lead to the singularity in the differential elastic cross-section at small angles:

$$\frac{d\sigma_e}{d\Omega} \propto \int d^3r\, C(r)\, e^{i\boldsymbol{q}\cdot\boldsymbol{r}} \propto \frac{1}{q^D} \tag{1.53}$$

where $q = 2\pi\lambda^{-1}\sin(\theta/2)$ and θ is the scattering angle. This singular behavior serves in some cases for the experimental determination of the fractal dimension D.

Voids in FAs (Eq. (1.45)) may create the conditions for the capture of light quanta inside them. Sometimes (under special resonance conditions) FAs consisting of weakly absorbed spherules are able to absorb light.

In their comparison of a fractal smoke optical model with light extinction measurements, Dobbins *et al.* [61] used the following expressions for the absorption and elastic scattering cross-sections:

$$\sigma_{abs} = \frac{6\pi E(m)}{\lambda\rho_p} \tag{1.54}$$

and

$$\sigma_{sca} = \frac{4\pi\overline{n}^{(2)} F(m)}{\lambda\rho_p\overline{n}^{(1)}}\left(1 + \frac{4}{3D}k^2R_g^2\right)^{-D/2} \tag{1.55}$$

with $\overline{n}^{(1,2)}$ being the first and second moments of the FA size distribution function, $k = 2\pi/\lambda$, $x_p = 2\pi r_0/\lambda$, λ being the wavelength of the incident light,

$$E(m) = \mathrm{Im}\left(\frac{m^2-1}{m^2+2}\right) \quad \text{and} \quad F(m) = \left|\frac{m^2-1}{m^2+2}\right| \tag{1.56}$$

These rather simple expressions were applied for the analysis of the results on the light extinction of aggregated soot aerosols with $D = 1.75$, and a reasonable agreement of predicted and measured values was found.

The paper by Berry and Persival [62] gave the starting push to the studies of the optics of FAs. The computational analysis of the *Rayleigh–Debye–Gans theory* performed by Farias *et al.* [63] (see also references therein) showed its applicability for soot FAs. The authors concluded that this theory should replace other approximations for the description of soot optical properties, such as Rayleigh scattering and Mie scattering for an equivalent sphere. Lushnikov and Maximenko [64] investigated the localization effects in FAs and found that FAs with $D < 3/2$ consisting of weakly absorbing materials may nevertheless be "black" due to the capture of the incident light quanta by voids inside the FAs. Other optical properties (hyper-combinational scattering, scattering at small angles, and photoabsorption) of FAs were also investigated [65]. Cabane and colleagues [66, 67] explained the contradictions between the results of polarization and photometric measurements of the upper layer of Titan's atmosphere by assuming that FA clouds are responsible for the light scattering effects.

1.5.6 Are Atmospheric Fractals Long-Lived?

The answer to this question depends on the mobility of FAs, which is expected to be much lower than that of compact particles of the same mass. The experimental and numerical studies of the mobilities of aggregated particles allow for some useful semiempirical relations to be established.

In the continuum regime, it was found [68] that the mobility diameter of an FA is

$$d_{\mathrm{mc}} = 2\beta R_{\mathrm{g}} = \beta d_1 \sqrt{\frac{D}{D+1}} g^{1/D} \tag{1.57}$$

with $\beta = 0.7\text{–}1.0$ and d_1 being the spherule diameter. In the free-molecule kinetic regime, $d_{\mathrm{mk}} \approx d_A$, where d_A is the radius of the equivalent projected sphere. The rather ancient "adjusted sphere" interpolation expression for the equivalent diameter by Dahneke (cited in [68]) has the form

$$\frac{d_{\mathrm{m}}}{C_{\mathrm{c}}(\mathrm{Kn}_{\mathrm{m}})} = \frac{d_{\mathrm{mc}}}{C_{\mathrm{c}}(\mathrm{Kn}_{\mathrm{ck}})} \tag{1.58}$$

where d_{m} is the transition mobility diameter, d_{mc} and d_{mk} are the kinetic and continuum regime mobility diameters defined above, $\mathrm{Kn}_{\mathrm{m}} = 2\lambda/d_{\mathrm{m}}$, and $\mathrm{Kn}_{\mathrm{ck}} = 2\lambda d_{\mathrm{mc}}/d_{\mathrm{mk}}^2$. The slip correction factor is introduced as

$$C_{\mathrm{c}}(\mathrm{Kn}) = 1 + \mathrm{Kn}\left[A + B\exp\left(\frac{-C}{\mathrm{Kn}}\right)\right] \tag{1.59}$$

with $A = 1.257$, $B = 0.4$, and $C = 1.1$; λ is the mean free path.

The sinks of FAs in the atmosphere are:

- diffusion deposition, which is smaller by $g^{2/D-2/3}$ in the free-molecule regime, and by $g^{1/D-1/3}$ in the continuum regime;
- sedimentation losses, which are smaller by $g^{1/D-1/3}$.

Other mechanisms are as follows:

- collapse by humidification, in which water condensation on atmospheric fractals may effectively enlarge their fractal dimensionality by 10–15%, making them more and more compact;
- water capture by (even hydrophobic) FAs;
- scavenging by rain- and snowfall.

The latter two mechanisms are likely the most effective.

Colbeck and Wu [69] used the relation $d_{\mathrm{m}} \propto d_V^{3(D-1)/2D}$ linking the mobility diameter d_{m} with the volume equivalent diameter d_V for determination of D of smoke FAs. They found D to lie within the interval 1.40–1.96. A useful relation linking the particle mobility with the optical diameter of soot aggregates ($d_{\mathrm{opt}} \propto d_{\mathrm{m}}^{1/3}$) has been reported [70]. Expressions for thermal and sedimentation velocities in terms of a geometric particle diameter have also been given [71]. Huang *et al.* [38] performed experiments with diesel engines that emit chain–agglomerate particles and found changes in D from 1.56 to 1.76 and from 1.40 to 1.54 depending on the sulfur content. The "rigidity" of the chains was demonstrated to grow on increasing the sulfur content.

1.5.7
Concluding Remarks

The fractal concept is undoubtedly fruitful for characterization of the present-day aerosol situation in the Earth's atmosphere. At the same time, it should be noted that the concept itself needs development when applied to atmospheric aerosols. The still not numerous observations of atmospheric fractal aggregates show that their sizes (better, the numbers of spherules comprising the aggregates) are not large enough to expose a well-developed fractal picture. Perhaps, distributions over D will be of use for their proper characterization.

The fractal aggregates manifest anomalous physico-chemical properties: their lifetimes are much longer than those of compact particles of the same mass (by 10–100 times in the case of atmospheric fractals); their light scattering and absorption cross-sections are higher by orders of magnitude than those of the equivalent collective non-aggregated spherules; and their chemical and catalytic activities are also enhanced by the specifics of the fractal aggregate morphology. This is why even a small admixture of fractal aggregates may seriously change the existing estimates of aerosol impact on the global radiation and chemical cycles in the atmosphere.

The condensation of atmospheric moisture on fractal aggregates was shown to restructure them, making the aggregates more compact. This process reduces their lifetimes. The recognition of this fact helps to answer the question of where to seek them. The upper layers of the atmosphere and near-space are the most probable places for the accumulation of fractal aggregates.

The fractals of the lower troposphere are mainly of anthropogenic origin and hardly to be thought as very desirable guests. Being good absorbers, they are able to accumulate harmful substances and radioactivity, and to transport them inside living organisms. Hence, the environmental aspects of the atmospheric fractal aggregates are of great importance.

1.6
Coagulation

Coagulation is a collective aerosol process. This means that the equation describing the kinetics of this process is nonlinear with respect to the particle size distribution (see Eq. (1.60) below). This section introduces the reader to some fairly new concepts that appeared not very long ago [54–58, 72, 73]. We begin with a short description of the coagulation process. More details can be found in the chapter written by Maisels in this book [74].

At first sight the coagulation process looks rather offenceless. A system of M monomeric objects begins to evolve by pair coalescence of g- and l-mers according to the scheme

$$(g) + (l) \longrightarrow (g + l) \tag{1.60}$$

It is easy to write down the kinetic equation governing this process as

$$\frac{dc_g(t)}{dt} = \frac{1}{2}\sum_{l=1}^{g-1} K(g-l, l)c_{g-l}(t)c_l(t)\,dl - c_g(t)\sum_{l=1}^{\infty} K(g, l)c_l(t)\,dl \tag{1.61}$$

This equation is known as the *Smoluchowski equation*. Here the coagulation kernel $K(g, l)$ is the transition rate for the process given by Eq. (1.60). The first term on the right-hand side of Eq. (1.61) describes the gain in the g-mer concentration $c_g(t)$ due to coalescence of $(g - l)$- and l-mers, while the second one is responsible for the losses of g-mers due to their sticking to all other particles. Equation (1.61) can be rewritten in the integral form (sums on the right-hand side of Eq. (1.61) are replaced with integrals)

$$\frac{\partial c(g, t)}{\partial t} = \frac{1}{2}\int_0^g K(g-l, l)c(g-l, t)c(l, t)\,dl - c(g, t)\int_0^{\infty} K(g, l)c(l, t)\,dl \tag{1.62}$$

Eqs. (1.61) and (1.62) should be supplemented with the initial conditions,

$$c_g(0) = c_g^0 \qquad \text{or} \qquad c(g, 0) = c^0(g) \tag{1.63}$$

where c^0 are known function of g.

There are a number of coagulation kernels that are commonly used in aerosol physics, and they look as follows:

1) Coagulation in the free-molecule regime:

$$K(g, l) = \pi a_0^2\sqrt{\frac{8kT}{\pi m_0}}(g^{1/3} + l^{1/3})^2\sqrt{g^{-1} + l^{-1}} \tag{1.64}$$

The physical meaning of this expression is apparent: it is just the geometrical cross-section of g- and l-mers times their mutual thermal velocity times their reduced mass. Here m_0 stands for the mass of the monomer. The analogy with the formula for the *condensation efficiency* of small particles is clearly seen.

2) Coagulation in the continuum regime:

$$K(g, l) = \frac{2kT}{3\rho\nu}(g^{1/3} + l^{1/3})(g^{-1/3} + l^{-1/3}) \tag{1.65}$$

3) Coagulation in laminar shear flow:

$$K(g, l) = \frac{4}{3}\Gamma m_0(g^{1/3} + l^{1/3})^3 \tag{1.66}$$

where Γ is the velocity gradient directed perpendicular to the flow of the carrier gas.

4) Coagulation in turbulent flow:

$$K(g,l) = \sqrt{\frac{\pi \epsilon}{120\nu}}\, 8a_0^3 (g^{1/3} + l^{1/3})^3 \tag{1.67}$$

where ϵ is the rate of dissipation of kinetic energy of the turbulent flow per unit mass.

It is important to emphasize that the above kernels are homogeneous functions of g and l, that is,

$$K(ag, al) = a^{\lambda} K(g,l) \tag{1.68}$$

where λ is the homogeneity exponent.

Equations (1.61) and (1.62) can be modified by adding a source of fresh particles (the term $If(g)$ on the right-hand side of these equations), a sink of particles (the term $\lambda_g c_g$), and the particle condensational growth (the term $\partial(\alpha_g c(g,t)/\partial g$ on the left-hand side of Eq. (1.62)). Then the full equation (the general dynamic equation in the terminology of Friedlander [34]) has the form

$$\frac{\partial c(g,t)}{\partial t} + \frac{\partial \alpha(g) c(g,t)}{\partial g} = I(g) + (\mathrm{Kcc})_g - \lambda(g) c(g,t) \tag{1.69}$$

where $(\mathrm{Kcc})_g$ stands for the right-hand side of Eq. (1.62):

$$(\mathrm{Kcc})_g = \frac{1}{2}\int_0^g K(g-l,l) c(g-l,t) c(l,t)\, \mathrm{d}l - c(g,t)\int_0^{\infty} K(g,l) c(l,t) \mathrm{d}l \tag{1.70}$$

1.6.1 Asymptotic Distributions in Coagulating Systems

In what follows, we will use the dimensionless version of this equation, that is, all the concentrations are measured in units of the initial monomer concentration $c_1(0)$ and time in units of $1/c_1(0)K(1,1)$. More details can be found in the review articles [75, 76].

Let us introduce a family of homogeneous kernels [72, 73, 77]

$$K(g,l) = \frac{1}{2}(g^{\alpha} l^{\beta} + l^{\alpha} g^{\beta}) \tag{1.71}$$

Then

$$\lambda = \alpha + \beta \tag{1.72}$$

We also introduce the exponent μ, as

$$\mu = |\alpha - \beta| \tag{1.73}$$

In addition, we assume that the condensation efficiency may be approximated by an algebraic function,

$$\alpha_g \propto g^{\gamma} \tag{1.74}$$

The late stages of the time evolution of disperse systems, when either coagulation alone governs the temporal changes of particle mass spectra or simultaneous condensation complicates the evolution process, are studied under the assumption that the condensation efficiencies and coagulation kernels are homogeneous functions of the particle masses, with γ and λ, respectively, being their homogeneity exponents. Three types of coagulating systems are considered: (i) free coagulating systems, where coagulation alone is responsible for disperse particle growth; (ii) source-enhanced coagulating systems, where an external spatially uniform source permanently adds fresh small particles, with the particle productivity being an algebraic function of time, $I(t) \propto t^s$; and (iii) coagulating-condensing systems, in which a condensation process accompanies the coagulation growth of disperse particles. The particle mass distributions of the form

$$c_A(g, t) = A(t)\psi(gB(t)) \tag{1.75}$$

are shown to describe the asymptotic regimes of particle growth in all the three types of coagulating systems (g is the particle mass).

Friedlander [78] was the first to introduce the self-preserving form of the mass spectra in free coagulating systems. According to the hypothesis of self-preservation $A(t) = N^2(t)$, $B(t) = N(t)$, with $N(t)$ being the total number concentration of the coagulating particles. However, the family of self-preserving spectra is much wider (see [72, 73] and references therein).

The functions $A(t)$ and $B(t)$ are normally algebraic functions of time whose power exponents are found for all possible regimes of coagulation and condensation as functions of λ and γ. The equations for the universality function $\psi(x)$ are formulated. It is shown that in many cases $\psi(x) \propto x^{-\sigma}$ ($\sigma > 1$) at small x, that is, the particle mass distributions are *singular*. The power exponent σ is expressed in terms of λ and γ.

We have given the classification of the singular self-preserving regimes in coagulating systems and have defined the conditions for their realization. They are listed below.

1) In the free coagulating systems $\psi(x) \propto 1/x^{1+\lambda}$ at $x \ll 1$, which corresponds to the mass distribution of the form:

$$c_A(g, t) \propto \frac{1}{g^{1+\lambda}t} \tag{1.76}$$

The condition for the realization of this asymptotics is $\alpha, \beta > 0$. At $\beta = 0$ the singularity is weaker, $\psi(x) \propto 1/x^{1+\gamma}$, where $0 < \gamma < \lambda$.
It is not so difficult to understand the physical meaning of this condition: the rate of interaction of small particles ($g \propto 1$) with large ones ($g \gg 1$) is on the order of $K(1, g) \propto g^{\alpha}$ and $K(g, g) \propto g^{\lambda}$, respectively, that is, the smaller particles interact with the larger ones much more slowly than the large ones between themselves ($\alpha \leq \lambda$). Strongly polydisperse mass spectra thus form, in which the role of larger particles is less than that of smaller ones.
The situation changes drastically at $\beta < 0$. In this case $K(1, g) \gg K(g, g)$, that is, the larger particles "eat" the smaller ones much faster than each other. A hump in the distribution at large masses develops, while the concentrations of

small particles drops with time. A singular and a non-singular distribution are shown in Figure 1 of [73].

2) The inequality $\lambda, \mu \leq 1$ defines the conditions for the singular distributions to exist in source-enhanced coagulating systems. It is simply the conditions for the convergence of the integral on the right-hand side of Eq. Eq. (1.62). The singularity of the mass spectra in the source-enhanced coagulating systems is $\psi(x) \propto x^{-(3+\lambda)/2}$ or, in terms of the particle masses,

$$c_A(g, t) \propto g^{-(3+\lambda)/2} t^{-(1-s)(1+\lambda)/2(1-\lambda)} \tag{1.77}$$

At $s = 1$ (a source that is constant in time) the time-dependent multiplier turns to unity. The mass spectrum has a steady-state left wing, that is, the spectrum of the highly disperse fraction is independent of time, although the source permanently supplies the system with fresh portions of small particles. These particles deposit mainly on the larger ones, providing the right wing of the spectrum to move to the right along the mass axis. The steady-state regimes of coagulation in source-enhanced systems have been investigated [79, 80] (see also [73] and references therein).

3) We have considered systems of coagulating particles in which a source that is constant in time produces a vapor of low volatility condensing onto the particle surfaces. The particle growth in such systems is similar in many respects to that in source-enhanced and (sometimes) free systems. Several regimes have been detected.

   a. The disperse phase consumes all the mass of the vapor. In this case $\psi(x) \propto 1/x^{2-\gamma+\lambda}$, or

$$c_A(g, t) \propto 1/t^{2\gamma-1-\lambda} g^{2-\gamma+\lambda} \tag{1.78}$$

   The conditions for realizing these distributions are: $\gamma < 1$ and $2\gamma > 1 + \lambda$. At $\lambda < \gamma < (\lambda + 1)/2$ the coagulating–condensing system behaves like a source-enhanced coagulating system with linearly growing mass concentration.

   b. When the mass of the disperse phase grows more slowly than t, the asymptotic mass distribution in coagulating–condensing systems is the same as in source-enhanced systems. The singular asymptotics, however, is never realized. At $\gamma \leq 2\lambda - 1$ condensation is so slow that the coagulating disperse system consumes only a finite part of the vapor and the coagulation process goes like in free coagulating systems.

Singular asymptotic distributions have been known since 1975 [81]. But what is especially wonderful is the fact that such distributions had appeared in the exactly solvable model $K(g, l) = g + l$ [82], but people (including me) did not want to notice them. A thorough numerical analysis by Lee [83] showed that the characteristic time for reaching the singular asymptotics is much longer than in the case of non-singular distributions. I did not cite here very many of my own papers on asymptotic distributions (a false modesty), but one can find almost a full list of these works in Lushnikov and Kulmala [72, 73].

1.6.2
Gelation in Coagulating Systems

A half a century ago it had become clear that there is something wrong with Eq. (1.61). An attempt by Melzak [84] to find an exact solution to this equation for the kernel proportional to the masses of coalescing particles

$$K(g, l) \propto gl \tag{1.79}$$

had led to a strange conclusion that the total mass concentration ceases to conserve after a finite time $t = t_c$ (in what follows t_c is referred to as the critical time) and the second moment of the particle mass distribution $\phi_2 = \sum g^2 c_g$ has a singularity,

$$\phi_2(t) \propto \frac{1}{t_c - t} \tag{1.80}$$

Even more strange is the fact that at $t = t_c$ nothing wrong happens either to the particle mass spectrum or the particle number concentration. The whole situation is displayed in Figure 1.4.

Immediately, the problems of the existence of the solution to Eq. (1.61) and of its uniqueness were posed and resolved [85–88]. But this did not help to answer the question of what does happen after $t = t_c$.

On the other hand, it is clear that, if we consider a finite coagulating system, then at any time we see a number of bigger and bigger particles whose total mass M cannot disappear somewhere. It is worthwhile to characterize such a system

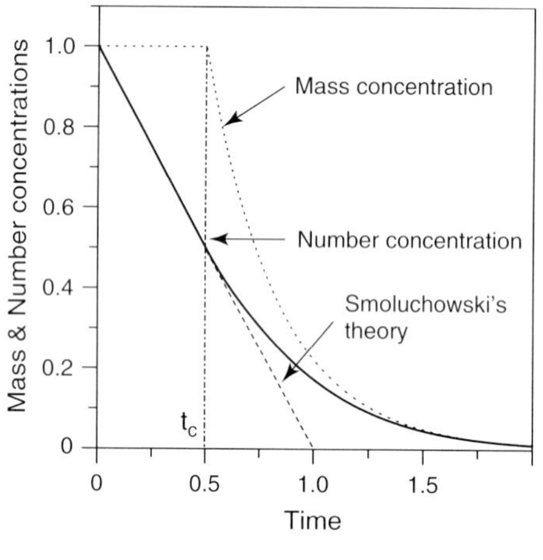

Figure 1.4 The total number and total mass concentrations of sol particles as functions of time (dimensionless units). After the critical time $t = t_c$ the mass concentration ceases to conserve, because a massive gel particle forms and begins to consume the mass of the sol. On the other hand, the number concentration does not feel the loss of one (although very big) gel particle. Still the post-critical behavior of the number concentration found from Eq. (1.61) differs from that predicted by the Smoluchowski equation ($n(t) = 1 - t$, dashed line).

by the set $\{n_g\}$ of occupation numbers of g-mers. Then it becomes possible to introduce the probability $W(\{n_g\}, t)$ for the realization of a given set at time t. Now the evolution of the coagulating system is fully described in terms of W. But a truncated description in terms of the average occupation numbers is also admissible:

$$\overline{n}_g(t) = \sum_{\{n_g\}} W(\{n_g\}, t) n_g \tag{1.81}$$

The particle concentrations are introduced as

$$c_g(t) = \frac{\overline{n}_g(t)}{V} \tag{1.82}$$

Here V is the total volume of the coagulating system.

Now we are ready to return to the question of what is going on in our system. The point is that the concentrations appearing in the Smoluchowski equation are defined as the thermodynamic limits of the ratios $\overline{n}_g/V$ (where $V \longrightarrow \infty, n_g \longrightarrow \infty$, and their ratio is finite), that is, if $\overline{n}_g$ are not large enough (for example, some of them $\propto V^{\alpha}, \alpha < 1$), then these particles are not "seen" in the thermodynamic limit, even if they exist. Still, these particles can have large masses, comparable to the mass of the entire system, and thus contribute to the mass balance. In coagulating systems, such particles can form *spontaneously* during a *finite* time (see Figure 1.5). This is gelation.

Two approaches have been applied for considering the *sol–gel transition*. The first approach does not conflict with the Smoluchowski description of gelling systems, that is, it starts with the Smoluchowski equation (see, for example, [89]). The mass deficiency appearing after the critical time is attributed to an *infinite* cluster (a gel), which is introduced "by hand" (its existence does not follow from the Smoluchowski equations) and serves only to restore mass conservation. The gel can be assumed to be either passive or active with respect to the sol fraction (defined as the collection of particles whose population numbers are macroscopically large). In the former case the gel grows due to a finite mass flux toward infinite particle sizes. The active gel grows, in addition, by consuming the sol particles. These two

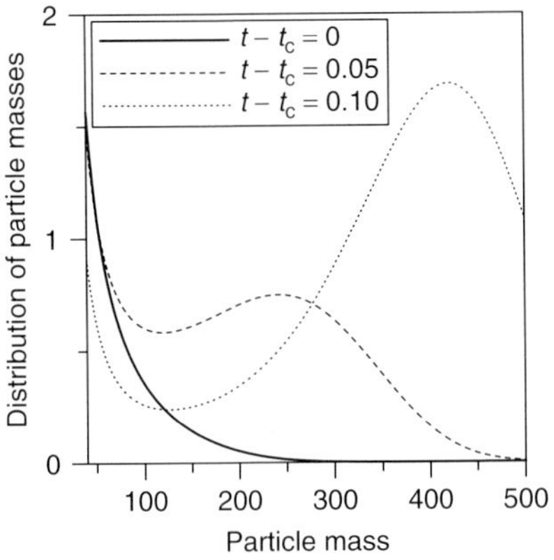

Figure 1.5 The particle mass spectrum at $t - t_c = 0$, 0.05, and 0.1. It is seen how the gel appears from nothing.

mechanisms are thoroughly discussed in [90], where the reader will find references to earlier works.

More accurately (but, again, within the Smoluchowski scheme) this process had been considered in [91], where an instantaneous sink of particles with masses exceeding a large one, G, was introduced. Then the kinetics of coagulation can be described by a truncated Smoluchowski equation, and no paradox with the total mass concentration comes up, for the mass excess is attributed to the deposit: the particles with masses $g > G$ consumed by the sink. Nevertheless, the difference between gelling and non-gelling systems manifests itself in the fact that during the whole pregelation period the mass is almost conserved, and only immediately after the critical time does a noticeable mass loss appear. The description of this model can be found in [58, 92].

The second, alternative, approach applying the Marcus [93] scheme to the gelation problem appeared earlier in [54–57, 94, 95]. The idea of this approach relies upon the consideration of finite coagulating systems. As mentioned above, this approach operates with the occupation numbers and the probability for the realization of a given set of occupation numbers. Within this scheme, the gel manifests itself as a narrow hump in the distribution of the average particle numbers over their masses. This hump appears after the critical time at macroscopically large mass $g \propto M$ and behaves like the active gel, that is, it influences the particle mass spectrum of the sol.

The master equation governing the time evolution of the probability is extremely complicated, but on replacing it by another one, for the generating functional of the probability W, it acquires a similarity with the Schrödinger equation for interacting quantum Bose fields. Although many features of the solution to the evolution equation for the generating functional were clear almost three decades ago, only very recently was I able to find the exact solution to this equation in a closed form and to analyze the behavior of the particle mass spectrum in the thermodynamic limit [54–58].

The description of the coagulation process in terms of occupation numbers (numbers of g-mers considered as random variables) was first introduced by [93]. This approach was then reformulated by me [52–58, 94, 95] in a form strongly reminiscent of the second quantization. Below I outline this approach.

The idea of this approach is very simple. Let us consider a process in which a pair of identical particles A, on colliding, produce one A particle (the process $A + A \longrightarrow A$). Let there be M such particles moving chaotically in the volume V. They collide and coalesce. Two particles produce one. This is exactly like in a coagulation process. The *collision rate* (the probability per unit time for a pair of particles to collide) is introduced as κ/V, where κ is the rate constant of the binary reaction $A + A \longrightarrow A$. The rate equation for the particle number concentration $c(t)$ describes the kinetics of the process:

$$\frac{\mathrm{d}c}{\mathrm{d}t} = -\kappa c^2 \tag{1.83}$$

However, we can choose an alternative route and introduce the probability $W(N, t)$ to find exactly N particles at time t in our system. It is also easy to guess that

$$\frac{dW(N,t)}{dt} = \frac{\kappa}{2V}[(N+1)NW(N+1,t) - N(N-1)W(N,t)] \tag{1.84}$$

The first term on the right-hand side of this equation gives the positive contribution to the rate $d_t W$ because of the coalescence of two particles ($(N+1)N/2$ is the number of ways to choose a pair of coalescing particles from $N+1$ particles). The second term describes the negative contribution to $d_t W$, because the particles continue to coalesce and transfer the system from the state with N particles to the state with $N-1$ particles.

Hence, a simple nonlinear equation (1.83) is replaced by a set of linear equations (1.84). However, we do not stop at this point and will make a step forward. We introduce the generating function for our probability,

$$\Psi(z,t) = \sum_N W(N,t) z^N \tag{1.85}$$

From Eq. (1.84) we can derive the equation for Ψ as

$$V\frac{\partial \Psi}{\partial t} = \frac{\kappa}{2}(z - z^2)\frac{\partial^2 \Psi}{\partial z^2} \tag{1.86}$$

This equation should be supplemented with the initial condition

$$\Psi(z,0) = \psi_0(z) \tag{1.87}$$

where $\psi_0(z)$ is a reasonable function (it should be analytical at $z = 0$). For example, if we fix the number of particles N_0 at the beginning of the process, then $\psi_0(z) = z^{N_0}$. Alternatively, the function $\psi_0(z) = e^{N_0(z-1)}$ corresponds to an initial Poisson distribution.

Two questions immediately come up: (i) Why should we introduce such a complex scheme for describing the kinetics of the reaction – why not use Eq. (1.83)? (ii) If the second scheme describes the same process as Eq. (1.83), then how do we derive Eq. (1.83) from Eq. (1.86)?

First, I answer the second question. Let us expand the right-hand side of Eq. (1.86) near $z = 1$, that is, we replace $z - z^2 \approx -2\xi$, where $\xi = z - 1 \ll 1$. We find from Eq. (1.86) that

$$V\frac{\partial \Psi}{\partial t} = -\kappa\xi\frac{\partial^2 \Psi}{\partial \xi^2} \tag{1.88}$$

Now it is easy to solve this equation to obtain

$$\Psi(z,t) = e^{c(t)V(z-1)} \tag{1.89}$$

On substituting this into Eq. (1.88) we come to the conclusion that the concentration $c(t)$ satisfies Eq. (1.83). Thus the probability has Poisson form. This approximation works well at very large N.

Now let us return to the first question. If we want to describe a *finite* system, then eventually we should use Eq. (1.84) or, better, Eq. (1.86). Because the description of a gel demands a step beyond the scope of the thermodynamic limit, I will use this very approach, although it requires much more serious efforts for operating and understanding the final results.

Here the "pathological" coagulating systems have been considered, that is, systems whose development in time leads to the formation of an object that is not provided for by the initial theoretical assumptions. In our case it is the gel whose appearance breaks the hypothesis that the kinetics of coagulation can be described in terms of the particle number concentrations defined as the thermodynamic limit of the ratio (occupation numbers)/volume.

The coagulating system with kernel proportional to the product of the masses of two colliding particles is the central object of the present study. Although the main decisive step in understanding the nature of the sol–gel transition in finite systems with $K \propto gl$ had been done long ago, only recently was I able to find the exact solution of this salient problem [54–58, 92, 96]. The central goal of this section was to introduce the reader to the main ideas of the approach that I so adore. Here I have demonstrated that this approach is eminently applicable to the solution of other problems, like the time evolution of random graphs or gelation in coagulating mixtures.

At first sight, the coagulation process cannot lead to something wrong. Indeed, let us consider a *finite* system of M monomers in the volume V. If the monomers move, collide, and coalesce on colliding, the coagulation process, after all, forms one giant particle of mass M. The concentration of this M-mer is small, $c_M \propto 1/M$. It is better to say that it is zero in the thermodynamic limit $V, M \longrightarrow \infty$, $M/V = m < \infty$. In other words, no particles exist in coagulating systems after a sufficiently long time. But still something unexpected goes on in gelling systems after a finite interval of time. The gel forms.

Two scenarios of gelation in coagulating systems have been considered in [54–58, 92]. The first one considers the coagulation process in a system of a finite number M of monomers enclosed in a finite volume V. In this case any losses of mass are excluded "by definition." The gel appears as a *single* giant particle of mass g comparable to the total mass M of the whole system.

What happens then in the system with $K(g, l) \propto gl$ in the thermodynamic limit? The answer is simple, although in no way apparent. In contrast to "normal" systems, where the time of formation of a large object grows with M, a giant object with a mass on the order of M forms during a *finite* (independent of V and M) time t_c. After $t = t_c$ this giant particle (gel) actively begins to absorb the smaller particles. Although the probability for any two particles to meet is generally small ($\propto K(g, l)/V$), in the case of $g \propto M$ this smallness is compensated by the large value of the coagulation kernel proportional to the particle mass M, which is, in turn, proportional to V. Hence, the gel whose concentration is zero in the thermodynamic limit can play a considerable role in the evolution of the whole system. The structure of the kernel is also the reason why only one gel particle can form. The point is that the time for the process $(l) + (m) \longrightarrow (l + m)$ is short for $l, m \propto M$: $\tau \propto V/K(l, m) \propto V/M^2 \propto 1/V \longrightarrow 0$ in the thermodynamic limit.

Of course, the Smoluchowski equation is not able to detect particles with zero concentration.

As mentioned above, the total mass concentration of the spectrum $\overline{n}_g^{(s)}(t)$ is not conserved at $t > t_c$. It is easy to show [54–57] that the deficit of the mass concentration after the critical time t_c is

$$2t = \frac{1}{\mu_c(t)} \ln\left(\frac{1}{1-\mu_c(t)}\right) \quad \text{or} \quad \mu_c = 1 - e^{-2\mu_c t} \tag{1.90}$$

This equation has only one root $\mu_c(t) = 0$ at $t < t_c$ and two roots at $t > t_c$. It is clear why we should choose the positive non-zero root after the critical time.

The mass distribution in the variables g, ϵ has the form (see also [54])

$$\overline{n}_g(t) = C(g,\epsilon)\exp\left(-\frac{g^3}{8M^2} + \epsilon\frac{g^2}{M} - 2g\epsilon^2\right) \tag{1.91}$$

Unfortunately, our asymptotic analysis does not allow for restoring the normalization factor $C(g,\epsilon)$. Still, some conclusions on its form can be retrieved from the mass conservation,

$$C(g,\epsilon) = \frac{M}{\sqrt{2\pi g^5}} + \frac{\sqrt{\epsilon}\,\theta(\epsilon)}{\sqrt{2\pi}\,M} \tag{1.92}$$

with $\theta(\epsilon)$ being the Heaviside step function. Indeed, below the transition point the total mass is conserved and the asymptotic mass spectrum is known. Equations (1.91) and (1.92) reproduce the latter at $g \ll M$. Above the transition point the second term normalizes the peak appearing at $g = \mu_- M$ to unity.

Now it becomes possible to describe what is going on. Below the transition point (at $\epsilon < 0$) the mass spectrum drops exponentially with increasing g. The terms containing M in the denominators (see Eq. (1.91)) play a role only at $g \propto M$. At these masses, the particle concentrations are exponentially small. In short, in the thermodynamic limit and at $\epsilon < 0$ the first two terms in the exponent on the right-hand side of Eq. (1.91) can be ignored. The spectrum is thus given by the equation

$$\overline{n}_g(t) = \frac{M}{\sqrt{2\pi g^5}} e^{-2g\epsilon^2} \tag{1.93}$$

At the critical point ($t = t_c$ or $\epsilon = 0$) the spectrum acquires the form

$$\overline{n}_g(t) = \frac{M}{\sqrt{2\pi g^5}} e^{-g^3/8M^2} \tag{1.94}$$

Although the expression in the exponent contains M in the denominator, we have no right to ignore it, for this exponential factor provides the convergence of the integral for the second moment $\phi_2 = M^{-1}\sum g^2 \overline{n}_g$ in the limit $M \longrightarrow \infty$. We thus have

$$\phi_2(t_c) = \frac{1}{\sqrt{2\pi}} \int_0^M \frac{e^{-g^3/8M^2}\,dg}{\sqrt{g}} \approx \frac{1}{3\sqrt{\pi}}\Gamma\left(\frac{1}{6}\right) M^{1/3} \tag{1.95}$$

Here $\Gamma(x)$ is the Euler gamma function.

The second (and the most widespread) scenario assumes that after the critical time the coagulation process instantly transfers large particles to a gel state, the latter being defined as an infinite cluster. This gel can be either passive (it does not interact with the coagulating particles) or active (coagulating particles can stick to the gel). In the latter case, the gel should be taken into account in the mass balance and no paradox with the loss of total mass comes up (see [52, 53, 94, 95]). Still, neither this definition nor the post-gel solutions to the Smoluchowski equation give a clear answer to the question of what the gel is.

The situation becomes more clear on considering a class of so-called truncated models (Section 1.5). In these models a cutoff particle mass G is introduced. The truncation is treated as an instant sink removing very heavy particles with masses $g > G$ from the system. So we sacrifice mass conservation from the very beginning. The particles whose mass exceeds G form a deposit (gel) and do not contribute to the mass balance. Of course, the total mass of the active particles plus deposit is conserved. The time evolution of the spectrum of active particles (with masses $g < G$) is described by the Smoluchowski equation as before, with the limit ∞ in the loss term being replaced with the cutoff mass G. The set of kinetic equations then becomes finite and no catastrophe is expected to come up. We have shown that, indeed, nothing wrong happens even for the coagulation kernel $K \propto gl$. The total mass concentration of active particles drops with time, as it should, because the largest particles settle out to deposit. But as $G \longrightarrow \infty$ the total mass concentration of active particles is almost conserved at $t < t_c$ and only after the critical time ($t_c - t \propto G^{-1/2}$) does the deposit begin to form and the mass to drop down with time.

The question immediately comes up: What kernels are pathological? In 1973 I tried to answer this question. A primitive analysis of [97] (see also [73]) shows that for homogeneous kernels $K(ag, al) = a^{\lambda} K(g, l)$ the self-preserving asymptotic solution to the Smoluchowski equation should have the form:

$$c_g(t) \approx t^{-2/(1-\lambda)} \psi(gt^{-1/(1-\lambda)}) \tag{1.96}$$

This asymptotic formula shows that something wrong should happen at $\lambda > 1$. So the kernels with $\lambda > 1$ occurred under suspicion. This opinion has survived until now. People continue to attack this problem, but the problem remains too hard. What is known up to now? For the kernels $K_\alpha(g, l) = g^\alpha l^\alpha$ it has been proved that the sol–gel transition exists [76]. At $\alpha > 1$ a gel appears already from the very beginning of the coagulation process ($t_c = 0$). So we know very little.

The model $K \propto gl$ considered in detail here admits an exact solution. As has been shown, this solution is not so simple, especially if one tries to consider a finite system. More general models are less pleasant in this respect. Still, the approaches described above can help to answer quantitatively what is going on in more general systems. For example, the truncated models can be analyzed numerically. We can find the time behavior of particle mass concentration, to detect the gelation point (there is a chance that the mass ceases to be conserved very near the critical point), and to try to look for a solution that decreases as t^{-1} (post-critical behavior). Moreover, such attempts have been reported (see [76]).

1.7
Laser-Induced Aerosols

The idea to use powerful lasers for aerosol particle production appeared rather long ago (see [59] and references therein). Although the laser technologies are typically expensive and not very productive, their advantages are apparent: in irradiating the targets (solid, liquid or even gaseous), it is easy to reach the necessary regimes for particle formation by changing the parameters of the incident laser beam and the carrier gas inside the vessel containing the target.

The laser beam interacting with a solid target creates a heated spot erupting plasma consisting of ionized vapor molecules and molecules of the carrier gas. This plasma cloud screens partially (or even totally in the case of breakdown) the incident light, heating itself by photoabsorption. Simultaneously it begins to expand. On cooling enough, the vapor of the target material begins to form aerosol particles, that then grow by condensing vapor molecules and coagulation. The balance of the characteristic times of all these processes defines the characteristics of the produced aerosol particles: number concentration, the shape of the particle size distribution, degree of agglomeration, and so on.

The chain of events leading to aerosol formation looks as follows:

$$\begin{aligned}&\text{incident beam} \longrightarrow \text{plasma eruption} \longrightarrow \text{expansion of plasma cloud}\\&\longrightarrow \text{nucleation} \longrightarrow \text{condensational growth plus expansion of cloud}\\&\longrightarrow \text{coagulation} \longrightarrow \text{fractals} \longrightarrow \text{aggregation} \longrightarrow \text{gelation.}\end{aligned}$$

The latter stage goes only in specially chosen conditions. Below, all the stages of the *laser-induced aerosols* are considered step by step.

1.7.1
Formation of Plasma Cloud

Focused laser irradiation heats the spot on the target surface up to temperatures above the boiling point. At the characteristic pulse energies of our experiment, the time for heat propagation inside the target body is much longer than the evaporation time, that is, all *absorbed* energy is spent on evaporating the atoms of the target material. The pressure of the vapor in the plasma cloud formed in this way is typically a little below the saturation pressure. The erupted vapor forms a one-dimensional plasma beam, the front of which propagates with the sound speed, has a temperature of $T_\text{f} \approx 0.7T_0$ (T_0 is the spot temperature), and has a density of $\rho_\text{f} = 0.25\rho_\text{s}$. The flux of evaporated atoms is $J \propto v_T N_0$. The vapor plasma is strongly non-equilibrium and the density of ions exceeds its equilibrium value by several decimal orders and reaches the value $N_\text{i} \propto 10^{18}–10^{20}\ \text{cm}^{-3}$.

1.7.1.1 Nucleation plus Condensational Growth

The front temperature of the plasma cloud is sufficiently low for aerosol particles to form. The characteristic times for particle formation are typically on the order of 10^{-9} s, that is, much shorter than the cloud formation times (10^{-7} s). This fact means that all erupted vapor is spent for particle formation and allows one to evaluate the total particle number per pulse. Most likely, the number concentration of the aerosol particles is on the order of the ion number concentration, that is, heterogeneous nucleation on ions plays the central role. In this case the nucleation–condensation process should produce the particles containing 10^2–10^3 atoms or 1 nm in diameter. Experimentally observed particles are typically several times bigger, which likely means that not all ions are effective condensation nuclei. There is another explanation of this fact: not all particles are formed simultaneously. The earlier particles may then deplete the vapor and not permit the other smaller embryos to grow up to the particle size.

1.7.1.2 Coagulation

The concentration of forming particles is extremely high, so the characteristic coagulation times should be of the order of 10^{-5} s (this is the upper estimate). The coagulation process is thus longer than the lifetime of the plasma cloud, which means that already cooled (solid) particles enter the coagulation process, thereby forming rather loose fractal aggregates with fractal dimensionality close to the $D = 1.8$ characteristic for the coagulation process. It is not easy to treat respective experimental data, for measurements of the aggregate mass spectra are still impossible. Nor are the mechanisms of aggregate–aggregate interaction well recognized.

The resulting particle mass spectra depend strongly on the form of the coagulation kernels. Wu and Friedlander [47, 48] assumed that the extension of free-molecule coagulation is enough in order to describe the coagulation of fractal aggregates. They replaced the colliding particle radii by $r \propto M^{1/D}$ in the expression for the coagulation kernel and investigated the dependence of the asymptotic mass spectra on D. The latter was shown to become broader with decreasing D. Another assumption was made by Lushnikov *et al.* [59]; the coagulation kernel is proportional to $(r_1 + r_2)^{\alpha}$ with $\alpha = 2$ or 3. In this case the homogeneity exponent of the coagulation kernel exceeds 1, and a gelation process should occur [59].

1.7.2
Laser-Induced Gelation

A strong laser beam hitting a solid target produces an eruption of vapor. Subsequent condensation of the cooling vapor gives nanometric particles that are able to form very crumbly fractal aggregates of micrometer size. In turn, aerosols consisting of such aggregates continue to coagulate. The most enigmatic is the final stage of the ageing process: the fractal aggregates form a web-like structure of macroscopic size. The experimental observation of this effect was reported by Lushnikov, Negin and Pakhomov [59] (hereafter LNP). Later, similar experiments were performed by

Friedlander's group, which investigated in addition some properties of the *fractal filaments* [98]. A theoretical explanation of this effect appeared much earlier [52, 53, 94, 95], who gave an exact analysis of the *Flory–Stockmeyer model* of polymerization (coagulation kernel $K \propto gl$ with g and l being the masses of colliding particles) and showed that *one* giant object (superparticle) should appear after a *finite* interval of time. This effect was then studied theoretically by a number of authors (for citations see the book *Fractals in Physics* [31]) who gave some evidence in favor of the fact that such a phenomenon is not a rarity and is of great significance for understanding the processes of fractal structure growth.

Let us consider the coagulation of particles placed in a uniform electric field of strength E. This field induces the dipole moment $\boldsymbol{d} = \alpha \boldsymbol{E}$, with α being the particle polarizability. The interaction energy of two particles is proportional to $\boldsymbol{d}_1\boldsymbol{d}_2$ and maximal for two aligned dipoles. The latter fact means that the coagulating particle should form a needle-like structure as observed in LNP. Below, we assume that the mutual particle motion is due to their dipole–dipole interaction. The kinetics of needle-like particle formation is thus described by the Smoluchowski equation, with the coagulation kernel proportional to the scalar product of the field-induced dipole moments: $K(g, l) \propto \boldsymbol{d}_1\boldsymbol{d}_2$. The polarizability of each needle is proportional to its maximal size, which, in turn, is proportional to the total mass of the needle g. This means that the coagulation kernel has the form

$$K = k_0 E^2 gl \tag{1.97}$$

with k_0 being a proportionality coefficient.

The equation for the second moment of the particle mass spectrum, $\phi_2(t) = \sum g^2 c_g(t)$, describes the kinetics of the pre-gelation period $t < t_c$:

$$\phi_2(t) = [Kc_0(t_c - t)]^{-1} \tag{1.98}$$

where the critical time

$$t_c = [k_0 E^2 \phi_2(0)]^{-1} \tag{1.99}$$

depends on the strength of the external electric field. Equations (1.98) and (1.99) were checked experimentally. To this end Fe, Ti, Ag, and Al targets placed in a specially designed vessel filled with inert carrier gas at different pressures were irradiated by pulsed ruby and CO_2 lasers. The time dependence of the scattered light signal was recorded and the dependence of t_c on E was found.

The comparison of the theoretical and experimental results clearly favors the coagulation mechanism of gelation. The coagulation kernel is proportional to the product of the dipole moments of the coagulating particles. Such a mechanism evidently produces needle-like structures. The specifics of the coagulation interaction eventually lead to the phase transition: during a finite period of time, a macroscopic object forms (in this particular case, it is a set of fractal filaments).

Fractal filaments that result from the aerosol–aerogel transition are of interest for those who develop the technology of nanomaterials. Although the laser method has many disadvantages (it is expensive and complicated), it does allow the production of aerogels made out of any thermostable material. There are many other processes giving fractal filaments: plasma discharge (for example, electric sparks), and thermal or chemical decomposition (for example, it is no problem to produce aerogels by thermal decomposition of $Fe(CO)_5$). The application of fractal filaments is a matter for the future.

1.8 Conclusion

Aerosol science does not belong to the group of sciences that are based on one equation or principle, like, for example, classical mechanics (Newton's equation), quantum mechanics (Schrödinger's equation), classical electrodynamics (Maxwell's equations), and so on. Rather, aerosol science applies the results and methods adopted from all other sciences. In particular, Newton's equation applies in aerosol mechanics, Maxwell's equations are used in the theory of light scattering by aerosols, quantum-mechanical approaches are needed for studying the structure of small clusters, and even quantum field ideas have been used in aerosol science – the theory of gelation, the derivation of Mie theory from quantum principles [99], and the study of inelastic electromagnetic processes on aerosol particles [100]. This rather speckled structure of aerosol science makes it difficult to write a review enveloping all branches of aerosol science. Here I have restricted myself to the problems of kinetics of aerosols.

Coagulation Aerosol particles are not simply suspended in a carrier gas. They always move due to their collisions with the carrier gas molecules. On colliding, the moving particles coalesce, forming a new daughter particle with mass equal to the total mass of the parent particles. This process continues until very few particles remain in the system.

There are two problems in the theory of coagulation: (i) how to find the efficiency of particle collisions; and (ii) how to describe the time evolution of coagulating aerosols, once the collision rates are known functions of the particle sizes. Both problems have been considered. Sometimes coagulation leads to gelation. I have explained in short the conditions under which the gel can form (without entering into the heavy mathematical details).

Charging of particles Charging of aerosol particles is of undoubted importance. An aerosol particle can carry from one to thousands of elementary charges. I have discussed the kinetics of particle charging. My starting point was the flux matching theory of charge transport in the carrier gas. The Coulomb and image forces make the problem extremely complicated, especially because the image forces are singular at the particle surface. Without derivation, I exposed the final results for the charging efficiency of aerosol particles.

Drag on aerosol particle In this section I discussed the drag force on an aerosol particle of a given radius. The central result is the formula proposed by Millikan and derived by Phillips from the numerical analysis of the Boltzmann equation. Mobility, electromobility, and diffusivity are discussed in the light of this formula.

Condensation, evaporation, nucleation These problems are discussed in a separate chapter in this book [101]. The situation with nucleation remains unsatisfactory. As for condensation, I showed that semiempirical formulas by Fuchs and Sutugin and by Dahneke give results very close to the expression derived theoretically by Lushnikov and Kulmala. The advantage of the latter approach is the possibility to extend it to more complicated situation when a single particle–molecule interaction is switched on.

References

1. Lushnikov, A.A. and Kulmala, M. (2000) Foreign aerosols in nucleating vapour. *J. Aerosol Sci.*, **31**, 651.
2. Lushnikov, A.A. and Kulmala, M. (2000) Nucleation burst in a coagulating system. *Phys. Rev. E*, **62**, 4932.
3. Aitchison, J. and Brown, J.A.C. (1957) *The lognormal distribution*, Cambridge, University Press, Cambridge.
4. Phillips, W.F. (1975) Drag on a small sphere moving through a gas, *Phys. Fluids*, **18**, 1089–1093.
5. Lushnikov, A.A. and Kulmala, M. (2005) A kinetic theory of particle charging in the free–molecule regime, *J. Aerosol Sci.*, **36**, 1069.
6. Lushnikov, A.A. and Kulmala, M. (2004) Flux matching theory of particle charging *Phys. Rev. E*, **70**, 046413.
7. Reist, P.C. (1984) *Introduction to Aerosol Science*, Macmillan, New York.
8. Smirnov, B.M. (2000) *Clusters and Small Particles in Gases*, Springer, New York.
9. Smirnov, B.M. (2000) Cluster plasma. *Phys. Usp.*, **170**, 495–534.
10. Stratton, J.A. (1941) *Electromagnetic Theory*, McGraw-Hill, New York.
11. Natanson, G.L. (1959) On the theory of volume ion recombination *Sov. Phys. – Tech. Phys.*, **29**, 1373–1380.
12. Natanson, G.L. (1960) On the theory of the charging of amicroscopic aerosol particles as a result of capture of gas ions. *Sov. Phys. – Tech. Phys.*, **30**, 573–588.
13. Gentry, J.W. and Brock, J.R. (1967) Unipolar diffusion charging of small aerosol particles. *J. Chem. Phys.*, **47**, 64.
14. Keefe, D., Nolan, P.J., and Scott, J.A. (1968) Influence of Coulomb and image forces on combination in aerosols. *Proc. R. Irish Acad.*, **66A**, 17–29.
15. Brock, J.R. (1970) Aerosol charging: the role of the image force. *J. Appl. Phys.*, **41**, 843–844.
16. Marlow, W.H. and Brock, J.R. (1975) Unipolar charging of small aerosol particles. *J. Colloid Interface Sci.*, **50**, 32–38.
17. Marlow, W.H. (1980) Derivation of the collision rates for singular attractive contact potentials. *J. Chem. Phys.*, **73**, 6284–6287.
18. Huang, D.D., Seinfeld, J.H., and Marlow, W.H. (1990) BGK equation solution for large Knudsen number aerosol with a singular attractive contact potential. *J. Colloid Interface Sci.*, **140**, 258–276.
19. Huang, D.D., Seinfeld, J.H., and Okuyama, K. (1991) Image potential between a charged particle and an uncharged particle in aerosol coagulation-enhancement in all size regimes and interplay with van der

Waals forces. *J. Colloid Interface Sci.*, **141**, 191–198.
20. Hoppel, W.A. and Frick, G.M. (1986) Ion–aerosol attachment coefficients and the steady-state charge distribution on aerosols in a bipolar ion environment. *J. Aerosol Sci. Technol.*, **5**, 1–21.
21. Lushnikov, A.A. and Kulmala, M. (2004), Charging of aerosol particles in the near free-molecule regime. *Eur. Phys. J. D*, **29**, 345–355.
22. Fuchs, N.A. (1964) *The Mechanics of Aerosols*, transl. ed. C.N. Davies, Macmillan, New York.
23. Burtcher, H. and Schmidt-Ott, A. (1985) Experiments on small particles in gas suspension. *Surface Sci.*, **156**, 735–740.
24. Hussin, A., Scheibel, H.G., Becker, K.H., and Porstendorfer, J. (1983) Bipolar diffusion charging of aerosol particles – I. Experimental results within the diameter range 4–30 nm. *J. Aerosol Sci.*, **14**, 671–677.
25. Liu, B.Y.H. (ed.) (1976) *Fine Particles: Aerosol Generation, Measurement, Sampling, and Analysis*, Academic Press, New York.
26. Pui, D.Y.H., Fruin, S., and McMurry, P.H. (1988) Unipolar diffusion charging of ultrafine aerosols. *J. Aerosol Sci. Technol.*, **8**, 173–187.
27. Romay, F.J. and Pui, D.Y.H. (1992) On the combination coefficient of positive ions with ultrafine neutral particles in the transition and free-molecule regimes. *J. Aerosol Sci. Technol.*, **17**, 134–137.
28. Whitby, K.T. (1976) Electrical measurement of aerosols, in *Fine Particles: Aerosol Generation, Measurement, Sampling, and Analysis*, ed. B.Y.H. Liu, Academic Press, New York.
29. Mandelbrot, B.B. (1977) *Fractals: Form, Chance and Dimension*, W.H. Freeman, San Francisco, CA.
30. Avnir, D. (ed.) (1989) *The Fractal Approach to Heterogeneous Chemistry*, John Wiley & Sons, Inc., New York.
31. Pietronero, L. and Tosatti, E. (eds) (1986) Fractals in physics, in *Proceedings of the VIth Trieste International Symposium on Fractals in Physics*, North-Holland, Amsterdam.
32. Feder, J. (1988) *Fractals*, Plenum, New York.
33. Colbeck, I., Eleftheriadis, K., and Simons, S. (1989) The dynamics and structure of smoke aerosols. *J. Aerosol Sci.*, **20**, 875–878.
34. Friedlander, S.K. (2000) *Smokes, Haze, Mist*, John Wiley & Sons, Inc., New York.
35. Katrinak, K.A., Rez, P., Perkes, P.R., and Buseck, P.R. (1993) Fractal geometry of carbonaceous aggregates from an urban aerosol. *Environ. Sci. Technol.*, **27**, 539–547.
36. Klinger, H.-J. and Roth, P. (1989) Size analysis and fractal dimension of diesel particles based on REM measurements with an automatic imaging system. *J. Aerosol Sci.*, **20**, 861–864.
37. Lesaffre, F. (1989) Characterization of aerosol aggregates through fractal parameters. Effects due to humidity. *J. Aerosol Sci.*, **20**, 857–859.
38. Huang, P.F., Turpin, B.J., Pipho, M.J., Kittelson, D.B., and McMurry, P.H. (1994) Effects of water condensation and evaporation on diesel chain-agglomerate morphology. *J. Aerosol Sci.*, **25**, 447–459.
39. Nyeki, S. and Colbeck, I. (1995) Fractal dimension analysis of single, in-situ, restructured carbonaceous aggregates. *Aerosol Sci. Technol.*, **23**, 109–120.
40. Nyeki, S. and Colbeck, I. (1995) An assessment of relevance of soot aggregate humidity cycling in the atmosphere. *J. Aerosol Sci.*, **26**, (Suppl.), S509–S510.
41. Ramachandran, G. and Reist, P.C. (1995) Characterization of morphological changes in agglomerates subject to condensation and evaporation using multiple fractal dimensions. *Aerosol Sci. Technol.*, **23**, 431–442.
42. Eltekova, N.A., Razdyakonova, G.I., and Eltekov, Y.A. (1993) Fractals in geometry of carbon-black. *Pure Appl. Chem.*, **65**, 2217–2221.
43. Nyeki, S. and Colbeck, I. (1994) Determination of the power law prefactor for fractal agglomerates. *J. Aerosol Sci.*, **25**, (Suppl. 1), S403–S404.

44. Smirnov, B.M. (1993) Physics of ball-lightning. *Phys. Rep.*, **224**, 151–236.
45. Meakin, P. (1989) Simulation of aggregation processes, in *The Fractal Approach to Heterogeneous Chemistry*, ed. D. Avnir, John Wiley & Sons, Inc., New York.
46. Lushnikov, A.A. and Kulmala, M. (1995) Source-enhanced condensation in disperse systems. *Phys. Rev. E*, **52**, 1658–1668.
47. Wu, M.K. and Friedlander, S.K. (1993) Note on the power-law equation for fractal-like aerosol agglomerates. *J. Colloid Interface Sci.*, **159**, 246–248.
48. Wu, M.K. and Friedlander, S.K. (1993) Enhanced power-law agglomerate growth in the free-molecule regime. *J. Aerosol Sci.*, **24**, 273–282.
49. Vemury, S. and Prastinis, S.E. (1994) Self-preserving distributions of agglomerates. *J. Aerosol Sci.*, **25**, (Suppl. 1), S305–S306.
50. Wu, Z.F., Colbeck, I., and Simons, S. (1994) Determination of the fractal dimension of aerosols from kinetic coagulation. *J. Phys. D*, **27**, 2291–2296.
51. Smirnov, B.M. (1990) The properties of fractal clusters. *Phys. Rep.*, **188**, 3–78.
52. Lushnikov, A.A. (1978) Coagulation in finite systems. *J. Colloid Interface Sci.*, **65**, 276–285.
53. Lushnikov, A.A. (1978) Some new aspects of coagulation theory. *Izv. Acad. Sci. USSR, Ser. Phys. Atmos. Oceans*, **14**, 1046–1054.
54. Lushnikov, A.A. (2004) From sol to gel exactly. *Phys. Rev. Lett.*, **93**, 198302.
55. Lushnikov, A.A. (2005) Exact kinetics of the sol–gel transition. *Phys. Rev. E*, **71**, 046129.
56. Lushnikov, A.A. (2005) Exact particle mass spectrum in a gelling system. *J. Phys. A: Math. Gen.*, **38**, L35–L39.
57. Lushnikov, A.A. (2005) Sol–gel transition in a coagulating mixture. *J. Phys. A: Math. Gen.*, **38**, L383–L387.
58. Lushnikov, A.A. (2006) Exact kinetics of sol–gel transition in a coagulating mixture. *Phys. Rev. E*, **73**, 036111.
59. Lushnikov, A.A., Negin, A.E., and Pakhomov, A.V. (1990) Experimental observation of aerosol–aerogel transition. *Chem Phys. Lett.*, **175**, 138–142.
60. Lushnikov, A.A., Negin, A.E., Pakhomov, A.V., and Smirnov, B.M. (1991) Aerogel structures in gas. *Sov. Phys. Uspekhi*, **161**, 113–123.
61. Dobbins, R.A., Mulholland, G.W., and Bryner, N.P. (1994) Comparison of a fractal smoke optics model with light extinction measurements. *Atmos. Environ.*, **28**, 889–897.
62. Berry, M.V. and Persival, I.C. (1986) Optics of fractal clusters such as smoke. *Opt. Acta*, **33**, 577–591.
63. Farias, T.L., Carvalho, M.G., Koylu, U.O., and Faeth, G.M. (1995) Computational evaluation of approximate Rayleigh–Debye–Gans fractal-aggregate theory for the absorption and scattering properties of soot. *J. Heat Transf. Trans. ASME*, **117**, 152–159.
64. Lushnikov, A.A. and Maximenko, V.V. (1991) Light absorption by fractal clusters. *J. Aerosol Sci.*, **22**, (Suppl. 1), S395–S398.
65. Lushnikov, A.A., Maximenko, V.V., and Pakhomov, A.V. (1989) Fractal aggregates from laser plasma. *J. Aerosol Sci.*, **20**, 987–991.
66. Cabane, M., Rannou, P., Chassefiere, E., and Israel, G. (1993) Fractal aggregates in Titan atmosphere. *Planet. Space Sci.*, **41**, 257–267.
67. Chassefiere, E. and Cabane, M. (1995) Two formation regions for Titan's hazes – indirect clues and possible synthesis mechanisms. *Planet. Space Sci.*, **43**, 91–103.
68. Rogak, S.N., Flagan, R.C., and Nguyen, H.V. (1993) The mobility and structure of aerosol agglomerates. *Aerosol Sci. Technol.*, **18**, 25–47.
69. Colbeck, I. and Wu, Z.F. (1994) Measurement of the fractal dimensions of smoke aggregates. *J. Phys. D*, **27**, 670–675.
70. Schmidt-Ott, A. and Wüstenberg, J. (1995) Equivalent diameters of non-spherical particles. *J. Aerosol Sci.*, **26**, (Suppl.), S923–S924.
71. Magill, J. (1991) Fractal dimension and aerosol particle dynamics. *J. Aerosol Sci.*, **22**, (Suppl. 1), S165–S168.

72. Lushnikov, A.A. and Kulmala, M. (2001) Kinetics of nucleation controlled formation and condensational growth of disperse particles. *Phys. Rev. E*, **63**, 061109.
73. Lushnikov, A.A. and Kulmala, M. (2002) Singular self-preserving regimes of coagulation–condensation process. *Phys. Rev. E*, **65**, 041604.
74. Maisels, A. (2010) High-temperature aerosol systems, in *Aerosol Science – A Comprehensive Handbook*, ed. I.E. Agranovski, Wiley-VCH, Weinheim, Chapter 2 (this book).
75. Drake, R.L. (1972) A general mathematical survey of the coagulation equation, in *Topics in Current Aerosol Research*, eds. G.M. Hidy and J.R. Brock, Pergamon, New York, vol. 2, pp. 201–376.
76. Leyvraz, F. (2003) Scaling theory and exactly solved models in the kinetics of irreversible aggregation. *Phys. Rep.*, **383**, 95–212.
77. Lushnikov, A.A. (1974) Evolution of coagulating systems. II. Asymptotic size distribution and analytic properties of generating functions, *J. Colloid Interface Sci.*, **48**, 400.
78. Friedlander, S.K. (1960) On the particle size spectrum of atmospheric aerosols. *J. Meteorol.*, **17**, 479.
79. Lushnikov, A.A. and Smirnov, V.I. (1975) Steady–state coagulation and formation of particle size distributions in the atmosphere. *Phys. Atmos. and Ocean.*, **11**, 139.
80. Lushnikov, A.A. and Piskunov, V.N. (1977) Asymptotic regimes in source–enhanced coagulating systems. *Kolloidn. Zh.*, **39**, 1076.
81. Lushnikov, A.A. and Piskunov, V.N. (1975) A class of solvable models in the coagulation theory. *Kolloidn. Zh.*, **37**, 285.
82. Scott, W.T. (1968) Analytic studies on cloud droplet coalescence. *J. Atmos. Sci.*, **25**, 54.
83. Lee, M.H. (2001) A survey of numerical solutions to the coagulation equation. *J. Phys. A*, **34**, 10219.
84. Melzak, Z.A. (1953) The effects of coalescence in certain collision processes. *Q. J. Appl. Math.*, **11**, 231–236.
85. McLeod, J.B. (1962) On an infinite set of nonlinear differential equations. *Q. J. Math. Oxford*, **2**, 119–124.
86. McLeod, J.B. (1962) On an infinite set of nonlinear differential equations. *Q. J. Math. Oxford*, **2**, 192–196.
87. Lushnikov, A.A. (2006) Gelation in coagulating systems. *Physica D*, **222**, 37–53.
88. Aldous, D.J. (1999) Deterministic and stochastic models for coalescence (aggregation, coagulation; review of the mean-field theory for probabilists). *Bernoulli*, **5**, 3–122.
89. Heindrics, E.M., Ernst, M.H. and Ziff, R.M. (1983) Coagulation equation with gelation. *J. Stat. Phys.*, **31**, 519.
90. Ziff, R.M. and Stell, G. (1980) Kinetics of polymer gelation. *J. Chem. Phys.*, **73**, 3492.
91. Lushnikov, A.A. and Piskunov, V.N. (1983) Analytical solutions in the theory of coagulating systems with sinks. *Applied Mathematics and Mechanics*, **47**, 931.
92. Lushnikov, A.A. (2007) Critical behavior of the particle mass spectra in a family of gelling systems. *Phys. Rev. E*, **76**, 011120.
93. Marcus, A.H. (1968) Stochastic coalescence. *Technometrics*, **10**, 133–143.
94. Lushnikov, A.A. (1977) Coagulation in systems with distributed initial conditions. *Dokl. Akad. Nauk SSSR*, **236**, 673–676.
95. Lushnikov, A.A. (1977) Some exactly solvable models of stochastic coagulation. *Dokl. Akad. Nauk SSSR*, **237**, 1122–1125.
96. Lushnikov, A.A., (2008) Exact post-critical behavior of a source-enhanced gelling system. *J. Phys. A*, **41**, 072001.
97. Lushnikov, A.A. (1973) Evolution of coagulating systems. *J. Colloid Interface Sci.*, **45**, 549–559.

98. Schleicher, B. and Friedlander, S.K. (1995) Characterization of Nanostructured Carbon Microwires Fabricated by an Aerosol Process. *Journal of Applied Physics*, **78**(10): 6046–6049.

99. Lushnikov, A.A. and Maximenko, V.V. (1993) Quantum optics of a metallic particle. *J. Expt. Theor. Phys.*, **103**, 1010.

100. Sipila, M., Lushnikov, A.A., Khriachtchev, L., Kulmala, M. and Rasanen, M. (2007) Experimental observation of two-photon photoelectric effect from silver aerosol nanoparticles. *New Journal of Physics*, **9**, 368.

101. Lushnikov, A.A. (2010) Condensation, evaporation, nucleation, in *Aerosol Science – A Comprehensive Handbook*, ed. I.E. Agranovski, Wiley-VCH, Weinheim, Chapter 4 (this book).

Part I
Aerosol Formation

Aerosols – Science and Technology. Edited by Igor Agranovski

ISBN: 978-3-527-32660-0

2
High-Temperature Aerosol Systems

Arkadi Maisels

2.1
Introduction

Among the different sources of *aerosol particles*, *high-temperature processes* are very important both in nature and in industry. Indeed, huge amounts of aerosols are emitted into the atmosphere during *combustion*, *explosions*, and *volcanic eruptions*. On the other hand, *industrial manufacture* of *powder materials* is often based on the implementation of different high-temperature processes for *aerosol generation*. Millions of tonnes of *carbon black*, silica, titanium, and so on are manufactured annually via aerosol-based technologies. Therefore, an understanding of aerosol formation and dynamics in high-temperature processes is of immense environmental and industrial importance.

In this chapter, an overview is given of different high-temperature routes for *aerosol synthesis*. The properties of the resulting particles are considered with respect to *particle design*. Besides engineering of the aerosol particle, the main *dynamic processes* are described.

2.2
Main High-Temperature Processes for Aerosol Formation

Aerosol formation at high temperatures results from either physical vapor synthesis (PVS) or chemical vapor synthesis (CVS). In the case of PVS, the precursor of the *particulate matter* is obtained by physical methods such as *evaporation*, *laser ablation*, *sputtering*, and so on. No chemical conversion takes place. In contrast, CVS is based on chemical reactions in the vapor phase, which leads to liquid or solid products. Particulate matter in both PVS and CVS arises in the form of tiny particles due to nucleation as soon as supersaturated conditions are reached. This could be caused either by cooling or by increase in vapor pressure due to increase of the concentration of particulate precursor. Since aerosols are highly dynamic systems, particles are subject to the continual influence of numerous dynamic processes, such as *coagulation*, *condensation*, evaporation, *sintering*, and so

Aerosols – Science and Technology. Edited by Igor Agranovski

ISBN: 978-3-527-32660-0

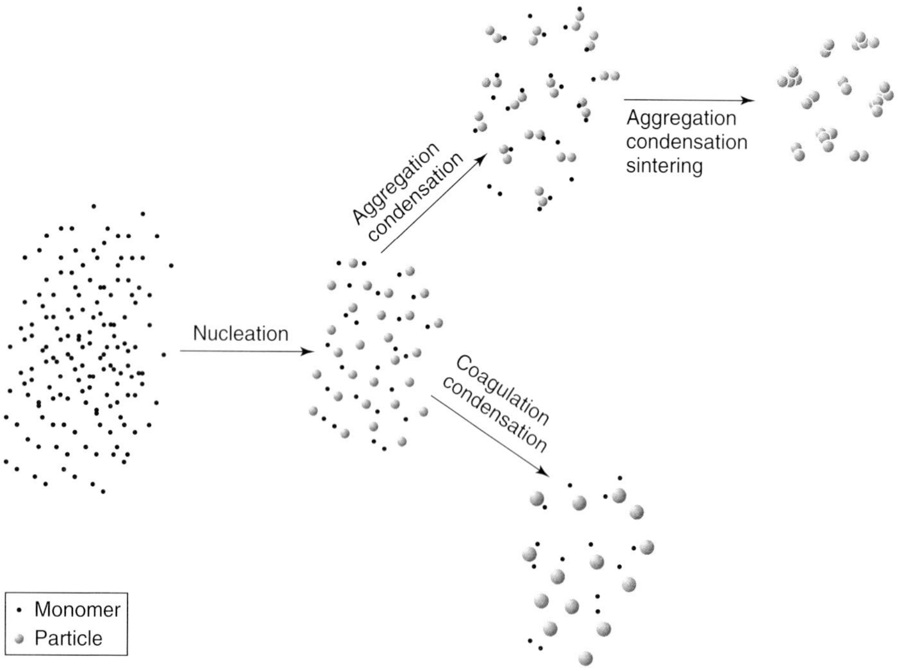

Figure 2.1 Mechanisms of particle formation in gas-phase processes.

on (Figure 2.1). Indeed, collisions lead to formation of new particles. Depending on the morphology of the resulting particles, *inter-particle collisions* are distinguished as either coagulation or *aggregation* processes. Inter-particle collisions are called coagulation if the colliding particles cannot be distinguished in the resulting particle, or aggregation if the colliding particles can be distinguished in the resulting particle after a suitable significant time interval. Coagulation leads to spherical or approximately spherical particles if the colliding particles are liquid in nature, or if the temperature is high enough for rapid sintering. Aggregation leads to the generation of aggregates of complex morphology, which is usually characterized by means of *fractal dimension* (see Chapter 21). Particle morphology in high-temperature processes is influenced by the condensation of vapor on the particle surface, which corresponds to some increase in the particulate *surface area* and mass. High temperature is responsible for the appearance of various *electrostatic effects* due to possible emission of charge by various objects, including aerosol particles, gas molecules, solid materials, or embedding gas flow, and due to charge collection by aerosol particles. High temperature also influences aggregate morphology, as it creates favorable conditions for particle sintering and coalescence.

The main advantage of high-temperature aerosol synthesis is the high purity of the resulting materials, as compared to wet routes of particle synthesis. Other advantages of high-temperature aerosol synthesis are equal processability of organic and *inorganic precursors*, simplicity, and uniformity of chemical composition. Different methods of high-temperature aerosol synthesis can be distinguished according to the kind of energy supply.

2.2.1 Flame Processes

Flame processes are very common for industrial manufacturing of powder materials via the high-temperature aerosol route. The main source of the particulate matter is usually injected into the tubular reactor either in the flame zone or shortly after it. The high temperature of the surrounding gas initiates chemical reactions of precursor gases, which lead to synthesis of the product vapor.

One of the major applications of *flame reactors* is the synthesis of fumed oxides (SiO_2, TiO_2, Al_2O_3, ZrO_2, and so on [1–4]). In industrial processes, vaporized *metal chloride* is introduced into the high-temperature hydrogen–oxygen flame and reacts with the hot water vapor to produce a *metal oxide* and HCl as follows:

$$MeCl_{2n} + nH_2O \rightarrow MeO_n + 2nHCl$$

Rapid chemical reaction leads to the formation of supersaturated oxide vapor, which nucleates to nanometer-sized nearly spherical particles. Inter-particle collisions lead to the formation of larger particles or so-called aggregates (Figure 2.2). *Surface reactions* with vapor species present (material additives from the gas phase) also increase the particle size. If aggregates remain at high temperatures then partial sintering of the composite primary particles may occur. Sinter necks determine

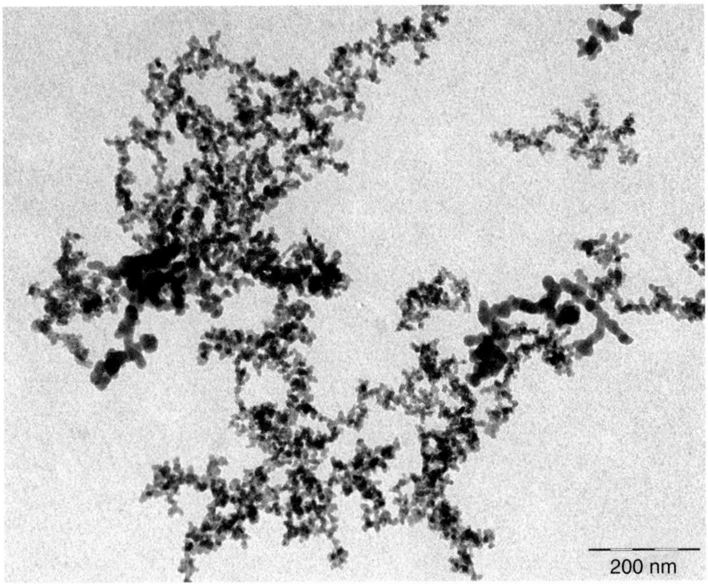

Figure 2.2 Transmission electron micrograph (TEM) of SiO_2 nanoparticle aggregate.

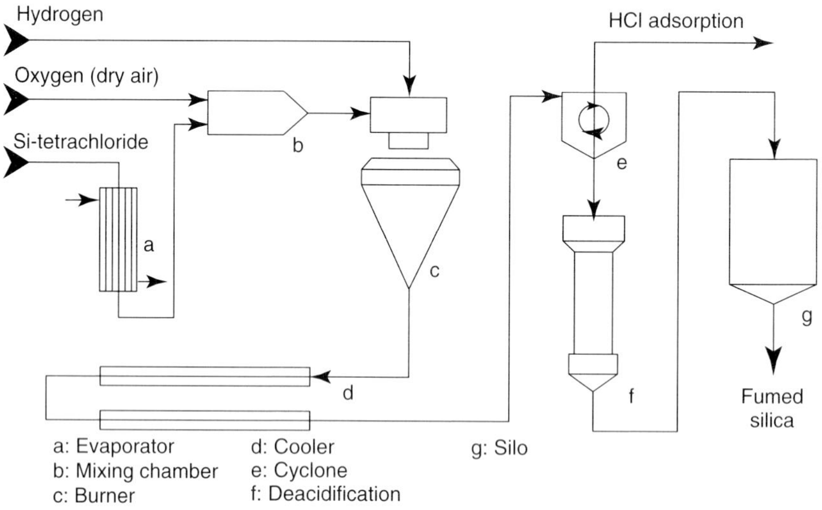

Figure 2.3 Industrial process for manufacturing of fumed silica.

the mechanical strength of aggregates and, consequently, their dispersability in liquids and polymers. In addition to particle removal from the aerosol flow, it is also important in industrial processes to reuse HCl for manufacturing metal chlorides. The entire industrial process is schematically depicted in Figure 2.3. The same idea is utilized in various scientific studies dedicated to the synthesis of oxides [5–8].

Carbon blacks represent probably the oldest and the largest group of aerosol-synthesized particles. Though several different manufacturing routes exist, most carbon blacks are made in flame reactors, where organic precursors are pyrolyzed by hot combustion gases to form the carbon black. The process is quite similar to the manufacture of metal oxides, with slightly different use of process gases – in manufacturing carbon black, hydrogen is usually burned in power plants [9]. The carbon black product is made of aggregates containing numerous elementary spherical particles.

Similar processes could also be used for the manufacture of *metal particles* from *metal–organic precursors* [10]. Flame reactors could be used for synthesis of composite materials [11, 12] by introducing precursors of different elements into the reaction zone.

Depending on the conditions of temperature, precursor concentration, and residence time, different sizes of primary particles and aggregates can be obtained. Fine particles, for example, could be made at short residence times and high temperatures by adjusting the air flow rate and the length-to-diameter ratio of the furnace [13]. The aggregation of particles can also be controlled by the addition of traces of potassium salts into the flames [14, 15]. The crystalline structure of the particles can be controlled by adding dopant vapors into the feed stream and by tailoring the temperature history of the reactor [16].

2.2.2
Hot-Wall Processes

In *hot-wall processes*, energy required for particle formation is supplied to a gas flow from outside through the reactor walls. This allows a more homogeneous temperature profile to be provided, compared to flame processes. On the other hand, temperatures in hot-wall processes cannot reach values comparable to those in flame processes. In industry, the hot-wall process is used, for example, for thermal synthesis of carbon black [9] and for the production of fine powders of silicon [17] or iron.

In hot-wall processes, precursor vapor is either decomposed under controlled temperature conditions in a tubular reactor heated from the outside [18–20], or produced by evaporation of solid material deposited in the heated zone of the reactor. Depending on the precursors used and the products desired, additional reactants can be introduced into the reactor. Similarly to flame reactors, hot-wall reactors can be used for the synthesis of composite [21] or doped [22] materials. Even solid precursors in the form of tiny particles can be introduced into hot-wall reactors to synthesize composite materials [23]. Well-defined temperature conditions within hot-wall reactors can be utilized for controlled synthesis of *fullerene*-like structures [24] and *carbon nanotubes* [25].

The dynamic processes responsible for particle formation and growth in hot-wall reactors are the same as in flame reactors. However, differences in temperature conditions could lead to slight differences in *particle morphology*, compared to that for flame reactors operated at similar flame temperatures. The resulting particles are usually aggregates of numerous nearly spherical primary particles with sizes in the range 5–50 nm.

2.2.3
Plasma Processes

In *plasma processes*, the energy of a plasma is used for particle formation [26–32]. Owing to the strongly dissimilar mobilities of plasma ions, aerosol particles become unipolarly charged during their time in the plasma zone. This feature allows the synthesis of aerosols with narrow *size distributions*.

Similarly to flame reactors, the variety of materials synthesized by different plasma-related techniques is also very broad. They include pure metals [29, 33, 34], oxides [35], nitrides [36], *composite materials* [37], and so on.

Depending on the method of introduction of the reactants into the apparatus, there is a range of plasma reactors available in industry. The operating temperatures in these devices (300–25 000 °C) are usually higher than in other aerosol reactors, and complete destruction of the reactants is common. This allows usage of molecular and solid feed streams so, in principle, any material can be processed. All plasma reactors produce product species that nucleate to form particles in the cooling zone immediately upon exit from the plasma-containing areas, with the following standard pathways of gas-to-particle aerosol formation. Two broad classes of thermal *plasma reactors* are used: *DC arc jets*, and high-frequency (microwave or radio-frequency) induction systems. In the case of a DC arc jet, current is supplied to the ionized gas (plasma) by physical contact with a metallic electrode surface. This system is relatively simple and inexpensive [38]. However, the electrodes are consumed and may contaminate the product. In the case of a high-frequency induction plasma (microwave or radio-frequency plasma reactor), there is no contact between the plasma and its power source. The induction coil lies outside the reactor walls like the electrical thermal elements in furnace reactors. Energy transfer takes place through the electromagnetic field of the induction coil, so there is no contamination of the product. A good overview of different plasma techniques is given by Vollath [39]. Variables that can be controlled are plasma composition and frequency. The most common case is an argon plasma operated at 200 kHz to 20 MHz with typical temperatures about 15 000 °C. As a result, plasma reactors can handle materials with high melting point and solid powder feeds.

2.2.4 Laser-Induced Processes

High-energy laser beams can also be used for particle production. In this case, tiny clusters are knocked out of the substrate by energy-reach laser beam. Laser ablation is usually a low-pressure technique with throughput high enough for scientific research but definitely not sufficient for industrial manufacturing. Laser ablation can be applied for synthesis of silicon [40], metals [41], metal oxides, and even complex oxides [42], which can hardly be obtained by other processes.

2.2.5 Gas Dynamically Induced Particle Formation

Gas dynamically induced particle formation is a rather new concept involving a high-throughput process for gas-phase particle synthesis. The initiation of the chemical reaction is realized by a stationary shock system. The quenching of the high-temperature gas flow is achieved by gas dynamic quenching, that is, accelerating the flow from subsonic to supersonic speed, which decreases the static temperature below the sinter temperature. The homogeneous flow field and high heating and quenching rates lead to narrow particle size distributions and low aggregation.

2.3 Basic Dynamic Processes in High-Temperature Aerosol Systems

Aerosols are physically unstable systems experiencing the continual influence of various dynamic processes. These processes change aerosol properties, such as particle size and *charge distributions*, particle morphology, and so on. Here, a short description of the main dynamic processes is presented.

The behavior of a system of spherical unequally sized particles undergoing *nucleation*, coagulation, and *surface growth* can be described by the general dynamics equation (GDE) in either continuous [43] or discrete [44] form. If the discrete form is chosen, the changes of the fraction $N_k(t)$ (dimension 1/kg) of particles containing k monomers can be described as follows:

$$\frac{\mathrm{d}N_k}{\mathrm{d}t} = \underbrace{\frac{1}{2}\rho_g \sum_{j=1}^{k-1} \beta_{k-j,j} N_{k-j} N_j - \rho_g \beta_k N_k \sum_{j=1}^{\infty} N_j}_{\text{coagulation}} + \underbrace{J(t)\delta_{i,k}}_{\text{nucleation}}$$
$$+ \underbrace{\rho_g \left(\beta'_{1,k-1} N'_1(t) N_{k-1} - \beta'_{1,k} N'_1(t) N_k\right)}_{\text{surface growth (condensation)}} \tag{2.1}$$

Here β is the *coagulation kernel* (coefficient) of two colliding particles (m^3/s); $J(t)$ is the *nucleation rate* (1/kg s); δ is the *Kronecker delta* ($\delta_{i,k} = 1$ for $i = k$; $\delta_{i,k} = 0$ otherwise); ρ_g is the *carrier gas density*; $N'_1(t)$ denotes the *number concentration* of condensing monomers (1/kg); while β' is the *collision frequency* of particles and monomers (m^3/s). These monomers are in the gas phase and should be distinguished from the *particle fraction* N_1, which represents condensed-matter particles of the smallest size. Particle interactions with gas-phase monomers result in particle growth but do not affect the total particle number concentration. The particle size is the only parameter taken into account in Eq. (2.1).

The dynamics of the entire system is described by a set of equations similar to Eq. (2.1), each of which corresponds to a specified k. The balance equation for the total *particle number concentration* N (1/kg of carrier gas) is obtained by summing over all size fractions:

$$\frac{\mathrm{d}N}{\mathrm{d}t} = -\frac{1}{2}\rho_g \sum_{k=1}^{\infty}\sum_{j=1}^{\infty} \beta_{k,j} N_k N_j + \sum_{k=1}^{\infty} J(t)\delta_{i,k} \tag{2.2}$$

The changes in the *total particle area* S (in 1 m^2/kg of carrier gas) and the *total particle mass* M (in 1 kg of carrier gas) are described by

$$\frac{\mathrm{d}S}{\mathrm{d}t} = \rho_g \sum_{k=1}^{\infty} s_k J(t)\delta_{i,k} + \pi\rho_g \sum_{k=1}^{\infty}\left(\frac{1}{2}\sum_{k=1}^{\infty} \beta_{k-j,j} N_{k-j} N_j (d_k^2 - d_{k-j}^2 - d_j^2) - \beta_k N_k \right.$$
$$\left. \times \sum_{k=1}^{\infty} N_j (d_{k+j}^2 - d_k^2 - d_j^2)\right) + \pi\rho_g \sum_{k=1}^{\infty} \beta_{1,k}\left((d_k^3 + (d_1)^3)^{\frac{2}{3}} - d_k^2\right) N'_1(t) N_k \tag{2.3}$$

and

$$\frac{\mathrm{d}M}{\mathrm{d}t} = \rho_g \sum_{k=1}^{\infty} m_k J(t)\delta_{i,k} + m'_1 \rho_g \sum_{k=1}^{\infty} \beta'_{1,k} N'_1(t) N_k \tag{2.4}$$

In Eq. (2.3) d_k and d'_1 denote the diameters of a particle in size class k and monomer, respectively, while s denotes the particle surface. Similarly, in Eq. (2.4), m denotes the mass of a particle or monomer. Since Eqs. (2.1)–(2.4) are written for spherical coalescing particles, processes of particle restructuring or sintering are not considered. These processes are considered in Section 2.3.4.

2.3.1 Nucleation

Both PVS and CVS create saturated and supersaturated atmospheres. The commencement of nucleation depends on a level of supersaturation and *residence time*. Nucleation is often presented as a dynamic process in which vapor molecules build clusters and dissociate up to the moment when stable clusters (critical nuclei) arise [45]. The nucleation theory is described in detail in Chapter 4. With respect to the dynamics of high-temperature aerosols, several approaches can be followed to describe nucleation rate as a function of temperature, pressure, vapor, and so on. One could follow *Becker–Döring nucleation theory*, which is widely supported by experimental studies [46]. According to this theory,

$$J = \rho_g N_s^2 s_1 \sqrt{\frac{k_B T}{2\pi m_1}} S^2 \sqrt[3]{\frac{2}{9\pi}} \Sigma^{1/2} \exp\left(-k^* \ln\left(\frac{S}{2}\right)\right) \tag{2.5}$$

Here, $S = N_1'/N_s$ is the *saturation ratio*, determined as a ratio of the actual monomer concentration to the monomer concentration at saturation conditions, s_1 and m_1 denote the monomer surface area and mass, respectively, Σ is a dimensionless *surface tension* parameter, and

$$k^* = \frac{\pi}{6}\left(\frac{4\Sigma}{\ln S}\right)^3$$

is the number of condensable monomers in the critical size nucleus. Stable nuclei undergo other dynamic processes such as coagulation and surface growth.

Since the saturation concentration of monomers is not known for most particulate systems, calculation of the nucleation rate according to Eq. (2.5) is not always unequivocal. This problem can be overcome if the monomers themselves are considered as nuclei, while the concentration of monomers can be calculated from the kinetics of the corresponding chemical reactions [47]. In this case $N_1 = N_1'$ and the last two terms in Eq. (2.1) corresponding to surface growth can be omitted.

2.3.2 Coagulation/Aggregation

Owing to strong *van der Waals forces*, collisions between nanometer-sized particles lead to particle sticking and the formation of new particles. This process is called coagulation if the resulting particles are spherical, which means that the colliding particles are liquid or complete sintering is very fast. Equation (2.1) can then be rewritten in the form [48]:

$$\frac{dN_k}{dt} = \frac{1}{2}\rho_g \sum_{j=1}^{k-1} \beta_{k-j,j} N_{k-j} N_j - \rho_g \beta_k \sum_{j=1}^{\infty} \beta_{k,j} N_j \tag{2.6}$$

Analytical solutions of Eq. (2.4) for total particle number concentration and particle size distribution can be found only for comparatively simple forms of β [49, 50].

Depending on the particle size, coagulation or aggregation is caused by *Brownian motion*, by *particle diffusion*, or by a combination of these mechanisms (see also Chapter 1). Collisions of small particles (*Knudsen number* $\mathrm{Kn} = 2\lambda_g/d > 10$, with λ_g being the *mean free path* of gas molecules, d the particle diameter) are caused mainly by ballistic reasons, and the coagulation kernel for two spherical particles with diameters d_i and d_j is determined as the product of their relative thermal velocity $v_{i,j}$ and projected surface:

$$\beta_{i,j} = \frac{1}{4}\pi v_{i,j}(d_i + d_j)^2 \tag{2.7}$$

In Eq. (2.5)

$$\overline{v}_{i,j} = \sqrt{\overline{v}_i^2 + \overline{v}_j^2} \quad \text{and} \quad \overline{v}_i = \sqrt{\frac{8k_B T}{\pi m_i}}$$

k_B is the *Boltzmann constant*, T is the fluid temperature, and m is the particle mass. Equation (2.7) is also known as the coagulation kernel for the free molecular regime. For $\mathrm{Kn} \leq 1$ (*near-continuum regime*), particle collisions are caused by diffusion, and the coagulation kernel can be calculated as

$$\beta_{i,j} = 2\pi(D_i + D_j)(d_i + d_j) \tag{2.8}$$

where D_i is the particle *diffusion coefficient* for a spherical particle of diameter d_i, which is determined as

$$D_i = \frac{C_S k_B T}{3\pi\eta d_i}$$

η is the *dynamic gas viscosity*, C_S is the *slip correction factor* [51], and

$$C_S = 1 + \mathrm{Kn}\left(A_1 + A_2 \exp\left(-\frac{2A_3}{\mathrm{Kn}}\right)\right)$$

Allen and Raabe [52] used an improved *Millikan apparatus* and determined the values $A_1 = 1.142$, $A_2 = 0.558$, and $A_3 = 0.4995$. For $\mathrm{Kn} \leq 0.1$, the slip correction factor $C_S \approx 1$ (*continuum regime*).

Equations (2.7) and (2.8) leave a gap in the calculation of the coagulation kernel for the particle size range $1 < \mathrm{Kn} < 10$. This gap is filled by a well-established *interpolation formula* [53] in aerosol science, which shows good accuracy across the entire particle size range:

$$\beta_{i,j} = \pi D_{i,j}(d_i + d_j)\left(\frac{d_i + d_j}{d_i + d_j + 2\Delta_{i,j}} + \frac{8D_{i,j}}{\overline{v}_{i,j}(d_i + d_j)}\right)^{-1} \tag{2.9}$$

where

$$\Delta_{i,j} = \sqrt{\Delta_i^2 + \Delta_j^2}, \quad \Delta_i = \left(\frac{1}{3d_i\lambda_i}\right)\left((d_i + \lambda_i)^3 - (d_i^2 + \lambda_i^2)^{3/2}\right) - d_i,$$

$$\lambda_i = \frac{8D_i}{\pi\overline{v}_i}, \quad \text{and} \quad D_{i,j} = D_i + D_j.$$

Equation (2.7) can be rearranged in the form

$$\frac{1}{\beta_{i,j}} = \underbrace{\frac{1}{\frac{2k_\mathrm{B}T}{3\eta}\left(\frac{C_{\mathrm{S},i}}{d_i}+\frac{C_{\mathrm{S},j}}{d_j}\right)(d_i+d_j+2\Delta_{i,j})}}_{\text{modified near - continuum}} + \underbrace{\frac{1}{\frac{1}{4}\pi v_{i,j}(d_i+d_j)^2}}_{\text{free molecular}} \tag{2.10}$$

which shows its interpolation nature much more clearly. Indeed, *Fuchs' expression* is a harmonic mean built by the coagulation kernel for the free molecular regime and by the modified coagulation kernel for the near-continuum regime. To simplify calculations, some authors [46] use the harmonic mean of the coagulation kernels according to Eqs. (2.7) and (2.8) without taking into account modifications. The error in calculation of the coagulation kernel is demonstrated in Figure 2.4. A more detailed analysis on the accuracy of different expressions for coagulation kernel was presented by Otto *et al.* [54].

For electrically charged particles, the *Coulomb forces* have to be considered and the coagulation kernel ought to be corrected by

$$\beta_{i,j;\mathrm{el}} = \frac{\beta_{i,j}}{W_{i,j}^{p,q}} \tag{2.11}$$

where $W_{i,j}^{p,q}$ is the *stability function* [53]

$$W_{i,j}^{p,q} = -\frac{1-\exp(\kappa)}{\kappa}, \quad \text{with} \quad \kappa = \frac{2pqe^2}{4\pi\varepsilon\varepsilon_0(d_i+d_j)k_\mathrm{B}T} \tag{2.12}$$

and p and q are the numbers of *elementary charges* on particles i and j, respectively.

Expressions (2.7)–(2.12) have been derived for spherical particles. Though Eq. (2.9) can be extended to the case of elliptically shaped particles, particle

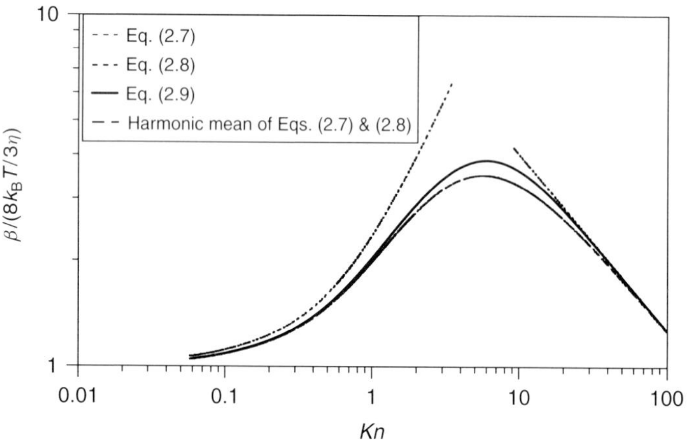

Figure 2.4 Coagulation kernel for equally sized spherical particles ($T = 1273$ K; $\eta = 3.8 \times 10^{-5}$ Pa s; particle density 1000 kg/m^3; $\lambda_g = 292$ nm).

morphology in nature and technology is much more complex (see Chapter 1). As already mentioned, solid particles originating from high-temperature processes are usually aggregates composed of numerous nearly spherical primary particles. If size uniformity of primary particles is assumed, the diameter of the aggregate can be expressed by means of the fractal dimension D_f as [55]

$$B\left(\frac{d}{d_0}\right)^{D_\mathrm{f}} = n_\mathrm{p} \tag{2.13}$$

where N_p is the number of primary particles with diameter d_0, and B is a constant ($1 \leq B \leq 1.4$). Though experimental studies point to $B = 1.23$, the value $B = 1$ is taken in most cases. Since the coagulation kernel in Eqs. (2.7)–(2.11) is a function of the diameters and masses of the colliding particles, for a fractal aggregate i with a number of primary particles $n_{\mathrm{p},i}$ and density of particulate material ρ_p, one can write

$$m_i = \frac{\pi}{6}\rho_\mathrm{p} n_{\mathrm{p},i} d_0^3 \quad \text{and} \quad d_i = d_0\left(\frac{n_{\mathrm{p},i}}{B}\right)^{1/D_\mathrm{f}} \tag{2.14}$$

Fractal aggregates with diameters calculated according to Eq. (2.14) have higher coagulation rates as compared to spherical particles of the same volume.

Equations (2.7)–(2.10) are valid for diluted aerosol. This means that particle mass concentrations are far below 1% of the mass of the carrier gas. This assumption allows the influence of the particle motion on the momentum of the gas molecules to be neglected. As shown by numerical simulations [56], the coagulation kernel in concentrated aerosols depends on the loading and exceeds the values calculated according to Eqs. (2.7)–(2.9).

2.3.3
Surface Growth Due to Condensation

If the GDE in the form of Eq. (2.1) is used, there is no special need for a detailed description of the surface growth mechanism. Indeed, the last two terms in Eq. (2.1) are very similar to coagulation terms. Calculation of the collision frequency for particles and monomers β' can be performed according to Eq. (2.7), if monomer size and mass are used instead of the size and mass of one of the colliding particles.

2.3.4
Sintering

Equation (2.1) is written for spherical coalescing particles, which do not change their shape. However, solid fractal-like particles can change their shape at elevated temperatures [57, 58]. These changes in particle shape are commonly known as sintering or *restructuring*. As depicted in Figure 2.5, there are two different mechanisms taking place within aggregates at elevated temperatures. The first mechanism is characterized by restructuring of the aggregate without changing the surface area of the composing particles and the entire aggregate. The second mechanism is characterized by slow coalescence of the composing primary

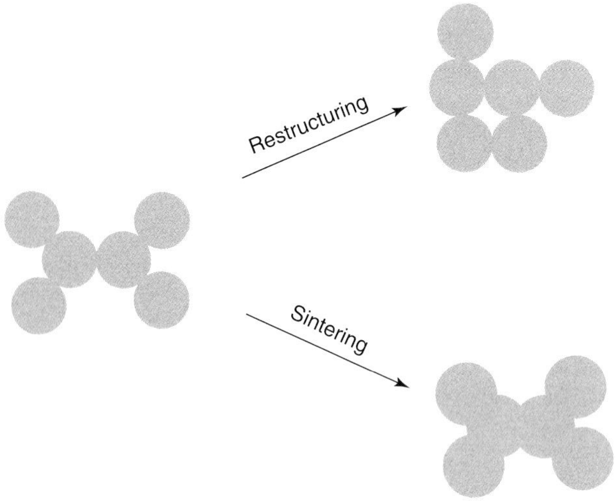

Figure 2.5 Particle restructuring and sintering.

particles, with a corresponding change in the aggregate surface area. For an aggregate of two spherical primary particles, the change in the particle surface area due to the sintering can be described by the following equation:

$$\frac{\mathrm{d}s}{\mathrm{d}t} = -\frac{1}{\tau_\mathrm{S}}(s - s_\mathrm{CS}) \tag{2.15}$$

where s is the particle surface area, s_SC is the surface area of the completely sintered particle (volume-equivalent sphere), and τ_S is the *characteristic sintering time*. The characteristic sintering time depends on the diameter of the primary particles, and on the particle temperature and material [59]:

$$\tau_\mathrm{S} = D_\mathrm{S} d_0^m \exp\left(\frac{E_\mathrm{A}}{RT}\right) \tag{2.16}$$

where D_S and E_A are diffusion coefficient and *activation energy*, respectively. The factor m in Eq. (2.16) depends on the sintering mechanism. For example, for fumed silica, Xiong *et al.* [60] determined $m = 1$, while for fumed titania $m = 4$ was reported by Kobata *et al.* [61].

Under the assumption of fractal aggregates undergoing sintering, Eq. (2.3) could be rewritten as

$$\frac{\mathrm{d}S}{\mathrm{d}t} = \rho_\mathrm{g} \sum_{k=1}^{\infty} a_k J(t) \delta_{i,k} - \frac{1}{\tau_\mathrm{S}}\left(S - \sum_{k=1}^{\infty} s_{\mathrm{CS},k}\right) + \pi\rho_\mathrm{g} \sum_{k=1}^{\infty} \beta'_{1,k} n_{\mathrm{p},k} \left(\left(d_0^3 + (d'_1)^3\right)^{2/3} - d_0^2\right) N'_1(t) N_k \tag{2.17}$$

Of course, Eq. (2.17) can be split into equations for each size fraction in a similar way to Eq. (2.1).

2.3.5 Charging

Heating of various materials is often accompanied by *charge emission*. Depending on the composition of the material, high-temperature aerosols can emit numerous elementary charges to an ambient gas due to *thermionic emission*. These elementary charges are picked up by the gas molecules, which form gas ions. A strong *electrostatic interaction* between highly charged particles and oppositely charged ions determines the intensive dynamics of such a system. The influence of electrical effects on the dynamics of high-temperature aerosols was reported by Vemury and Pratsinis [62] and by Katzer *et al.* [63]. The first detailed study on thermionic charging of aerosol particles was presented by Schiel *et al.* [64]. Under the assumption that thermionic emission by particles is the only ion source, the dynamic behavior of the charging process for a polydisperse aerosol can be described by the following system of differential equations (see also [65]):

$$\frac{dn_{\text{ion}}^-}{dt} = \sum_i \sum_{q=q_{\min}}^{q=q_{\max}} \vartheta_i^{q\to q+1} N_i^q - \rho_g n_{\text{ion}}^- \sum_i \sum_{q=q_{\min}}^{q=q_{\max}} \beta_i^{q\to q-1} N_i^q \tag{2.18}$$

$$\frac{\mathrm{d}N_i^q}{\mathrm{d}t} = \vartheta_i^{q-1\to q} N_i^{q-1} - \vartheta_i^{q\to q+1} N_i^q + \rho_g n_{\text{ion}}^- \beta_i^{q+1\to q} N_i^{q+1} - \rho_g n_{\text{ion}}^- \beta_i^{q\to q-1} N_i^q \tag{2.19}$$

Equations (2.18) and (2.19) represent the balance equation for the concentrations of negative ions n_{ion}^-. It is assumed that all ions have one negative elementary charge and equal mass. Equation (2.20) is the balance equation for the aerosol fraction N_i^q with particle diameter d_i and charge q. All terms on the right-hand side of Eq. (2.19) account for changes in fraction N_i^q due to different charging processes. The frequency of *ion–particle collisions* is determined by the *ion concentrations* and the *ion attachment coefficient* β_i^q. Ion production by particles is expressed by the *combination coefficient* ϑ_i^q. Coagulation, *electrostatic dispersions*, and so on are not considered in Eqs. (2.18) and (2.19) (see [66]).

For spherical particles of radius r_p, the ion attachment coefficient $\beta^{q\to q-1}$ is determined as [67]:

$$\beta^{q\to q-1} = \frac{4\pi r_p D_{\text{ion}}}{\dfrac{4 r_p D_{\text{ion}}}{\delta^2 \overline{v}_{\text{ion}} \alpha_{\text{coll}}} \exp\left(\dfrac{\phi(\delta, q)}{k_B T}\right) + \displaystyle\int_0^{r_p/\delta} \exp\left(\dfrac{\phi(r_p/y, q)}{k_B T}\right) \mathrm{d}y} \tag{2.20}$$

Here D_{ion} is the diffusion coefficient for ions, $\overline{v}_{\text{ion}}$ is the *mean ion thermal velocity*, and α_{coll} is the *collision parameter* [67]. The parameter $\phi(l, q)$ is the *interaction potential* between an ion and a q-charged particle, which is determined by superposition of the Coulomb and *image forces*. For negative ions,

$$\phi(l, q) = \frac{e^2}{4\pi\varepsilon_0}\left(-\frac{q}{l} - K(l)\frac{r_p^3}{2l^2\left(l^2 - r_p^2\right)}\right) \tag{2.21}$$

where K is a material-dependent function and l is the distance from the particle. Parameter δ in Eq. (2.20) is the radius of the adsorbing sphere [67]:

$$\delta = \frac{r_\mathrm{p}}{\mathrm{Kn}_\mathrm{ion}^2}\left(\frac{(1+\mathrm{Kn}_\mathrm{ion})^5}{5} - \frac{\left(1+\mathrm{Kn}_\mathrm{ion}^2\right)(1+\mathrm{Kn}_\mathrm{ion})^3}{3} + \frac{2\left(1+\mathrm{Kn}_\mathrm{ion}^2\right)^{5/2}}{15}\right) \tag{2.22}$$

in which Kn_ion is the Knudsen number for ions.

The combination coefficient ϑ_i^q is a function of the particle size, charge, and material. According to the *Richardson equation*, for spherical particles of radius r_p the combination coefficient $\vartheta^{q\to q+1}$ is determined as

$$\vartheta^{q\to q+1} = 4\pi r_\mathrm{p}^2 A T^2 \exp\left(-\frac{\Phi^{q\to q+1}}{k_\mathrm{B}T}\right) \tag{2.23}$$

where $\Phi^{q\to q+1}$ is the *work function* that must be overcome by an electron to escape from the surface of a spherical particle with q elementary charges [68]:

$$\Phi^{q\to q+1} = \Phi_\infty + \frac{e^2(q+1)}{4\pi\varepsilon_0 r_\mathrm{p}} - \frac{5}{8}\frac{e^2}{4\pi\varepsilon_0 r_\mathrm{p}} \tag{2.24}$$

The constant A in Eq. (2.23) is determined as

$$A = \frac{4\pi m_\mathrm{el} k_\mathrm{B}^2}{h^3} = 7.49\times 10^{24}\ \mathrm{m}^{-2}\ \mathrm{s}^{-1}\ \mathrm{K}^{-2}$$

with m_{el} being the mass of the electron and h the *Planck constant*.

Equations (2.18) and (2.19) with the combination coefficient and ion attachment coefficient according to Eqs. (2.20) and (2.23) can only be solved numerically if no simplifying assumptions are made. So, if stationary conditions are reached, for any size fraction i, the charge fractions N_i^q are related to each other by

$$\frac{N_i^q}{N_i^{q-1}} = \frac{\vartheta_i^{q-1\to q}}{n_\mathrm{ion}^- \beta_i^{q\to q-1}} \tag{2.25}$$

The expression for the maximum possible charge achievable by a monodisperse aerosol with particle radius r_p at isothermal conditions without external ion sources after making simplifying assumptions for the combination coefficient and ion attachment coefficient can be written as [69]

$$\frac{q_\mathrm{max} N D_\mathrm{ion} e^2}{\varepsilon_0 k_\mathrm{B} T} = 4\pi r_\mathrm{p}^2 A T^2 \exp\left(-\frac{\Phi_\infty + q_\mathrm{max}e^2/(4\pi\varepsilon_0 r_\mathrm{p})}{k_\mathrm{B}T}\right) \tag{2.26}$$

As follows from Eq. (2.26), the maximum possible charge state depends on the particle size, the *carrier gas temperature*, the particle material (work function), and the particle number concentration. The influences of the particle number concentration, particle size, and work function of the particle material are demonstrated in Figure 2.6 for a carrier gas temperature of 1500 K.

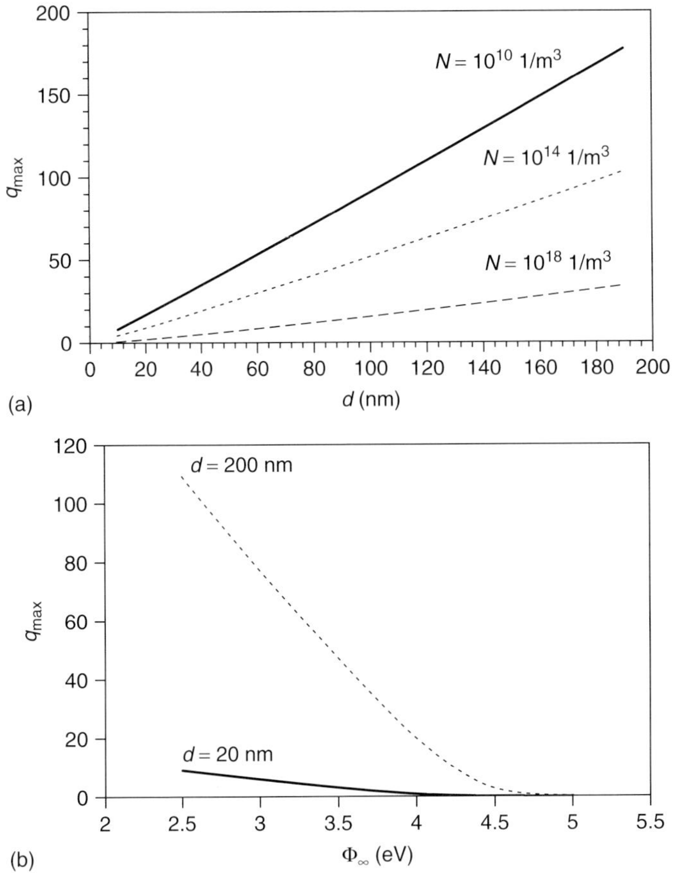

Figure 2.6 Maximum particle charge due to thermionic charging at $T = 1500$ K as a function of (a) particle size and (b) work function of particle material.

2.4 Particle Tailoring in High-Temperature Processes

As already mentioned, Eq. (2.1) can be solved analytically in only a few special cases, which are usually not representative of real conditions. Various techniques for simulation of particle dynamics have been developed over the past 20 years [70, 71], but implementation of these methods is quite a challenging task, especially in the case of multivariate systems and different simultaneous dynamic processes. However, the estimation of the expected tendencies for the resulting particle size distributions after changing parameters such as precursor concentration, carrier gas temperature, residence time in high-temperature zone or cooling rate can be done by comparatively simple methods. As a result, Eqs. (2.2) and (2.4) can be rewritten under the assumption of constant coagulation kernel and constant nucleation rate, while nucleation takes place only during some restricted time interval, and surface growth by condensation and sintering is neglected:

$$\frac{\mathrm{d}N}{\mathrm{d}t} = -\frac{1}{2}\rho_g \beta N^2 + J_0, \quad \text{if } t < t_{\text{nucl}}; \quad \frac{\mathrm{d}N}{\mathrm{d}t} = -\frac{1}{2}\rho_g \beta N^2, \quad \text{if } t \geq t_{\text{nucl}} \tag{2.27}$$

$$\frac{\mathrm{d}M}{\mathrm{d}t} = \rho_g m_0 J_0, \quad \text{if } t < t_{\text{nucl}}; \quad \frac{\mathrm{d}M}{\mathrm{d}t} = 0, \quad \text{if } t \geq t_{\text{nucl}}; \tag{2.28}$$

$$N_{t=0} = 0; \quad M_{t=0} = 0 \tag{2.29}$$

From these equations, the mean particle mass can be calculated as follows:

$$\langle m \rangle = \frac{M}{N};$$

$$\langle m \rangle = \frac{1}{2} m_0 \frac{t}{\sqrt{2\beta J_0}\tan\mathrm{h}\left(t/\sqrt{2\beta J_0}\right)}, \quad \text{if } t < t_{\text{nucl}};$$

$$\langle m \rangle = \frac{M_{\text{nucl}}}{N_{\text{nucl}}} \left(1 + \frac{1}{2}\beta N_{\text{nucl}}\left(t - t_{\text{nucl}}\right)\right), \text{if } t \geq t_{\text{nucl}}$$

where $M_{\text{nucl}} = M(t = t_{\text{nucl}})$, and $N_{\text{nucl}} = N(t = t_{\text{nucl}})$.

Values of $\langle m \rangle$ for different nucleation rates J_0 and for different values of β are plotted in Figure 2.7. The increase of J_0 leads to some increase in the mean particle size if the same time interval is considered. The same effect is observed for β. From the practical point of view, an increase in the nucleation rate can be interpreted as an increase in precursor concentration, while an increase in the coagulation kernel can be achieved by increasing the carrier gas temperature. Remarkably, the same tendencies were observed in various experimental studies [8, 72]. Some analysis of Eqs. (2.27) and (2.28) allows several important tendencies for particle design to be deduced.

The mean particle size increases with increasing precursor concentration. The precursor concentration has a direct impact on the nucleation rate and, consequently, on the amount of particles produced. Higher nucleation rates lead to a higher concentration of critical nuclei, while the size of the critical nuclei does not vary much. This fact leads to more intensive coagulation and to more rapid particle growth.

The mean particle size also increases with increasing temperatures and residence times, which arises from increased particle coagulation kernels. If aggregated particles are considered, the increase in the carrier gas temperature reduces the characteristic sintering times, which leads to an increase in the primary particle size. The same effect is also related to the residence time at higher temperatures. In contrast, rapid cooling rates weaken the influence of the sintering process.

The influence of coagulation can be reduced by electrical repulsion if a significant part of the aerosol is charged unipolarly [62]. Though thermionic emission can lead to such a unipolar charge distribution, rapid discharging at lower temperatures has to be taken into account.

The understanding of basic dynamic processes in high-temperature aerosol systems and their interdependence allows the tailoring of particulate products

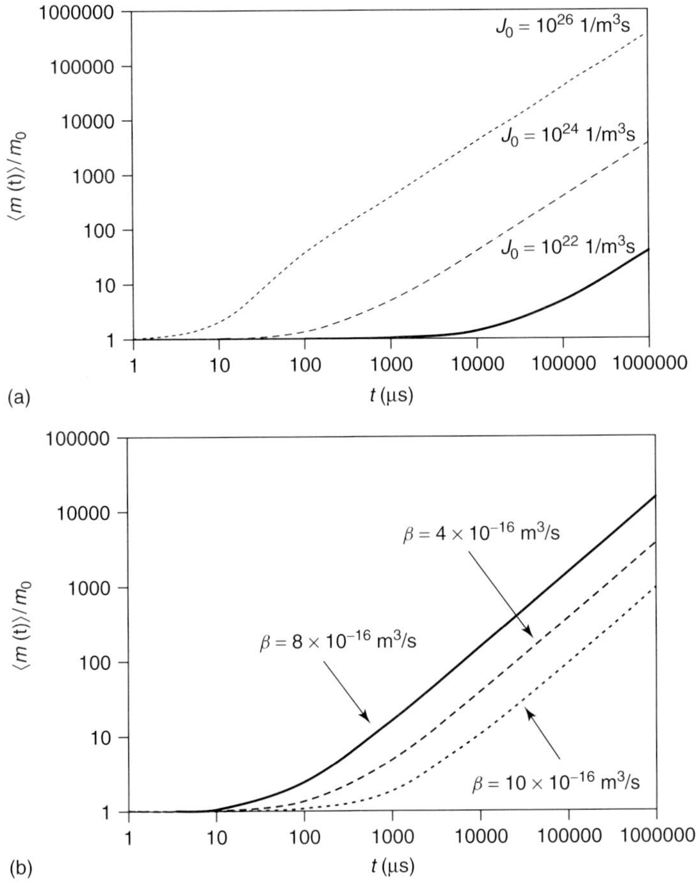

Figure 2.7 Solution of Eqs. (2.27)–(2.29) for (a) different nucleation rates and (b) different coagulation kernels ($T = 1273$ K; $\eta = 3.8 \times 10^{-5}$ Pa s; particle density 1000 kg/m^3; $\lambda_g = 292$ nm).

with respect to size, morphology, and composition. This opens possibilities for manufacturing of new products, including new composites, and their applications.

References

1. Wagner, E. and Brünner, H. (1960) Aerosil, Herstellung, Eigenschaften und Verhalten in organischen Flüssigkeiten. *Angew. Chem.*, **72**, 744–750.
2. Kriechbaum, G. and Kleinschmidt, P. (1989) Superfine oxide powders: flame hydrolysis and hydrothermal synthesis. *Angew. Chem. Adv. Mater.*, **10**, 330–337.
3. Basic Characteristics of AEROSIL® Fumed Silica. Technical Bulletin Fine Particles. Published by Evonik Degussa GmbH, Frankfurt am Main, Germany 2006.
4. Wagner, E. and Brünner, H. (1960) Aerosil, Herstellung, Eigenschaften und Verhalten in organischen

Flüssigkeiten. *Angewandte Chemie*, **72**, 744–750.
5. Biswas, P., Li, X., and Pratsinis, S.E. (1989) Optical waveguide preform fabrication: silica formation and growth in a high-temperature aerosol reactor. *J. Appl. Phys.*, **65**, 2445–2450.
6. Akhtar, M.K., Xiong, Y., and Pratsinis, S.E. (1991) Vapor synthesis of titania powder by oxidation of titanium tetrachloride. *AIChE J.*, **37**, 1561–1570.
7. Mädler, L., Kammler, H., Mueller, R., and Pratsinis, S.E. (2002) Controlled synthesis of nanostructured particles by flame spray pyrolysis. *J. Aerosol Sci.*, **33**, 369–389.
8. Mueller, R., Jossen, R., and Pratsinis, S.E. (2004) Zirconia nanopowders made in spray flames at high production rates. *J. Am. Ceram. Soc.*, **87**, 197–202.
9. What is Carbon Black? Published by Evonik Degussa GmbH, Frankfurt am Main, Germany, 2008.
10. Makela, J.M., Keskinen, H., Forsblom, T., and Keskinen, J. (2004) Generation of metal and metal oxide nanoparticles by liquid flame spray process. *J. Mater. Sci.*, **39**, 2783–2788.
11. Knipping, J., Pridöhl, M., Roth, P., and Zimmermann, G. (2003) Superparamagnetic nanocomposite materials. *Magnetohydrodynamics*, **39**, 71–76.
12. Athanassiou, E.K., Grass, R.N., and Stark, W.J. (2006) Large scale production of carbon coated copper nanoparticles. *Nanotechnology*, **17**, 1668–1673.
13. Gutsch, A., Mühlenweg, H., and Krämer, M. (2005) Tailor-made nanoparticles via gas-phase synthesis. *Small*, **1**, 30–46.
14. Dannenberg, E.M. (1971) Progress in carbon black technology. *J. Inst. Rubber Ind.*, **5**, 190–195.
15. Wu, M.K., Windeler, R.S., Steiner, C.K.R., Börs, T., and Friedlander, S.K. (1993) Controlled synthesis of nanosized particles by aerosol processes. *Aerosol Sci. Technol.*, **19**, 527–548.
16. Mezey, E.J. (1966) in *Vapor Deposition* (eds C.F.Powell, J.H. Oxley, and J.M. Blocher Jr.), John Wiley & Sons, Inc., New York.
17. WO 2005/049492 A1, 2005.
18. Alam, M.K. and Flagan, R.C. (1986) Controlled nucleation aerosol reactors: production of bulk silicon. *Aerosol Sci. Technol.*, **5**, 237–248.
19. Okuyama, K., Kousaka, Y., Tohge, N., Yamamoto, S., Wu, J.J., Flagan, R.C., and Seinfeld, J.H. (1986) Production of ultrafine metal oxide aerosol particles by thermal decomposition of metal alkoxide vapors. *AIChE J.*, **32**, 2010–2019.
20. Kim, K.S. (1997) Analysis on SiO_2 particle generation and deposition using furnace reactor. *AICHE J.*, **43**, 2679–2687.
21. Shi, L., Li, C., Chen, A., Zhu, Y., and Fang, D. (2001) Morphological structure of nanometer $TiO_2-Al_2O_3$ composite powders synthesized in high temperature gas phase reactor. *Chem. Eng. J.*, **84**, 405–411.
22. Huang, Y., Ho, W., Ai, Z., Song, X., Zhang, L., and Lee, S. (2009) Aerosol-assisted flow synthesis of B-doped, Ni-doped and B–Ni-codoped TiO_2 solid and hollow microspheres for photocatalytic removal of NO. *Appl. Catal. B*, **89**, 398–405.
23. Moore, K., Caserano, J., Smith, D.M., and Kodas, T. (1992) Synthesis of submicrometer mullite powder via high-temperature aerosol decomposition. *J. Am. Ceram. Soc.*, **75**, 213–215.
24. Zak, A., Feldman, Y., Alperovich, V., Rosentsveig, R., and Tenne, R. (2000) Growth mechanism of MoS_2 fullerene-like nanoparticles by gas-phase synthesis. *J. Am. Chem. Soc.*, **122**, 11108–11116.
25. Nasibulin, A.G., Moisala, A., Brown, D.P., and Kauppinen, E.I. (2003) Carbon nanotubes and onions from carbon monoxide using $Ni(acac)_2$ and $Cu(acac)_2$ as catalyst precursors. *Carbon*, **41**, 2711–2724.
26. Gani, M. and McPherson, R. (1980) The structure of plasma-prepared Al_2O_3 and TiO_2 powders. *J. Mater. Sci.*, **15**, 1915–1925.
27. Girshick, S.L. and Chiu, C.-P. (1989) Homogeneous nucleation of particles from the vapor phase in thermal plasma synthesis. *Plasma Chem. Plasma Process.*, **9**, 355–369.
28. Barry, T.I., Bayliss, R.K., and Lay, L.A. (1968) Mixed oxides prepared with an

induction plasma torch. *J. Mater. Sci.*, **3**, 229–238.

29. Yoshida, T. and Akashi, K. (1981) Preparation of ultrafine iron particles using an RF plasma. *Trans. Jpn. Inst. Metals*, **22**, 371–378.
30. Canteloup, J. and Mocellin, A. (1976) Ultrafine TaC powders prepared in a high frequency plasma. *J. Mater. Sci.*, **11**, 2353.
31. Kumar, R., Cheang, P., and Khor, K.A. (2001) RF plasma processing of ultrafine hydroxyaptite powders. *J. Mater. Processing Technol.*, **113**, 456–462.
32. Fabry, F., Flamant, G., and Fulcheri, L. (2001) Carbon black processing by thermal plasma. Analysis of the particle formation mechanism. *Chem. Eng. Sci.*, **56**, 2123–2132.
33. Weigle, J.C., Luhrs, C.C., Chen, C.K., Perry, W.L., Mang, J.T., Nemer, M.B., Lopez, G.P., and Phillips, J. (2004) Generation of aluminum nanoparticles using an atmospheric pressure plasma torch. *J. Phys. Chem. B*, **108**, 18601–18607.
34. Dutta, J., Hofmann, H., Houriet, R., Valmalette, J.-C., and Hofmeister, H. (1997) Crystallization of nanosized silicon powder prepared by plasma-induced clustering reactions. *AIChE J.*, **43**, 2610–2615.
35. Suzuki, M., Kagawa, M., Syono, Y., and Hirai, T. (1992) Synthesis of ultrafine single-component oxide particles by the spray-ICP technique. *J. Mater. Sci.*, **27**, 679–684.
36. Singh, R. and Doherty, R. (1990) Synthesis of TiN powders under glow discharge plasma. *Mater. Lett.*, **9**, 87–89.
37. Bystrzejewski, M., Karoly, Z., Szepvolgyi, J., Kaszuwara, W., Huczko, A., and Lange, H. (2008) Continuous synthesis of carbon-encapsulated magnetic nanoparticles with a minimum production of amorphous carbon. *Carbon*, **47**, 2040–2048.
38. Rao, N., Girshick, S., Heberlein, J., McMurry, P., Jones, S., Hansen, D., and Micheel, B. (1995) Nanoparticle formation using a plasma expansion process. *Plasma Chem. Plasma Process.*, **15**, 581–606.
39. Vollath, D. (2008) Plasma synthesis of nanopowders. *J. Nanopart. Res.*, **10**, 39–57.
40. Orii, T., Hirasawa, M., and Seto, T. (2003) Tunable, narrow-band light emission from size-selected Si nanoparticles produced by pulsed-laser ablation. *Appl. Phys. Lett.*, **83**, 3395–3397.
41. Cai, H., Chaudhary, N., Lee, J., Becker, M.F., Brock, J.R., and Keto, J.W. (1998) Generation of metal nanoparticles by laser ablation of microspheres. *J. Aerosol Sci.*, **29**, 627–636.
42. Sasaki, T., Terauchi, S., Koshizaki, N., and Umehara, H. (1997) Preparation of nanoparticles by excimer laser ablation of calcium iron complex oxide. *AIChE J.*, **43**, 2636–2640.
43. Gelbard, F. and Seinfeld, J.H. (1979) The general dynamic equation for aerosols – theory and application to aerosol formation and growth. *J. Colloid Interface Sci.*, **68**, 363–382.
44. Seinfeld, J.H. (1986) *Atmospheric Chemistry and Physics of Air Pollution*, John Wiley & Sons, Inc., New York.
45. Friedlander, S.K. (1977) *Smoke, Dust and Haze*, Wiley Interscience, New York.
46. Pratsinis, S.E. (1988) Simultaneous nucleation, condensation and coagulation in aerosol reactors. *J. Colloid Interface Sci.*, **124**, 416–427.
47. Tsantilis, S. and Pratsinis, S.E. (2004) Soft- and hard-agglomerate aerosols made at high temperatures. *Langmuir*, **20**, 5933–5939.
48. Smoluchowski, M.V. (1917) Versuch einer mathematischen Theorie der Koagulationskinetik kolloider Lösungen. *Z. Phys. Chem.*, **XCII**, 129–168.
49. Lushnikov, A.A. and Piskunov, V.N. (1975) A class of solvable models in the theory of coagulation. *Kolloidn Zh.*, **37**, 285–291.
50. Spouge, J.L. (1983) Solutions and critical times for the polydisperse coagulation equation when $a(x, y) = A + B(x + y) + Cxy$. *J. Phys. A: Math. Gen.*, **16**, 3172–3132.
51. Hinds, W.C. (1982) *Aerosol Technology*, John Wiley & Sons, Inc., New York.
52. Allen, M.D. and Raabe, O.G. (1985) Slip correction measurements of spherical

aerosol particles in an improved Millikan apparatus. *Aerosol Sci. Technol.*, **4**, 269–286.

53. Fuchs, N.A. (1964) *The Mechanics of Aerosols*, Pergamon Press, Elmsford, NY.
54. Otto, E., Fissan, H., Park, S.H., and Lee, K.W. (1999) The log normal size distribution theory of Brownian aerosol coagulation for the entire particle size range. Part II – Analytical solution using Dahneke's coagulation kernel. *J. Aerosol Sci.*, **30**, 17–34.
55. Matsoukas, T. and Friadlander, S.K. (1991) Dynamics of aerosol agglomerate growth. *J. Colloid and Interface Sci.*, **146**, 495–506.
56. Buesser, B., Heine, M.C., and Pratsinis, S.E. (2009) Coagulation of highly concentrated aerosols. *J. Aerosol Sci.*, **40**, 89–100.
57. Koch, W. and Friedlander, S.K. (1990) The effect of particle coalescence on the surface area of a coagulating aerosol. *J. Colloid Interface Sci.*, **140**, 419–427.
58. Weber, A.P. and Friedlander, S.K. (1997) *In situ* determination of the activation energy for restructuring of nanometer aerosol agglomerates. *J. Aerosol Sci.*, **28**, 179–192.
59. Shimada, M., Seto, T., and Okuyama, K. (1994) Size change of very fine silver agglomerates by sintering in a heated flow. *J. Chem. Eng. of Japan*, **27**, 795–802.
60. Xiong, Y., Akhtar, M.K., and Pratsinis, S.E. (1993) Formation of agglomerate particles by coagulation and sintering. 2. The evolution of the morphology of aerosol-made titania, silica and silica-doped titania powders. *J. Aerosol Sci.*, **24**, 301–313.
61. Kobata, A., Kusakabe, K., and Morooka, S. (1991) Growth and transformation of TiO_2 crystallites in aerosol reactor. *AIChE J.*, **37**, 347–359.
62. Vemury, S. and Pratsinis, S.E. (1996) Charging and coagulation during flame synthesis of silica. *J. Aerosol Sci.*, **27**, 951–966.
63. Katzer, M., Weber, A.P., and Kasper, G. (2001) Collision kinetics and electrostatic dispersion of airborne submicrometer fractal agglomerates. *J. Colloid Interface Sci.*, **240**, 67–77.
64. Schiel, A., Weber, A.P., Kasper, G., and Schmid, H.-J. (2002) *Part. Part. Syst. Char.*, **19**, 410–418.
65. Jiang, J., Lee, M.-H., and Biswas, P. (2007) Model for nanoparticle charging by diffusion, direct photoionization, and thermionization mechanisms. *J. Electrostat.*, **65**, 209–220.
66. Oron, A. and Seinfeld, J. (1989) The dynamic behavior of charged aerosols. III. Simultaneous charging and coagulation. *J. Colloid Interface Sci.*, **133**, 80.
67. Fuchs, N.A. (1963) On the stationary charge distribution on aerosol particles in a bipolar ionic atmosphere. *Geophys. Pura E Appl.*, **56**, 185.
68. Wood, M.D. (1981) Classical size dependence of the work function of small metallic spheres. *Phys. Rev. Lett.*, **21**, 623–636.
69. Maisels, A., Jordan, F., and Fissan, H. (2002) Dynamics of the aerosol particle photocharging process. *J. Appl. Phys.*, **91**, 3377–3383.
70. Landgrebe, J.D. and Pratsinis, S.E. (1990) A discrete-sectional model for powder production by gas chemical reaction and aerosol coagulation in the free-molecular regime. *J. Colloid Interface Sci.*, **139**, 315–325.
71. Zhao, H., Maisels, A., Matsoukas, T., and Zheng, C. (2007) *Powder Technol.*, **173**, 38–50.
72. Wegner, K. and Pratsinis, S.E. (2003) Scale-up of nanoparticle synthesis in diffusion flame reactors. *Chem. Eng. Sci.*, **58**, 4581–4589.

3
Aerosol Synthesis of Single-Walled Carbon Nanotubes

Albert G. Nasibulin and Sergey D. Shandakov

3.1
Introduction

3.1.1
Carbon Nanotubes as Unique Aerosol Particles

Carbon nanotubes (CNTs) are a unique family of materials exhibiting diverse useful chemical and *physical properties* [1, 2]. The CNTs and especially single-walled carbon nanotubes (SWCNTs) were found to have exceptional mechanical, thermal, and electronic properties [1–3], which are strongly determined by their chiralities. A brief summary of the physical properties of SWCNTs and materials made of SWCNTs are collected in Table 3.1. SWCNTs are the strongest known material, with exceptionally high *Young modulus* of elasticity and *tensile strength* [4, 5]. Both *thermal conductivity* and *electrical conductivity* show remarkably high values. Nanotubes in a polymer matrix significantly improve the thermal and *mechanical properties* of the matrix material [6]. Taking into account the very high *current density* that the tubes can withstand without destruction, up to 10^9 A/cm^2, SWCNTs are believed to be an ideal material to replace copper and aluminum in *integrated circuits*. Semiconducting SWCNTs have electrical properties that are better than those of any known *semiconductors*. The *hole mobility* in SWCNTs is higher than in silicon metal–oxide–semiconductor field-effect transistors (FETs) and comparable to the *in-plane mobility* of *graphene* [7]. The applications of SWCNTs in *microelectronics* are not limited to the utilization of individual CNTs. Sub-monolayer random networks or aligned arrays of SWCNTs could behave as a *thin-film semiconductor* with a *charge mobility* of up to $\sim$2500 cm^{2}V^{-1}s^{-1} [8]. An obvious advantage of CNT utilization in electronics is the possibility to create flexible and transparent components [8, 9]. Owing to their high *specific surface area* and developed microporous structure, CNT materials can be widely applied as filters, electrodes, effective adsorbents and absorbents, and so on [10].

CNTs have very interesting *molecular structure*. A SWCNT can be schematically created by rolling up into a cylinder a single graphene layer, which consists of carbon atoms packed in a honeycomb crystal lattice (Figure 3.1). Depending on

Aerosols – Science and Technology. Edited by Igor Agranovski

ISBN: 978-3-527-32660-0

Table 3.1 Physical properties of individual SWCNTs and materials made of SWCNTs.

	Individual SWCNTs	References
Young modulus of elasticity	300–1470 GPa (compared to 200 GPa for high-strength steel)	[4]
Tensile strength	30–200 GPa (compared to 1–2 GPa for high-strength steel)	[4, 5]
Thermal conductivity along the tube	~6600 W/m K (twice as high as that of diamond)	[11]
Electrical resistance of metallic CNTs	3×10^{-6} Ω cm at 300 K (compared to 2.82×10^{-6} Ω cm for Al and 1.72×10^{-6} Ω cm for Cu)	[12, 13]
Maximum current density (without destruction)	10^9 A/cm^2 at theoretical limit of 10^{13} A/cm^2 (compared to 10^7 A/cm^2 for copper wire of 100 nm diameter)	[14–16]
Hole and electron mobilities	(2–6) $\times 10^4$ cm^2/V s (compared to 450 and 1400 cm^2/V s for Si at 300 K)	[17, 18]
	Materials made of SWCNTs	
Microporous structures	Porous diameter: ~0.7 to ~2 nm	
Specific surface area	1300 m^2/g for closed CNTs; >2000 m^2/g for open CNTs	[10]
Young modulus of composites	2.5 GPa for SWCNT–poly(ε-caprolactone) composite (compared to 0.38 GPa for pure polymer)	[8]
Composite electrical resistance	10^6 Ω cm at 300 K (compared to 10^{15} Ω cm for pure polymer)	[19, 20]
Composite heat conductivity	10 W/m K (compared to 0.1–1 W/m K for pure polymers)	[19]

the method of rolling, one can get CNT structures with different helicities: achiral (zig-zag and armchair) or chiral (Figure 3.2). The zig-zag and armchair nanotubes correspond to chiral angles of 0° and 30°, respectively. There is a simple rule for the chiral indices (n, m), which allows the prediction of the *electronic properties* of SWCNTs: if the indices are equal $(n = m)$, then the tube has armchair structure and possesses metallic properties; if $(n - m)/3$ gives an integer value, then the CNT is semimetallic; otherwise, it is semiconducting, with a bandgap of $E_g = 0.9/D$ eV, where D is the tube diameter in nanometers [7].

The aerosol route is the most popular way to produce CNTs. All synthesis methods can be classified according to the carbon atomization as physical or chemical techniques. The physical method is associated with a high energy input to the carbon source (graphite or carbon black) by either arc discharge,

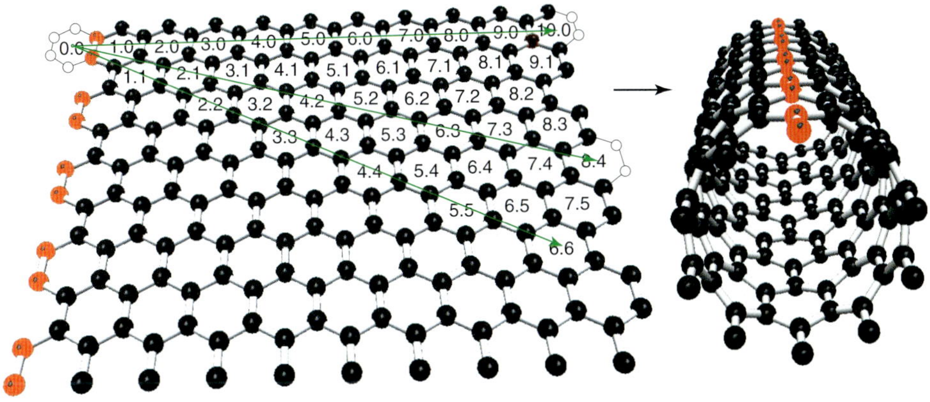

Figure 3.1 A schematic representation of the construction of a zig-zag (10, 0) CNT by rolling a cylinder joining hexagon (0, 0) and (10, 0). Lines also indicate the possibility of creating the armchair (6, 6) and chiral (8, 4) tubes shown in Figure 3.2.

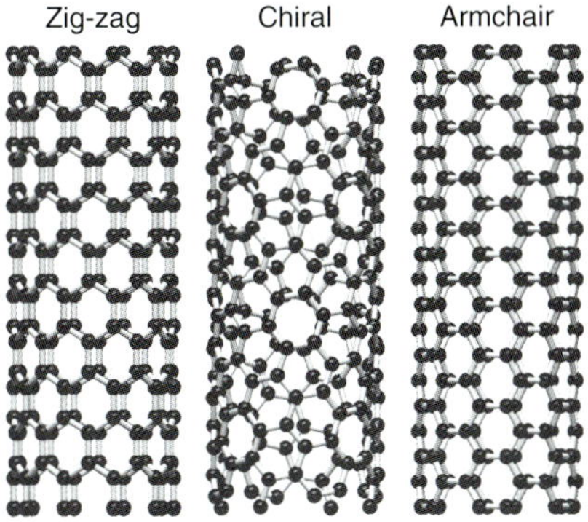

Figure 3.2 Examples of CNTs with different chiralities: zig-zag (10, 0), chiral (8, 4), and armchair (6, 6).

laser, or induction heating evaporation [21–23]. For the synthesis of SWCNTs, a small amount of *metal catalyst* is usually required. Chemical production is based on the *catalytic decomposition* of carbon-containing precursors. An obvious advantage of this method is the possibility of producing CNTs at relatively low temperatures. The synthesis methods can be further divided into *substrate-supported* chemical vapor deposition (*CVD*) [24–26] and aerosol-unsupported (free-floating catalyst) CVD [27–30]. In the substrate-supported CVD process, the *carbon precursor* decomposition and CNT formation take place on the surface of *catalyst* particles that are supported on a substrate. The substrates can be aerosolized. In the *free-floating catalyst* or aerosol-unsupported CVD method, the whole process takes place in the gas phase or on the surface of catalyst particles suspended in the gas. As a catalyst, typically *transition metals* such as Fe, Co, or Ni are used.

3.1.2
History and Perspectives of CNT Synthesis

Intensive CNT research was initiated in 1991 by Sumio Iijima after his famous publication [31], in which he showed clear structures of double-walled and *multi-walled CNTs*. *Single-walled CNTs* were independently discovered by Iijima and Ichihashi [32] and Bethune *et al.* [21] in 1993. Nevertheless, investigations on CNTs (carbon filaments or fibers) had been performed for many years before the described events. The earliest paper that the authors could find, related to the synthesis and observation of tubular carbon structures with a diameter of around 100 nm, was published in 1946 by Watson and Kaufmann [33]. They examined *cuprene*, the product of C_2H_2 polymerization over fine copper oxide catalyst below 300 °C. Cuprene, the chemical formula of which can be represented as C_xH_y, was likely an intermediate product between the carbon precursor and CNTs. In 1952 Radushkevich and Lukyanovich published the first clear transmission electron microscope (TEM) images of 30–50 nm thick CNTs [34]. In 1960 Bacon investigated the growth, structure, and properties of coiled *graphite whiskers* grown by arc discharge [35]. In 1975 Baker *et al.* finished their series of studies with a description of the first mechanism of CNT growth, examining hydrocarbon decomposition over Pt–Fe particles [36]. Oberlin *et al.* produced hollow *carbon fibers* with a diameter below 10 nm (admittedly a single- or double-walled CNT) using an aerosol growth technique [37, 38]. The crystallographic relationship between catalyst (FeCo and FeNi) particles and grown carbon nanofibers was investigated in 1981 [39]. The same year, a group of Soviet scientists reported the results of TEM observations of the product of carbon monoxide (CO) disproportionation on an iron catalyst and suggested the first chirality model of CNTs [40]. In 1984 Tibbets attempted to answer the question "Why are carbon filaments tubular?" He considered the energy change during filament growth and explained the tubular structure formation as due to *free-energy minimization* [41]. Further investigations of the growth [42] and applications [43] of the CNTs and filaments proceeded intermittently until the boom of CNT research.

The wide interest in CNTs was caused by their diverse useful properties, which provided various applications in many fields, such as emission technologies, *nanoelectronics*, *superstrong fibers*, composite materials, catalysts, *molecular wires*, straws, gears, switches, and photonic materials, and so on. It is believed that many new applications for CNTs will be proposed in the very near future when this material can be supplied on an industrial scale. So far, the ordinary utilization of CNTs has been limited by their very high price [44]. Nowadays, the price for purified SWCNTs remains quite high for many bulk applications (40 USD/g from Kunming Guorui Nanotechnologies Co., China). In order to reduce the cost significantly and to provide satisfactory quantities, development of the available methods for CNT mass production is needed. The available techniques have been successfully utilized to synthesize these materials in laboratory-scale quantities, and only a few methods are able to manufacture SWCNTs in gram quantities.

The first success in producing substantial amounts of SWCNTs was achieved by Bethune *et al.* in 1993 by the arc discharge method [21]. They used graphitic carbon with a small amount of cobalt catalyst. Smalley's group succeeded in growing high-quality SWCNTs at the 1–10 g scale by using laser ablation of a graphite target placed in an oven [13]. Later, the same group developed another method called the *HiPco process* [45], which will be discussed later. In 2000, the group led by D.E. Resasco from the University of Oklahoma developed the *CoMoCat method* utilizing Co- and Mo-impregnated fume silica particles in a CO atmosphere in a *fluidized bed reactor* [46], commercialized by SouthWest Nanotechnologies, Inc. One of the successful approaches for high-yield synthesis is based on the combustion of hydrocarbon–oxygen–catalyst precursors. This method has been extensively applied by different groups to synthesize SWCNTs [23, 47–50]. A Canadian group from the University of Sherbrooke has recently developed a large-scale system for the synthesis of SWCNTs using induction thermal plasma technology [23]. The method is based on the direct evaporation of carbon black and metallic catalyst mixtures in a radio-frequency plasma torch reaching extremely high temperature (~15 000 K). The 40 wt% purity SWCNT samples can be continuously synthesized at a production rate of 100 g/h.

Analyzing the various techniques for the synthesis of SWCNTs, we believe that aerosol-unsupported CVD methods have many advantages over other methods. Arc discharge and laser ablation methods rely on evaporation of carbon from solid carbon sources at temperature of 3000–4000 °C and therefore are very energy-consuming methods. Additionally, the nanotubes synthesized by carbon evaporation are in tangled forms, difficult to unbundle and purify. The CVD methods are operated at essentially low temperatures (about 600–1000 °C). From an industrial point of view, for many applications it is desirable to produce and/or directly deposit CNTs onto the required substrates, so that time-consuming steps of CNT purification from the catalyst and support, dispersion, and deposition processes are avoided. Therefore, the aerosol-unsupported CVD technique, which allows the production of high-quality clean SWCNTs, is more preferable. Supplementary advantages of the aerosol-unsupported method are the possibilities for on-line control of CNT quality and separate individual and bundled CNTs. The continuous CVD process is one of the most promising and powerful methods for high-yield synthesis under controlled conditions.

We have restricted this chapter by considering only SWCNTs produced by aerosol-unsupported CVD synthesis methods. Even though practically all synthesis techniques (except substrate-supported CVD) deal with *aerosolized catalysts* and CNTs, investigations of SWCNTs in the gas phase have widely been ignored. Practically all methods are limited to CNT synthesis and the characterization of the product. Evolution of CNT concentrations and aerosol particle diameters in the gas phase during growth have very rarely been discussed. The first reactors that allowed controlled aerosol-unsupported growth with subsequent investigations of CNTs as an aerosol object were built in 2002–2003 in the NanoMaterials Group (Helsinki University of Technology). In this chapter, two synthesis methods for controlled production of SWCNTs, that is, the hot-wire generator (HWG) [29] and the ferrocene-based method [30], will be described in detail.

3.2
Aerosol-Unsupported Chemical Vapor Deposition Methods

Many groups have successfully utilized the aerosol-unsupported CVD method for the laboratory-scale synthesis of SWCNTs. One of the most common ways to produce SWCNTs is based on the thermal decomposition of *ferrocene* (usually along with *thiophene*) dissolved in different carbon sources. A prototype of the reactor was built and used for the growth of carbon fibers in the 1980s by Endo [43]. An aerosol-unsupported continuous CVD method, in which SWCNTs are grown from a benzene–ferrocene–thiophene mixture at a temperature of 1200 °C, was developed in 1998 by Cheng *et al.* [51]. Rao and co-workers [52] and Bladh *et al.* [28] produced SWCNTs by pyrolyzing C_2H_2 and CH_4, respectively, in the temperature range of 800–1200 °C. Iijima's group at Japan's National Institute of Advanced Industrial Science and Technology has been actively working on the growth of SWCNTs utilizing other different aromatic hydrocarbons [53]. Recently, a simple technique based on spray pyrolysis of ferrocene dissolved in alcohols has been reported [54]. A similar method has been utilized by a group led by A. Windle from Cambridge University for spinning fibers from single-, double-, and multi-walled CNTs [55]. Synthesis of macro-scale amounts of SWCNTs from a mixture of ferrocene–sulfur–acetylene was achieved by Xie Sishen and colleagues at the Institute of Physics (Chinese Academy of Sciences) [56].

For many purposes, the presence of impurities is the main concern limiting the application of SWCNTs. Usually, as-produced SWCNTs contain undesired by-products in the form of soot particles, amorphous carbon or unused catalyst. Purification of the CNTs is very expensive and sometimes as expensive as the production of the tubes. The amount of undesirable products can be decreased by proper selection of experimental conditions and precursors. In particular, non-graphitic carbon impurities can be avoided by utilizing CO as the carbon source, which is known to disintegrate only on the surface of catalyst. Here, three experimental techniques based on CO and utilizing different catalyst sources, that is, *iron pentacarbonyl* (HiPco process) [45], ferrocene [28], and *supersaturated iron vapor* in the HWG method [29] will be briefly described. In order to demonstrate aerosol methods for the synthesis of SWCNTs, on-line monitoring of the SWCNT products, gas-phase separation of the products, and studies of the SWCNT growth mechanism, only the ferrocene-based and HWG reactors will be considered.

3.2.1
The HiPco Process

The HiPco (high pressure CO) process was developed by Nikolaev *et al.* at Rice University in 1999 [57]. Nowadays, this method is a well-known technique to grow bulk quantities of SWCNTs. HiPco is based on CO disproportionation (*Boudouard*

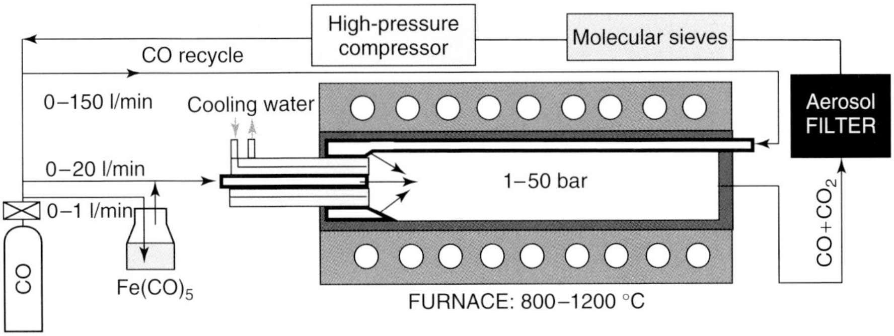

Figure 3.3 Schematic representation of the HiPco experimental setup.

reaction) according to

$$2CO \rightleftharpoons CO_2 + C_{(s)} \tag{3.1}$$

on the surface of iron particles. The catalyst is generated *in situ* by the thermal decomposition of iron pentacarbonyl ($Fe(CO)_5$) in the reactor heated to 1000–1200 °C. The HiPco process occurs at high CO pressures (up to 50 bar), which significantly increases the CO disproportionation rate and thus enhances the SWCNT yield.

A schematic of the HiPco reactor is shown in Figure 3.3. A flow of cold $Fe(CO)_5$ vapor containing CO is introduced into the reactor through a nozzle and turbulently mixed with a hot CO flow. The SWCNTs produced are collected by filtering the gas flow. The gaseous product, CO_2, is removed by passing the flow through molecular sieves. The CO gas is purified, compressed, and introduced into the reactor. Recirculation is needed due to the very low CO conversion – typically the mole fraction of CO_2 does not exceed 0.005% at the outlet of the reactor. The HiPco process allows the production of SWCNT material with a purity up to 97% at a rate of about 500 mg/h [58]. Recently, the HiPco process was applied to produce SWCNTs by thermal decomposition of ferrocene vapor without additional carbon sources [59].

3.2.2
Ferrocene-Based Method

The ferrocene-based method is also based on the catalytic CO disproportionation reaction (3.1). The main differences between this method and the HiPco process are the catalyst precursor source, the ambient pressure, and the laminar flow in the reactor. Utilization of ferrocene as a catalyst precursor allows high-quality SWCNTs to be produced in the temperature range of 600–1150 °C [30]. Ferrocene is vaporized at room temperature by passing CO (at a flow rate of 300 cm^3/min) through a cartridge filled with ferrocene powder (Figure 3.4). The flow containing ferrocene vapor (0.8 Pa) is then introduced into the high-temperature zone of the ceramic tube (internal diameter 22 mm) reactor through a water-cooled probe and

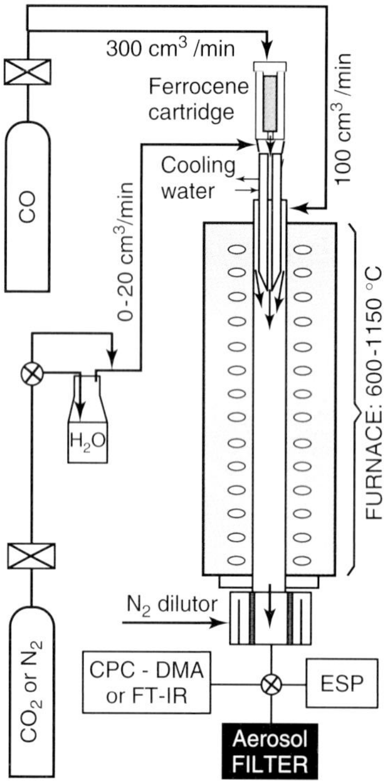

Figure 3.4 Schematic representation of the ferrocene-based reactor.

mixed with an additional CO flow (100 cm^3/min). The outlet of the water-cooled probe is at the wall temperature of around 700 °C, which is needed for fast heating of the vapor–gas mixture and the production of tiny catalyst particles. In order to enhance the yield, usually a small amount of water vapor (around 50 ppm) or CO_2 (around 600 ppm) is introduced into the reactor. The SWCNTs are collected either by filtering the flow for macroscopic investigations or by an electrostatic precipitator (ESP) for observation by transmission electron microscopy (TEM) or scanning electron microscopy (SEM). The distribution of aerosol CNT particles can be measured by a system comprising a differential mobility analyzer and a condensation particle counter. The gaseous composition of the products is measured by a Fourier-transform infrared (FTIR) spectroscopy system.

Raman measurements and TEM observations detected high-purity SWCNTs. As an example, Raman measurements carried out with 633 and 488 nm excitation wavelength lasers from a sample collected at 800 °C are shown in Figure 3.5. The existence of a radial breathing mode (RBM) in the low-frequency region (100–200 cm^{-1}) and the strong G-band (with peak at 1592 cm^{-1}) in the collected Raman spectrum indicate the formation of SWCNTs. A notable characteristic of the collected spectrum is a weak D-band (around 1300 cm^{-1}), indicating a low fraction of disordered carbon in the product. Depending on the experimental conditions, the diameters of the CNTs produced by this method can be varied from 1.1 to 2.0 nm.

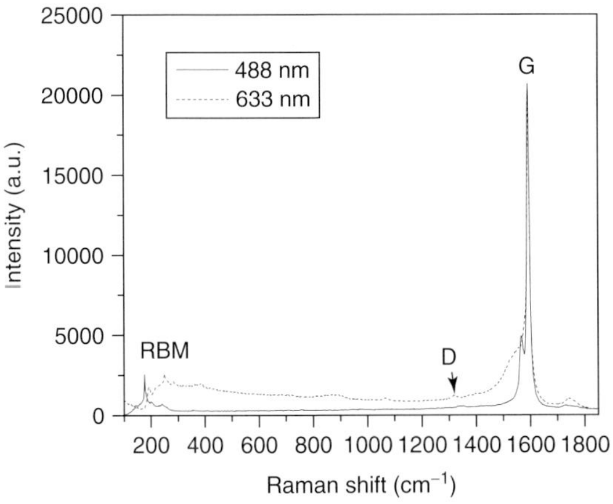

Figure 3.5 Raman spectra of SWCNTs synthesized in the ferrocene-based reactor.

3.2.3
Hot-Wire Generator

The HWG method is an original method developed in the NanoMaterials Group (Helsinki University of Technology) for the growth of SWCNTs [29] and multi-walled CNTs [60]. This method is based on the introduction of pre-made catalyst particles into the reactor and mixing them with the carbon source. A ceramic tube, with 22 mm internal diameter, inserted inside a furnace, is used as the laminar flow reactor (Figure 3.6). The HWG, which is a resistively heated thin iron wire (0.25 mm diameter), was placed inside a ceramic tube (with external and internal diameters of 13 and 9 mm) inserted inside the reactor to maintain inert conditions for catalyst particle formation. The outlet of the HWG tube was placed at the location with a wall temperature of around 400 °C. The iron particles produced in the HWG tube were carried in a N_2/H_2 (or Ar/H_2) mixture with mole component ratio of 93/7 at a flow rate of 400 cm^3/min and mixed with the outer 400 cm^3/min CO flow. Inside the reactor, in addition to the Boudouard reaction (3.1), the following CO hydrogenation reaction occurs:

$$CO + H_2 \rightleftharpoons H_2O + C_{(s)} \tag{3.2}$$

Downstream of the reactor, a porous tube dilutor (12 l/min) was used to decrease CNT bundling and to prevent product deposition on the walls by thermophoresis and diffusion. The reactor is operated at ambient pressure. The average residence time inside the reactor is about 2–3 s. In order to preserve the conditions for SWCNT formation, a small amount of etching agents, such as water vapor or carbon dioxide, can be added [61]. An important characteristic of this process is the efficiency of utilization of catalyst material. Practically all the catalyst particles initiate the growth of CNTs. The concentration of CNTs grown in the reactor

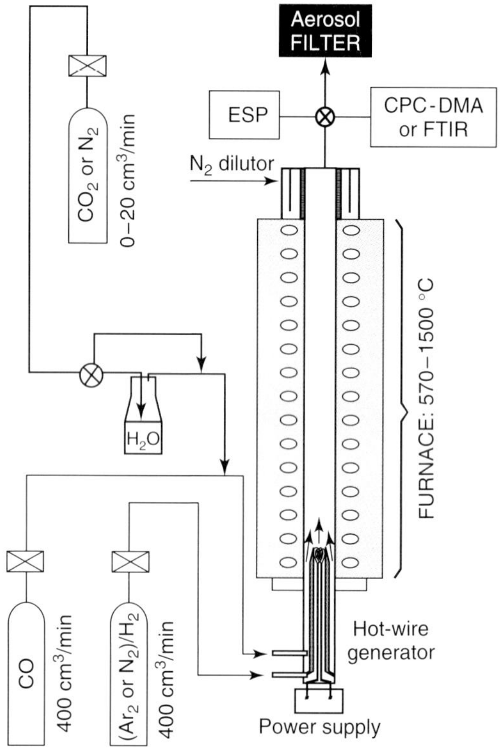

Figure 3.6 Schematic representation of the HWG experimental setup.

as well as the morphology of the products can be easily varied by changing the concentration or *activation degree* of the catalyst particles [60]. Depending on the experimental conditions, the diameter of the CNTs is determined by the catalyst particle size and varies from 1.1 to 1.4 nm [62].

3.3
Control and Optimization of Aerosol Synthesis

3.3.1
On-Line Monitoring of CNT Synthesis

Despite progress in the synthesis of CNTs, their detection has typically been realized only by SEM, TEM, or spectroscopy methods (for example, Raman, photoluminescence or optical absorption). These techniques generally require additional work, they are time- and resource-consuming, and they do not give direct feedback to the synthesis process. In order to provide on-line information on the reactor process conditions, differential mobility analysis (DMA) technology was adapted [63].

The DMA technique is widely used in the field of aerosol science for particle number size distribution (NSD) measurements and is based on the size classification of charged aerosol particles according to their electrical mobility in an electric field [64]. Particles with selected *electrical mobility* after the DMA are introduced to the condensation particle counter (CPC) for the concentration measurements. Scanning the voltage (typically applied to the internal electrode) enables the selected particle size to be changed within the range regulated by DMA sheath flow rate, voltage, and geometry.

Figure 3.7 shows an example of the NSDs of the product coming from the reactor under conditions in which the product consists of either inactive catalyst particles or bundles of SWCNTs. A clear difference in the NSDs can be seen in both ferrocene-based and HWG reactors. The total aerosol concentration drops about two to three times and the geometric mean diameter is shifted to larger sizes when the reactor parameters change from inactive to active conditions for SWCNT growth. This on-line technique significantly facilitates the execution of experimental work, especially when the parameters are tuned to find appropriate conditions for SWCNT growth. It is worth noting that the geometric mean diameters corresponding to the conditions of CNT growth in ferrocene and HWG reactors (100 and 35 nm, respectively) are determined by the geometry of the product (typically bundles) and do not give direct information about the length and diameter of the SWCNTs produced. The description of the theoretical approach to the mobility of non-spherical particles in an electric field has been reported elsewhere [63]. The correlation between electrical mobility size measured with the DMA and physical size of high-aspect-ratio objects such as multiwalled CNTs and nanowires was considered in [65].

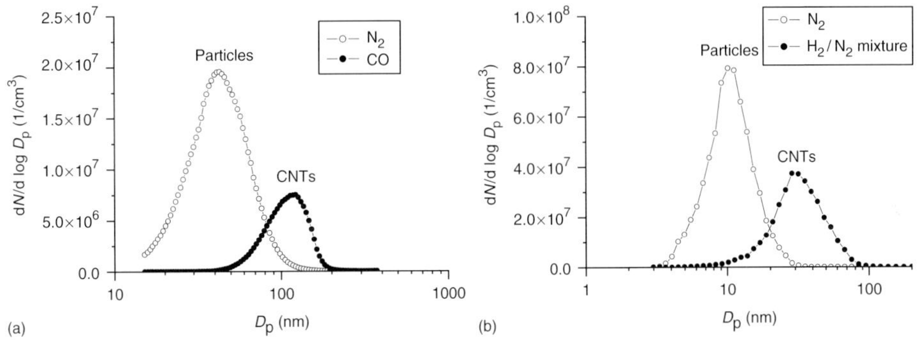

Figure 3.7 NSDs of the product formed in active and inactive CNT growth conditions in (a) the ferrocene-based reactor and (b) the HWG reactor using different carrier gases through the HWG tube.

3.3.2 Individual CNTs and Bundle Separation

Ideally, even one SWCNT with a certain chirality placed in a certain location is sufficient for many applications, such as an FET, memory device, quantum wire, or logic gate circuit. Therefore, methods for the controllable synthesis of individual CNTs are extremely desirable. However, this is still a challenging task, since CNTs tend to bundle spontaneously, and, as a result, most of the CNT synthesis methods produce bundled tubes. In order to exfoliate the bundles, additional steps of CNT functionalization, ultrasonication, and deposition are required. Isolated and individual CNTs can be synthesized by the substrate CVD method. Nevertheless, the requirement to use high growth temperatures (between 400 and 1000 °C) inevitably limits the utilization of temperature-sensitive substrate materials and the simple integration of CNTs into nano-scale electronic devices.

A one-step process for the gas-phase CVD synthesis, *in situ* separation, and deposition of individual SWCNTs on a wide variety of substrates at ambient temperature has been developed. This approach is based on spontaneous charging of CNTs during their bundling in the gas phase [66, 67]. Charged CNTs were removed from the gas by passing them via an electrostatic filter (ESF). The remaining non-charged fraction was found to consist of individual CNTs, while the filtered charged fraction of the CNTs were bundled. The collection of the individual CNTs can be realized by thermophoresis or by utilizing a corona charger ESP. This opens new avenues for the direct integration of individual CNTs into molecular electronics based on both conventional oxidized silicon substrates and temperature-sensitive materials, for example, for flexible electronics.

3.3.3 CNT Property Control and Nanobud Production

Initial experimental investigations revealed unstable CNT production in the HWG and ferrocene-based reactors. Analysis of the experimental data showed that the difficulties were associated with the reactor wall conditions. It was found that, in order to provide stable CNT synthesis, the walls needed to be saturated by the catalyst material. Practically, this could be achieved either by deposition of the catalyst particles on the reactor walls or simply by using a reactor tube made of the catalyst material. Further investigations revealed the essential need for the presence of small amounts of etching agents (CO_2 or H_2O vapor) for the successful synthesis of CNTs [61, 68].

By varying the concentration of H_2O vapor or CO_2 introduced into the CNT reactor, SWCNTs covered with covalently attached fullerenes were synthesized. This material was termed *carbon nanobuds*, since the fullerenes on the surface of the SWCNTs were reminiscent of buds on a branch [69]. The introduction of H_2O and CO_2 into the ferrocene reactor revealed that the optimum reagent

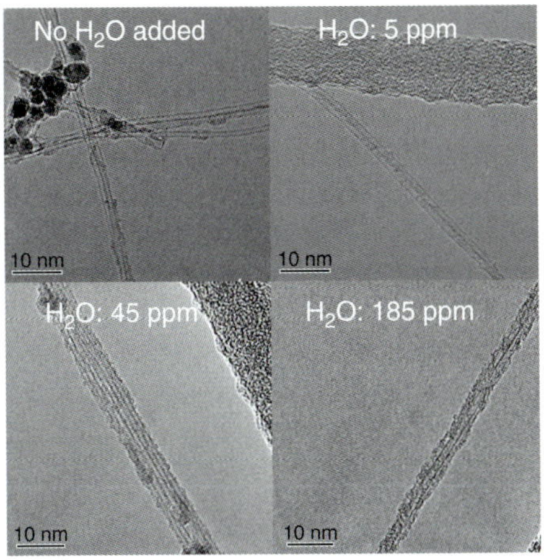

Figure 3.8 Effect of water vapor concentration in the ferrocene reactor at 1000 °C.

concentrations were between 45 and 245 ppm for H_2O (see Figure 3.8) and between 2000 and 6000 ppm for CO_2. *Atomistic calculations* based on *density functional theory* showed the possibility of the existence of fullerenes covalently bonded to the SWCNTs by [2+2] and [4+4] *cycloaddition* as well as the formation of one-body *SWCNT–fullerene* hybrid structures (Figure 3.9). Nanobuds possess

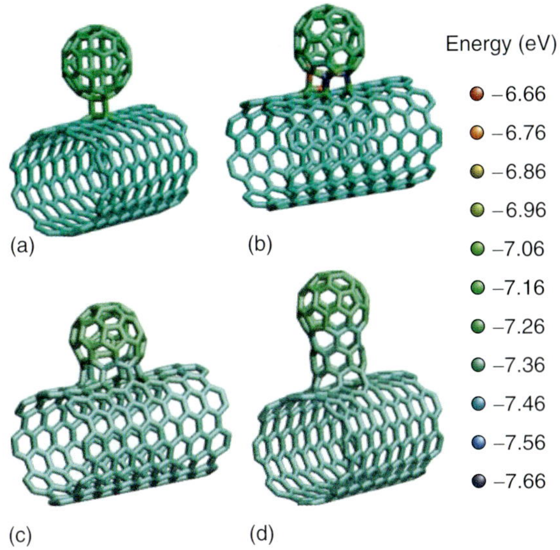

Figure 3.9 Bonding scenarios of fullerenes with SWCNTs: (a) [2+2] and (b) [4+4] cycloaddition; (c) and (d) SWCNT–fullerene hybrid structures. The relative binding energy of the individual atoms is reflected in the color for (a)–(d). (Courtesy of Dr A.V. Krasheninnikov.)

advantageous properties compared to SWCNTs or fullerenes alone, or in their non-bonded configurations. For instance, this structural arrangement of highly curved fullerenes and inert, but thermally and electrically conductive, CNTs was shown to exhibit enhanced cold electron field-emission properties [69]. Additionally, higher reactivity due to the presence of fullerenes opens new possibilities for the functionalization of SWCNTs. Furthermore, the attached fullerene molecules could be used as molecular anchors to prevent slipping of SWCNTs in composites. Owing to the charge transport between SWCNTs and functionalizing fullerenes, both the electrical and optical properties of the material can be tuned.

3.4
Carbon Nanotube Bundling and Growth Mechanisms

3.4.1
Bundle Charging

Commonly, for the DMA measurements, a radioactive bipolar charger is utilized for the artificial charging of the aerosol particles. It was observed that CNTs synthesized in both the ferrocene and HWG systems were naturally charged. Aerosol mobility size measurements were presented in two different ways: as distributions and as spectra. The *mobility size distributions* were measured by passing the aerosol-containing flow through a radioactive charger, and then a typical inversion procedure was performed to calculate the real aerosol concentration, assuming equilibrium charging in the charger [70]. Spectra in which the concentration of the naturally charged aerosol was not subjected to the inversion procedure were obtained without the charger. The *mobility diameter, D*, was calculated assuming a spherical shape and a single charge of aerosol particles on the basis of the *Millikan equation* [71].

Aerosol size distribution measurements of charged and non-charged CNTs showed that, depending on the concentration of CNTs in the HWG reactor, a certain fraction of the CNTs are charged. The higher the total concentration, the higher is the fraction of charged CNTs [67]. The tubes produced in the ferrocene reactor were practically all charged due to higher CNT concentration. The fraction of charged CNTs was determined on the basis of DMA size mobility measurements using an ^{85}Kr charger. Electrically neutral and all (neutral and charged) aerosol CNTs were measured after the ESF, where the potential was or was not, respectively, applied. The concentration of charged CNTs was detected to be 92% at 800 °C, 99% at 1000 °C, and 98% at 1150 °C. An example of the NSD and spectra obtained after the reactor at 1150 °C is shown in Figure 3.10. The CNTs were found to be charged both positively and negatively. At a temperature of 800 °C the concentration of negatively charged ions was about six times higher than that of positive CNTs. Increasing the reactor temperature to 1000 °C resulted in an increase of the fraction of positively charged CNTs: the difference decreased to a factor of 2. At 1150 °C, the spectra of both negative and positive polarities were very similar.

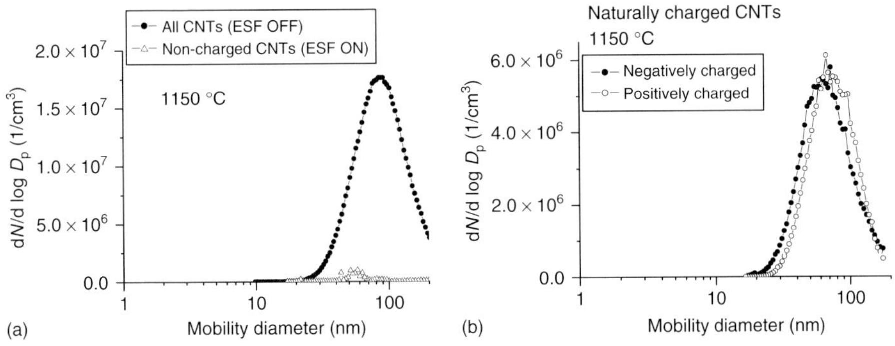

Figure 3.10 (a) Number size distributions of all and non-charged fraction of CNTs. (b) Mobility spectra of negatively and positively naturally charged CNTs. The measurements were carried out after formation in the ferrocene reactor operating at 1150 °C.

In order to detect the charging degree of the naturally charged CNTs, two different DMAs in tandem were utilized in the aerosol experiments [72]. The results of the Gaussian function fittings for 80 nm mobility-selected CNTS showed that the CNTs possessed from one to five elementary charges (Figure 3.11), which indicated non-equilibrium charging of the CNTs.

In order to examine the nature of the ions that were emitted from the CNTs and thereby were responsible for CNT ionization, laser desorption ionization time-of-flight (LDI-TOF) measurements of the CNT samples were carried out. Carbon-containing ions ($C_6OH_3^+$, $C_7O_2H_3^+$, and $C_9O_2H_9^+$) were found to be responsible for the negative charging of the CNTs. The positive charging occurs because of electron emission. The bundle can become charged due to the emission

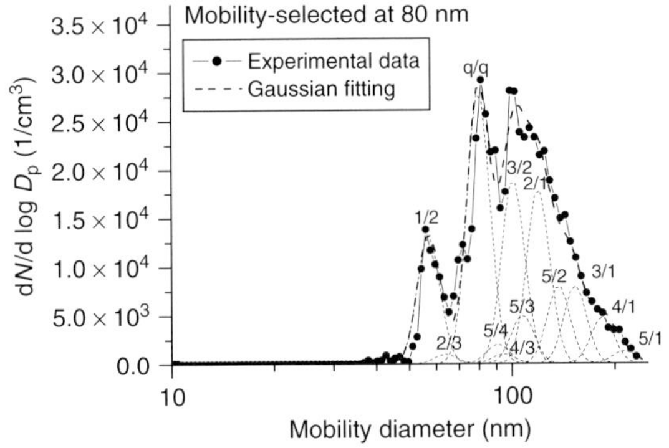

Figure 3.11 Results of tandem mobility measurements of naturally negatively charged mobility-selected CNTs at 80 nm. Charging states are represented as follows: original number of charges/number of charges after passing through the charger.

of electrons and ions by dissipation of the released *van der Waals energy* during CNT bundling. For instance, the bundling of two armchair (10, 10) nanotubes leads to the total energy decrease as high as 0.95 eV/nm. Considering that the average length of our bundles is larger than 100 nm, bundling is expected to result in a very high energy release.

3.4.2 Growth Mechanism

The mechanism of CNT growth is usually investigated *ex situ* by studying the products prepared under different experimental conditions. Another approach to examine the mechanism of SWCNT formation, which can be easily realized in the aerosol technique, is to sample *in situ* from different locations in the reactor [68, 73]. These allow one to follow the changes in product morphology, state, chemical composition of catalyst particles, and length of the CNTs [73].

For *in situ* sampling, a stainless-steel rod with a TEM grid attached to the top was rapidly inserted into the reactor and held in a steady position for 30 s. The product was collected due to the thermophoretic forces between the hot reactor's atmosphere and the cold sampling rod. Aerosol product samples were taken from different positions inside the reactor. Multiple measurements made from TEM images determined the average length of the SWCNTs at different locations in the reactor. On the basis of the CNT lengths, temperature, and residence time in the reactor, the average growth rate of the SWCNTs was calculated. The calculations showed that the growth rate of the CNTs varied from 0.67 to 2.7 μm/s when the temperature changed from 804 to 915 °C. The activation energy was found to be $E_a = 134$ kJ/mol, which was close to the values reported in the literature for multi-walled CNTs, and could be attributed to the activation energy for the diffusion of carbon atoms in bulk solid iron with concentrations of carbon from 0.1 to 1 mass% [73].

Even though the catalyst particles are very small, the fact that they are in the solid state can be confirmed by calculations on the basis of the *Kelvin equation*:

$$T_m = T_0 \exp\left(-\frac{2\sigma_{sl}V}{r\Delta H_{fus}}\right)$$

Here T_m is the *melting temperature* for a given particle of radius r; T_0 is the bulk melting temperature (1535 °C); ΔH_{fus} is the *latent heat of fusion* (13.8 kJ/mol); V is the volume of a metal molecule, which can be calculated from the density (7.87×10^3 kg/m^3); and σ_{sl} is the surface tension between the liquid and solid (0.86 J/m^2). As can be seen from Figure 3.12, the catalyst particles of around 4.5 nm in the temperature range of 804–915 °C should be in the solid state [73].

The shape of the catalyst particle determines the probability of the carbon atoms being on the surface. The particle shape is not always ideally spherical, and in addition to being convex can also contain concave regions, where the particle surface curvature is negative. In general, a system adopts a configuration that minimizes the *surface energy*. Owing to the diffusion process, the system continually changes

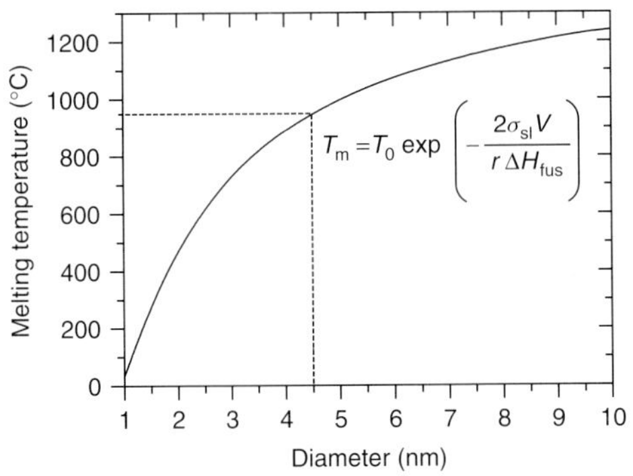

Figure 3.12 The effect of particle curvature on the melting temperature and carbon solubility in iron.

and tends to occupy lower potential energy level. Therefore, the probability of carbon atoms being on the negative-curvature surface is higher since then carbon would get more neighbors compared to on convex regions. Therefore, excess carbon dissolved in a particle can preferentially segregate to regions with negative curvature. The existence of concave regions in the particles and the segregation conditions were observed during our TEM investigations and could be easily seen during *in situ* TEM studies of CNT growth [74, 75].

The results of the kinetic studies, which determined the limiting stage for CNT growth (carbon diffusion through the solid catalyst particle), TEM observations, and calculations allowed us to propose a mechanism for CNT nucleation and growth (Figure 3.13). The nucleation of CNTs is believed to occur from solid iron supersaturated by carbon. A concave region, where carbon segregation can occur preferentially, is created in the particle (Figure 3.13a). Carbon released on the surface forms a graphitic cap (Figure 3.13b), a CNT nucleation site. The CNT starts to grow by feeding carbon into the concave regions, where the carbon atoms are consumed due to incorporation into the hexagonal CNT network. This leads to the creation of a carbon concentration gradient in the particle, which provides a continuous flux of carbon atoms through the catalyst particles from the surface to

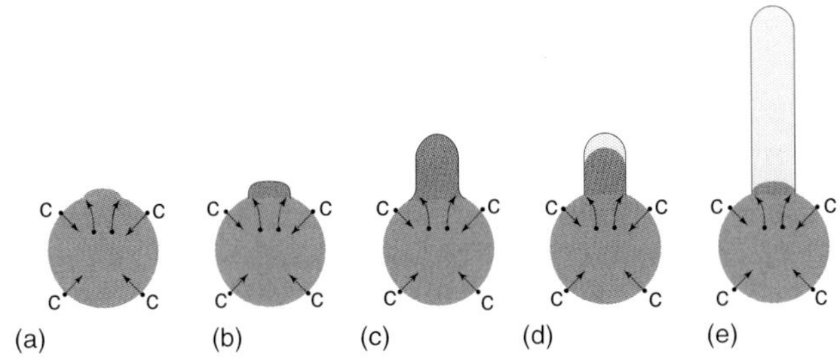

Figure 3.13 Schematic representation of CNT nucleation and growth. (Courtesy of Anton Anisimov.)

the region of CNT growth. The concave region is preserved by the growing CNT, since the CNT walls are much stiffer than the iron particle. Part of the particle embedded inside the growing CNT in the initial stage (Figure 3.13c) is pulled out from inside of the CNT due to the surface tension of the catalyst particle (Figure 3.13d). Figure 3.13e corresponds to the conditions of steady-state growth of CNTs. It is worth mentioning that the proposed mechanism (Figure 3.13) is valid not only for SWCNTs, but also for the growth of multi-walled CNTs synthesized by the substrate CVD method [74, 75].

3.5 Integration of the Carbon Nanotubes

The main advantage of the considered aerosol methods is the possibility to directly integrate the CNTs into some applications without time-consuming sample purification, dispersion, and subsequent deposition. Since the CNTs are clean, they can be directly utilized in the form in which they come from the reactor. Since the CNTs are collected at room temperature, they can be deposited onto any substrate, including temperature-sensitive polymers. CNTs can be directly deposited onto the desired substrate by applying either thermophoretic or electrophoretic forces, or they can be simply filtered and subsequently transferred onto the secondary support.

Collection of CNTs from the gas phase can also be carried out using a thermophoretic precipitator. Thermophoresis is a physical phenomenon in which aerosol particles, subjected to a temperature gradient, move from high- to low-temperature zones [76]. A thermophoretic precipitator employs an electrically heated top plate (120 °C) and a water-cooled bottom plate (10 °C) with an aerosol sample flow through a gap between the plates [77]. Substrates for SWCNT collection are placed on the cold plate. Collection of both neutral and electrically charged CNTs from the gas phase onto substrates can also be carried out at ambient temperature using a point-to-plate ESP (Intox Products, New Mexico, USA). In this device, CNTs are charged in a corona discharge that ionizes the gas and creates a small current between the two plates. Charged SWCNTs can also be collected on a substrate to which a certain potential is applied. This type of ESP is a cylindrical chamber with an inner horizontally positioned metal electrode, where the substrate is placed [78].

Deposited individual CNTs have been utilized to create high-efficiency FETs. Transistors made out of CNTs are known to have very high mobilities, but usually exhibit large hysteresis [79]. Nevertheless, the unwanted hysteresis in FETs could be successfully utilized in memory devices. Rinkiö *et al.* [80] showed high-yield memory elements with an ON/OFF ratio up to 10^5 using a nanometer-thick gate dielectric deposited by atomic layer deposition.

Devices built on individual CNTs allow the study of the properties of tubes and are very important from the scientific point of view. However, they are resource- and time-consuming to manufacture and very difficult to manipulate. Usually, electron-beam and multiple-step nanolithography processes are required. However, for many applications, group properties of CNTs can be utilized. The uniformity of CNT properties is achieved by statistical averaging over a number of individual tubes comprising the network. This gives reproducible electrical behavior over larger length scales in comparison to individual CNTs, whose tube-to-tube variations in chirality, position, and orientation constitute a major fabrication obstacle [83]. Recently, a fabrication method for high-performance FETs based on SWCNTs deposited from an aerosol reactor was developed [83]. Two types of thin-film transistor structures have been fabricated to evaluate the FET performance of SWCNT films: bottom-gate transistors on Si/SiO_2 substrates, and top-gate transistors on polymer substrates. Devices exhibited ON/OFF ratios of up to 10^5 and field-effect mobilities of up to 4 cm^2/V s.

Filtering, which is the easiest way to gather aerosol particles from the gas phase, can also be utilized for some applications. Figure 3.14 shows the SWCNTs collected from a ferrocene reactor by filtering downstream of the reactor through 2.45 cm diameter nitrocellulose (or silver) disk filters (Millipore Corp., USA). Depending on the desired film thickness, the deposition time could be varied from a few seconds to several hours. A simple thermo-compression method for integrating SWCNT films of adjustable thickness, transparency, and conductivity into polymer films was proposed in [79]. Produced *SWCNT–polyethylene* composite films have exhibited good optical transparency and conductivity as well as high mechanical flexibility. It was found that the electrical conductivity of the SWCNT films could be significantly improved by ethanol densification. SWCNT–polyethylene thin films demonstrated excellent cold electron field-emission properties [79]. CNTs can also be transferred onto a mirror for laser mode locking [80] and onto a polyethylene terephthalate (PET) substrate for electrochemical applications [84] or for transparent and flexible electrodes for replacement of indium tin oxide (ITO) in flat displays. Various other devices and components based on CNT networks have already been successfully demonstrated, including diodes, logic circuit elements, solar cells, displays, and sensors.

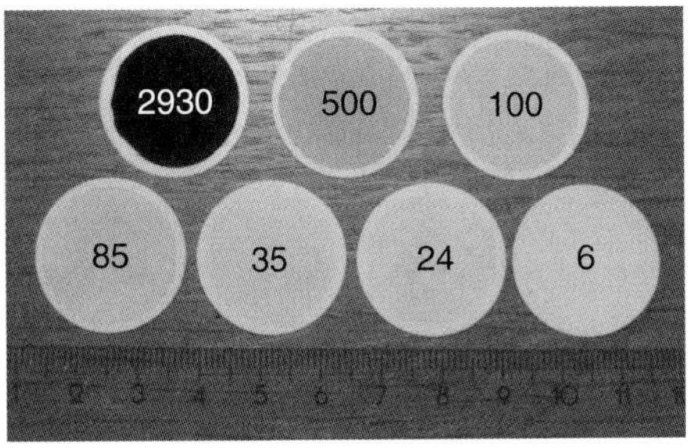

Figure 3.14 SWCNT film collected on nitrocellulose filters. (The numbers show the thickness value in nanometers.)

3.6 Summary

In this chapter, research in the field of CNTs concerning their discovery, intriguing properties, and applications has been briefly reviewed. A comparison of different synthesis methods for both laboratory and industrial production was presented. Special attention was devoted to the aerosol-unsupported CVD method and its advantages in the controlled production and integration of CNTs. On-line monitoring of CNT fabrication using DMA measurements and separation of individual CNTs based on spontaneous charging of CNTs were considered. Direct integration of the CNTs produced by the aerosol methods into different applications, especially for high-performance flexible electronics, was briefly discussed.

Acknowledgements

We thank Prof. E.I. Kauppinen, and the present and former members of the NanoMaterials Group, namely Dr A. Moisala, A.S. Anisimov, Dr H. Jiang, Dr D.P. Brown, Dr D. Gonzalez, Dr P. Queipo, and M.Y. Zavodchikova, who contributed to the original papers used as the basis of this review. Also, we are grateful to Dr P. Pikhitsa, Dr A.V. Krasheninnikov, Prof. D. Resasco, Prof. O. Tolochko, Dr D. Tománek, Y. Tian, L.I. Nasibulina, and Dr I.V. Anoshkin. This work was supported by the Academy of Finland, the European Commission through a Marie Curie Individual Fellowship (No. MIF1-CT-2005-022110), and partly by the Federal Agency for Science and Innovation of Russia Federation.

References

1. Dresselhaus, M.S., Dresselhaus, G., and Eklund, P.C. (1996) *Science of Fullerenes and Carbon Nanotubes*, Academic Press, San Diego.
2. Reich, S., Thomsen, C., and Maultzsch, J. (2004) *Carbon Nanotubes*, Wiley-VCH Verlag GmbH, Weinheim.
3. Rotkin, S.V. and Subramoney, S. (2005) *Applied Physics of Carbon Nanotubes: Fundamentals of Theory, Optics and Transport Devices*, Springer, Berlin.
4. Yu, M.-F., Bradley, S.F., Arepalli, S., and Ruoff, R.S. (2000) Tensile loading of ropes of single wall carbon nanotubes and their mechanical properties. *Phys. Rev. Lett.*, **84**, 5552–5555.
5. Lau, K.T. and Hui, D. (2002) The revolutionary creation of new advanced materials – carbon nanotubes composites. *Composites B*, **33** (4), 263–277.
6. Chatterjee, T., Mitchell, C.A., Hadjiev, V.G., and Krishnamoorti, R. (2007) Hierarchical polymer–nanotube composites. *Adv. Mater.*, **19**, 3850–3853.
7. McEuen, P.L. and Parl, J.-Y. (2004) Electron transport in single-walled carbon nanotubes. *MRS Bull.*, **25**, 272–275.
8. Cao, Q., Kim, H., Pimparkar, N., Kulkarni, J.P., Wang, C., Shim, M., Roy, K., Alam, M.A., and Rogers, J.A. (2008) Medium-scale carbon nanotube thin-film integrated circuits on flexible plastic substrates. *Nature*, **454**, 495–500.
9. Nasibulin, A.G., Ollikainen, A., Anisimov, A.S., Brown, D.P., Pikhitsa, P.V., Holopainen, S., Penttilä, J.S., Helistö, P., Ruokolainen, J., Choi, M., and Kauppinen, E.I. (2008) Integration of single-walled carbon nanotubes into polymer films by thermo-compression.

Chem. Eng. J., **136** (2–3), 409–413.

10. Fonseca, A. and Nagi, J.B. (2001) in *Carbon Filaments and Nanotubes: Common Origins, Differing Applications?* (eds L.P. Biró, C.A. Bernardo, G.G. Tibbetts, and Ph. Lambin), NATO Science Series E, Vol. 372, Kluwer Academic, Dordrecht, pp. 75–84.
11. Berber, S., Kwon, Y.-K., and Tomanek, D. (2000) Unusually high thermal conductivity of carbon nanotubes. *Phys. Rev. Lett.*, **84** (20), 4613–4616.
12. Osawa, E. (ed.) (2001) *Perspectives of Fullerene Nanotechnology*, Kluwer Academic, Dordrecht.
13. Thess, A., Lee, R., Nikolaev, P., Dai, H.J., Petit, P., Robert, J., Xu, C.H., Lee, Y.H., Kim, S.G., Rinzler, A.G., Colbert, D.T., Scuseria, G.E., Tomanek, D., Fischer, J.E., and Smalley, R.E. (1996) Crystalline ropes of metallic carbon nanotubes. *Science*, **273**, 483–487.
14. Ebbesen, T.W., Lezec, H.J., Hiura, H., Bennett, J.W., Ghaemi, H.F., and Thio, T. (1996) Electrical conductivity of individual carbon nanotubes. *Nature*, **382**, 54–56.
15. Wei, B.Q., Vajtai, R., and Ajayan, P.M. (2001) Reliability and current carrying capacity of carbon nanotubes. *Appl. Phys. Lett.*, **79**, 1172–1174.
16. Moulin, J., Woytasik, M., Grandchamp, J.-P., Dufour-Gergam, E., and Bosseboeuf, A. (2006) High current densities in copper microcoils: influence of substrate on failure mode, in *Symposium on Design, Test, Integration and Packaging of MEMS/MOEMS*, TIMA Editions, Grenoble.
17. Hull, R. (ed.) (1999) *Properties of Crystalline Silicon*, INSPEC, The Institution of Electrical Engineers, London, pp. 430–443.
18. (a) Tameev, A.R., Jiménez, L.L., Pereshivko, L.Y., Rychwalski, R.W., and Vannikov, A.V. (2007) Charge carrier mobility in films of carbon-nanotube–polymer composites. *J. Phys.: Conf. Ser.*, **61**, 1152–1156; (b) Martel, R., Schmidt, T., Shea, H.R., Hertel, T., and Avouris, P. (1998) Single- and multi-wall carbon nanotube field-effect transistors. *Appl. Phys. Lett.*, **73**, 2447–2449.
19. Choi, E.S., Brooks, J.S., Eaton, D.L., Al-Haik, M.S., Hussaini, M.Y., Garmestani, H., Li, D., and Dahmen, K. (2003) Enhancement of thermal and electrical properties of carbon nanotube polymer composites by magnetic field processing. *J. Appl. Phys.*, **94** (9), 6034–6039.
20. Strümpler, R. and Glatz-Reichenbach, J. (1999) Conducting polymer composites. *J. Electroceram.*, **3** (4), 329–346.
21. Bethune, D.S., Kiang, C.H., de Vries, M.S., Gorman, G., Savoy, R., Vazquez, J., and Beyers, R. (1993) Cobalt-catalysed growth of carbon nanotubes with single-atomic-layer walls. *Nature*, **363**, 605–607.
22. Guo, T., Nikolaev, P., Thess, A., Colbert, D.T., and Smalley, R.E. (1995) Catalytic growth of single-walled nanotubes by laser vaporization. *Chem. Phys. Lett.*, **243**, 49–54.
23. Kim, K.S., Cota-Sanchez, G., Kingston, C.T., Imris, M., Simard, B., and Soucy, G. (2007) Large-scale production of single-walled carbon nanotubes by induction thermal plasma. *J. Phys. D: Appl. Phys.*, **40**, 2375–2387.
24. Dai, H., Rinzler, A.G., Nikolaev, P., Thess, A., Colbert, D.T., and Smalley, R.E. (1996) Single-wall nanotubes by metal-catalyzed disproportionation of carbon monoxide. *Chem. Phys. Lett.*, **260**, 471–475.
25. Bachilo, S.M., Balzano, L., Herrera, J.E., Pompeo, F., Resasco, D.E., and Weisman, R.B. (2003) Narrow (n,m)-distribution of single-walled carbon nanotubes grown using a solid supported catalyst. *J. Am. Chem. Soc.*, **125**, 11186–11187.
26. Miyauchi, Y., Chiashi, S., Murakami, Y., Hayashida, Y., and Maruyama, S. (2004) Fluorescence spectroscopy of single-walled carbon nanotubes synthesized from alcohol. *Chem. Phys. Lett.*, **387** (1-3), 198–203.
27. Zhou, Z., Ci, L., Chen, X., Tang, D., Yan, X., Liu, D., Liang, Y., Yuan, H., Zhou, W., Wang, G., and Xie, S. (2003) Controllable growth of

double wall carbon nanotubes in a floating catalytic system. *Carbon*, **41**, 337–342.

28. Bladh, K., Falk, L.K.L., and Rohmund, F. (2000) On the gas phase growth of single-walled carbon nanotubes and encapsulated metal particles in the gas phase. *Appl. Phys. A*, **70**, 317–322.
29. Nasibulin, A.G., Moisala, A., Brown, D.P., Jiang, H., and Kauppinen, E.I. (2005) A novel aerosol method for single walled carbon nanotube synthesis. *Chem. Phys. Lett.*, **402**, 227–232.
30. Moisala, A., Nasibulin, A.G., Brown, D., Jiang, H., Khriachtchev, L., and Kauppinen, E.I. (2006) Single-walled carbon nanotube synthesis using ferrocene and iron pentacarbonyl in a laminar flow reactor. *Chem. Eng. Sci.*, **61**, 4393–4402.
31. Iijima, S. (1991) Helical microtubules of graphitic carbon. *Nature (London)*, **354**, 56.
32. Iijama, S. and Ichihashi, T. (1993) Single-shell carbon nanotubes of 1-nm diameter. *Nature*, **363**, 603–605.
33. Watson, J.H.L. and Kaufmann, K. (1946) Electron microscope examination of the microphysical properties of the polymer cuprene. *J. Appl. Phys.*, **17**, 996–1005.
34. Radushkevich, L.V. and Lukyanovich, V.M. (1952) Carbon nanostructure for alternative power engineering. *J. Phys. Chem.*, **26** (1), 88–95.
35. Bacon, R. (1960) Growth, structure, and properties of graphite whiskers. *J. Appl. Phys.*, **31** (2), 283–290.
36. Baker, R.T.K. and Waite, R.J. (1975) Formation of carbonaceous deposits from the platinum–iron catalyzed decomposition of acetylene. *J. Catal.*, **37**, 101–105.
37. Oberlin, A., Endo, M., and Koyama, T. (1976) High resolution electron microscope observations of graphitized carbon fibers. *Carbon*, **14**, 133–135.
38. Oberlin, A., Endo, M., and Koyama, T. (1976) Filamentous growth of carbon through benzene decomposition. *J. Cryst. Growth*, **32**, 335–349.
39. Audier, M., Oberlin, A., and Coulon, M. (1981) Crystallographic orientations of catalytic particles in filamentous carbon: case of simple conical particles. *J. Cryst. Growth*, **55**, 549–556.
40. Kolesnik, N.F., Nesterenko, A.M., Akhmatov, Y.S., Suhomlin, V.I., and Prilutskii, O.V. (1982) Phase composition and structure peculiarities of products of NiO and Fe_2O_3 and carbon monoxide interactions. *Izv. Akad. Nauk. SSSR, Met.*, **3**, 12–17 (in Russian).
41. Tibbets, G.C. (1984) Why are carbon filaments tubular? *J. Cryst. Growth*, **66**, 632–638.
42. Sacco, A., Thacker, P., Chang, T.N., and Chiang, A.T.S. Jr. (1984) The initiation and growth of filamentous carbon from α-iron in H_2, CH_4, H_2O, CO_2, and CO gas mixtures. *J. Catal.*, **85**, 224–236.
43. Endo, M. (1988) Grow carbon fibers in the vapor phase. *ChemTech*, **18**, 568–576.
44. *www.mercorp.com*, *www.sesres.com* (last accessed 31 March 2010).
45. Nikolaev, P.M., Bronikowski, J., Bradley, R.K., Rohmund, F., Colbert, D.T., Smith, K.A., and Smalley, R.E. (1999) Gas-phase catalytic growth of single-walled carbon nanotubes from carbon monoxide. *Chem. Phys. Lett.*, **313**, 91–97.
46. Kitiyanan, B., Alvarez, W.E., Harwell, J.H., and Resasco, D.E. (2000) Controlled production of single-wall carbon nanotubes by catalytic decomposition of CO on bimetallic CoMo catalysts. *Chem. Phys. Lett.*, **317**, 497–503.
47. Height, M.J., Howard, J.B., Tester, J.W., and Sande, J.B.V. (2004) Flame synthesis of single-walled carbon nanotubes. *Carbon*, **42**, 2295–2307.
48. Vander Wal, R.L., Ticich, T.M., and Curtis, V.E. (2000) Diffusion flame synthesis of single-walled carbon nanotubes. *Chem. Phys. Lett.*, **323** (3–4), 217–223.
49. Vander Wal, R.L., Hall, L.J., and Berger, G.M. (2002) Optimization of flame synthesis for carbon nanotubes using supported catalyst. *J. Phys. Chem. B*, **106** (51), 13122–13132.

50. Yuan, L.M., Saito, K., Pan, C.X., Williams, F.A., and Gordon, A.S. (2001) Nanotubes from methane flames. *Chem. Phys. Lett.*, **340** (3–4), 237–241.
51. Cheng, H., Li, F., Su, G., Pan, H., and Dresselhaus, M. (1998) Large-scale and low-cost synthesis of single-walled carbon nanotubes by the catalytic pyrolysis of hydrocarbons. *Appl. Phys. Lett.*, **72**, 3282–3284.
52. Satishkumar, B.C., Govindaraj, A., Sen, R., and Rao, S.N.R. (1998) Single-walled nanotubes by the pyrolysis of acetylene–organometallic mixtures. *Chem. Phys. Lett.*, **293**, 47–52.
53. Saito, T., Xu., W.-C., Ohshima, S., Ago, H., Yumura, M., and Iijima, S. (2006) Supramolecular catalysts for the gas-phase synthesis of single-walled carbon nanotubes. *J. Phys. Chem. B*, **110**, 5849–5853.
54. Su, L.F., Wang, J.N., Yu, F., Sheng, Z.M., Chang, H., and Pak, C. (2006) Continuous production of single-wall carbon nanotubes by spray pyrolysis of alcohol with dissolved ferrocene. *Chem. Phys. Lett.*, **420**, 421–425.
55. Motta, M., Moisala, A., Kinloch, I.A., and Windle, A.H. (2007) High performance fibres from 'dog bone' carbon nanotubes. *Adv. Mater.*, **19**, 3721–3726.
56. Song, L., Ci, L., Lv, L., Zhou, Z., Yan, X., Liu, D., Yuan, H., Gao, Y., Wang, J., Liu, L., Zhao, X., Zhang, Z., Dou, X., Zhou, W., Wang, G., Wang, C., and Xie, S. (2004) Direct synthesis of a macroscale single-walled carbon nanotube non-woven material. *Adv. Mater.*, **16**, 1529–1534.
57. Nikolaev, P. (2004) Gas-phase production of single-walled carbon nanotubes from carbon monoxide: a review of the HiPco process. *J. Nanosci. Nanotechnol.*, **4**, 307–316.
58. Bronikowski, M.J., Willis, P.A., Colbert, D.T., Smith, K.A., and Smalley, R.E. (2001) Gas-phase production of carbon single-walled nanotubes from carbon monoxide via the HiPco process: a parametric study. *J. Vac. Sci. Technol. A*, **19** (4), 1800–1805.
59. Barreiro, A., Hampel, S., Rümmeli, M.H., Kramberger, C., Grüneis, A., Biedermann, K., Leonhardt, A., Gemming, T., Büchner, B., Bachtold, A., and Pichler, T. (2006) Thermal decomposition of ferrocene as a method for production of single-walled carbon nanotubes without additional carbon sources. *J. Phys. Chem. B*, **110**, 20973–20977.
60. Nasibulin, A.G., Moisala, A., Jiang, H., and Kauppinen, E.I. (2006) Carbon nanotube synthesis from alcohols by a novel aerosol method. *J. Nanopart. Res.*, **8**, 465–475.
61. Nasibulin, A.G., Brown, D.P., Queipo, P., Gonzalez, D., Jiang, H., and Kauppinen, E.I. (2006) An essential role of CO_2 and H_2O during single-walled CNT synthesis from carbon monoxide. *Chem. Phys. Lett.*, **417**, 179–184.
62. Nasibulin, A.G., Pikhitsa, P.V., Jiang, H., and Kauppinen, E.I. (2005) Correlation between catalyst particle and single-walled carbon nanotube diameters. *Carbon*, **43** (11), 2251–2257.
63. Moisala, A., Nasibulin, A.G., Shandakov, S.D., Jiang, H., and Kauppinen, E.I. (2005) On-line detection of single-walled carbon nanotube formation during aerosol synthesis methods. *Carbon*, **43**, 2066–2074.
64. Knutson, E. and Whitby, K. (1975) Aerosol classification by electric mobility: apparatus, theory and applications. *J. Aerosol Sci.*, **6**, 443–451.
65. Kim, S.H., Mulholland, G.W., and Zachariah, M.R. (2007) Understanding ion-mobility and transport properties of aerosol nanowires. *J. Aerosol Sci.*, **38**, 823–842.
66. Gonzalez, D., Nasibulin, A.G., Shandakov, S.D., Jiang, H., Queipo, P., and Kauppinen, E.I. (2006) Spontaneous charging of single-walled carbon nanotubes in the gas phase. *Carbon*, **44**, 2099–2101.
67. Gonzalez, D., Nasibulin, A.G., Shandakov, S.D., Jiang, H., Queipo, P., Anisimov, A.S., Tsuneta, T., and Kauppinen, E.I. (2006) Spontaneous

charging of single-walled carbon nanotubes: a novel method for the selective substrate deposition of individual tubes at ambient temperature. *Chem. Mater.*, **18**, 5052–5057.

68. Nasibulin, A.G., Queipo, P., Shandakov, S.D., Brown, D.P., Jiang, H., Pikhitsa, P.V., Tolochko, O.V., and Kauppinen, E.I. (2006) Studies on mechanism of single-walled carbon nanotube formation. *J. Nanosci. Nanotechnol.*, **6**, 1233–1246.

69. Nasibulin, A.G., Pikhitsa, P.V., Jiang, H., Brown, D.P., Krasheninnikov, A.V., Anisimov, A.S., Queipo, P., Moisala, A., Gonzalez, D., Lientschnig, G., Hassanien, A., Shandakov, S.D., Lolli, G., Resasco, D.E., Choi, M., Tománek, D., and Kauppinen, E.I. (2007) A novel hybrid carbon nanomaterial. *Nat. Nanotechnol.*, **2**, 156–161.

70. Wiedensohler, A. (1988) An approximation of the bipolar charge distribution for particles in the submicron size range. *J. Aerosol Sci.*, **19**, 387–389.

71. Friedlander, S.K. (2000) *Smoke, Dust, and Haze. Fundamentals of Aerosol Dynamics*, Oxford University Press, New York.

72. Nasibulin, A.G., Shandakov, S.D., Anisimov, A.S., Gonzalez, D., Jiang, H., Pudas, M., Queipo, P., and Kauppinen, E.I. (2008) Charging of aerosol products during ferrocene vapor decomposition in N_2 and CO atmospheres. *J. Phys. Chem. C*, **112** (15), 5762–5769.

73. Anisimov, A.S., Nasibulin, A.G., Jiang, H., Launois, P., Cambedouzou, J., Shandakov, S.D., and Kauppinen, E.I. (2010) Mechanistic investigations of single-walled carbon nanotube synthesis by ferrocene vapor decomposition in carbon monoxide. *Carbon*, **48** (2), 380–388.

74. Yoshida, H., Takeda, S., Uchiyama, T., Kohno, H., and Homma, Y. (2008) Atomic-scale in-situ observation of carbon nanotube growth from solid state iron carbide nanoparticles. *Nano Lett.*, **8**, 2082–2086.

75. Helveg, S., López-Cartes, C., Sehested, J., Hansen, P.L., Clausen, B.S., Rostrup-Nielsen, J.R., Abild-Pedersen, F., and Nørskov, J.K. (2004) Atomic-scale imaging of carbon nanofibre growth. *Nature*, **427**, 426–429.

76. Kodas, T. and Hampden-Smith, M. (1999) *Aerosol Processing of Materials*, John Wiley & Sons, Inc., New York.

77. Gonzalez, D., Nasibulin, G.A., Baklanov, A.M., Shandakov, S.D., Brown, D.P., Queipo, P., and Kauppinen, E.I. (2005) A new thermophoretic precipitator for collection of nanometer-sized aerosol particles. *Aerosol Sci. Technol.*, **39**, 1064–1071.

78. Krinke, T.G., Deppert, K., Magnusson, M.H., Schmidt, F., and Fissan, H. (2002) Microscopic aspects of the deposition of nanoparticles from the gas phase. *J. Aerosol Sci.*, **33**, 1341–1359.

79. Nasibulin, A.G., Ollikainen, A., Anisimov, A.S., Brown, D.P., Pikhitsa, P.V., Holopainen, S., Penttilä, J.S., Helistö, P., Ruokolainen, J., Choi, M., and Kauppinen, E.I. (2008) Integration of single-walled carbon nanotubes into polymer films by thermo-compression. *Chem. Eng. J.*, **136**, 409–413.

80. Kivistö, S., Hakulinen, T., Kaskela, A., Aitchison, B., Brown, D., Nasibulin, A., Kauppinen, E., Härkönen, A., and Okhotnikov, O.G. (2009) Carbon nanotube films for ultrafast broadband technology. *Opt. Express*, **17**, 2358–2363.

81. Durkop, T., Getty, S.A., Cobas, E., and Fuhrer, M.S. (2004) Extraordinary mobility in semiconducting carbon nanotubes. *Nano Lett.*, **4**, 35–39.

82. Rinkiö, M., Johansson, A., Zavodchikova, M.Y., Toppari, J., Nasibulin, A.G., Kauppinen, E.I., and Törmä, P. (2008) High-yield of memory elements from carbon nanotube field-effect transistors with atomic layer deposited gate dielectric. *New J. Phys.*, **10**, 103019.

83. Zavodchikova, M.Y., Kulmala, T., Nasibulin, A.G., Ermolov, V., Franssila, S., Grigoras, K., and

Kauppinen, E.I. (2009) Carbon nanotube thin film transistors based on aerosol methods. *Nanotechnology*, **20**, 085201.

84. Heras, A., Colia, A., López-Palacios, J., Kaskela, A., Nasibulin, A.G., Ruiz, V., and Kauppinen, E.I. (2009) Flexible optically transparent single-walled carbon nanotube electrodes for UV–Vis absorption spectroelectrochemistry. *Electrochem. Commun.*, **11**, 442–445.

4
Condensation, Evaporation, Nucleation

Alexey A. Lushnikov

4.1
Introduction

An aerosol is a collection of particles suspended in a gaseous medium (carrier gas, in what follows). In many cases, the aerosol reveals itself as a collection of particles, but sometimes it leaves the possibility to study the individual characteristics of particles. Then the collective properties of the aerosol can be considered as the sum of the individual properties of the particles comprising it.

This chapter considers the most important phenomena in the life of aerosol particles: nucleation, condensation, and evaporation.

Nucleation is responsible for the production of the tiniest particles due to gas-to-particle conversion. Condensable vapors cannot exist in the vapor phase under certain conditions, normally when the vapor pressure exceeds the saturation value, which, in turn, depends on the temperature. Statistical mechanics (see, for example, [1–9]) predicts the phase transition in these cases, but it does not answer the question of how it goes on. For example, the vapor can condense on the walls of the vessel containing the gas–vapor mixture, or on foreign aerosol particles suspended in the carrier gas, or it can form the particles itself without any help from other external factors. The latter case is referred to as *spontaneous nucleation* [1]. There are some principal difficulties in the theoretical description of the nucleation process. I will try to elucidate the nature of these difficulties and to outline ways to avoid them.

Although everybody has seen how a kettle boils, producing visible vapor, quantitative measurements on nucleation are very far from being simple. The point is that the various nucleation theories are applicable for steady-state and very clean (no foreign condensation nuclei) conditions. It is not an easy task to provide such conditions [3].

At first sight, condensation is a simpler process. A vapor molecule reaches the particle surface and sticks to it. But even in this case one encounters the necessity to solve the kinetic *Boltzmann equation* in order to describe the transport of molecules toward the particle surface [10]. More problems come up if we want to take into account the surface processes related to latent heat release or

Aerosols – Science and Technology. Edited by Igor Agranovski

ISBN: 978-3-527-32660-0

surface chemical reactions. Experiments on condensation are more accessible than nucleation experiments. Their interpretation is also more straightforward [11].

Evaporation – the escape of a molecule from the particle – is much less understandable than condensation. This is the reason why one tries to express the *evaporation rate* through the *equilibrium vapor pressure* over the particle surface and the condensation efficiency. But this is not always a valid procedure. It may be better to try to attack the problem from the "inside," that is, to consider a heated particle and to calculate the flux of the molecules from the particle surface [4, 12].

4.2
Condensation

Let us consider a *gas–vapor mixture*. The process

$$(g) + (1) \longrightarrow (g+1) \tag{4.1}$$

is referred to as condensation. A g-mer just joins the molecules of condensable vapor one by one. Let n_∞ be the number concentration of vapor molecules far away from the particle. The particle absorbs the molecules and thus the vapor concentration at the particle surface becomes equal to n_a. The concentration difference $n_\infty - n_a$ drives a flux toward the particle surface. The particle begins to grow. The rate of change in the number of the vapor molecules in the particle is proportional to the difference $n_\infty - n_a$, that is,

$$\frac{dg}{dt} = \alpha(a)(n_\infty - n_a) \tag{4.2}$$

where the particle radius is $a = a_0(4\pi g/3)^{1/3}$, with a_0 being the radius of a molecule of the condensable substance. The coefficient $\alpha(a)$ is called the condensation efficiency. It has dimension $[L^3/T]$. The central problem of the theory of condensation is to find $\alpha(a)$.

Very simple dimensional considerations allow us to establish a general form of the condensation efficiency. There are three parameters that govern the *condensation kinetics*: the *particle radius a*, the *thermal velocity* of the condensable gas molecules $v_T = \sqrt{8kT/\pi m}$, and their *diffusivity D*. Their units are: a [cm], v_T [cm/s], and D [cm^2/s]. Because the units of $\alpha(a)$ are [cm^3/s], we can write

$$\alpha(a) = \pi a^2 v_T f(a v_T/D) \tag{4.3}$$

The multiplier π normalizes $f(0)$ to unity and $f(0) = 1$ (see Eq. (4.10) below). The function $f(x)$ is not yet known. In order to find this function, one should solve the kinetic Boltzmann equation that describes the time evolution of the coordinate–velocity distribution of the condensing molecules, then find the flux of the condensing molecules toward the particle, and then extract $\alpha(a)$. This is not easy to do in general form. However, the limiting situations have been analyzed [10, 13–16].

4.2.1
Continuum Transport

Let the growing particle be large enough so that the diffusion equation is applicable for the description of the molecular transport of the condensable gas toward the particle surface. Then, if we know the vapor concentration n_∞ far away from the particle, the solution to the steady-state diffusion equation has the form

$$n(r) = n_\infty - \frac{J}{4\pi Dr} \tag{4.4}$$

Here the total flux J of the condensable vapor is introduced. It does not depend on the coordinates (the condensable vapor does not disappear somewhere on its way from infinity to the particle surface). We can thus define the value of J from a boundary condition fixing the concentration at the particle surface $n(a) = n_a$. This condition defines the expression for the total flux,

$$J = 4\pi Da(n_\infty - n_a) \tag{4.5}$$

One sees that Eq. (4.5) has the form predicted by Eq. (4.3). The function $f(x)$ in this case is just

$$f(x) \approx \frac{4}{x} \tag{4.6}$$

This result is understandable: the thermal velocity v_T should not enter the final expression in the continuum limit, because the vapor transport is governed by the diffusion process of the vapor molecules. In this case the thermal motion provides only very tiny jumps at a distance on the order of the molecular mean free path. The *diffusion flux* arises at larger distances, here on the order of the particle size. If we remove v_T from the list of parameters responsible for molecular transport, then only two parameters should enter the final result for α, namely, D and a. The only combination with the correct dimension is $\alpha \propto Da$.

4.2.2
Free-Molecule Transport

When the particle size is much smaller than the mean free path of the vapor molecules, then the diffusion coefficient does not define the vapor transport toward the particle. There is no diffusion. The free molecules reach the particle surface moving with velocity v_T. Hence, the diffusivity does not enter the set of parameters defining the particle transport. We then have only one combination of correct dimension, $a^2 v_T$. Hence,

$$\alpha(a) \propto a^2 v_T \tag{4.7}$$

The proportionality factor can be found from a simple kinetic consideration. The free-molecule flux is the product of the particle flux $n_\infty \boldsymbol{v}$ times the differential of the particle surface $d\mathbf{S}$. Integrating over all molecular velocities and elements of the particle surface directed toward the particle center yields

$$J = \frac{n_\infty}{Z} \int (\boldsymbol{v} \cdot \mathrm{d}\boldsymbol{S})\, \mathrm{e}^{-E/kT}\, \mathrm{d}^3 v \tag{4.8}$$

Here

$$Z = \int \mathrm{e}^{-E/kT}\, \mathrm{d}^3 v \tag{4.9}$$

is the *partition function* for a single vapor molecule, E is its kinetic energy, and $\mathrm{d}\boldsymbol{S}$ stands for the element of the particle surface. The integration is trivial, and its details can be found elsewhere. The final result is

$$\alpha(a) = \pi a^2 v_T \tag{4.10}$$

Then $f(x) = 1$.

4.3
Condensation in the Transition Regime

The first simple theories applied the continuum models of condensation (the particle radius a greatly exceeds the condensing molecule mean free path lv [17]). Such models were not able to describe very small particles with sizes less than l. It was quite natural, therefore, to try to attack the problem by starting with the free-molecule limit, that is, to consider a collisionless motion of condensing molecules. Various expressions for the condensation efficiencies were derived [10, 13, 15, 18–21]. The important step directed to reconciliation of these two limiting cases was done by Fuchs [22], who invented the *flux-matching theory*.

The *flux-matching theories* are well adapted for studying the behavior of aerosol particles in the transition regime. Although these theories mostly did not have a firm theoretical basis, they successfully served for systematizing numerous experiments on the growth of aerosol particles. Up to the present day, these theories remain rather effective and very practical tools for studying the kinetics of aerosol particles in the transition regime (see [9, 10, 13]). On the other hand, these theories are always semiempirical, that is, they contain a parameter that should be taken from somewhere else, not from the theory itself.

We introduce the reader to the ideology of the flux-matching theories by considering the condensation of a non-volatile vapor onto the surface of an aerosol particle. The central idea of the flux-matching procedure is a hybridization of the diffusion and the free-molecule approaches. The concentration profile of a condensing vapor far away from the particle is described by the diffusion equation. This profile coincides with the real one down to distances on the order of the vapor molecule mean free path. A limiting sphere is then introduced, wherein the free-molecule kinetics governs the vapor transport. The equality of the fluxes in both zones and the continuity of the concentration profile at the surface of the limiting sphere define the flux and the condensing vapor concentration at the particle surface. The third parameter, the radius of the limiting sphere, cannot be found from such a consideration.

We apply a more sophisticated approach [23]. We also introduce a limiting sphere outside of which the density profile of the condensing vapor can be described by the diffusion equation. Inside the limiting sphere, we solve the collisionless Boltzmann equation subject to a given boundary condition at the particle surface and introduce an additional condition: the vapor concentration at the surface of the limiting sphere coincides with that found from the solution of the diffusion equation. Even in the absence of any potential created by the particle, the vapor profile in the free-molecule zone depends on the radial coordinate, because the particle surface adsorbs all incoming molecules. We thus gain the possibility to call for the continuity of the first derivatives of the profile on both sides of the limiting sphere. This additional condition defines the radius of the limiting sphere.

4.3.1
Flux-Matching Theory

As mentioned above, the steady-state molecular flux $J(a)$ onto the surface of a particle of radius a can always be written as

$$J(a) = \alpha(a)(n_\infty - n_a) \tag{4.11}$$

that is, the flux is proportional to the difference of the vapor density n_∞ far away from the particle and at the particle surface. The proportionality coefficient $\alpha(a)$ is referred to as the condensation efficiency. The problem is to find $\alpha(a)$.

Below, we adapt the flux-matching theory of [23] (see also Chapter 1 in this book [24]) to the case of condensation of neutral molecules onto the particle surface.

To this end, we generalize Eq. (4.11) as follows:

$$J(a, R, n_R) = \alpha(a, R)(n_R - n_a) \tag{4.12}$$

where n_R is the vapor concentration at a distance R from the particle's center. It is important to emphasize that n_R is (still) an *arbitrary* value introduced as a boundary condition at distance R (also arbitrary) to a kinetic equation that it is necessary to solve for defining $\alpha(a, R)$.

The flux defined by Eq. (4.11) is thus

$$J(a) = J(a, \infty, n_\infty) \qquad \text{and} \qquad \alpha(a) = \alpha(a, \infty) \tag{4.13}$$

The value of $\alpha(a, R)$ does not depend on $n_R - n_a$ because of the linearity of the problem.

Assuming that we know the exact vapor concentration profile $n_{\text{exact}}(r)$ corresponding to the flux $J(a)$ from infinity (see Eqs. (4.11) and (4.12)), we can express $J(a)$ in terms of n_{exact} as follows:

$$J(a) = J(a, R, n_{\text{exact}}(R)) = \alpha(a, R)(n_{\text{exact}}(R) - n_a) \tag{4.14}$$

If we choose R sufficiently large, then the diffusion approximation reproduces the exact vapor concentration profile,

$$n_{\text{exact}}(R) = n^{(J(a))}(R) = -\frac{J(a)}{4\pi DR} + n_\infty \tag{4.15}$$

with $n^{(J)}(r)$ being the steady-state vapor concentration profile corresponding to a given total molecular flux $J(a)$.

On combining Eqs. (4.14) and (4.15) with Eq. (4.13) we obtain

$$J(a) = \alpha(a, R)\left(n_\infty - n_a - \frac{J(a)}{4\pi DR}\right) \tag{4.16}$$

On solving this equation with respect to $J(a)$, one finds $\alpha(a)$ as

$$\alpha(a) = \frac{\alpha(a, R)}{1 + [\alpha(a, R)/4\pi DR]} \tag{4.17}$$

Equation (4.16) is exact if $R \gg l$. In order to find $\alpha(a, R)$ and R, we must call on approximations.

4.3.2 Approximations

All the currently used approximations for α can be derived from Eq. (4.17)

4.3.2.1 The Fuchs Approximation

This approximation follows from Eq. (4.17) if one puts $\alpha(a, R) = \alpha_{\text{fm}}(a) = \pi a^2 v_T$ and $R = a + \Delta$, where the distance $\Delta \propto l$, with l being the mean free path of the condensing molecule in the carrier gas. Fuchs does not specify the value of Δ and suggests that it be used as a fitting parameter. There were other suggestions: $\Delta = 0$, $\Delta = l$, and $\Delta = 2D/v_T$.

4.3.2.2 The Fuchs–Sutugin Approximation

These authors [19] used the Sahni [25] numerical solution of the Bhatnagar–Gross–Krook (BGK) equation [26] in order to fit the parameter Δ in the Fuchs theory. Their final expression for the condensation efficiency looks as follows:

$$\alpha = 4\pi Da \frac{1 + \text{Kn}}{1 + 1.71\,\text{Kn} + 1.33\,\text{Kn}^2} \tag{4.18}$$

Here $\text{Kn} = l/a$ is the Knudsen number and $l = 3D/v_T$. At the present time Eq. (4.18) is the most popular one. The function $f(x)$ in Eq. (4.3) is then

$$f(x) = 4\frac{x + 3}{x^2 + 5.13x + 12} \tag{4.19}$$

4.3.2.3 The Lushnikov–Kulmala Approximation

These authors [23] called upon two approximations:

1) The free-molecule expression approximates $\alpha(a, R)$ as

$$\alpha(a, R) \approx \alpha_{\text{fm}}(a, R) \tag{4.20}$$

where

$$\alpha_{\text{fm}}(a, R) = \frac{2\pi a^2 v_T}{1 + \sqrt{1 - a^2/R^2}} \tag{4.21}$$

2) The radius R of the limiting sphere is found from the condition that "the diffusion flux from the diffusion zone is equal to the diffusion flux from the free-molecule zone." The diffusion flux is defined from Fick's law. Hence,

$$\mathrm{d}_r n_{\mathrm{fm}}(r)|_{r=R} = \mathrm{d}_r n^{(J(a))}(r)|_{r=R} \tag{4.22}$$

where

$$n_{\mathrm{fm}}(r) = (n_R - n_a)\frac{1+\sqrt{1-a^2/r^2}}{1+\sqrt{1-a^2/R^2}} + n_a \tag{4.23}$$

is the vapor concentration profile found in the free-molecule zone for $a < r < R$. The distance R separates the zones of the free-molecule and the continuum regimes. If $n_a > n_R$ then

$$R = \sqrt{a^2 + \left(\frac{2D}{v_T}\right)^2} \tag{4.24}$$

The value $2D/v_T$ has the order of the molecular mean free path l.

The final result is

$$\alpha(a) = \frac{2a^2 v_T}{1+\sqrt{1+(a v_T/2D)^2}} \tag{4.25}$$

The function $f(x)$ in this case looks as follows:

$$f(x) = \frac{2}{1+\sqrt{1+0.25x^2}} \tag{4.26}$$

The approximate $f(x)$ are presented in Figure 4.1 (the curve from Dahneke [27] is added).

4.3.3 More Sophisticated Approaches

As mentioned above, the analysis of the mass transport of condensable vapors should rely upon the solution of the Boltzmann equation written down for the vapor–carrier gas mixture. Considerable success in this direction was achieved by Sitarski and Novakovski [28] and by Loyalka and co-workers [10, 29–32] in the 1980s. These authors developed methods for the numerical solution of the Boltzmann equation and suggested several interpolation formulas (see [13]). Their results demonstrated good agreement with the measurements of Ray *et al.* [11].

4.4 Evaporation

The difference between the evaporation and condensation processes is in the sign of the difference $n_\infty - n_a$. In the case of evaporation, this difference is negative,

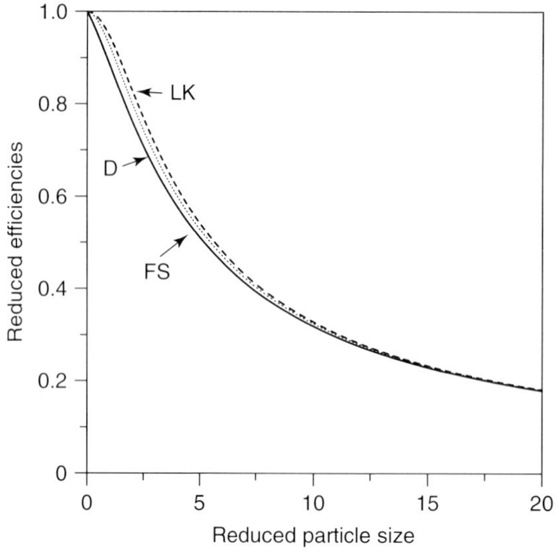

Figure 4.1 Condensation efficiency versus particle size. The condensation efficiency depends on the particle size as follows: $\alpha(a) = \pi a^2 v_T f(a v_T/D)$. Three approximate $f(x)$ are plotted in this figure: FS, Fuchs and Sutugin [19]; D, Dahneke [27]; and LK, Lushnikov and Kulmala [23]. These three approximations give almost coincident results (within 5%).

and the flux is directed from the particle surface to infinity [20]. If the evaporation rate is so low that the vapor concentration over the particle surface is equal to the equilibrium vapor pressure given by the Kelvin formula, then

$$n_a = n_s \exp\left(\frac{2\sigma v_1}{kTa}\right) \tag{4.27}$$

where n_s is the equilibrium concentration of vapor molecules over the planar surface of the liquid, σ is the surface tension, and v_1 is the molecular volume. In principle, the pressure over the particle surface is always lower and depends on the rates of the transport processes.

There have been several attempts to attack the problem of evaporation rate from first principles. The idea of Schenter *et al.* [12] is to apply the variational theory of transition states in order to formulate the expression for the evaporation rate. The evaporation rate is defined as the number of molecules leaving the particle surface per unit time, that is,

$$\gamma_g = \frac{1}{Z_g}\frac{1}{g}\sum_{k=1}^{g}\int v_k \mu \mathrm{e}^{-H/kT}\delta(R - r_k)\theta(v_k\mu_k)\,\mathrm{d}\Gamma \tag{4.28}$$

where H is the classical Hamiltonian of g molecules in the embryo, $\delta(x)$ is the Dirac delta function, R is the radius of the constraining sphere (the distance from the embryo center to the points where the potential created by the embryo does not affect the escaped molecule), $\boldsymbol{v}_k = \boldsymbol{p}_k/m$ are the molecular velocities, $\mu = \cos\phi$,

ϕ is the angle between the velocity and the radial direction, and $d\Gamma$ stands for the integration over all phase space of g molecules. The theta function (the Heaviside step function) forbids integration over velocities directed toward the center of the embryo. The delta function on the right-hand side fixes the positions of outwardly moving molecules at the surface of the constraining sphere. The normalization multiplier $1/Z_g$ is just the reciprocal partition function of g molecules inside the sphere:

$$Z_g = \sum_{k=1}^{g} \int e^{-H/kT} \prod\nolimits_{k=1}^{g} \theta(R - r_k)\, d\Gamma \tag{4.29}$$

Here the Heaviside step function $\theta(R - r_k)$ restricts the integration over coordinates to the part of the coordinate space lying inside the constraining sphere.

The most attractive feature of this theory is the fact that the parameter R can be found from the condition for the minimum of the evaporation rate. Next, we can perform the integration over velocities in the numerator and denominator and present the result in the form

$$\beta = -\pi R^2 v_T \frac{d}{dV_R} \ln Z_g \tag{4.30}$$

Here V_R is the volume of the constraining sphere and the value $-d \ln Z_g/dV_R = P^{\text{int}}/kT$, with P_{int} being the internal pressure at the embryo surface.

However, all is not so simple. The point is that we again encounter the problem of how to calculate Z_g, the partition function of g molecules. Once again, we encounter the need to solve a many-body problem. Although *molecular dynamics* methods permit such calculations to be performed, it is not very convenient to have to do so every time it is needed. Another drawback of this approach is our lack of knowledge about the intermolecular potentials, which also limits the precision of the molecular dynamics calculations.

4.5
Uptake

Chemical processes at the particle surface and in its volume create additional complications related to the need to describe *diffusional transport* of reactants through the interface and in the volume of the particle. Trace gases are commonly recognized to react actively with the aerosol component of the Earth's atmosphere. Substantial changes to atmospheric chemical cycles due to the presence of aerosol particles in the atmosphere make us look more attentively at the nature of the processes stipulated by the activity of atmospheric aerosols (see [13, 18, 33–36]). The process of *gas–particle interaction* is usually a first-order chemical reaction going along the route

$$X + AP \longrightarrow (APX) \tag{4.31}$$

where X, AP, and (APX) stand, respectively, for a reactant molecule, an aerosol particle, and the final product resulting from the reaction.

Some important aspects of aerosol heterogeneous chemistry are still not so well studied, among them being the interconnection between uptake and mass accommodation efficiencies. There still exist discrepancies between the results of different authors because of different understandings of the meaning of the *uptake coefficient*. In 2001 an entire issue of *Journal of Aerosol Science* (volume 32, issue 7) was devoted to the problems of *gas–aerosol interaction* in the atmosphere.

Below, I wish to outline my point of view. Let a particle of radius a initially comprising N_B molecules of a substance B be embedded into the atmosphere containing a reactant A. The reactant A is assumed to be able to dissolve in the host particle material and to react with B. The particle will begin to absorb A and will do this until the pressure of A over the particle surface is enough to block the diffusion process. Our task is to find the consumption rate of the reactant A as a function of time. Next, we focus on sufficiently small particles whose size is comparable to or less than the mean free path of the reactant molecules in the carrier gas. The mass transfer to such particles is known to depend strongly on the dynamics of the interaction between the incident molecules and the particle surface. In particular, the value of the probability β for a molecule to stick to the particle surface is suspected to strongly affect the uptake kinetics. The question "How?" has not yet found a full resolution.

Below, we shall try to answer this question starting with a simple analysis of the boundary condition to the kinetic equation for the molecules of A.

4.5.1
Getting Started

Let an aerosol particle be put in the atmosphere containing a reacting gas admixture A. The molecules of A are assumed to react with a guest reactant B dissolved in a host material of the particle (in principle, B itself can be the host material itself). The reactant A is assumed to react with the reactant B along the route:

$$A + B \longrightarrow C \tag{4.32}$$

The particle initially containing no molecules of A begins to consume those crossing the particle–carrier gas interface. Our goal is to investigate the kinetics of this process.

Our basic integral principle asserts that

$$\text{flux of A from outside} = \text{total consumption of A inside} \tag{4.33}$$

We consider four stages of the uptake process:

- diffusion of A toward the particle;
- crossing the particle air interface;
- diffusion reaction process inside the particle; and
- accumulation of non-reacted A molecules in the particle.

Balancing the fluxes gives the equation for the uptake rate.

4.5.2
Hierarchy of Times

In principle, the consideration of uptake requires a solution of the time-dependent transport problem [37]. Here we give some order-of-magnitude estimations allowing for a correct statement of the problem in realistic conditions of the Earth's atmosphere. Our idea is to get rid of the non-stationarity wherever possible. Below, we use the notation D_X ($X = \mathrm{A, B}$) and D for the diffusivity of the reactant molecules inside the particle and in the gas phase, respectively.

The characteristic time of the non-stationarity in the gas phase is estimated as $\tau_g \propto a^2/D$, where a is the particle radius. This time is extremely short. For $D = 0.1$ $\mathrm{cm^2/s}$ and $a = 1\ \mu\mathrm{m}$, we obtain $\tau_g \propto 10^{-7}$ s. So the transport in the gas phase can be considered in the steady-state limit. The diffusion process in the liquid phase is much slower. Its characteristic time is $\tau_l \propto a^2/D_X \propto 10^{-3}$ s for micrometer-sized particles and $D_X = 10^{-5}\ \mathrm{cm^2/s}$.

The time for the chemical reaction of A molecules in the liquid phase is estimated as $\tau_A \propto \kappa n_B$, where κ is the binary reaction rate constant for the reaction given by Eq. (4.32). The maximal value of $\kappa \propto D_X a_m \propto 10^{-13}\ \mathrm{cm^2/s}$ for the molecular radius $a_m \propto 10^{-8}$ cm. The book of Seinfeld and Pandis [13] cites values within the interval $\kappa = 10^{-11}$–$10^{-18}\ \mathrm{cm^3/s}$. The estimate of the characteristic time τ_A depends on the value of n_B. If the gaseous reactant reacts with the host material, the characteristic reaction times are very short ($\tau_{chem} \propto 10^{-9}$ s) for diffusion-controlled reactions and much longer (up to seconds or even minutes) for other types of chemical processes.

The characteristic transport times should be compared to the characteristic times of substantial chemical changes inside the particle that are limited by the flux of A from outside. These times are on the order of $\tau_{changes} \propto 1/j_A$, where j_A is the total flux of A molecules trapped by the particle. This is the characteristic time for one molecule of A to attach to the particle surface. Actually, this time depends on the sticking probability β and can reach tens of seconds.

These estimates show (see also [13]) that all the characteristic times for the transient processes inside and outside the particle are much shorter that the characteristic time for the particle to change its chemical composition due to uptake. This means that a quasi-steady-state approximation can be used for the description of very slowly changing parameters such as the total number of molecules inside the particle or its size. Fast transport processes establish instantly the steady-state concentration profiles.

4.5.3
Diffusion in the Gas Phase

The flux j_A of A in the carrier gas should be found from the solution of the kinetic Boltzmann equation, for we want to consider all regimes of reactant transport, $0 < \mathrm{Kn} < \infty$, where the Knudsen number is defined as $\mathrm{Kn} = l/a$, with $l = 3D/v_T$ [13]. Here D is the diffusion coefficient of A in air and v_T is the thermal velocity of an A molecule. Actually, we do not solve the kinetic equation. Instead, we analyze its boundary condition and express the flux of A in the gas phase in terms of the parameters entering the boundary condition. This step is of primary importance for deriving the correct dependence of the *uptake efficiency* on the *sticking probability*. As was mentioned, this question has been raised many times – for example, see the discussion in the 2001 issue of *Journal of Aerosol Science* (volume 32, issue 7) mentioned above – but no unified opinion as yet exists.

Let us first consider the condensation of A molecules onto a spherical particle of A liquid. In the gas phase, the distribution function f_A of A molecules over coordinates and velocities satisfies the Maxwell boundary condition [38]

$$f_A^- = (1-\beta)f_A^+ + \frac{\beta}{2\pi}n_{Ae} \tag{4.34}$$

where β is the sticking probability, f^- is the velocity distribution function of molecules flying outward from the particle, f^+ is the same for molecules flying toward the particle surface, and n_{Ae} is the equilibrium concentration of A molecules over the particle surface. The first term on the right-hand side of this equation describes the mirror reflection of A molecules from the particle surface. The second term gives the density of A molecules first captured and then emitted from the surface. The coefficient $1/(2\pi)$ reflects the fact that the molecules fly only in the outward direction. At $f^- = f^+ = n_e/(2\pi)$ (full thermodynamic equilibrium), Eq. (4.34) is satisfied automatically.

The total flux of A is expressed as

$$J_A = \alpha(a)(n_{A\infty} - n_{Ae}) \tag{4.35}$$

Here $n_{A\infty}$ is the concentration of A far away from the particle.

The solution of the kinetic equation defines the concrete form of the dependence of the condensation efficiency $\alpha(a)$ on the particle radius and β. The form of the second multiplier is *universal* and it is not related in any way to any approximation.

Now we shall return to uptake. In this case the inward flux inside the particles makes the concentration n_A^+ of particles flying outward lower than n_{Ae}. The boundary condition in Eq. (4.6) is replaced by

$$f_A^- = (1-\beta)f_A^+ + \frac{1}{2\pi}n_A^+ \tag{4.36}$$

with the value of the reactant concentration n_A^+ being determined from flux balance. This is the principal point of our further consideration.

Because the value n^+/β replaces n_e in Eqs. (4.34) and (4.35), the flux of the reactant toward the particle can be written as

$$j_A = \alpha(a)\left(n_{A\infty} - \frac{n_A^+}{\beta}\right) \tag{4.37}$$

Instead of solving the kinetic equation, we will use the semiempirical Fuchs–Sutugin expression for $\alpha(a)$, that is,

$$\alpha(a) = \frac{\alpha_{fm}(a)}{1 + \beta S(\mathrm{Kn})} \tag{4.38}$$

where

$$\alpha_{\text{fm}}(a) = \beta \pi v_T a^2 \tag{4.39}$$

is the condensation efficiency in the *free-molecule regime* and

$$S(x) = \frac{3}{2}\left(\frac{1}{2x} - \frac{0.311}{x+1}\right) \tag{4.40}$$

4.5.4
Crossing the Interface

Let n_A^* be the concentration of the reactant in the liquid phase immediately beneath the surface and n_A^+ its concentration immediately above the particle surface. In equilibrium, the concentrations n_A^+ and n_A^* are linked by Henry law as

$$n_A^* = H n_A^+ \tag{4.41}$$

where H is the dimensionless *Henry's constant*. This can be expressed in terms of the commonly accepted one as

$$H = 0.8159 \times 10^{-7} T H_S \tag{4.42}$$

where H_S is the Henry's constant (in units of M/atm) defined in the book of Seinfeld and Pandis [13], and T is the absolute temperature (in K). The largest cited $H_S = 2.1 \times 10^3$ is for the system $NO_3 + H_2O$, so the value of H is on the order of unity (see also [39]).

4.5.5
Transport and Reaction in the Liquid Phase

Inside the particle, the diffusion–reaction process [40] settles the concentration profile $n_X(r)$,

$$\frac{\partial n_X}{\partial t} - D_X \Delta n_X = -\kappa n_A n_B \tag{4.43}$$

The boundary condition for the component B corresponds to its zero flux through the interface, that is,

$$\left.\frac{\partial n_B}{\partial r}\right|_{r=a} = 0 \tag{4.44}$$

Instead of the boundary condition for A, we use the integral principle,

$$\frac{dN_A}{dt} = J_A - \kappa n_A^* n_B^* \int_V f_A(r) f_B(r)\, d^3 r \tag{4.45}$$

where we have introduced the notation $n_A(r,t) = n_A^*(t) f_A(r,t)$, $n_B(r,t) = n_B^*(t) f_B(r,t)$, $n_B^*(t) = n_B(0,t)$, and

$$N_X(t) = n_X^*(t) Q_X(t) \tag{4.46}$$

where $Q_X(t) = \int_V f_X(r,t)\, d^3 r$. The values $n_X^*(t)$ are the maximal concentrations of the reactants.

4.6 Balancing Fluxes

4.6.1 No Chemical Interaction

In this case the component A dissolves and accumulates inside the particle. Because the diffusion process in the liquid phase is very fast, the concentration profile of A is constant inside the particle – this is the well-stirred reactor approximation (WSRA). Hence

$$\frac{dN_A}{dt} = \alpha(a)\left(n_{A\infty} - \frac{n_A^*}{H\beta}\right) \tag{4.47}$$

The newly arriving molecules of A increase the particle volume,

$$V(t) = V_0 + N_A v_a \tag{4.48}$$

Here V_0 is the initial particle volume and v_a is the volume per added molecule of A. Next,

$$n_A^* = \frac{N_A}{H(N_A v_a + V_0)} \tag{4.49}$$

Finally, we find the equation for N_A as

$$\frac{dN_A}{dt} = \alpha(a)\left(n_{A\infty} - \frac{N_A}{H\beta(V_0 + N_A v_a)}\right) \tag{4.50}$$

The link between $a(t)$ and N_A simplifies Eq. (4.50) to

$$\frac{4\pi}{3}a^3 = V_0 + N_A v_a \tag{4.51}$$

If we introduce the scales

$$t_0 = \frac{\beta H V_0}{\alpha(a_0)} \quad \text{and} \quad N_{A\infty} = \beta H n_{A\infty} V_0 \tag{4.52}$$

and the dimensionless functions and the variable

$$X = \frac{N_A}{N_{A\infty}} \quad Y = \frac{V}{V_0} \quad \text{and} \quad \tau = \frac{t}{t_0} \tag{4.53}$$

then the set of Eqs. (4.50) and (4.51) takes a more observable form:

$$\frac{dX}{d\tau} = \frac{\alpha(a_0 Y^{1/3})}{\alpha(a_0)}\left(1 - \frac{X}{Y}\right) \tag{4.54}$$

$$Y = 1 + \zeta X \tag{4.55}$$

where $\zeta = \beta H n_{A\infty} v_0$.

We solve Eq. (4.50) in the limiting case when the growth of the particle volume can be ignored. Equation (4.50) then takes the form

$$\frac{dN_A}{dt} = \alpha(a)\left(n_\infty - \frac{N_A}{\beta H V_0}\right) \tag{4.56}$$

The solution to this equation is readily found as

$$N_{\mathrm{A}}(t) = n_\infty \beta H V_0 (1 - \mathrm{e}^{-\alpha(a)t/\beta H V_0}) \tag{4.57}$$

It is remarkable that the relaxation time is independent of β at small β. The maximal number of trapped molecules is

$$N_{\mathrm{A}}(\infty) \approx n_\infty \beta H V_0 \tag{4.58}$$

which corresponds to $n^- = n_\infty$ as it should. At this stage we ignore the changes of the particle volume in the course of uptake. Then the kinetics of the uptake process is defined by the simple exponential law

$$N_{\mathrm{A}}(t) = \beta H V_0 n_\infty (1 - \mathrm{e}^{-t/t_0}) \tag{4.59}$$

where the time of the uptake process is

$$t_0 = \frac{4Ha[1 + \beta S(\mathrm{Kn})]}{3v_T} \tag{4.60}$$

Two remarkable facts should be emphasized: (i) t_0 remains finite at $\beta = 0$; and (ii) the β dependence of t_0 is linear. Figure 4.2 displays the dependence of t_0 on the particle size (actually on Kn). The kinetic curves $N_{\mathrm{A}}(t)$ are shown for several values of Kn and β.

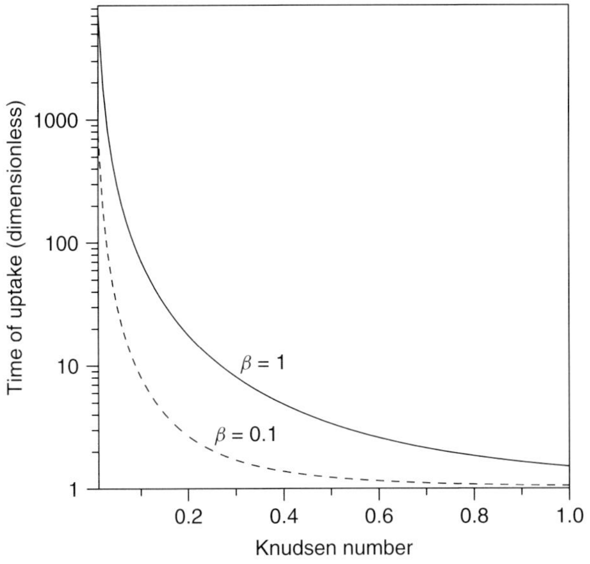

Figure 4.2 Time of uptake. The molecules of a trace gas A reach the inner part of the particles, thus increasing the A concentration over the particle surface. The uptake process ceases as soon as the A concentration over the particle surface becomes equal to the A concentration far away from the particle. The saturation time in units of $4Ha/3v_T$ is plotted as a function of the Knudsen number.

4.6.2
Second-Order Kinetics

Let us apply the WSRA to a consideration of the general case. We assume that both reactants are well stirred (the chemical reaction goes too slow, so the diffusion process has time to smooth over the reactant profiles). Let N_A and N_B be the total number of molecules of the reactants, and N_C be the number of the molecules of the reaction product. We also introduce $v_{a,b,c}$ as the molecular volume of each reactant. Then the following set of equations governs the process:

$$\frac{dN_A}{dt} = \alpha(a)\left(n_{A\infty} - \frac{N_A}{H\beta V}\right) - \frac{\kappa}{V} N_A N_B \tag{4.61}$$

$$\frac{dN_B}{dt} = -\frac{\kappa}{V} N_A N_B \tag{4.62}$$

and

$$V(t) = v_a N_A + v_b N_B + v_c N_C \tag{4.63}$$

In order to introduce the evaporation of C, we should add the outgoing flux of C in the flux balance,

$$\frac{dN_C}{dt} = -\alpha(a)\frac{N_C}{H_C \beta_C V} + \frac{\kappa}{V} N_A N_B \tag{4.64}$$

where H_C is Henry's constant for the reaction product C, and β_C is the sticking probability of C molecules. No C molecules are assumed to be present far away from the particles.

Let us introduce the scales

$$t_0 = \frac{\beta H V_0}{\alpha_0} \qquad N_{A\infty} = n_{A\infty}\beta H V_0 \qquad t_A = \frac{1}{\kappa n_{B0}} \qquad \text{and} \qquad t_B = \frac{V_0}{\kappa N_{A\infty}}$$

Then the set of Eqs. (4.61) and (4.62) takes the form

$$\frac{dX}{d\tau} = \frac{\alpha(a)}{\alpha(a_0)}\left(1 - \frac{X}{V}\right) - \xi\frac{XY}{V} \qquad \frac{dY}{d\tau} = -\eta\frac{XY}{V} \tag{4.65}$$

where $\tau = t/t_0$, $\xi = t_0/t_A$, and $\eta = t_0/t_B$. Equation (4.59) is also simplified to

$$\frac{dZ}{d\tau} = -\frac{t_0}{t_C}Z + \frac{XY}{V} \tag{4.66}$$

where $t_C = H_C\beta_C V_0/\alpha_0$ and $N_{C,0} = \kappa N_{A\infty} N_B(t_0/V_0) = N_A(t_0/t_A)$. The dimensionless volume is introduced as

$$V = \frac{1}{V_0}(v_a N_A + v_b N_B + v_c N_C) \tag{4.67}$$

where $V_0 = v_b N_{B0}$. Hence, the dimensionless volume V satisfies the equation

$$V = Y + \mu X + \nu Y \tag{4.68}$$

with $\mu = N_{A\infty} v_a/V_0$ and $\nu = N_{C0} v_c/V_0$. Figure 4.3 displays the solution of the set of Eqs. (4.65), (4.66) and (4.68).

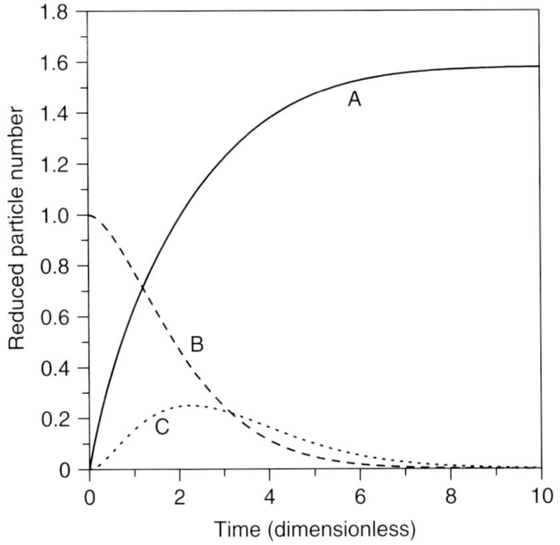

Figure 4.3 Second-order chemistry inside the particle. The particle consisting of the substance A traps molecules of reactant B from the surrounding atmosphere. A chemical reaction between A and B produces a neutral substance C that leaves the particle. Shown is the time evolution of the total molecule numbers of A, B, and C. Results are presented in dimensionless units.

We have considered the uptake of gaseous reactants by aerosol particles. The decisive point of our consideration is Eq. (4.35) linking the flux of the reactant toward particle with the concentration of molecules n^+ emitted from the particles. The value of n^+ is fixed by the condition that the external flux is equal to the consumption of the gas inside the particle (because of chemical reaction, for example). Actually, this means that we have ignored very fast transient processes related to diffusion. However, the time dependence of more lengthy processes that change the particle composition is retained. The central attention has been paid to the dependence of the kinetics of uptake on the sticking probability β for a reactant molecule to stick to the particle surface. This sticking probability is unambiguously introduced by the boundary condition imposed on the distribution function of reactant molecules.

Any uptake process continues for a finite time until the moment either when the reactant concentration inside the particle becomes sufficient to stop the inward diffusion flux or when chemical reactions entirely change the particle composition (see Figure 4.2). We found the dependence of the characteristic times on the sticking probability. Of course, the kinetics of the process has been investigated.

Because our major goal is the study of uptake kinetics on the sticking probability, we consider only the Knudsen and free-molecule aerosols (sufficiently large Knudsen numbers), otherwise the reactant flux toward the particle is independent of the sticking probability. Still the β dependence can give information about the uptake kinetics even in the case of very small Knudsen numbers if β is also very small but the ratio β/Kn is finite.

4.7 Nucleation

For many decades there has been an irreconcilable conflict between the theoretical predictions and the experimental data on the spontaneous nucleation of vapors. The reason for this is quite clear: the problem itself is far from being simple either theoretically or experimentally. This opinion is commonly recognized and shared by most researchers investigating the nucleation of vapors [1, 4, 8, 9, 41–43]. The diversity of the results of nucleation measurements made me think that there exist a number of principally different scenarios of the nucleation process. The goal of this section is to introduce the reader to some of them.

Almost all modern theories about forming a new phase from supersaturated vapors rely upon the model that considers the supersaturated vapor to comprise monomeric molecules and clusters (g-mers) suspended in a carrier gas. The clusters either grow on colliding or decay into fragments, forming new smaller clusters along the way:

$$(g_1) + (g_2) \longrightarrow (g_1 + g_2) \longrightarrow (g_1') + (g_2') + \cdots + (g_k') \tag{4.69}$$

Mass conservation is assumed:

$$g_1 + g_2 = \sum_{s=1}^{k} g_s' \tag{4.70}$$

At first sight such an approach looks very natural. On the other hand, it is not easy to define what this cluster is.

Indeed, the calculation of the cluster free energies demands the representation of the total partition function as a product of the individual cluster partition functions. The cluster is an equilibrium system. Its partition function includes the summation over states that contain spatially separated objects. For example, the partition function of a g-mer contains the summation over states where $g - 1$ molecules are bounded and a monomer is located beyond the zone of action of the intermolecular forces. The history of this problem already has a rather venerable age exceeding half a century (see Reiss *et al.* [41] and references therein). But it has not been fully resolved hitherto. Simply put, people silently came to the consensus that this scheme was correct and that it was permissible to consider the equilibrium of an individual g-mer suspended in a gas–condensable vapor mixture.

The theory of nucleation aims at predicting the rate of new phase formation $\mathcal{J}$, that is, the rate of production of new stable particles in unit volume as a function of temperature, supersaturation, and some other external parameters. The free energy of a g-mer grows at small g, passes through a maximum, and then diminishes with growing particle size. Clusters whose size corresponds to maximal free energy are referred to as *critical clusters*. These clusters are able to grow, giving life to the new liquid (solid) phase. Below I will explain how this scheme works.

4.7.1
The Szilard–Farkas Scheme

Current kinetic approaches to the problem of the nucleation of supersaturated vapors assume particle formation to go along a simpler scheme than that given by Eq. (4.69) (the *Szilard–Farkas scheme*):

$$(g) + (1) \rightleftharpoons (g+1) \tag{4.71}$$

with the kinetic coefficients (forward $\alpha_g c_1$ and backward β_g rates) being known functions of g – the number of condensing vapor molecules (monomers) – and the parameters of the carrier gas. This scheme thus assumes that g-mers interact only with the monomers.

Let n_g be the concentrations of clusters of mass g and C be the vapor concentration. Then the steady-state rate $\mathcal{J}$ of new particle production can be written as follows:

$$\mathcal{J} = \alpha_{g-1} C n_{g-1} - \beta_g n_g \tag{4.72}$$

The right-hand side of this equation is just the number of g-mers crossing the point g along the mass axis. In the steady state this number is independent of g and thus equal to the productivity of supercritical embryos $\mathcal{J}$. Equation (4.72) is just the set of non-homogeneous linear algebraic equations with respect to n_g. The solution to this equation can be readily found. We introduce the auxiliary concentrations n_g^0 that are the solutions to the homogeneous equation,

$$0 = \alpha_{g-1} C n_{g-1}^0 - \beta_g n_g^0 \tag{4.73}$$

Many authors consider Eq. (4.73) as a formulation of the principle of *detailed balance* expressing n_g^0 in terms of the equilibrium characteristics of the embryos. It was emphasized long ago by Lushnikov and Sutugin [44] that Eq. (4.73) *does not bear on the equilibrium.* Equation (4.73) is just a formal trick that solves the non-homogeneous linear set Eq. (4.72).

From Eq. (4.72) we find

$$n_g^0 = A \frac{\alpha_{g-1}\alpha_{g-1}\ldots\alpha_1}{\beta_g\beta_{g-1}\ldots\beta_2} \tag{4.74}$$

and look for the solution to Eq. (4.72) in the form $c_g = A_g n_g^0$. One recognizes the method of variable constants in this trick. If we put the boundary condition $n_G = 0$ at some very large G, we find the following:

1) The nucleation rate in terms of the condensation and evaporation coefficients and the monomer concentration (see [1] for an overview and discussion of the various approaches) looks like

$$\frac{\mathcal{J}_G}{\mathcal{J}_2} = \frac{1}{1 + x_2 + x_2x_3 + \cdots + x_2x_3\ldots x_{G-1}} \tag{4.75}$$

where $\mathcal{J}_2(c_1) = \frac{1}{2}\alpha_1 C^2$ is the rate of dimerization (the reaction (1) + (1) $\longrightarrow$ (2)), C is the monomer number concentration, and

$$x_g = \frac{\beta_g}{\alpha_g C} = \frac{n_g^*}{C} \tag{4.76}$$

with $n_g^* = \beta_g/\alpha_g$

2) The mass spectrum in the nucleation mass interval $1 < g < G$ is

$$c_g = \frac{\alpha_1}{2\alpha_g} C \frac{1 + x_{g+1} + x_{g+1}x_{g+2} + \cdots + x_{g+1}x_{g+2} \ldots x_{G-1}}{1 + x_2 + x_2 x_3 + \cdots + x_2 x_3 \ldots x_{G-1}} \tag{4.77}$$

4.7.2
Condensation and Evaporation Rates

The rates β_g and γ_g can be found from some very simple but not always correct considerations. For example, the condensation rate is simply replaced by the product of geometrical cross-section times the thermal velocity of vapor molecules. In this case,

$$\beta_g = \pi (a_1 + a_g)^2 v_T \tag{4.78}$$

where a_g is the radius of the g-mer, and v_T is the monomer–g-mer relative thermal velocity,

$$v_T = \sqrt{\frac{\pi k T}{8\mu_{1,g}}} \qquad \mu_{1,g} = \frac{m_1 m_g}{m_1 + m_g} \tag{4.79}$$

with m_g the g-mer mass.

The evaporation rate is expressed in terms of the equilibrium distribution of vapor clusters at saturation by using a detailed balance consideration. Next, this distribution is either calculated starting with the principles of statistical mechanics or expressed in terms of the physico-chemical constants of bulk liquids: surface tension, bulk density, and so on [1, 4, 8, 9, 43].

The latter approach, although being the most widespread one, is essentially restricted in describing the properties of the smallest clusters, where macroscopic notions like surface tension or liquid-state density do not work. Meanwhile, the initial stages of the nucleation process either appreciably affect the nucleation rate or even totally define the latter as, for example, in the case of small "magic" embryos whose ability to decay is suppressed by their very high binding energies [45]. There are other situations where the initial steps of the new phase formation are of primary importance: at very high supersaturations the mass of a critical embryo is small and the formation process is regulated only by the dimer stage.

Equation (4.75) clearly manifests the role of the initial stages of the nucleation process, as follows:

1) The nucleation rate is proportional to α_1 – the rate of dimer formation. This is not such a simple process because no stable dimers can form without a third

participant whose role is just to remove the energy excess and allow the dimer to transfer to a bound state.

2) Mixed dimers consisting of a molecule of condensing vapor and a molecule of the carrier gas should exist in the nucleating vapor. The formation of the dimers of two vapor molecules proceeds easier via the exchange reaction when the incident vapor molecule replaces the carrier gas one.
3) Each term in the denominator on the right-hand side of Eq. (4.75) (except 1, which is typically neglected) begins with the multipliers x_2, x_3, ... containing the condensation and evaporation rates of the smallest clusters. In particular, if there are anomalously stable small clusters at $g = g_a$ whose evaporation rate is negligibly low, the sums in the denominators of Eqs. (4.75) and (4.77) are cut off at $g = g_a$. The value of g_a is in no way connected with the supersaturation ratio and may be arbitrarily smaller than the mass of the critical embryo. In this case the critical embryos do not play a decisive role in the nucleation process.

The above list demonstrates clearly that sometimes there are no ways to avoid a detailed knowledge of the kinetics of the smallest cluster formation in treating the data on spontaneous nucleation.

4.7.3 Thermodynamically Controlled Nucleation

The first attempts to describe the nucleation process were based on the thermodynamic theory of fluctuations. The idea was (and is) quite clear and transparent: the molecules of a condensing vapor should occur in a volume where the intermolecular interaction is sufficiently strong to keep them together. In order to find the probability of forming such a cluster (critical or supercritical embryo), it is necessary to find the free energy of the cluster and exponentiate it. The problem is just how to find this free energy. It was quite natural to assume that this free energy is identical to the free energy of a liquid (solid) droplet of the condensing substance. This assumption is not so bad for large droplets; but if the embryo is small, doubts about the validity of such an approach can arise. Of course, it is possible to try to find the free energy starting with the micro-principles, that is, to calculate the free energy of the cluster by solving the respective classical or quantum-mechanical problem. At present the most widespread expression for the nucleation rate is [13]

$$J = \left(\frac{2\sigma}{\pi m_1}\right)^{1/2} \frac{v_1 N_1^2}{S} \exp\left(-\frac{16\pi}{3}\frac{v_1^2\sigma^3}{(kT)^3(\ln S)^2}\right) \tag{4.80}$$

4.7.4 Kinetically Controlled Nucleation

It is very difficult to imagine that a very slow thermodynamically controlled nucleation scenario can be realized in the atmosphere. More likely is another (kinetically controlled) scenario that assumes the formation of the critical embryo after one successful collision with another vapor molecule. In this case a dimer forms. This dimer can grow and change to a trimer, and so on, once the latter is stable. The formation of the dimer requires the presence of a third body that takes away the excess energy appearing after the formation of the bound state of two molecules. The nucleation rate in this case is proportional to the second power of the vapor concentration.

This scenario has been investigated in detail in [24]. It has been shown that only the bound states of the dimer contribute to the growth process, which means that thermodynamic equilibrium is never attained in this case. The states in the continuous spectrum are short-lived and thus should be ignored in the kinetic consideration of the process. The kinetic approach allows one to assume that the supercritical embryo can form via the formation of mixed clusters comprising molecules of the condensing vapor and molecules of the carrier gas. When such a cluster grows and reaches a thermodynamically controllable size, the molecules of the carrier gas escape (they evaporate back from the droplet). The mixed states of the growing embryo are apparently non-equilibrium and thus cannot be predicted within the scope of the thermodynamically controlled scenario. As far as I know, nobody has yet tried to consider such a type of nucleation.

Meanwhile, the existing experiments sometimes display the dependence of the nucleation rate on the pressure of the carrier gas [43]. The kinetics of the particle formation–growth process is described by the Szilard–Farkas scheme, which assumes that the vapor molecules can join (or escape) one by one until the growing particle reaches the critical mass. This kinetic scenario produces the well-known chain of equations for the concentrations of growing particles. This set of equations can be solved in the steady-state limit and gives the expression for the nucleation rate in terms of the evaporation and condensation efficiencies. If one assumes the principle of detailed balance to be valid, the evaporation efficiencies can be expressed in terms of the equilibrium concentrations and condensation efficiencies. Then the Szilard–Farkas scheme leads to the well-known expression for the nucleation rate. The kinetic approach denies this step. Now it has become evident that the secondary atmospheric aerosols in most cases form according to the kinetic scenario.

Let us consider the dimer stage of nucleation and focus on the kinetics of dimer formation. This means that the sum in the denominator of Eq. (4.75) is restricted to the first two terms. This approximation is valid at sufficiently high supersaturations.

On the other hand, dimers give us a very nice opportunity to perform an explicit analysis, that is, to calculate exactly some values of interest and to see what kind of approximations underlie the derivation of the expressions for condensation–evaporation rates.

We start with the definition of the dimer. The most natural way to do this is to consider the *bound states* of two particles as dimers [46]. It is this definition that has been adopted in all *ab initio* calculations.

But even in the case of dimers it is not easy to answer the question: What does "bound state" mean? Do we consider as dimers the quasi-stationary states with positive energies belonging to the continuum spectrum (states below the centrifugal barrier are meant)? The answer to the latter question is positive, for the probability of under-barrier penetration is typically small and the corresponding decay rates are comparable to or lower than that of the direct processes of breaking the dimers by incident carrier gas molecules. Fortunately, as our estimations show, the role of these states is negligible for sufficiently smooth intermolecular potentials.

The dimers result from three-body processes (two condensing molecules and a molecule of the carrier gas). Our consideration does not go far beyond traditional approaches replacing real cross-sections by geometrical ones. Still, a short revision of this point is done in order to specify the values entering into the expression of the condensation coefficient. The decay rate of dimers can be found from a detailed balance consideration. The classical approach and semiclassical quantization rule are applied for calculating the energy density of bound (and quasi-bound) states of two molecules and then the probability for a dimer to exist. After this, the equilibrium concentration of dimers is found and the equilibrium detailed balance consideration yields the dimer evaporation rate. This is where the expression for the equilibrium concentration of mixed dimers (a molecule of vapor plus a molecule of carrier gas) is given. Although this concentration is small (several percent), the role of mixed dimers should not be underestimated.

The kinetics of nucleation via the dimer stage was considered in [46]. We formulated and solved the set of differential equations describing the time evolution of the nucleating system. Attention was focused on the time dependence of the vapor concentration that practically defined all the rest: particle size spectra, and number and mass concentrations of the disperse phase. The most essential conclusion inferred from this consideration was that, under a wide range of external conditions, the kinetic coefficients rescale the time–concentration plane, leaving unchanged the dimensionless shape of the kinetic curve.

4.7.5
Fluctuation-Controlled Nucleation

Even very strongly supersaturated vapors consist of independent (non-interacting) molecules (the interaction time is much shorter than the free flight time). However, in such vapors, the nucleation rate is extremely high. Moderately supersaturated vapor, in principle, always contains very highly supersaturated areas forming due to density fluctuations. The nucleation process within these areas goes very quickly, so the rate of nucleation is limited by the rate of formation of such fluctuation areas. Therefore, the details of the nucleation process inside these highly supersaturated areas are not important – it happens instantly.

Here I return to the old idea on the role of fluctuations in the nucleation process. But in contrast to the classical thermodynamic approach, I consider the fluctuation areas wherein the molecules do not (yet) interact. The vapor density is sufficient for creating high supersaturation, but it consists of *non-interacting molecules*. It is clear that such a scenario can be described by the scheme, $1 + 1 + 1 + \cdots + 1 \longrightarrow G^*$. The rate of the process is proportional to the vapor concentration to the power

G^*. This very type of nucleation has been introduced and investigated in [46]. It is important to emphasize that the formation of particles by nucleation alone cannot be observed directly. The point is that the nucleation process is accompanied by coagulation, and the latter process is also very swift. So just formed (by nucleation) particles coagulate, and we observe only the final result of this process. This means that the observed nucleation rate is slower than that predicted by the nucleation theory alone.

The nucleation rate for fluctuation-controlled nucleation is

$$\mathcal{J} = AC^{G*} \tag{4.81}$$

This very expression was used in [46–50] for studying *nucleation-controlled processes* of particle formation–growth (see the next sections).

4.8 Nucleation-Controlled Processes

In this section we consider the nucleation-controlled processes of particle formation and growth [48–52]. A nucleation process produces particles that are able to consume the vapor and remove the supersaturation. After this, the nucleation process ceases, but the particles already formed continue to grow by condensation and coagulation. This effect is referred to as the *nucleation burst*.

4.8.1 Nucleation Bursts

Here we consider the development of the aerosol state, making the following assumptions:

1) At the initial moment of time, the source of a condensable vapor is switched on. The productivity of the source (the number of particles produced per unit volume per unit time) is $I(t) = I_0 i(t)$, where I_0 is the dimensionality carrier and the dimensionless function $i(t)$ describes the time (t) dependence of the productivity.
2) At $t = 0$ there exists a foreign aerosol with known size distribution, distributed over particle masses g as $N(g, 0) = N_0 n(g, 0)$, where $N_0 = \sum N(g, 0)$ is the particle number concentration and the particle mass g is measured in units of the masses of the condensable molecules.

The qualitative picture of the development of the situation looks as follows (Figure 4.4). At the initial period, the concentration grows linearly with time (for simplicity we consider $i = 1$), and then it begins to bend because part of the vapor condenses onto foreign particles. At $t = t^*$, when the vapor concentration reaches a sufficiently high level for the spontaneous nucleation process to start, newly born particles appear and also begin to consume the vapor. The vapor concentration thus passes through a maximum whose value is determined either

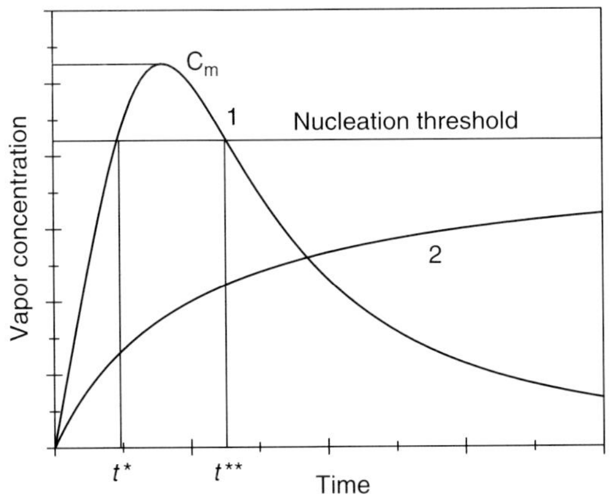

Figure 4.4 Schematic picture of the nucleation burst. A source of productivity J produces a condensable gas that at some critical concentration is able to nucleate and produce aerosol particles (curve 1). These aerosol particles, in turn, serve as a sink for the condensable molecules. The particles thus remove the supersaturation and lower the gas concentration below the threshold of nucleation. The nucleation ceases, but the newly formed particles continue to grow by consuming the condensable molecules. This very short nucleation event is referred to as the nucleation burst. A foreign aerosol presenting in the carrier gas is able to prevent the nucleation burst by consuming the excess of the condensable vapor (curve 2).

by the nucleation rate or by the rate of vapor loss by condensation onto foreign particles. The nucleation process stops at $t = t^{**}$ when the vapor concentration level drops below the critical value c^* and results in the formation of fine particles. The number concentration of the latter depends on the characteristics of foreign particles: their number concentration and some geometrical characteristics (more complex than simply average particle radius or average particle surface [49]). In the post-nucleation stage, the evolution continues mainly due to the growth of foreign particles by simultaneous joining of the vapor molecules and the smaller particles to larger ones.

4.8.2
Nucleation-Controlled Condensation

New phase formation in first-order phase transitions often goes through a disperse phase state, that is, the new phase appears as disperse particles suspended in a carrier medium. The processes responsible for disperse particle formation are:

- formation of stable embryos by nucleation,
- their growth by condensation, and
- coagulation aging of the thus-formed disperse system.

The kinetics of nucleation-controlled formation and condensational growth of disperse particles is considered under the following two assumptions:

1) Only a small amount of condensable substance nucleates and forms particles that grow by condensing the rest of the substance.
2) The condensation efficiency is a power function of the particle mass.

A non-trivial perturbation theory with respect to the smallness parameter

$$\mu = (\text{mass of nucleated matter})/(\text{total mass of condensable matter})$$

is developed, allowing one to describe the source-enhanced and free (no source) condensation processes in terms of two universal functions: the *particle mass spectrum*, and the concentration of condensable matter. The theory relies upon a renormalization group transformation that either totally removes the smallness parameter from the evolution equations (if the nucleation rate is an algebraic function of the concentration of condensable matter) or leaves it in the expression for the nucleation rate, where this parameter defines only the concentration scale of the nucleation process (for nucleation rates of general form). The theory is illustrated by the exact analytical solutions of the nucleation–condensation kinetic equations for three practically important cases:

- gas-to-particle conversion in the free-molecule regime,
- formation and diffusion-controlled condensational growth of islands on surfaces, and
- formation and diffusion-controlled growth of disperse particles in the continuum regime.

The analytical expressions for mass spectra of growing particles are found in the case of free condensing particles. The final mass spectra in free condensing systems display rather unusual behavior: they are either singular at small particle masses or not, depending on the value of the power exponent in the mass dependence of the condensation rate.

Here we return to a simple model of particle formation–growth already studied in [46]. The model considers particle formation by nucleation and their subsequent condensational growth in a spatially uniform gas–vapor mixture. A simple power dependence of the condensation efficiency $\alpha(g)$ on the particle mass is assumed, $\alpha(g) = \alpha g^{\gamma}$, where g is the number of molecules in the particle, and α and γ are constants. In [46] it was shown that, at $\gamma = n/(n+1)$, with n being an integer, the condensation stage can be described in terms of $n+1$ moments of the particle mass distribution, and that all the kinetic curves are universal functions of a specially defined universal variable playing the role of time. One question immediately arises: Does this universality hold for other γ? Here we give a positive answer to this question and propose a renormalization procedure valid in the general case.

The rescalings of the vapor concentration $C(t) = C_0 c(\theta)$, time $t = t_0\theta$, particle mass $g = g_0 y$, and particle mass spectrum $N(g,t) = N_0 n(y,\theta)$, with

$$C_0 = t_0 = \mu^{-(1-\gamma)/(4-2\gamma)} \qquad g_0 = \mu^{-1/(2-\gamma)} \qquad N_0 = \mu^{(3+\gamma)/(4-2\gamma)} \tag{4.82}$$

for source-enhanced condensation, or

$$t_0 = \mu^{-(1-\gamma)/(2-\gamma)} \qquad g_0 = \mu^{-1/(2-\gamma)} \qquad N_0 = \mu^{2/(2-\gamma)} \tag{4.83}$$

for free (no source) condensation, remove the smallness parameter μ from the governing equations and allow for the (already dimensionless) evolution equations for the vapor concentration and particle mass spectrum to be cast into a universal form.

The final expression for the particle mass spectrum is found in the form:

$$n(y, \tau) = \frac{1}{y^{\gamma}} \frac{j(x)}{c(x)} \Theta(x) \tag{4.84}$$

where $x = \tau - (1-\gamma)^{-1} y^{1-\gamma}$, $j(x) = j(c(x)/c_c)$ is the dimensionless nucleation rate, c/c_c is the supersaturation, $\tau = \int_0^t c(t')\,dt'$, and $\Theta(x)$ is the Heaviside step function.

The dimensionless concentration satisfies the equation

$$c\frac{dc}{d\tau} = 1 - (1-\gamma)^{\gamma/(1-\gamma)} c \int_0^{\tau} (\tau - \zeta)^{\gamma/(1-\gamma)} \frac{j(\zeta)}{c(\zeta)}\, d\zeta \tag{4.85}$$

It is not a problem to solve this equation numerically or even analytically. In particular, for free particle growth in the free-molecule regime and barrierless nucleation, one finds the final mass spectrum in the form

$$n(y) = y^{-2/3} \cosh(\tfrac{1}{2}\pi - 3p_0 y^{1/3}) \sin(3p_0 y^{1/3}) \Theta(\tfrac{1}{2}\pi - 3p_0 y^{1/3}) \tag{4.86}$$

where $p_0 = (18)^{-1/4}$. The spectrum is thus stretched from $y = 0$ to $y = y_{max} = (\pi/6p_0)^3$.

The particle mass spectrum is characterized by only one mass parameter $\mu^{-3/4}$ and cannot be described by a narrow curve. The characteristic size scale is $a_0\mu^{-1/4}$, where a_0 is the molecule size (on the order of 0.1 nm). The value of maximal particle radius is thus on the order of 10 nm at $\mu = 10^{-8}$.

The most essential step made here and in earlier papers [47–49] is the effective use of the relative smallness of the vapor mass spent in nucleation compared to the total vapor mass. This parameter is believed to be almost always small: the total mass of the aerosol comes from the vapor condensation onto the particles formed by nucleation. This fact immensely simplifies the consideration of the kinetics of nucleation–condensation processes.

4.8.3 Nucleation-Controlled Growth by Coagulation

Here we present a simple model of the particle formation–growth process that takes into account nucleation, condensational growth, and coagulation [48, 49, 53–55]. The particles are assumed to form in the free-molecule regime. We consider the free-molecule regime not only because of its practical importance, but also because the simple and specific dependence of the condensation efficiency on the particle mass, $\alpha(g) = \alpha g^{2/3}$ (g being the number of molecules in the particle), allows the whole consideration to be restricted to three moments of the particle mass distribution. These moments satisfy the set of four first-order differential equations together with the vapor concentration. Some complications appear if the process involves other moments. In this case the assumption on log-normal shape of the particle mass distribution function saves the simplicity of the scheme.

We assume next that there is a spatially uniform source of condensable vapor of productivity I, and only a small amount of condensable vapor is spent on nucleation. The newly born particles grow after the nucleation burst by condensing the non-volatile vapor onto their surfaces and changing their total number concentration by coagulation.

Although we consider here the barrierless nucleation (the nucleation rate J is proportional to the squared vapor concentration C), the results can be easily extended to arbitrary dependence of the nucleation rate on supersaturation. This is absolutely clear, because the time of the nucleation burst is much shorter than other characteristic time scales. On the other hand, the barrierless nucleation is very often met in the processes of formation of nanomaterials and functionally is rather simple to operate with. So we assume $\mathcal{J} = AC^2$.

Our model uses the moment method, which is very well suited for considering the particle formation–growth process in the free-molecule regime. Lushnikov and Kulmala [46–50] discussed the application of this method to growth processes in the free-molecule regime and found that three moments of the particle size distribution and the concentration of condensable vapor can be described in terms of universal functions, with all details of the process being hidden in the scales of the time and concentration axes.

This approach can be extended by including the coagulation process into consideration. We show that there are two different scales of time, the shorter of which defines the dynamics of the nucleation–condensation stage, while the longer one scales coagulation aging. It is found that each stage is described by a set of four universal functions that satisfy four (different for each stage) first-order differential equations, the right-hand sides of which contain coagulation integrals. These integrals are evaluated and expressed in terms of the parameters of the log-normal particle mass distribution. In contrast to commonly accepted approaches, these parameters include the particle number concentration, and two moments of the order of 1/3 and 2/3. This step allows one to formulate the close set of equations for these three values and the vapor concentration. This set contains the smallness parameter $\mu = A/\alpha$, which cannot be treated by a straightforward application of perturbation theory. However, two rather non-trivial rescalings allow for the separation of the nucleation–condensation and coagulation–condensation stages of the particle formation–growth process and the formulation of two closed sets of equations not containing the smallness parameter at all.

It is shown that the time for the condensation–nucleation stage is longer than the characteristic condensation time $1/\sqrt{I\alpha}$ by the factor $\mu^{-1/8}$. The particle number concentration contains the smallness parameter to the power 5/8: $\phi_0 \propto \mu^{5/8}\sqrt{I\alpha}$.

The coagulation stage is longer than the condensation stage by $\mu^{-3/16}$. Asymptotic analysis shows that the moments and particle number concentration are algebraic functions of time: $\phi_{1/3} \propto t^{-3/5}$, $\phi_{2/3} \propto t^{1/5}$, and $\phi_0 \propto t^{-7/5}$. These values of the exponents correspond to the predictions of the self-preservation theory for source-enhanced coagulation in the free-molecule regime. Numerical analysis confirms these power laws and gives the values of the constants before the powers. The above asymptotic dependences correspond to a constant width of the log-normal function ($s = 0.69$ for the source-enhanced growth process).

4.8.4
Nucleation Bursts in the Atmosphere

Regular production of non-volatile species of anthropogenic or natural origin in the atmosphere eventually leads to their nucleation, formation of tiny aerosol particles, and subsequent growth. The aerosol thus formed is able to inhibit the nucleation process because of condensation of non-volatile substances onto the surfaces of newly born particle surfaces. This process is referred to as the nucleation burst [9, 56].

The dynamics of atmospheric nucleation bursts possesses its own specifics – in particular, particle production and growth are suppressed mainly by pre-existing aerosols rather than by freshly formed particles of nucleation mode [21, 48, 49, 57, 58]. In many cases the nucleation bursts have a heterogeneous nature. The smallest (undetectable) particles accumulated during the night-time begin to grow at day-time because of sunlight-driven photochemical cycles producing low-volatility (but not nucleating) substances that are able to activate the aerosol particles [59]. Stable sulfate clusters [60–62] can serve as heterogeneous embryos provoking the nucleation bursts.

Nucleation bursts have been observed regularly in atmospheric conditions and shown to serve as an essential source of cloud condensation nuclei [61, 63].

Now it becomes more and more evident that the nucleation bursts in the atmosphere can contribute substantially to the production of cloud condensation nuclei [64] and can thus affect the climate and weather conditions on our planet (see, for example, [65, 66] and references therein). Present opinion connects the nucleation bursts with the additional production of non-volatile substances that can then nucleate, producing new aerosol particles, and/or condense onto the surfaces of newly born particles, foreign aerosols or atmospheric ions. The production of non-volatile substances, in turn, demands some special conditions to be imposed on (i) the emission rates of volatile organic compounds from vegetation, (ii) the current chemical content of the atmosphere, (iii) the rates of stirring and exchange processes between the lower and upper atmosphere, (iv) the presence of foreign aerosols serving as condensational sinks for trace gases (accumulation mode) and as coagulation sinks for the particles (nucleation mode), and (v) the interactions with air masses from contaminated or clean regions [58, 61–63, 67–72]. Such a plethora of very diverse factors, most of which have a stochastic nature, prevents direct attacks on this effect. A huge amount of field measurements of nucleation burst dynamics have appeared during the past decade [58, 61, 63, 67, 69–81].

Attempts to model this important and still enigmatic process also appeared rather long ago. Here we avoid the long history of this problem and cite only the recent models that have appeared in the twenty-first century [65, 81–92]. Extensive earlier citations can be found in these papers. All models (with no exception) start from the commonly accepted point of view that the chemical reactions of trace gases are responsible for the formation of non-volatile precursors, which then give rise to subnanometric and nanoparticles in the atmosphere. In their turn, these particles are considered as active participants of the atmospheric chemical cycles leading to particle formation [76, 93–95]. Hence, any model of nucleation bursts included (and includes) *coupled* chemical and aerosol blocks. This coupling leads to strong nonlinearities, which means that all intra-atmospheric chemical processes (not all of which are, in addition, firmly established) are described by a set of nonlinear equations, and there is no assurance that we know all the participants of the chemical cycles leading to the production of low-volatility gas constituents that then convert to the tiniest aerosol particles. The results of modeling the nucleation bursts in the atmosphere are shown in Figure 4.5.

The special significance of atmospheric sulfuric acid was emphasized in [96]. Although its total concentration is not enough to provide the observed particle growth, the primary role of H_2SO_4 in the processes of atmospheric nucleation has been proven in a number of works.

4.9
Conclusion

In this chapter we have considered condensation, evaporation, and nucleation. These three processes together with coagulation are responsible for particle formation and growth processes. We have emphasized that aerosols are not static objects. In principle, they cannot exist in the equilibrium state. They always change (sometimes very slowly). Evaporation decreases the particle size because of the loss of molecules from the particle surface. On the contrary, condensation increases the particle mass and size by joining molecules of condensable vapors from the carrier gas. These two effects can be described within the *single-particle model*, that is, the collection of aerosol particles does not affect these processes. Coagulation of aerosols is principally different in this respect. The aerosol particles are able to coalesce on colliding, thus producing larger daughter particles. The role of the particle collection is of principal importance in this case.

Condensation has been considered as particle growth due to collisions between particles and condensable molecules. A molecule hitting a particle surface is captured by the particle. The rate of this process is proportional to the difference in the number concentrations of the condensable molecules far away from the particle and at the particle surface. The proportionality coefficient (referred to as the condensation efficiency) has been found for three regimes of condensable molecule motion toward the particle surface. I have repeated the derivation of the expressions for the condensation efficiency for the free-molecule and continuum regimes (small and large particles compared to the mean free path of the condensable molecules). Then the transition regime of condensation has been discussed in detail. The main result of this section is the derivation of the formal expression for $\alpha(a)$. This expression sheds new light on the nature of the well-known and widely used approximations currently used. In

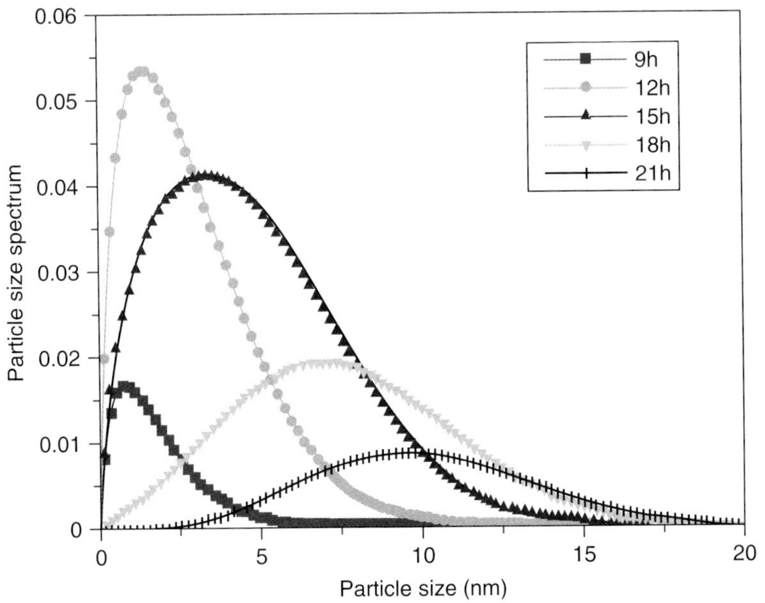

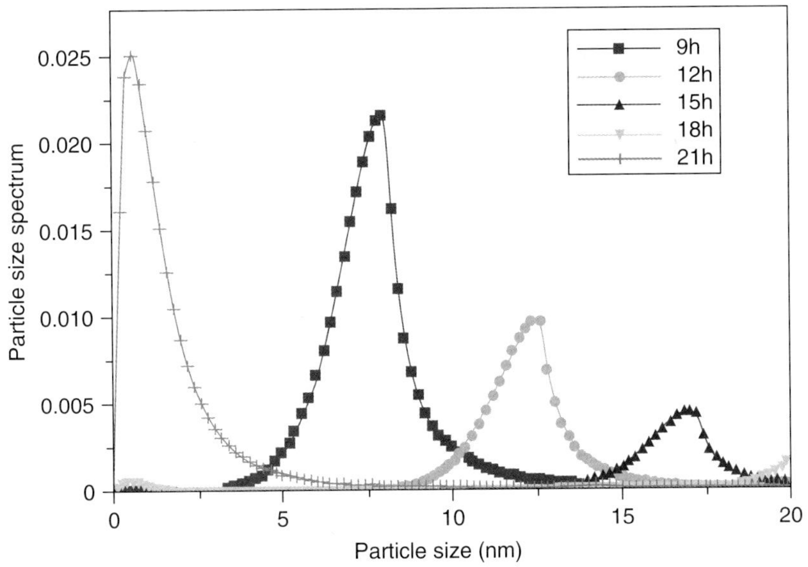

Figure 4.5 Nucleation bursts in the atmosphere. Time evolutions of the particle size spectrum. (a) An example of day-time nucleation. The source of tiny particles begins to work simultaneously with the photochemical production of low-volatility substances providing particle growth. (b) An example of night-time nucleation and subsequent growth. The maximal intensity of the source is shifted back by 12 h. The source does not work during the day-time and the particles produced during the night-time grow in the free (no source) regime.

addition I have demonstrated the derivation of a new approximate formula for $\alpha(a)$.

Evaporation is the loss of molecules from the particle. This process is always possible because of thermal fluctuations of single-molecule energies. When the molecular energy exceeds the potential barrier created by other molecules in the particle, the respective molecule is able to escape from the particle. The rate of evaporation can be expressed through the equilibrium vapor pressure and the condensation rate. There are other approaches that consider the escape of molecules from particles starting from first principles.

The chemical content of aerosol particles is important for the evolution of the aerosol. It affects the rates of physico-chemical processes and plays an important role in establishing the chemical state of the environment. I have classified all (known to me) chemical processes that can influence the fate of the aerosol particles. In particular, the chemical content of particles defines the equilibrium pressure over the particle surface and thus the rate of condensational growth. In this section I have investigated the role of the mass accommodation efficiency in the processes of particle growth and evaporation. The consideration of this section allows one to understand better the processes accompanying condensation and evaporation. It has become clear that evaporation is just the diffusion of foreign molecules in the potential well created by the molecules of the host material and subsequent diffusion in the gas phase (on crossing the potential barrier at the particle surface). The pressure of the evaporating molecules is limited by the transport processes inside and outside the particle.

After a short discussion on what the term "nucleation" means, I have considered three scenarios of nucleation: (i) thermodynamically controlled nucleation, (ii) kinetically controlled nucleation, and (iii) fluctuation-controlled nucleation. The main task of this consideration has been the derivation and discussion of the expression for the nucleation rate. The nucleation problem is not yet resolved.

References

1. Abraham, F.F. (1974) *Homogeneous Nucleation Theory*, Academic Press, New York.
2. Landau, L.D. and Lifshits, E.M. (1998) *Statistical Physics*, Nauka, Moscow.
3. Anisimov, M.A. (2003) Nucleation: theory and experiment. *Russ. Chem. Rev.*, 72, 591.
4. Ford, I.J. and Harris, S.A. (2004) Molecular cluster decay viewed as escape from a potential of mean force. *J. Chem. Phys.*, **120**, 4428.
5. Fuchs, N.A. (2002) *Mechanics of Aerosols*, Pergamon, New York.
6. Gross, D.H.E. (2001) *Microcanonical Thermodynamics*, World Scientific, Singapore.
7. Hill, T. (1956) *Statistical Mechanics, Principles and Selected Applications*, McGraw-Hill, New York.
8. Kaschiev, D. (2000) *Nucleation: Basic Principles and Application*, Butterworth, Heinemann, Oxford.
9. Friedlander, S.K. (2000) *Smokes, Haze, Mist*, John Wiley & Sons, Inc., New York.
10. Williams, M.M.R. and Loyalka, S.K. (1991) *Aerosol Science, Theory and Practice*, Pergamon, Oxford.

11. Ray, A.K., Lee, A.J., and Tilley, H.L. (1988) Direct measurements of evaporation rates of single droplets at large Knudsen numbers. *Langmuir*, **4**, 631.
12. Schenter, G.K., Katzmann, S.M., and Garrett, B.C. (1999) Dynamical nucleation theory: a new molecular approach. *Phys. Rev. Lett.*, **82**, 3484.
13. Seinfeld, J.H. and Pandis, S.N. (1998) *Atmospheric Chemistry and Physics*, John Wiley & Sons, Inc., New York.
14. Friedlander 2001 to come.
15. Reist, P.C. (1984) *Introduction to Aerosol Science*, Macmillan, New York.
16. Hidy, J.M. and Brock, J.R. *The Dynamics of Aerocolloidal Systems*, Pergamon, Oxford.
17. Fuchs, N.A. (1959) *Evaporation and Droplet Growth in Gaseous Media*, Pergamon, London.
18. Davis, E.J. (1983) Transport phenomena with single aerosol particle. *Aerosol Sci. Technol.*, **2**, 121.
19. Fuchs, N.A. and Sutugin, A.G. (1971) High-dispersed aerosols, in *Topics in Current Aerosol Research*, eds. G.M. Hidy and J.R. Brock, Pergamon, Oxford, vol. 2, pp. 1–60.
20. Li, W. and Davis, E.J. (1995) Aerosol evaporation in the transition regime. *Aerosol Sci. Technol.*, **25**, 11.
21. McGraw, R. and Marlow, W.H. (1983) The multistate kinetics of nucleation in the presence of an aerosol. *J. Chem. Phys.*, **78**, 2542.
22. Fuchs 1964 to come.
23. Lushnikov, A.A. and Kulmala, M. (2004) Flux-matching theory of particle charging. *Phys. Rev. E*, **70**, 046413.
24. Lushnikov, A.A. (2010) Introduction to aerosols, in *Aerosol Science–A Comprehensive Handbook*, ed. I.E. Agranovski, Wiley-VCH, Weinheim, Chapter 1 (this book).
25. Sahni, D.C. (1966) The effect of black sphere on the flux distribution of an infinite moderator. *J. Nucl. Energy*, **20**, 915–920.
26. Bhatnagar, P.L., Gross, E.P., and Krook, M. (1954) A model for collision processes in gases. I. Small amplitude processes in charge and neutral one-component systems. *Phys. Rev.*, **94**, 511–525.
27. Dahneke, B. (1983) Simple kinetic theory of Brownian diffusion in vapors and aerosols, in *Theory of Dispersed Multiphase Flows*, ed. R.E. Meyer, Academic Press, New York, pp. 97–133.
28. Sitarski, M. and Novakovski, B. (1979) Condensation rate of trace vapor on Knudsen aerosol from solution of the Boltzmann equation. *J. Colloid Interface Sci.*, **72**, 113–122.
29. Loyalka, S.K. (1986) Rarefied gas dynamics problems in environmental sciences, in *Rarefied Gas Dynamics XV*, vol. 1, eds V. Boffi and C. Cercignani, B.G. Tuebner, Stuttgart.
30. Loyalka, S.K. and Park, J.W. (1988) Aerosol growth by condensation: a generalization of Mason's formula. *J. Colloid Interface Sci.*, **125**, 712.
31. Loyalka, S.K., Hamoodi, S.A., and Tompson, R.V. (1989) Isothermal condensation on a spherical particle. *Phys. Fluids A*, **1**, 358.
32. Tompson, R.V. and Loyalka, S.K. (1986) Condensational growth of a spherical droplet: free molecule limit. *J. Aerosol Sci.*, **17**, 723.
33. Charlson, R.J. and Heitzenberg, R.L. (1995) *Aerosol Forcing of Climate*, John Wiley & Sons, Ltd, Chichester.
34. Charlson, R.J., Schwartz, S.E., Hales, J.M., Cess, R.D., Coakley, J.A. Jr., Hansen, J.E., and Hofmann, D.J. (1992) Climate forcing by anthropogenic aerosols. *Science*, **255**, 423.
35. Davidovits, P., Hu, J.H., Worsnop, D.R., Zahnister, M.S., and Kolb, C.E. (1995) Entry of gas molecules into liquids. *Faraday Discuss.*, **100**, 65.
36. Clement, C.F., Kulmala, M., and Vesala, T. (1996) Theoretical consideration on sticking probabilities. *J. Aerosol Sci.*, **27**, 869–882.
37. McMurry, P.H. and Wilson, J.C. (1982) Growth laws for formation of secondary ambient aerosols: implication for chemical conversion mechanisms. *Atmos. Environ.*, **16**, 121.
38. Cercignani 1993 to come.
39. Natanson, G.M., Davidovits, P., Worsnop, D.R., and Kolb, C.E. (1996) Dynamics and kinetics at the gas–liquid interface. *J. Phys. Chem.*, **100**, 13007.

40. Shi, B. and Seinfeld, J.H. (1991) On mass transport limitation to the rate of reaction of gases in liquid droplets. *Atmos. Environ.*, **25A**, 2371.
41. Reiss, H., Tabahzade, A., and Talbot, J. (1990) Molecular theory of vapor phase nucleation: the physically consistent cluster. *J. Chem. Phys.*, **92**, 1266.
42. Laaksonen, A., Talanquer, V., and Oxtoby, D.W. (1995) Nucleation: measurements, theory, and atmospheric applications. *Annu. Rev. Phys. Chem.*, **46**, 189.
43. Anisimov 2004 to come.
44. Lushnikov and Sutugin 1976 to come
45. Smirnov, B.M. (2000) *Clusters and Small Particles in Gases and Plasma*, Springer, New York.
46. Lushnikov, A.A. and Kulmala, M. (1998) Dimer in nucleating vapor. *Phys. Rev. E*, **58**, 3157.
47. Lushnikov, A.A. and Kulmala, M. (1998) Nucleation controlled formation and growth of disperse particles. *Phys. Rev. Lett.*, **81**, 5165.
48. Lushnikov, A.A. and Kulmala, M. (2000) Foreign aerosols in nucleating vapour. *J. Aerosol Sci.*, **31**, 651.
49. Lushnikov, A.A. and Kulmala, M. (2000) Nucleation burst in a coagulating system. *Phys. Rev. E*, **62**, 4932.
50. Lushnikov, A.A. and Kulmala, M. (2001) Kinetics of nucleation controlled formation and condensational growth of disperse particles. *Phys. Rev. E*, **63**, 061109.
51. Lushnikov, A.A. and Kulmala, M. (1995) Source enhanced condensation in monocomponent disperse systems. *Phys. Rev. E*, **52**, 1658.
52. Lushnikov, A.A., Kulmala, M., Arstila, H., and Zapadinskii, E.L. (1996) Source enhanced condensation of a single component vapor in the transition regime. *J. Aerosol Sci.*, **27**, 853.
53. Warren, D.R. and Seinfeld, J.H. (1985) Simulation of aerosol size distribution evolution in systems with simultaneous nucleation, condensation and coagulation. *Aerosol Sci. Technol.*, **4**, 31.
54. Pratsinis, S.E., Friedlander, S.K., and Pearlstein, A.J. (1986) Aerosol reactor theory: stability and dynamics of a continuous stirred tank aerosol reactor. *AIChE J.*, **32**, 177.
55. Pratsinis, S.E. (1988) Simultaneous nucleation, condensation and coagulation in aerosol reactors. *J. Colloid Interface Sci.*, **124**, 416–427.
56. Friedlander, S.K. (1983) Dynamics of aerosol formation by chemical reactions. *Ann. N.Y. Acad. Sci.*, **354**, 1.
57. Kerminen, V.-M., Pirjola, L., and Kulmala, M. (2001) How significantly does coagulation scavenging limit atmospheric particle production? *J. Geophys. Res.* **106**, (D20), 24119–24125.
58. Dal Maso, M., Kulmala, M., Lehtinen, K.E.J., Mäkelä, J.M., Aalto, P., and O'Dowd, C.D. (2002) Condensation and coagulation sinks and formation of nucleation mode particles in coastal and boreal boundary layers. *J. Geophys. Res.*, **107**, (D19), 8097.
59. Kulmala, M., Lehtinen, K.E.J., and Laaksonen, A. (2006) Cluster activation theory as an explanation of the linear dependence between formation rate of 3 nm particles and sulphuric acid concentration. *Atmos. Chem. Phys.*, **6**, 787–793.
60. Kulmala, M., Pirjola, L., and Mäkelä, J.M. (2000) Stable sulfate clusters as a source of new atmospheric particles. *Nature*, **404**, 66–69.
61. Kulmala, M., Vehkamäki, H., Petäjä, T., Dal Maso, M., Lauri, A., Kerminen, V.M., Birmili, W., and McMurry, P.H. (2004) Formation and growth rates of ultrafine atmospheric particles: a review of observations. *J. Aerosol Sci.*, **35**, 143–176.
62. Kulmala, M., Hari, P., Laaksonen, A., and Viisanen, Y. (2005) Research Unit of Physics, Chemistry and Biology of Atmospheric Composition and Climate Change: overview of recent results. *Boreal Environ. Res.*, **10**, 459–478.
63. Kulmala, M. (2003) How particles nucleate and grow. *Science*, **302**, 1000–1001.
64. Pruppacher, H.R. and Klett, J.D. (1980) *Microphysics of Clouds and Precipitation*, Reidel, Dordrecht.
65. Spracklen, D.V., Carslaw, K.S., Kulmala, M., Kerminen, V.M.,

Mann, G,V., and Sihto, S.-L. (2006) The contribution of boundary layer nucleation events to total particle concentration on regional and global scales. *Atmos. Chem. Phys.*, **6**, 5631–5648.

66. Taylor, F.W. (2002) The greenhouse effect and climate change revisited. *Rep. Prog. Phys.*, **65**, 1.

67. Kerminen, V.-M., Virkkula, A., Hillamo, R., Wexler, A.S., and Kulmala, M. (2000) Secondary organics and atmospheric cloud condensation nuclei production. *J. Geophys. Res.*, **105**, 9255–9264.

68. Adams, P.J. and Seinfeld, J.H. (2002) Predicting global aerosol size distribution in general circulation models. *J. Geophys. Res.*, **107**, (D19), 4370.

69. Boy, M. and Kulmala, M. (2002) Nucleation events on the continental boundary layer: influence of physical and meteorological parameters. *Atmos. Chem. Phys.*, **2**, 1–16.

70. Kulmala, M., Kerminen, V.-M., Anttila, T., Laaksonen, A., and O'Dowd, C.D. (2004) Organic aerosol formation via sulfate cluster activation. *J. Geophys. Res.*, **109**, (D4), 4205.

71. Kulmala, M., Laakso, L., Lehtinen, K.E.J., Riipinen, I., Dal Maso, M., Anttila, T., Kerminen, V.-M., and Horrak, U. (2004) Initial steps of aerosol growth. *Atmos. Chem. Phys.*, **4**, 2553–2560.

72. Kerminen, V.-M., Anttila, T., Lehtinen, K.E.J., and Kulmala, M. (2004) Parametrization for atmospheric new-particle formation: application to a system involving sulfuric acid and condensable water-soluble organic vapors. *Aerosol Sci Technol.*, **38**, 1001–1008.

73. Kavouras, I.G., Mihalopoulos, N., and Stephanou, E.G. (1998) Formation of atmospheric particles from organic acids produced by forests. *Nature*, **395**, 683–686.

74. Kulmala, M., Dal Maso, M., Mäkelä, J.M., Pirjola, L., Väkevä, M., Aalto, P., Miikkulainen, P., Hämeri, K., and O'Dowd, C. (2001) On the formation, growth and composition of nucleation mode particles. *Tellus B*, **53**, 479–490.

75. Aalto, P., Hämeri, K., Becker, I., Weber, R., Salm, J., Mäkelä, J.M., Hoell, C., O'Dowd, C.D., Karlsson, H., Hansson, H.-C., Väkevä, M., Koponen, I.K., Buzorius, G., and Kulmala, M. (2001) Physical characterization of aerosol particles during nucleation events. *Tellus B*, **53**, 344–358.

76. Janson, R., Rozman, K., Karlsson, A., and Hansson, H.-C. (2001) Biogenic emission and gaseous precursor to forest aerosols. *Tellus B*, **53**, 423–440.

77. O'Dowd, C.D., Aalto, P., Hämeri, K., Kulmala, M., and Hoffmann, T. (2002) Aerosol formation: atmospheric particles from organic vapors. *Nature*, **416**, 497–498.

78. Boy, M., Rannik, U., Lehtinen, K.E., Tarvainen, V., Hakola, H., and Kulmala, M. (2003) Nucleation events in the continental boundary layer: long-term statistical analysis of aerosol relevant characteristics. *J. Geophys. Res.*, **108**, (D21), 4667–4675.

79. Lyubovtseva, Yu.S., Sogacheva, L., Dal Maso, M., Bonn, B., Keronen, P., and Kulmala, M. (2005) Seasonal variations of trace gases, meteorological parameters, and formation of aerosols in boreal forests. *Boreal Environ. Res.*, **10**, 493–510.

80. Kerminen, V.-M., Lehtinen, K., Anttila, T., and Kulmala, M. (2004) Dynamics of atmospheric nucleation mode particles: timescale analysis. *Tellus B*, **56**, 135–146.

81. Stolzenburg, M.R., McMurry, P.H., Sakurai, H., Smith, J.N., Mauldin, R.L., Eisele, F.L., and Clement, C.F. (2005) Growth rates of freshly nucleated atmospheric particles in Atlanta. *J. Geophys. Res.*, **110**, (D22), D22S05.

82. Barrett, J.C. and Clement, C.F. (1991) Aerosol concentrations from a burst of nucleation. *J. Aerosol Sci.*, **22**, 327–335.

83. Clement, C.F. and Ford, I.J. (1999) Gas to particle conversion in the atmosphere: II. Analytic models of nucleation bursts. *Atmos. Environ.*, **33**, 489–499.

84. Clement, C.F., Pirjola, L., Twohy, C.H., Ford, I.J., and Kulmala, M. (2006) Analytic and numerical calculations of the formation of a sulfuric acid aerosol in

the upper troposphere. *J. Aerosol Sci.*, **37**, 1717–1729.

85. Adams *et al* 2002 to come.
86. Korhonen, H., Lehtinen, K.E.J., Pirjola, L., Napari, I., Vehkamaki, H., Noppel, M., and Kulmala, M. (2003) Simulation of atmospheric nucleation mode: a comparison of nucleation models and size distribution representations. *J. Geophys. Res.*, **108**, (D15), 4471.
87. Lehtinen, K.E.J. and Kulmala, M. (2003) A model for particle formation and growth in the atmosphere with molecular resolution in size. *Atmos. Chem. Phys.*, **3**, 251–257.
88. Easter, R.C., Ghan, S.J., Zhan, Y., Sailor, R.D., Chapman, E.G., Laulainen, L.S., Abdul-Razzak, H., Leung, L.R., Bian, X., and Zaveri, R.A. (2004) MIRAGE: model description and evaluation of aerosols and trace gases. *J. Geophys. Res.*, **109**, D20210.
89. Anttila, T., Kerminen, V.-M., Kulmala, M., Laaksonen, A., and O'Dowd, C.D. (2004) Modelling the formation of organic particles in the atmosphere. *Atmos. Chem. Phys.*, **4**, 1071–1083.
90. Korhonen, H., Lehtinen, K.E.J., and Kulmala, M. (2004) Multicomponent aerosol dynamic model UHMA: model development and validation. *Atmos. Chem. Phys. Discuss.*, **4**, 471–506.
91. Jacobson, M.Z. and Turko, R.P. (1995) Simulating condensational growth, evaporation and coagulation of aerosols using a combined moving and stationary size grid. *Aerosol Sci. Technol.*, **22**, 73.
92. Grini, A., Korhonen, H., Lehtinen, K., Isaksen, I., and Kulmala, M. (2005) A combined photochemistry/aerosol dynamics model: model development and a study of new particle formation. *Boreal Environ. Res.*, **10**, 525–541.
93. Hofmann, T., Odum, J., Bowman, F., Collins, D., Klockow, D., Flagan, R., and Seinfeld, J. (1997) Formation of organic aerosols from the oxidation of biogenic hydrocarbons. *J. Atmos. Chem.*, **26**, 189.
94. Griffin, R., Cocker, D.R., Flagan, R., and Seinfeld, J.H. (1999) Organic aerosol formation from the oxidation of biogenic hydrocarbons. *J. Geophys. Res.*, **104**, 3555.
95. Griffin, R., Dabdub, D., and Seinfeld, J.H. (2002) Secondary organic aerosol: I. Atmospherical chemical mechanism for production of molecular constituents. *J. Geophys. Res.*, **107**, (D17), 4332.
96. Kulmala, M., Kerminen, V.-M., and Laaksonen, A. (1995) Simulation on the effect of sulfuric acid formation on atmospheric aerosol concentration. *Atmos. Environ.*, **29**, 377–382.

Further Reading

Adams, P.J. and Seinfeld, J.H. (2003) Disproportionate impact of particulate emissions on global cloud condensation nuclei concentration. *Geophys. Res. Lett.*, **30**, 1239.

Cercignani, C. (1975) *Theory and Application of the Boltzmann Equation*, Scottish Academic Press, Edinburgh.

Dal Maso, M., Kulmala, M., Riippinen, I., Hussein, T., Wagner, R., Aalto, P.P., and Lehtinen, K.E.J. (2005) Formation and growth of fresh atmospheric aerosols: eight years of aerosol size distribution data from SMEAR II, Hyytiälä, Finland. *Boreal Environ. Res.*, **10**, 323–336.

Eisele, F.L. and Hanson, D.R. (2000) First measurements of prenucleation molecular clusters. *J. Phys. Chem.*, **104**, 830–836.

5
Combustion-Derived Carbonaceous Aerosols (Soot) in the Atmosphere: Water Interaction and Climate Effects

Olga B. Popovicheva

5.1
Black Carbon Aerosols in the Atmosphere: Emissions and Climate Effects

In recent years, significant man-made *climate changes* have become increasingly evident. *Anthropogenic aerosols* from combustion sources have substantially increased the global burden of aerosols with respect to those of pre-industrial times. Global emission estimates for submicrometer *carbonaceous aerosols* are as high as 8–17 Tg/year depending on the various combustion sources considered, emission factors, and fuel use [1, 2]. The emissions from fuel oil combustion (industrial and residential), transportation (highway, aviation, and shipping), and biomass burning (wildfires and domestic wood burning) account for approximately 25% of all anthropogenic fine aerosol emissions. Black carbon (BC) aerosol is a ubiquitous substance that is produced by incomplete combustion of hydrocarbon fuels. It may be found in the atmosphere, soils, and sediments as a long-term geochemical sink of carbon and is a tracer of anthropogenic activity in urban regions. The contribution of *fossil fuel* is estimated as 38% of the global inventory of combustion aerosol emission [2].

Over the past years, *combustion-derived aerosols* have gradually received wider attention, owing to their environmental and climate impacts on the global scale and to health risks on the local scale. The physical, chemical, and toxicological characteristics of BC particles may vary significantly between urban and rural regions, and between continental and marine environments. In urban areas, the aerosol emitted by diesel engines is considered to be a dangerous pollutant affecting human health, especially near regions of public transportation (roads and harbors). Such aerosol may aggravate respiratory, cardiovascular, and allergic diseases [3].

Combustion-generated aerosols are currently attracting increasing attention because of the possible impacts on the radiation budget and clouds. The Intergovernmental Panel on Climate Change (IPCC) has stressed the importance of producing comprehensive estimates of anthropogenic aerosol "forcing" in order to gain an understanding of the most important aspects of climate change [1]. The radiative effect is typically expressed in terms of "forcing": the change in net

Aerosols – Science and Technology. Edited by Igor Agranovski

ISBN: 978-3-527-32660-0

radiation flux at the tropopause, which has units of watts per square meter. The *greenhouse effect* of heat-trapping gases such as CO_2 is the best known of the global atmospheric changes, being the largest in terms of global average, at a forcing of about $+2.5\,W\,m^{-2}$ [1]. An opposing effect, cooling the atmosphere, is provided by increases in scattering or reflective aerosols, primarily sulfates: their current forcing is near $-0.4\,W\,m^{-2}$. While many aerosols reflect light back to space, only carbonaceous aerosols (also termed *black carbon* in optical studies) dominate light absorption by aerosols in many regions. A direct radiative warming effect due to fossil-fuel BC is estimated by a forcing up to $0.4\,W\,m^{-2}$, while higher estimates have resulted from considering biomass burning near the most polluted areas in Europe, a maximum forcing of $2\,W\,m^{-2}$ [4]. Anthropogenic BC is predicted to raise globally and annually averaged equilibrium surface air temperature by 0.2 K, and even by as much as 0.4 K if BC is assumed to be internally mixed with sulfate aerosols [5].

One of the most pressing problems of research priority in climate estimates of BC aerosol impacts is their influence on cloud radiative properties [6]. Scheme 5.1 presents the major climate effects of BC aerosols. The first indirect effect is that BC aerosols increase the *cloud droplet concentration* serving as cloud condensation nuclei (CCN) and thereby decrease the cloud effective radius (the "radius" effect). Well-absorbing particles may induce a specific semi-direct effect relating to the heating of the air and cloud evaporation (the "heating" effect), which can change the vertical temperature profile and the dynamic structure of clouds. The second indirect effect is that a decreased cloud droplet effective radius decreases the

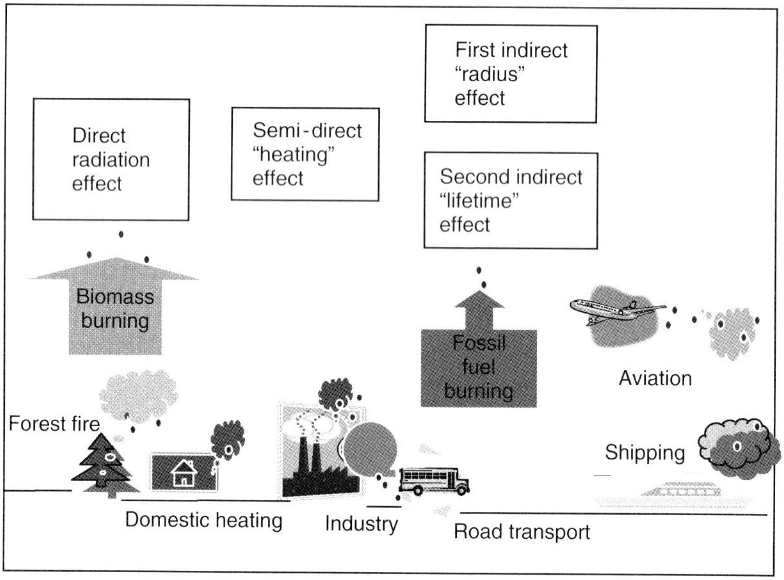

Scheme 5.1 Climate effects of BC aerosols.

rate of precipitation, causing longer cloud lifetime and higher cloud amount (the "lifetime" effect). The indirect effect of BC anthropogenic aerosols on changing the cloud optical properties is the most uncertain component of climate forcing over the past 100 years [7].

Currently, emissions of particulate matter from ships is gaining increasing attention because shipping may contribute significantly to the total global BC budget, but its emissions are poorly known in comparison with those from land-based transport [8]. Observations confirm that diesel-engine-powered ships burning heavy fuel oil may contribute to CCN in marine stratus clouds and produce ship tracks that change the *cloud reflectance*, while the impact of ships burning diesel distilled fuel is considerably less [9].

Industry, road transport, and residential wood burning compose the BC sources located over land. Even if they do not produce visible cloudiness, their particulate emission is lifted by convection and could substantially impact the tropospheric BC aerosol loading [10]. However, while diesel particulate emission in urban areas has been of great concern in the past years, the other emissions are still not well described.

A number of global climate and *radiative transfer models* [6, 7, 10] have estimated the first indirect BC aerosol effect, providing a value near $-1.5\ \mathrm{W\,m^{-2}}$. However, the uncertainty in their estimations remains one of the highest in climate studies today. This is due to the high inhomogeneity of BC aerosol emission and burden, and to the complex relationship between aerosol *physico-chemical properties* and cloud microphysics. The situation is complicated by: (i) the wide variety of combustion sources burning gaseous, liquid, and solid fuels in different engines, boilers, stoves, and ovens; (ii) the source-dependent properties of the emitted soot particles; and (iii) an incomplete database for the behavior of soot aerosols in the humid atmosphere. The advanced global climate models currently simplify the situation by assuming that BC particles are initially hydrophobic and act as CCN only after they have become internally mixed with sulfates, thus leaving out any consideration of the hydration properties of the original combustion aerosols.

Remote-sensing observations and theoretical studies of the Earth's climate prove that clouds provide a strong influence on the radiative energy balance. In this context, *ice-nucleating aerosols* have special significance because they may allow ice nucleation at lower supersaturations than those required for homogeneous freezing, resulting in an increase of the *cirrus cloud* coverage and changing its microphysics and optical properties [11]. Analysis of ice-nucleating aerosols taken in cirrus clouds showed that they are dominated by carbonaceous particles [12]. However, there is considerable uncertainty regarding the quantitative estimate of the *ice-nucleating ability* of combustion-derived aerosols in the upper *troposphere* because using *in situ* measurements it is difficult to obtain unambiguous evidence that BC aerosols are directly involved in cloud formation.

A potentially strong indirect aerosol effect on ice crystals has been inferred by including heterogeneous ice nucleation on carbonaceous aerosols along with homogeneous freezing of sulfate aerosols into the microphysical models of cirrus clouds [11, 13, 14]. Cirrus clouds appear to be far more sensitive to the impact of heterogeneous freezing ice nuclei (IN) increases than they are to increasing numbers of sulfate aerosol particles. Adding even a small number of, but efficient, heterogeneous IN to a region where ice crystals form primarily through homogeneous freezing can lead to a marked suppression of the relative humidity (RH) with respect to ice (RH_i) and can thereby reduce cirrus ice crystal number densities. Moreover, the importance of the heterogeneous freezing compared to the homogeneous freezing nucleation process depends greatly on the particle size and the mass percent in the droplet [11]. The presence of insoluble particulates at sizes greater than 0.5 μm within just a few percent of the CCN population lowers the ice crystal concentration by an order of magnitude in comparison with smaller size particles of 0.08 μm. Less frequent observations of RH_i above 130% outside of clouds in the Northern Hemisphere compared to the Southern Hemisphere may support the assumption that heterogeneous freezing on combustion aerosols occurs at least in some polluted regions of Northern Hemisphere midlatitude cirrus clouds [15]. However, in the background atmosphere, BC aerosols should be present mixed with other aerosols, whereas the fraction of particles possessing specific heterogeneous ice-nucleation activity may be relatively low except perhaps in relatively undiluted source plumes (for example, aircraft exhausts or biomass burning plume).

Recent campaigns of BC particle measurement in the upper troposphere have reported their mean number density to be on the order $0.1\ cm^{-3}$, co-varying in density with commercial air traffic fuel consumption [1]. The highest BC concentrations in the upper troposphere can be found in regions over the USA and Europe. Aviation can cause large-scale increases in the BC particle number concentration of more than 30% in regions highly frequented by aircraft [10]. Although the amount of soot, sulfur, metal particles, and nitrogen oxides emitted from airplanes is small compared to the total loading in the atmosphere, at the altitude of aircraft flight these emissions may represent a dominant source.

Aviation-produced soot aerosols are suspected to enhance contrails and cirrus formation [16], thus giving rise to a positive radiative forcing [17]. A major source of uncertainty in assessing the impact of aircraft-emitted soot aerosols on climate change is their role in contrail formation and secondary effects on cirrus formation through potential action as IN. Enhancement in the frequency of occurrence of cirrus clouds up to 10% per decade strongly suggests that emitted soot may act as IN for cirrus formation. The maximum estimate of the number concentration of aircraft-emitted soot particles results in an increase of the potential IN up to 50% at the main aircraft flight altitude [10].

The phenomenon of aircraft contrail formation has attracted much attention as a visible and direct anthropogenic impact upon the atmosphere because: (i) field observations in aircraft plumes indicate that jet exhaust aerosols with diameters $D > 0.01$ μm mainly consist of BC particles [18]; (ii) the sampling of ice residuals in contrails [19] and in ice-nucleating aerosols from aircraft plumes [12] shows that they are mostly composed of BC with small amounts of oxygen, sulfur, and metals; and finally (iii) optical observations of an internal mixture of ice–BC aerosols in plumes [20] directly confirm theoretical model predictions showing [21] that exhaust soot particles may serve as IN. Other studies suggest that the soot aerosols from evaporated contrails make their way into natural cirrus clouds and may affect their microphysical properties [22].

The quantification of aircraft-emitted soot impact has been advanced, but the state of scientific understanding is still poor, mainly because of insufficient studies of the ice-nucleating ability of atmospheric soot aerosols forming the contrails and cirrus clouds. Indeed, sparse documentation exists on the CCN and ice-nucleating properties of soot particles and on the relationship between the physico-chemical characteristics of these particles and their ability to take up water. Laboratory studies can provide relevant information concerning soot CCN and IN behaviors. However, the literature concerning soot studies in the laboratory shows the use of a great variety of combustion sources and soot surrogates for atmospheric studies. Commercially available soots [23, 24], spark discharge soot [25, 26], and samples produced by different burners and fuels [27–31] have all been used to study the CCN and/or IN activity of soot particles. These different particles represent a variety of physico-chemical properties, leading to differences in their water/ice nucleation ability. Combustion aerosols generated by burning acetylene fuel in a welding torch under high oxygen conditions as well as by flaming wood combustion have demonstrated the highest *CCN activity*: the ratio of the number of CCN to the total number of condensation nuclei (CN) is near 0.49 and 0.72 for these aerosols, respectively, at 1% supersaturation [31]. In contrast, the open burning of JP-4 aviation fuel produced soot particles of weak CCN activity (CCN/CN $\approx 8 \times 10^{-3}$) for similar conditions. Unfortunately, not all such studies have characterized the properties responsible for soot *hygroscopicity* and CCN activity such as composition and water-soluble fraction. The great variety of physico-chemical properties of soots produced by different laboratory combustion and commercial sources remains the problem in the application of laboratory-made soot for atmospheric studies [32].

The lack of experimental data on the microstructure, composition, and hygroscopicity of original soot from aircraft engines has led some investigators [21] to assume that all exhaust soot particles are hydrophobic. Moreover, Kärcher *et al.* [21] assumed that the properties of aircraft-engine-generated soot are similar to those of graphitized soot, having a perfectly homogeneous surface and an extremely low density of chemical heterogeneities (surface *functional groups*). In that case, to activate *water adsorption*, a hygroscopic coverage (for example, of sulfuric acid) should be formed on the surface of the emitted soot particles by interaction with gas and gaseous particles in the plume. However, such assumptions did not allow a few general questions arising from the observations to be answered, namely why supersaturation with respect to water is needed for visible contrail formation [33], and why changes in the fuel sulfur content have little impact on the threshold for contrail formation [34]. Therefore, it is reasonable to reconsider the hypothesis about hydrophobic exhaust soot and to assume an initial heterogeneity of engine-generated soot particles with respect to their microstructure and chemical composition, which should lead to the ability to take up water in the condensation process.

The mechanism that could lead to heterogeneous ice nucleation by soot particles at the cirrus level and in lower tropospheric clouds is not clear. Numerical simulations [11] focus on immersion freezing nucleation because of the low likelihood of finding completely insoluble particles available for direct ice deposition nucleation. The lack of observation and laboratory data has led some investigators [13, 14] to evaluate the effectiveness of heterogeneous freezing versus homogeneous freezing, hypothetically assuming that soot particles act as immersion freezing nuclei at $RH_i = 130\%$. They conducted a scenario with hypothetically assumed heterogeneous freezing ability of soot particles, where the specific soot physico-chemical properties are not considered. As a result, a lot of uncertainties remain in estimates of the role of soot aerosols in cold cloud formation [35].

However, soot particles emitted into the atmosphere from a great variety of combustion sources have a wide range of natural variability with respect to physico-chemical properties, including composition and hygroscopicity. The situation is complicated by the limited possibilities for *in situ* exhaust measurement of soot characteristics, such as wetting, *water uptake,* and *surface chemistry*. Experiments in which actual combustion particles are collected and characterized offer a direct method for studying atmospheric soot–water interactions while mitigating the expense and difficulty of direct atmospheric sampling. Test experiments at ground aircraft engine facilities are an example of such characterization studies. In the PartEmis gas turbine engine measurement experiments [36, 37], the impact of the fuel sulfur content on the hygroscopic growth factors and on CCN activities of exhaust soot particles were reported. The morphology, microstructure, and chemical composition of original aircraft engine combustor (AEC) soot generated by burning TC1 aviation kerosene in a typical gas turbine engine combustor were described in Popovicheva *et al.* [38]. Possible pathways for CCN and IN activation of *AEC soot* through the condensation freezing mode have been suggested in Popovicheva *et al.* [38] and Shonija *et al.* [39].

Therefore, a starting point for research in order to evaluate the effects of anthropogenic aerosols on climate, and specifically on clouds, is to document the nature of the particulate emissions from major combustion sources and to quantify their interactions in the humid environment.

5.2
Physico-Chemical Properties of Black Carbon Aerosols

Carbonaceous aerosols are generated from traffic and industry burning fossil fuels, from outdoor fires, and from households burning coal and biomass fuels. Since the formation of carbonaceous particles depends on several factors (for example, type of fuel, temperature, and duration of combustion), its morphology and composition are also expected to vary considerably. BCs comprise distinct particle types. During combustion, small particle aggregates (soot) and/or large spherical or layered structural residuals (char/charcoal) are formed [40–42]. For wood burning, liquid-like agglutinated structures with no well-defined boundaries are typical [40]. Owing to contamination of fuel, some mineral compounds may pollute these particles or even form distinct ash residuals [43]. The basis feature that all BCs share is a matrix formed from layers of polycyclic aromatic rings. Features that are variable include the presence of associated organics, carbon–oxygen functionalities, and inorganic impurities.

The microstructure and composition of combustion-generated particulates could have significant effects on their roles and fates in the environment. A specific key property of combustion-derived aerosols, important to all major potential climate impacts of BCs, is their behavior in the humid atmosphere, their ability to act as CCN and IN. Indirect BC effects are critically dependent on the particle hydrophilic properties – cloud activation is determined solely by the ability of a particle to take up water in a supersaturated environment. Therefore, current investigations are focused on experimental characterizations of key physico-chemical properties of combustion particles relating to water uptake. The volatile, soluble organic, and water-soluble fractions are analyzed. Surface chemistry related to surface functional groups (functionalities) and ion composition are determined. Special attention is paid to the solubility in water of different classes of inorganic and organic surface compounds, since the water-soluble fractions play a crucial role in analyzing the ability of aerosol particles to serve as CCN/IN in the atmosphere.

Directly related to poor knowledge of identifying surface properties, uncertainties exist in understanding the detailed role of the composition of surface compounds in the interaction with water, and in the prediction of the effects of the BC particulates in a humid atmosphere. A major emphasis of the present section is to provide those combustion particle characteristics which are related to or responsible for environmental impact, more precisely for interaction with water vapor and freezing.

5.2.1 General Characteristics

Combustion-derived residuals are revealed in distinctive physical forms. As commonly accepted, soot is a product of incomplete burning of hydrocarbon fuels. However, sometimes the particles from spark discharge with graphite rods are also called soot. Commercial carbon black is a material generated by high-temperature combustion of petroleum feedstock and natural gas. Diesel soot is produced by the operation of heavy-duty truck diesel engines and corresponds to traffic-derived BC, as the particulate emission factors of gasoline engines are smaller by several orders of magnitude.

There are numerous articles and excellent reviews on the formation and structure of soot particles in diffusion flames and diesel environments (see, for example, [44]). First, hydrocarbon fuel degrades (fuel pyrolysis) into small radicals from which the formation of *aromatic rings* occurs mainly via the addition of acetylene and polyaromatic hydrocarbons (PAHs) in fuel-rich areas (particle inception). The increase of soot nuclei size is assumed to happen by both coalescence of large aromatic structures and surface growth. It is a stage in the formation of the elemental carbon (EC) soot structure. With the decrease of the temperature in the flame, the primary particles coagulate into chain- or grape-like (aciniform) aggregates and many of them may be subsequently oxidized. Hydrocarbons, soluble organics, and inorganic sulfur-containing materials of atomized fuel and evaporated lubrication oil appear in the plume as soluble organic compounds and sulfuric acid or sulfates. They may condense on the particle surface. Fuel metal impurities are found to exhibit catalytic activity for soot nuclei formation and lead to the formation of an inorganic ash fraction of soot particles.

Soot from any combustion source reveals surface irregularities over several length scales, which is a typical fractal agglomerated structure. Morphological studies on soot agglomerates have a long and elaborate history [45]. The degree of aggregate compactness, fractal dimension, varies from 1.7 to 2.4 for particles originating from diesel engine exhaust, from laboratory burning diesel fuel and natural gas, as well as from spark discharge generation [46–48]. But an often-cited soot structure consisting of point-contacting spheres assembled into branched fractals is rather the model assumption. In contrast, transport, commercial carbon black, and laboratory-generated soots typically appear as compact aggregates with primary particles that are not clearly defined [49, 50].

Figure 5.1a shows the typical morphology of aggregates of laboratory-made *kerosene flame soot* [51]. On a primary particle scale (see Figure 5.2a), they appear mostly as significantly merging spheroids with distortions and irregularities. Therefore, their size may be only roughly estimated. There is remarkable similarity in estimations of the primary particle size for carbon black and diesel soot: their

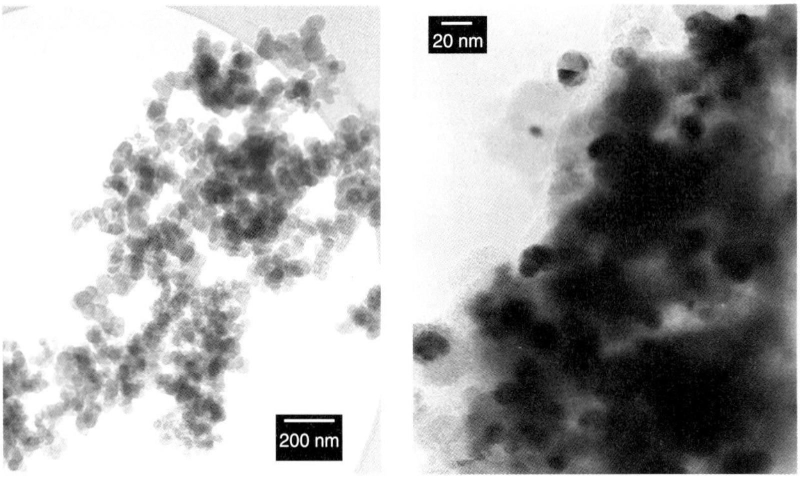

(a) (b)

Figure 5.1 Morphology of (a) aggregates of laboratory kerosene flame soot and (b) particles of aircraft combustor soot containing iron impurities.

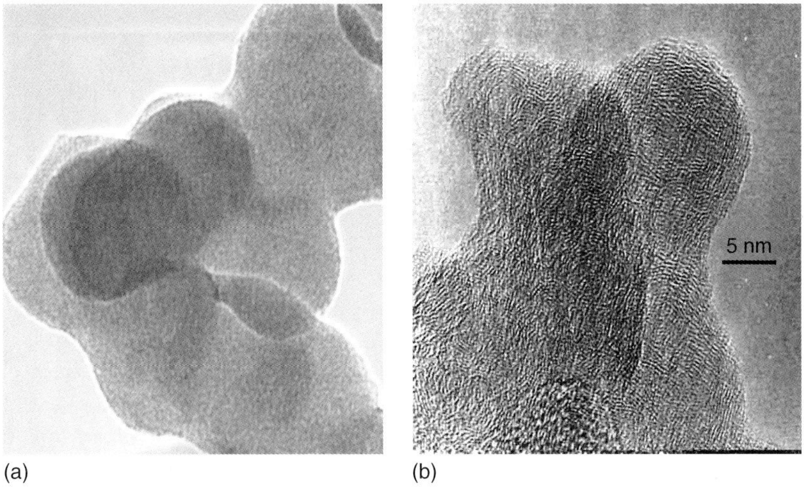

Figure 5.2 Primary particles of (a) amorphous nanostructure of laboratory kerosene flame soot and (b) onion-like nanostructure of aircraft combustor soot.

diameter, D, ranges from approximately 20 to 100 nm. The small size of soot primary particles determines the high specific surface area of soots, $\leq 100\ m^2\ g^{-1}$, related to the developed mesoporosity (inter-particle spaces).

Most of the agglomerated particles emitted by diesel and spark-ignition engine combustion are in the accumulation mode, $50\ nm < D < 1000\ nm$ diameter range [52]. The aircraft soot exhaust aerosol consists of a primary mode with a median $D \sim 35$ nm and agglomerated particle mode with a peak near 160 nm [18]. Aerosol measurements carried out in diesel engines burning heavy fuel oil demonstrate the bimodal particulate mass size distribution, with a main mode under 100 nm and a second mode at 7–10 μm [41, 43], providing strong evidence for a specific difference in morphology of carbonaceous particles in comparison with burning light diesel fuel or kerosene. This is the presence of large *char particles* originating from unburned fuel droplets by subsequent pyrolysis, oxidation, and carbonization. Char particles typically have a spherical morphology and are observed in aerosols emitted by marine ships [43, 53]. The transmission electron microscope (TEM) image of a char particle present in ship residues is shown together with soot particles in Figure 5.3a [54].

In terms of microstructure, there is a high similarity between transport exhaust soots, commercial carbon blacks, and laboratory-made soots. They consist of an EC matrix (nanocrystalline domains) and adsorbed or internally mixed materials, which represent organic carbon (OC) and inorganic compounds (amorphous domains). Numerous work has proved the presence of graphite microcrystallites inside soot particles, which consist of graphite planes rotated along axes perpendicular to the graphite planes [32, 50, 55]. Raman data indicate the highly disordered graphitic structure in almost all types of soots [56].

Figure 5.3 (a) Char and soot particles and (b) primary soot particles polluted by transition metals of marine heavy fuel oil combustion residuals.

Three distinct forms of EC may be observed at the microscopic level: amorphous, fullerenic, and graphitic types. For combustion-generated particles produced by diffusion flames, amorphous microstructure without any nanostructure is typical [51, 57], as shown in Figure 5.2a by the TEM image of kerosene flame soot produced in a common wick lamp. Conditions for soot formation in diffusion flames are much more favorable than in premixed flames where fuel and oxygen are intimately mixed prior to reaction.

The microcrystallites in soot particles can be aligned in different shapes, forming fullerenic turbostratic (onion-like) structure. This consists of a continuous network of concentric, size-limited graphene layers composed of several parallel planes of carbon atoms arranged in hexagonal arrays. The graphene layer planes show a wavy shape. This indicates that the atomic configuration of layer planes is far more random than that of graphite, involving numerous lattice defects and pores. A detailed analysis by electron microscopy reveals that the primary soot particles consist of two distinct parts: inner core and outer shell [57]. The inner core may be about 10 nm in diameter and contain several fine nuclei of 3–4 nm diameter, providing evidence of differently scaled agglomeration processes. Such fullerene-type carbon structures are typically observed in premixed flames and in diesel exhaust [50, 55].

The configuration of the combustor of an aircraft engine is defined by turbulent diffusion combustion processes, having the features of both diffusion and premixed flames. This is why the various microstructures can be found in engine *combustor soot* particles. Figure 5.2b shows the typical onion-like structure of primary particles of AEC soot [51]. Well-graphitized structures (nanotubes and flakes) may be caused by high-temperature graphitization of the soot particles inside the combustor, but they appear more rarely [32]. Moreover, the formation of highly ordered carbon nanostructures such as onions and tubes appears to require much longer residence times, perhaps seconds or minutes, in the flame environment.

The chemical composition of soot particles depends upon the type of engine, the combustion conditions, and fuel formulation. While carbon, hydrogen, and oxygen compose any combustion particles, sulfur and transition metals may accumulate on the surface and inside soot particles due to contamination in the fuel and engine. Heavy-duty diesel engine-emitted particles contain sulphuric acid or sulfate fraction roughly proportional to the fuel sulfur content, fuel additive ash, and heavy-metal compounds originating from the lubricating oil [52]. The iron oxides observed in aircraft-engine-emitted particles are due to corrosion processes in the combustor system [19, 38]. Figure 5.1b shows soot particles containing iron impurities [51], thus composing the fraction of impurities of AEC soot.

Soot particles in ship residuals produced by burning low-grade heavy fuel exhibit a similar morphology to any other soot particles, but the comparison of microstructure clearly indicates the presence of a dark core inside the particles (see Figure 5.3b). *Elemental analysis* confirms the large contamination of these particles by transition metals and *alkaline earth metals* (V, Ni, Ca, Fe, Na, Al, Si, and P). Diffraction analysis proves the presence of various chemical forms, such as Ni_3S_2, $Na_6(CO_3)(SO_4)$, Ni_2Fe, Ni_3Fe, NiO, V_2O_3, and NiS, probably forming cores inside particles [43, 54]. Hazardous constituents and their soluble or insoluble chemical forms (sulfides, sulfates, oxides, and carbides) are released by large ship- and land-based diesel engines into the atmosphere, together with emitted particles [41, 58].

The plant-derived residuals include straw and wood charcoal, vegetation and wood fire residuals, and chimney residuals. They exhibit various forms, including large spherical char and layered structures with hard edge boundaries, and liquid-like agglutinated structures with no well-defined boundaries [41]. Therefore, they have low specific surface area, typically $\leq 20\ m^2\ g^{-1}$. Aciniform carbon is a quite random fraction in biomass burning products, in respect of low EC content.

Discrepancies between the amount and the composition of condensed compounds on the surface of combustion-generated particles arise from their post-formation and sampling conditions as the exhaust is diluted and cooled, then nucleation and condensation transform volatile materials to solid matter [48]. The soluble organic fraction (SOF) is varied from $\leq 2\%$ for carbon blacks to $>13\%$ for diesel and chimney residuals [40]. The ratio of polar/non-polar SOF is generally $<7\%$ for fossil BC but $>30\%$ for plant-derived BC. The systems developed recently for the characterization of diesel particles – like for measurement of particle emission, number concentration, size distribution, EC and OC analysis – are described in Burtscher's review [48] and Gelencser [59].

5.2.2
Key Properties Responsible for Interaction with Water

Incorporation of soot-containing particles into cloud droplets depends on the critical water vapor supersaturation, at which a particle of given diameter activates and forms a cloud droplet [60]. BC is usually treated as an inert substance that contributes to the particle volume. In the absence of aqueous solution coverage, the classical nucleation theory is used to describe cloud droplet nucleation for insoluble particles. The theory considers the *energy of formation* of a critical cluster on a particle surface with *wettability* characterized by the *contact angle* of water with the surface [60]. Literature on the wettability of carbonaceous substances yields a consistent picture of hydrophobicity (see [61] and references therein). The measured water/ice contact angles on the soot surfaces vary in the range 40°–80°. Non-zero water droplet contact angle is traditionally associated with poor ability to take up water and low rate of heterogeneous nucleation. This is why there was a widespread assumption that, for instance, fresh aircraft-emitted soot particles are hydrophobic and should be a poor substrate for nucleating water embryos until they have undergone chemical activation through the accumulation of soluble species such as sulfuric acid [21, 62].

However, a water contact angle is a macroscopic measure of wettability. Measurements of water uptake from the vapor phase in a wide range of relative humidities may provide much more comprehensive information on the mechanism of water molecule interaction with the surface [63], allowing the classification of soot from hydrophobic to hydrophilic. Moreover, the critical supersaturation, S_c, is influenced by soot particle surface properties, microstructure, and the presence of water-soluble compounds on the surface. Classical *Köhler theory* calculates the *activation barrier* based on the saturation vapor pressure depression due to the dissolved solute [60]. It follows from the Köhler theory that the fraction of water-soluble compounds, ε, of the particle coverage may decrease S_c; therefore, ε may be proposed as a measure of soot hygroscopicity. But it should also be taken into account that water-insoluble particles are also assumed to serve as CCN in respect of the *Kelvin activation* if they are wettable. This is the case for $\varepsilon \rightarrow 0$. If soot particles are insoluble but wettable, what surface characteristics do they have? The answer to this question should be found by analyzing the water interaction mechanism at the microscopic level.

Water adsorption is proposed to be observed on soot particles due to interaction with active sites (surface functionalities) [64–66]. Moreover, water may condense on partly wettable particles even below saturation conditions due to capillary condensation into mesopores. For agglomerated soot particles, the classical nucleation theory is extended to account for mesopores, the spaces between the primary particles [67]. Irregular soot agglomerated structure is suggested to amplify ice nucleation due to condensation into soot inter-particle cavities and an inverse *Kelvin effect* [61]. Since soot aerosols typically consist of agglomerated chains of roughly spherical primary particles, restructuring may be observed during water condensation–evaporation cycles [46, 68].

Experimental characterization of original soot generated by burning sulfur-free hydrocarbon fuel in a typical gas turbine engine combustor and by rich and lean flames has shown that, due to the microporous structure and oxygen-containing surface functional groups, these soots may take up significantly more water than a reference graphite material [69, 70]. Such soot aerosols can acquire a substantial fraction of a water monolayers (MLs) even in the conditions of the young plume [71]. As a result, a sulfur-independent heterogeneous nucleation mode driven by the presence of water on such soot particle types was suggested.

A number of techniques useful for the analysis of the structure and chemical properties of soot have been applied. Measurements of surface area and texture parameters are made using N_2 adsorption, while net water uptake is assessed using H_2O adsorption [65, 71]. The combination of these methods provides information about the detailed mechanism of *water–soot interaction* at RH $\leq 100\%$, and clarification of the role of surface chemistry and micro- and mesoporosity. Additional information for understanding the behavior of atmospheric soot particles is obtained by analyzing the changes in size of free-flowing particles versus RH using a humidified tandem differential mobility analyzer (HTDMA) [36, 66, 68]. Such data also have direct relevance to warm cloud formation processes.

The most complex problem of CCN and IN activation of BC aerosols is determining the surface requirements for water/ice nucleation at the microscopic level. Observations from recent experiments concurrently utilizing the methodology and water uptake measurements described above allow the quantitative examination of the role of key properties of soot particles in the observed water interactions, such as water-soluble fraction, hydrophilic organic or inorganic surface compounds, surface-active functionalities, and porosity. Broad chemical features may be inferred from elemental analysis, analysis of impurities, polar and non-polar organic matter, ionic content, and volatile surface organic compounds. Chemical functionalities observed on the particle surface support the identification of hydrophobic and hydrophilic soot. These features are characterized by the methodology elaborated using TEM, Fourier-transform infrared (FTIR) spectroscopy, gas chromatography–mass spectrometry (GCMS), and ion chromatography to determine the composition features responsible for their interaction with water.

It is expected that soot aerosols may be incorporated into cloud droplets by the presence of water-soluble compounds. However, the content of OC and SOF is known to vary with engine operating conditions, and can represent 5–90% of the whole mass of particulate matter [52]. A comprehensive evaluation of organic coverage composition in urban areas and in diesel exhaust [72, 73] identified the presence of different classes of polar compounds (aromatic, alkanoic and alkanedioic acids, and phenols) as well as non-polar PAHs, alkanes, and alkenes. Published data provide evidence that polar organic compounds take up water (see [74] and references therein) and thus increase the cloud and fog formation ability of insoluble soot cores. Non-polar organic compounds do not generally take up water and also may form films that lower the accommodation coefficient of water on the particles, thereby inhibiting absorption of water on the surface.

The literature on the heterogeneous ice-forming ability of insoluble particles yields a number of requirements relating to size, microstructure, insolubility, chemical bonding, and surface-active sites [60]. Sites that are capable of adsorbing water molecules are also sites at which ice nucleation is likely to be initiated, while the number of critical embryos per unit area of the substrate is proportional to the number of water molecules adsorbed per unit area. Current studies seek direct evidence for CCN activity at environmentally relevant water supersaturation, and for ice-nucleation behavior at contrail and cirrus formation conditions, using a set of carefully selected soots, of varying physical and chemical characteristics [75]. Special attention is paid to a systematic comparative analysis of water uptake on combustion nanoparticles from transport systems and on the identification of water–soot interaction mechanisms in relation to the exhaust residual composition.

5.3 Water Uptake by Black Carbons

Long-term recommended research needs currently include the requirement for the development of new concepts with respect to soot impacts on CCN and ice cloud formation. The first actual need is the quantification of water uptake by BC aerosols, since the present atmospheric studies will remain relatively empirical until a *quantification measure* of the extent of BC hydrophilicity is defined. In adsorption science, the high fraction of the surface area covered by adsorbed water is proposed as a criterion for hydrophilicity. A hypothetical statistical water monolayer is assumed as a reference for the characterization of the level of surface polarity. A water film may originate due to cluster formation around active sites, which may eventually become connected if the number of active sites is relatively high. The formation of a water film on the soot surface releases the wetting phenomena at the microscopic level – that may represent a feature of hydrophilic soot. Additionally, it may serve as the necessary condition for the Kelvin activation of a soot particle. To quantify the conditions for water film formation, a comprehensive analysis of water uptake on soot particles with various surface chemical and structural properties is needed.

This section is devoted to a systematic comparative analysis of water uptake on laboratory soots proposed for atmospheric studies and on original transport-emitted residuals with the purpose of identifying water–black carbon interaction mechanisms in relation to particle composition [54, 75, 76]. The purpose is to describe a *quantification measure* for the separation between hygroscopic and non-hygroscopic BCs and for the identification of hydrophilic and hydrophobic particles within non-hygroscopic atmospheric aerosols. A few classical water adsorption models are applied for parameterization and quantitative comparison.

5.3.1 Fundamentals of Water Interaction with Black Carbons

The adsorption behavior of a solid is generally characterized by a plot of the amount of gas adsorbed as a function of gas pressure at constant temperature (*adsorption isotherm*). In the simplest *Langmuir model* the adsorbed film is assumed to be just one monolayer, owing to the very short range of the intermolecular forces and negligible interaction between the adsorbate molecules [63]. The assumption that all adsorption sites on the surface are equivalent to each other leads to a famous *Langmuir isotherm* for the amount of vapor adsorbed a as a function of the pressure p:

$$a = \frac{a_{\mathrm{m}} K p}{(1 + K p)} \tag{5.1}$$

where a_{m} is the monolayer coverage, and K is a constant for a given temperature and adsorbing material. The Brunauer–Emmett–Teller (BET) theory extends the Langmuir model to include multilayer adsorption and assumes that: (i) each molecule adsorbed in a particular layer is a possible site for adsorption of a molecule in the next layer; (ii) no horizontal interactions exist between adsorbed molecules; (iii) the *heat of adsorption* is the same for all molecules in any given adsorbed layer; and (iv) the heat of adsorption is equal to the *latent heat of vaporization* for all adsorbed layers except the first one. The *BET equation* is written as

$$a = \frac{a_{\mathrm{m}} C p/p_{\mathrm{s}}}{(1 - p/p_{\mathrm{s}})(1 + p/p_{\mathrm{s}}(C - 1))} \tag{5.2}$$

where p_{s} is the saturation vapor pressure for the vapor being adsorbed, and C is a constant for the *adsorbent–adsorbate interaction*. The BET equation (5.2) is very widely used for surface area estimations by determining a_{m}. However, usually it treats adsorption data in the range of p/p_{s} from 0.05 to 0.35 and at $C > 10$.

The lattice theory of Ono and Kondo was used by Aranovich [77] to improve the BET theory, since it allows the vertical *adsorbate–adsorbate interactions* to be taken into account. The two-parameter *Aranovich equation*

$$a = \frac{a_{\mathrm{m}} C p/p_{\mathrm{s}}}{(1 + C p/p_{\mathrm{s}})(1 - p/p_{\mathrm{s}})^{0.5}} \tag{5.3}$$

provides qualitatively correct results in the limiting cases and successfully simulates multilayer adsorption on various solids [78], extending the range of application of the equation up to $p/p_{\mathrm{s}} \sim 0.7 - 0.8$.

The classical model of BCs proposes that they are composed of misaligned graphite platelets that form a nanoporous network in which oxygenated functional groups are contained [45]. The peculiar nature of water adsorption on BCs is related to the relatively low dispersion energy between water molecules and graphite platelets. According to the fundamental mechanism [63, 79], initial water adsorption takes place on oxygen-containing groups, which may act as primary adsorption sites. At low relative pressures, the amount of water adsorbed is determined by the total density of primary adsorption sites (*Henry's law*); this amount gives the value of a_{m} in Eqs. (5.1)–(5.3). At higher relative pressures, *hydrogen bonding* between free and adsorbed water occurs, the primary adsorbed water molecules act as nucleation sites for further adsorption of water, and the formation of three-dimensional clusters begins [80, 81]. Both the density and the geometrical arrangement of the active sites have a pronounced effect on the value of $p_{\mathrm{c}}/p_{\mathrm{s}}$ (p_{c} indicates p relating to cooperative effect) at which water clusters form bridges (*fluid–fluid cooperative effect*), producing a film of bonded water molecules extending over the surface.

A secondary water adsorption mechanism realizing as the fluid–fluid cooperative effect is unique for the *water–carbon system*, so it is not considered by the BET theory. The theory of cooperative multi-molecular sorption (CMMS) (see [82] and references therein) has been elaborated to account for the influence of the adsorbed molecule promoting the entry of other molecules to adjacent sites. The CMMS theory is employed to account for the type II, III, V, and hybrid isotherms (in the BET classification scheme) for water–carbon systems.

The model of "long-winded" fluid formation has been proposed by Berezin *et al.* [83] to describe the water cluster confluence and liquid water phase origin on carbonaceous adsorbents. It is based on a thermodynamic analysis of multilayer water adsorption, which shows the existence of a fluid film long-winded by the surface forces [84]. The *Halsey–Hill equation* for the long-winded fluid film was used to obtain the isotherm equation, taking into account its correction on the curvature of particles [84]:

$$\ln(p_s/p) = b/\theta^3 - W/RT \tag{5.4}$$

Here θ is the adsorption coverage (in monolayers) calculated assuming one statistical monolayer of water estimated from the effective molecular cross-sectional area for the water molecule of 0.105 nm^2, W is the binding energy of the surface film, and b is a constant defined as $b = 2.5V\sigma/RTd$, where V and σ are the mole volume and surface tension, respectively, R is the gas constant, and d is a diameter of the adsorbate molecule. Assuming the Kelvin equation for the convex surface, W can be written as

$$W = 2\frac{V\sigma}{r} \tag{5.5}$$

where r is the radius of curvature of the particle. Finally, for a water film extended over a convex surface, we have the isotherm equation

$$\ln(p_s/p) = 4.34/\theta^3 - 1.08/r(\text{nm}) \tag{5.6}$$

which may be used to estimate p/p_s when there are bridges between water clusters producing a large film extended over the surface of a soot particle with radius r.

If the soot coverage thickness exceeds a few monolayers and the coverage material consists (even partially) of water-soluble materials ($\varepsilon > 0$), water uptake should be treated in terms of bulk dissolution, while adsorption should be assumed for interaction with the surface-active sites and with sites inside the water-soluble coverage material (volume sites). The general model of "dual sorption" should be useful in that case [85], where both the absorption and permeation processes of a vapor in a soluble material are taken into account. Then, the concentration of water molecules in the soot coverage a is the sum of dissolved molecules and the molecules absorbed on active sites, determined by Henry's law and from the Langmuir isotherm, similar to Eq. (5.1):

$$a = K_d p + \frac{a_v Kp}{(1 + Kp)} \tag{5.7}$$

where K_d is the dissolution coefficient, and a_v is the equilibrium concentration of vapor determined by volume adsorption sites (*sorption capacity*).

Micropores between the graphite platelets inside a soot particle and mesopores in the inter-particle cavities of soot agglomerates create the soot porous structure, which may impact upon water uptake in accordance with the pore size and volume. It is found that soot micropores may impact initial water uptake. Even if a single water molecule does not have sufficient dispersive force to adsorb inside the micropore, a cluster of a few water molecules produces sufficient forces to stay within the micropore [82]. Micropore filling by water may lead to swelling of the soot particles [66].

Finally, because of the complicated chemistry and various microstructures of the original combustion particles, there is no universal model of water–soot interaction. But the fundamentals of classical models may help us to perform a quantitative comparison analysis and to distinguish the different mechanisms of water interaction with soot.

5.3.2
Concept of Quantification

Identification of water uptake mechanisms and comparative analysis of isotherms for soots of various composition, from pure EC to complex composites with large water-soluble coverage, allow us to quantify water uptake. Figure 5.4 presents a schematic isotherm plot to demonstrate the *concept of quantification*. For clear atmospheric applications, the dependence on $RH = (p/p_s) \times 100\%$ is presented. The isotherms of the surface water film (Eq. (5.4)) are plotted for radius of soot primary particles of 5 and 125 nm (low and high boundaries of marked area, respectively). They are suggested as a *quantification measure*, which separates

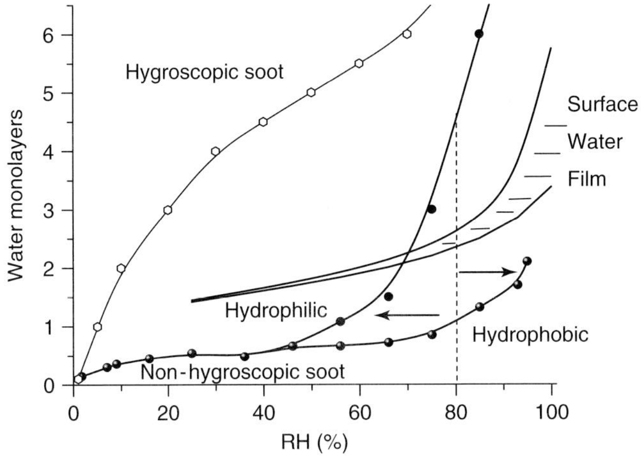

Figure 5.4 Schematic isotherm plot for a *concept of quantification*. The region of existence for a water film on the surface is indicated, with low and high boundaries for 5 nm and 125 nm radius particles, respectively.

hygroscopic from non-hygroscopic soot. If the soot water-soluble fraction is quite large, we may assume hygroscopic soot, the isotherm of which is higher than that for the surface water film on a particle of corresponding size. If the soot particles are made mostly from EC and/or the organic coverage has low water solubility, we may propose *non-hygroscopic soot*, the isotherm of which is less than or maybe just approaching the isotherm for the surface water film.

The number density of active sites on the surface may be a parameter for definition of hydrophilic and *hydrophobic soots* within non-hygroscopic ones. A high value of a_m on the surface of hydrophilic soot leads to water cluster confluence and the formation of a water film extended over the total soot surface. Therefore, the water adsorption isotherm for hydrophilic soot should approach the isotherm (Eq. (5.4)) for the surface water film (see Figure 5.4). To conclude about the hydrophilicity of primary soot particles, the surface water film should be formed at RH $<$ 80%, as typically the capillary condensation of water in inter-particle cavities and *polymolecular adsorption* happens at RH $>$ 80% [63]. But the isotherm of hydrophobic soot never approaches the isotherm for the surface water film due to the low value of a_m (see Figure 5.4); hydrophobic soot particles may be only partly covered by water molecules.

5.3.3
Laboratory Approach for Water Uptake Measurements

Because of the difficulties with *in situ* measurements of the water interaction with airborne soot particles, the laboratory approach is widely utilized. Various laboratory-produced soot surrogates have been introduced for atmospheric studies. Commercially available soots, namely FW1 [24] and lamp soot [23], have been used for CCN and IN studies. *Palas soot* generated by spark discharge between graphite electrodes has recently become popular among researchers because of the high reproducibility of the Palas GFG 1000 generator [25, 26, 87]. Combustion particles produced by burning various fuels in different burners are proposed for hydration and wetting studies [61, 64, 65, 70, 88]. Especially, *CAST soot* produced by the Combustion Aerosol STandard burner has currently become a subject of intensive studies [27], as the CAST generator allows the production of soot under controlled combustion conditions. But it is obvious that one of the best soot surrogates for upper troposphere research is soot produced by the combustor of a typical gas turbine engine operating at cruise conditions [32, 36, 38].

To perform the quantitative comparison analysis of water uptake and demonstrate the concept of quantification, a *set of soots* with a large variety of compositions and structural properties is proposed in this study. It includes spark discharge Palas soot analyzed previously in Kuznetsov *et al.* [87], commercial *thermal soot* produced by pyrolysis of natural gas (Electrougly Ltd, Russia) presented in Popovicheva *et al.* [66], and laboratory-made TC1 kerosene soot produced by burning the TC1 aviation kerosene in the usual type of oil lamp described in Popovicheva *et al.* [66]. To examine the hydrophilic properties of CAST soot, two samples were generated by burning propane at controlled flows of propane and synthetic air, namely at C/O ratio 0.29 and 0.4 as described in Möhler *et al.* [27]. We will term them CAST-4 and CAST-27, respectively, correlating with the OC content found in these samples (which will be defined below). To have the best surrogate for atmospheric applications, original AEC soot produced under typical cruise combustion conditions at the background facilities [54] is included in the set of soots. Additionally, combustor soot produced by burning propane or butane fuel in a gas turbine engine combustor [32] is taken to be available for comparison.

The mean diameter of primary particles, specific surface area, elemental composition, water-soluble fraction (WSF) by mass, and OC content is presented in Table 5.1. The oxygen content as well as the WSF is assumed to correlate with the water adsorption ability of carbonaceous adsorbents. With the purpose to identify the functional groups responsible for the presence of active sites on the soot surface, infrared spectra were recorded using FTIR spectroscopy. A large variety of $-CH$ and $-CH_2$ groups in aromatics and aliphatics as well as oxygen-containing functional groups such as $-C=O$ carbonyl, carboxyl, and $-OH$ hydroxyl were found; they may create hydrogen bonds of different strength, impacting soot hygroscopicity. Additionally, the peaks of HSO_4^- ions and organic sulfates were found in the spectra of AEC soot, which indicates the presence of really hydrophilic compounds on the surface of this soot [38].

The FTIR spectra of CAST soots for 4% and 27% OC content are plotted in Figure 5.5. Different functional groups such as polyaromatic at 836 and 880 cm^{-1}, carbonyl at 1400 cm^{-1}, carboxyl at 1485, 1604, and 1629 cm^{-1}, and hydroxyl at 3201 and 3402 cm^{-1} are observed for CAST soot containing a large amount of organics,

Table 5.1 Physico-chemical characteristics of soots.

Soot type	Mean diameter (nm)	Surface area (m^2 g^{-1})	Composition (wt%)	WSF (wt%)	OC content (%)
Palas	6.6	308	C 100	0	$\sim 0^c$
Thermal	246	10	O 0.4	0.5	0.6^c
TC1 flame kerosene	57	49	C 95	0.9	2.1 ± 1.4^d
			O 5		4^c
CAST-4	300	46	nd[a]	nd	4 ± 4^d
CAST-27	32; 90;140	21	nd	nd	27 ± 7^d
Aircraft EC (AEC)	30–50[b]	6	C	13.5	20 ± 1.5^d
			O, S, Fe, K		17^c
Combustor	30–50	54	C 95	2.0	nd
			O 5		

[a] Not determined.
[b] Two fractions are found in AEC soot; for details see Demirdjian *et al.* [51].
[c] Determined by volatility method [54].
[d] Determined by thermographic method [54].

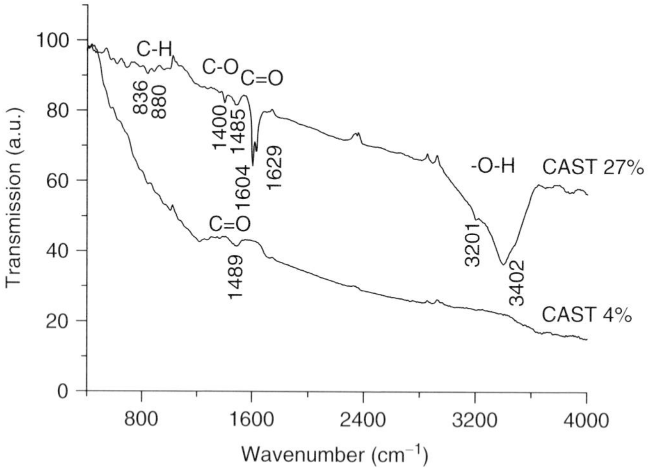

Figure 5.5 FTIR spectra of CAST soots with 4% and 27% of organic carbon. The peaks of intensive absorption bands corresponding to surface functional groups are indicated.

and only the carbonyl group at 1489 cm^{-1} is prominent on surfaces having low organic coverage.

5.3.4 Quantification of Water Uptake

5.3.4.1 Hydrophobic Soot

Water uptake by soot, a, as a function of RH is measured in millimoles per gram of soot by a gravimetric method. To build the absolute adsorption isotherm for comparative analysis, we recalculate the a value into the amount of statistical monolayers assuming the measured specific surface area, S (m^2 g^{-1}), for a given soot. One monolayer is estimated as $1\ ML \approx a \times 60/S$, assuming the cross-sectional area of a water molecule to be 0.105 nm^2.

Spark discharge Palas soot is found to consist of totally elemental carbon with a negligible amount of water-soluble material on the surface but still having some functional groups responsible for the interaction with water. The absolute isotherm of water adsorption on Palas soot is presented in Figure 5.6. The isotherm for Palas soot is always lower than the lower boundary of the marked area for the existence of the water surface film for particles of radius 5 nm marked in Figure 5.6. Therefore, we conclude that the initial water adsorption on the active sites of Palas soot may probably stimulate water cluster growth but cannot lead to the cluster confluence and the formation of a water film extended over the total soot surface. This is a feature of hydrophobic soot in respect of the concept of quantification.

For the quantitative description of the water–soot interaction mechanism as well as for parameterization of water uptake, the application of classical adsorption equations to experimental data is fruitful. For these purposes the BET equation

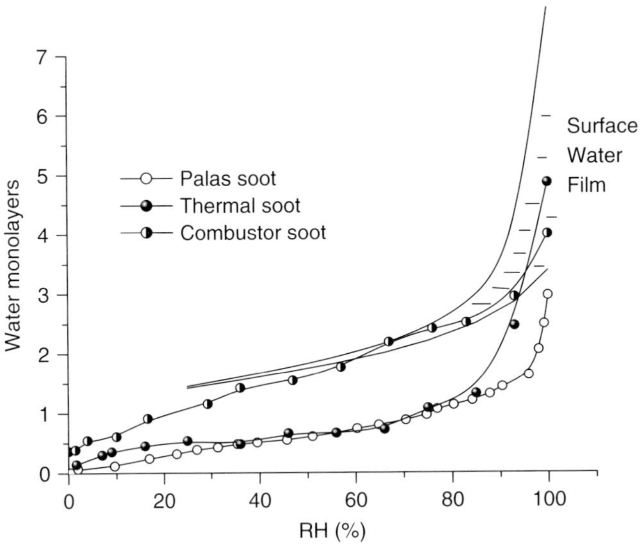

Figure 5.6 Isotherms of water adsorption on hydrophobic Palas and thermal soots, and on hydrophilic combustor soot. The region of existence for a water film on the surface is indicated, similar to Figure 5.4.

(5.2) and the Aranovich equation (5.3) are frequently used in their linear forms:

$$Y_{\mathrm{BET}} = \frac{p/p_\mathrm{s}}{a(1 - p/p_\mathrm{s})} = \frac{1}{a_\mathrm{m} C} + \frac{C-1}{a_\mathrm{m} C}\frac{p}{p_\mathrm{s}} \tag{5.8}$$

$$Y_{\mathrm{AR}} = \frac{p/p_\mathrm{s}}{a(1 - p/p_\mathrm{s})^{0.5}} = \frac{1}{a_\mathrm{m} C} + \frac{p/p_\mathrm{s}}{a_\mathrm{m}} \tag{5.9}$$

The parameters a_m and C found from the linear fits with the highest correlation R^2 as well as the ranges for good application of the BET and Aranovich theories are summarized in Table 5.2. The range for application of the BET equation is found to be up to 0.5, which is even higher than the range generally accepted for the BET theory (up to 0.35). The estimation of the amount of water molecules in a monolayer, a_m, allows the calculation of the surface area covered by water, S_{H_2O}, presented in Table 5.2. As one can see, both theories give values of S_{H_2O} less than the surface area measured by N_2 adsorption – this is a typical feature of hydrophobic soot.

Comparison of the experimental data for water uptake on Palas soot and the isotherms obtained from various equations with parameters a_m and C presented in Table 5.2 leads to the conclusion that, within experimental accuracy, the Aranovich equation gives a better representation of the experimental data over the entire p/p_s range than the BET equation, in accordance with the more developed consideration of adsorbate–adsorbate interactions in the Aranovich theory. Probably the vertical interactions taken into account in the Aranovich theory may address water cluster formation more comprehensively.

Table 5.2 Parameters for fitting of experimental data by the BET and Aranovich (AR) equations.

	Palas soot		Thermal soot		Combustor soot		TC1 soot		AEC soot	
Parameter	BET	AR	BET	AR	BET	AR	BET	AR	BET	AR
R^2	0.981	0.931	0.995	0.996	0.985	0.989	0.995	0.985	0.997	0.994
a_m (mmol g^{-1})	1.08	2.77	0.06	0.08	0.81	1.42	1.44	1.59	2.02	2.46
C	4.84	1.58	62.1	25.8	65.7	7.9	104	49.8	49.14	31.2
Range of application, up to	0.7	0.5	0.5	0.7	0.6	0.76	0.4	0.7	0.4	0.5
S_{H_2O} (m^2 g^{-1})	65	167	3.6	4.6	49	86	86.5	96	122	148

Additionally, the Langmuir isotherm was calculated from Eq. (5.1) using a similar linearization procedure as described above for the BET and *Aranovich isotherms*. It is found that the Langmuir equation does not describe water adsorption on Palas soot at p/p_s higher than 0.1. Therefore, there is no justification for using the Langmuir theory for the description of the water–soot interaction. This is why the attempt of Kärcher *et al.* [21] to apply the Langmuir equation for water uptake on hydrophobic aircraft-generated soot failed.

Gas pyrolysis thermal soot is characterized by the low oxygen content and WSF (see Table 5.1). Functional groups are found on its surface but, because of either their small amount or low strength, low water adsorption on thermal soot is observed, similar to Palas soot up to RH $\approx$ 80% (see Figure 5.6). The increase in adsorption observed at higher RH may take place due to polymolecular adsorption in the mesopores. The water film is never extended over the hydrophobic thermal soot surface because the isotherm does not approach the surface water film isotherm for particles of 250 nm diameter (the higher boundary of the marked area in Figure 5.6). The parameters a_m and C found from a linear fit of the experimental data by the BET equation (5.8) and Aranovich equation (5.9) are presented in Table 5.2.

5.3.4.2 **Hydrophilic Soot**

The oxygen content and water-soluble fraction are found to be higher for combustor soot produced by burning gaseous fuel in the combustor of a gas turbine engine in comparison with those for the hydrophobic soots (see Table 5.1). Accordingly, the isotherm of combustor soot may be perfectly treated in respect of the concept of quantification as a characteristic of hydrophilic soot (see Figure 5.6). It starts increasing at very low RH due to a significantly higher amount of active sites on its surface than on the surface of Palas or thermal soots. Water cluster confluence may be predicted at RH $\geq$ 60% where the isotherm approaches

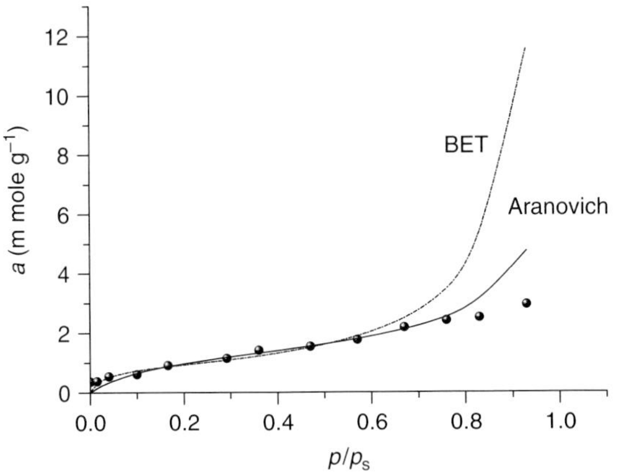

Figure 5.7 Adsorption isotherms of combustor soot: experimental data and fitting curves obtained from the BET and Aranovich equations.

water film formation for particles with 15–25 nm diameter (size for combustor soot particles). However, owing to the relatively small amount of water-soluble surface compounds and, therefore, relatively low water uptake, even in the better case of the formation of a water film on the surface, all Palas, thermal, and combustor soots belong to a class of non-hygroscopic soots, as marked in Figure 5.4.

Figure 5.7 shows the experimental data for water uptake on combustor soot and the fitting isotherms obtained from the BET and Aranovich equations. Equation (5.8) gives a good representation of the data only up to 0.6, while it appears that the Aranovich model is even valid in the range of p/p_S greater than 0.7. This finding proves that the secondary water adsorption mechanism on soot realized as the fluid–fluid cooperative effect can hardly be considered by the BET theory.

Further, the formal quantification may be done for water uptake on TC1 kerosene flame soot. It should be classified as really hydrophilic, or even slightly hygroscopic, soot with respect to the concept since its isotherm lies higher than the upper boundary of the marked area in Figure 5.8. However, there is one suspicious fact, in that the WSF of TC1 soot is noticeably less than the WSF of combustor soot (see Table 5.1). Probably, there is another mechanism impacting water uptake that is different from water solubility of the soot coverage. In the case of TC1 soot, it is the high (in comparison with other soots) microporosity, which, as proved in Popovicheva *et al.* [54], may increase the water adsorption. Formal treatment of experimental data by the BET and Aranovich equations demonstrates the better applicability of the latter equation and only a restricted range (up to $p/p_S \sim 0.3$) for the former. Therefore, we conclude that, for soot with higher hydrophilicity, the BET theory is less applicable.

Special attention should be paid to water uptake of CAST soot since it can be produced in the laboratory study with various and controlled amounts of

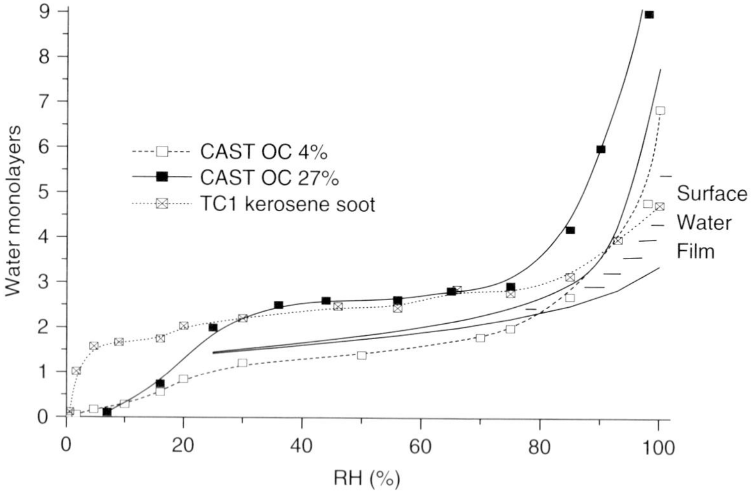

Figure 5.8 Isotherms of water adsorption of TC1 kerosene soot, and on CAST-4 and CAST-27 soots.

organics on the surface. Möhler *et al.* [27] concluded that the increase in the OC content of soot markedly suppresses the ice-nucleation ability of particles. This result led Kärcher *et al.* [35] to conclude that the increase in the OC content makes the flame soot particles more hydrophobic. Measurements of water uptake prove the increase of water adsorption with the increase in the organic content (see Figure 5.8). FTIR spectra (see Figure 5.5) clearly indicate more pronounced features of oxygen-containing groups on the surface with 27% of organics in correlation with the isotherm for CAST-27 soot being higher than the isotherm for CAST-4 soot.

The isotherm for CAST-4 soot with a mean particle diameter of 300 nm does not approach the higher boundary of the area for the existence of a water film indicated in Figure 5.8. Therefore, we should classify CAST-4 as hydrophobic soot. In contrast, a significant increase in water adsorption for CAST-27 soot is observed at RH $\leq 20\%$ (see Figure 5.8); this happens even before the threshold for water film formation. In this case we are definitely sure about the hydrophilic nature of the CAST-27 soot surface and maybe we should consider the hygroscopic character of water uptake in respect of the concept of quantification.

But in the range of initial RH (less than 10%) the isotherms of both CAST soots exhibit negligible uptake, even lower than on hydrophobic Palas and thermal soots. Such behavior is related to the presence of hydrophobic organics on the surface of CAST soots, which are to be replaced by water molecules before the following significant adsorption begins. It is worth noting that TC1 kerosene flame soot has less OC on the surface than CAST soots, only near 2 wt% (see Table 5.1). This demonstrates the significant uptake already at the initial RH, around 5%, and similar adsorption on CAST-27 soot in the RH range from 30 to 80% (see Figure 5.8), probably due to the similar oxidation level of condensable organic compounds on its surface. Such a complex mechanism of water uptake on CAST soots with large organic coverage is beyond the range of applicability of the BET and Aranovich equations; therefore, these equations describe experimental data poorly for values of $p/p_s \geq 0.2$.

5.3.4.3 Hygroscopic Soot

Finally, AEC soot produced by burning TC1 kerosene in the combustor of a gas turbine engine has a really high OC content, $\approx$ 20 wt%, as well as high water-soluble fraction, $\approx$ 13.5 wt%, including 3.5 wt% of sulfates. The water uptake by AEC soot is presented in Figure 5.9. Significant increase up to 10 monolayers does not leave any doubt about the dominant mechanism of water uptake being the dissolution of water into the soluble soot coverage and the formation of a thick solution film surrounding the soot particles.

The role of sulfates in hygroscopicity may be dominant. Thus, AEC soot is the best candidate for being the original hygroscopic soot with respect to the concept of quantification.

The dual sorption theory is proposed above for modeling water uptake on hygroscopic soot. The isotherm obtained using Eq. (5.7) by least-squares fitting has the following parameters: $a_v = 2.1$ mmol g^{-1}, $K = 15$ Torr^{-1}, and $K_d = 0.15$ mmol g^{-1} Torr; it treats measurements well up to 70% RH. The formal application of the BET and Aranovich equations (see Table 5.2) gives very high values of surface areas covered by water, tens of times more than the surface area measured by N_2 adsorption. This finding confirms the absorption mechanism of water uptake on hygroscopic AEC soot, while these classic theories are preferentially applied for adsorption on non-hygroscopic (hydrophobic or hydrophilic) soots.

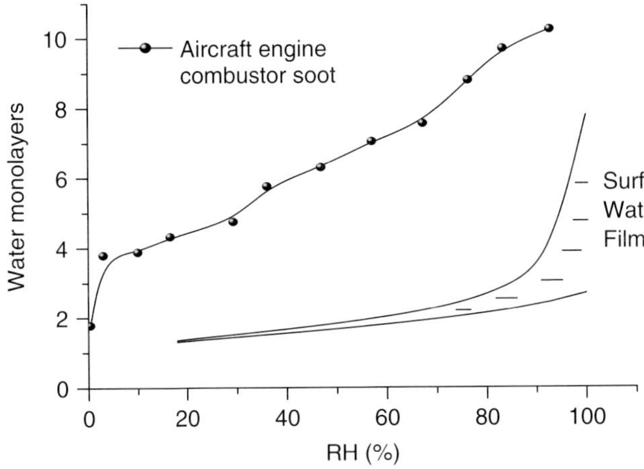

Figure 5.9 Isotherm of water absorption on hygroscopic aircraft engine combustor AEC soot.

5.4 Conclusions

Soot is a complex material composed of elemental carbon covered by organic and/or inorganic compounds. In the general case, two mechanisms of water interaction, namely, bulk dissolution into the water-soluble coverage (*ab*sorption mechanism) and water molecule adsorption on active sites (*ad*sorption mechanism), govern the water–soot interaction. Comparative analysis of water uptake on soots of various compositions, from ECs to complex composites with large water-soluble fractions, allows us to suggest a *concept of quantification*. The isotherm for a water film extended over the surface is proposed as a *quantification measure*, which separates hygroscopic from non-hygroscopic soot. Water uptake on hygroscopic soot significantly exceeds the surface water film formation and may reach tens of water monolayers. If soot particles made mostly from EC and/or the organic coverage are totally water-insoluble, we assume non-hygroscopic soot with water uptake less than (hydrophobic soot) or approaching (hydrophilic soot) the surface water film formation.

Water uptake measurements accompanied by comprehensive soot characterization have shown that AEC soot is representative of hygroscopic soots due to its high water-soluble fraction leading to a multilayer water uptake. Spark discharge Palas soot is a hydrophobic BC because of its chemically pure surface, which supports only low water adsorption due to few active sites. Laboratory-produced TC1 kerosene flame and CAST burner soots may be classified as hydrophilic within non-hygroscopic soots because of the presence of some water-soluble compounds and functionalities on their surface leading to cluster formation and confluence into the water film extended over the surface.

Classical models of water adsorption may be applied for parameterization of experimental data in some ranges of relative pressures. The more adsorbate–adsorbate interactions are developed in the model, the better is its application. But there is no single model that could be universally applied for water uptake on soot of arbitrary extent of hydrophilicity.

The concept of quantification of water uptake by soot particles developed in this study is useful for relating the isotherm data and the cloud condensation soot activity (CCN data). It may allow one to define which combustion soot may be activated with respect to the Kelvin barrier overcome. The formation of a water film extended over the hydrophilic surface is assumed to be needed for wetting that soot particle which becomes activated at water supersaturations in respect of the Kelvin theory. Whether the condition is achieved may be concluded from the adsorption data. In the case of hydrophobic soot, the low adsorption on just a few active sites is not indicative of CCN activation; such a soot will require really high supersaturations to become activated. Further measurements to prove this conclusion are required.

We believe that the concept of quantification of water uptake will also be useful for understanding ice nucleation on soot particles since water–soot interaction is the first step before freezing in any atmospheric systems, whether in a cooling plume or the background atmosphere. The amount of water adsorbed on soot particles and the fraction of frozen water may depend on temperature [38, 86]. But if surface water remains unfrozen at low temperatures, the mechanism of water uptake and water distribution over the surface may be concluded from the concept of quantification.

Acknowledgements

The author is very grateful to Dr N.K. Shonija and Dr N.M. Persiantseva at Moscow State University for a long and fruitful collaboration, numerous analyses, and interest in this work. The financial support from the Integrated Project QUANTIFY (Contract No. **003893 GOGE**), the project CRDF/RFBR 2949/09-05-92506, and the grant of the President of the Russian Federation **SS-3322.2010.2** are gratefully acknowledged.

References

1. International Panel on Climate Change (2001) *Climate Change 2001: The Scientific Basis*, Cambridge University Press, New York.
2. Bond, T., Streets, D., Yarber, K., Nelson, S., Woo, J.-H., and Klimont, Z. (2004) A technology-based global inventory of black and organic carbon emissions from combustion. *J. Geophys. Res.*, **109**, D14203. doi: 10.1029/2003JD003697
3. Bernstein, J.A., Alexis, N., Barnes, C., Bernstein, I.L., Bernstein, J.A., Nel, A., Peden, D., Diaz-Sanchez, D., Tarlo, S.M., and Williams, P.B. (2004) Health effects of air pollution. *J. Allergy Clin. Immunol.*, **114**, 1116–1123.
4. Kirkevag, A., Iversen, T., and Dahlback, A. (1999) On radiative effects of black carbon and sulphate aerosols. *Atmos. Environ.*, **33**, 2621–2635.
5. Chung, S. and Seinfeld, J. (2005) Climate response of direct radiative forcing of anthropogenic black carbon. *J. Geophys. Res.*, **110**, D11102. doi: 10.1029/2004JD005441
6. Chen, Y. and Penner, J. (2005) Uncertainty analysis for estimates of the first indirect aerosol affect. *Atmos. Chem. Phys.*, **5**, 2935–2948.
7. Lohmann, U., Feichter, J., Pebber, J., and Leaitch, R. (2000) Indirect effect of sulfate and carbonaceous aerosols: a mechanistic treatment. *J. Geophys. Res.*, **105** (D10), 12193–12206.
8. Lauer, A., Eyring, V., Hendricks, J., Jockel, P., and Lohmann, U. (2007) Global model simulations of the impact of ocean-going ships on aerosols, clouds, and the radiation budget. *Atmos. Chem. Phys.*, **7**, 5061–5079.
9. Hobbs, P. *et al.* (2000) Emissions from ships with respect to their effects in clouds. *J. Atmos. Sci.*, **57**, 2570–2590.
10. Hendricks, J., Kärcher, B., Dopelheuer, A., Feichter, J., Lohmann, U., and Baumgartner, D. (2004) Simulating the global atmospheric black carbon cycle: a revisit to the contribution of aircraft emissions. *Atmos. Chem. Phys.*, **4**, 2521–2541.
11. DeMott, P.J., Rogers, D.C., and Kreidenweis, S.M. (1997) The susceptibility of ice formation in upper tropospheric clouds to insoluble aerosol components. *J. Geophys. Res.*, **102**, 19575–19584.
12. Chen, Y., Kreidenweis, S.M., McInnes, L.M., Rogers, D.C., and DeMott, P.J. (1998) Single particle analysis of ice nucleating aerosols in the upper troposphere and lower stratosphere. *Geophys. Res. Lett.*, **25**, 1391–1394.

13. Gierens, K. (2003) On the transition between heterogeneous and homogeneous freezing. *Atmos. Chem. Phys.*, **3**, 437–446.
14. Penner, J.E., Chen, Y., Wang, M., and Liu, X. (2008) Possible influence of anthropogenic aerosols on cirrus clouds and anthropogenic forcing. *Atmos. Chem. Phys. Discuss.*, **8**, 13903–13942.
15. Haag, W., Kärcher, B., Strom, J., Minikin, A., Lohmann, U., Ovarlez, J., and Stohl, A. (2003) Freezing thresholds and cirrus formation mechanisms inferred from *in situ* measurements of relative humidity. *Atmos. Chem. Phys.*, **3**, 1791–1806.
16. Schumann, U. (2005) Formation, properties and climatic effects of contrails. *C. R. Phys.*, **6** (4–5), 549–565.
17. Seinfeld, J.H. (1998) Clouds, contrails and climate. *Nature*, **391**, 837–838.
18. Petzold, A., Dopelheuer, A., Brock, C.A., and Schröder, F. (1999) *In situ* observations and model calculations of black carbon emission by aircraft at cruise altitude. *J. Geophys. Res.*, **104** (D18), 22171–22181.
19. Petzold, A., Ström, J., Ohlsson, S., and Schroder, F.P. (1998) Elemental composition and morphology of ice-crystal residual particles in cirrus clouds and contrails. *Atmos. Res.*, **49**, 21–34.
20. Kuhn, M., Petzold, A., Baumgardner, D., and Schröder, F. (1998) Particle composition of a young condensation trail and of upper tropospheric aerosol. *Geophys. Res. Lett.*, **25**, 2679–2682.
21. Kärcher, B., Peter, T., Biermann, U.M., and Schumann, U. (1996) The initial composition of jet condensation trails. *J. Atmos. Sci.*, **53**, 3066–3083.
22. Ström, J. and Ohlsson, S. (1998) *In situ* measurements of enhanced crystal number densities in cirrus clouds caused by aircraft exhaust. *J. Geophys. Res.*, **103**, 11355–11361.
23. DeMott, P.J., Chen, Y., Kreidenweis, S.M., Rogers, D.C., and Sherman, D.E. (1999) Ice formation by black carbon particles. *Geophys. Res. Lett.*, **26**, 2429–2432.
24. Lammel, G. and Novakov, T. (1995) Water nucleation properties of carbon black and diesel soot particles. *Atmos. Environ.*, **29**, 813–823.
25. Kotzick, R. and Niessner, R. (1999) The effect of aging processes on critical supersaturation ratios of ultrafine carbon aerosols. *Atmos. Environ.*, **33**, 2669–2677.
26. Möhler, O., Buttner, S., Linke, C., Schnaiter, M., Saathoff, H., Stetzer, O., Wagner, R., Krämer, M., Mangold, A., Ebert, V., and Schurath, U. (2005) Effect of sulfuric acid coating on heterogeneous ice nucleation by soot aerosols particles. *J. Geophys. Res.*, **110**, D11210. doi: 10.1029/2004JD005169
27. Möhler, O., Linke, C., Saathoff, H., Schnaiter, M., Wagner, R., and Schurath, U. (2005) Ice nucleation on flame soot aerosol of different organic carbon content. *Meteorol. Z.*, **14**, 477–484.
28. DeMott, P. (1990) An exploratory study of ice nucleation by soot aerosols. *J. Appl. Meteorol.*, **29**, 1072–1079.
29. Diehl, K. and Mitra, S.K. (1998) A laboratory study of the effects of a kerosene-burner exhaust on ice nucleation and the evaporation rate of ice crystals. *Atmos. Environ.*, **32**, 3145–3151.
30. Dymaska, M., Murray, B.J., Sun, L., Eastwood, L., Knohf, D.A., and Bertram, A.K. (2006) Deposition ice nucleation on soot at temperatures relevant for the low troposphere. *J. Geophys. Res.*, **111**, D04204. doi: 10.1029/2005JD006627
31. Hallet, J., Hudson, G., and Rogers, C.F. (1989) Characterization of Combustion aerosols for haze and cloud formation. *Aerosol Sci. Technol.*, **10**, 70–83.
32. Popovicheva, O.B., Persiantseva, N.M., Kuznetsov, B.V., Rakhmanova, T.A., Shonija, N.K., Suzanne, J., and Ferry, D. (2003) Microstructure and water adsorbability of aircraft combustor and kerosene flame soots: toward an aircraft-generated soot laboratory surrogate. *J. Phys. Chem. A*, **107**, 10046–10054.
33. Kärcher, B., Busen, R., Petzold, A., Schröder, F., Schumann, U., and Jensen, E.J. (1998) Physicochemistry of aircraft-generated liquid aerosols, soot, and ice particles – 2. Comparison

with observations and sensitivity studies. *J. Geophys. Res.*, **103**, 17129–17147.

34. Schumann, U., Arnold, F., Busen, R., Curtius, J., Kärcher, B., Kiendler, A., Petzold, A., Schlager, H., Schröder, F., and Wohlfrom, K.H. (2002) Influence of fuel sulfur on the composition of aircraft exhaust plumes: the experiments SULFUR 1–7. *J. Geophys. Res.*, **107**, D15. doi: 10.1029/2001JD00813
35. Kärcher, B., Möhler, O., DeMott, P., Pechtl, S., and Yu, F. (2007) Insights into the role of soot aerosols in cirrus cloud formation. *Atmos. Chem. Phys. Discuss.*, **7**, 7843–7905.
36. Gysel, M., Nyeki, S., Weingartner, E., Baltensperger, U., Giebl, H., Hitzenberger, R., Petzold, A., and Wilson, C.W. (2003) Properties of jet engine combustion particles during the PartEmis experiment: hygroscopicity at subsaturated conditions. *Geophys. Res. Lett.*, **30**, 1566.
37. Hitzenberger, R., Giebl, H., Petzold, A., Gysel, M., Nyeki, S., Weingartner, E., Baltensperger, U., and Wilson, C.W. (2003) Properties of jet engine combustion particles during the PartEmis experiment. Hygroscopic growth at supersaturated conditions. *Geophys. Res. Lett.*, **30**, 1779. doi: 10.1029/2003GL017294
38. Popovicheva, O.B., Persiantseva, N.M., Lukhovitskaya, E.E., Shonija, N.K., Zubareva, N.A., Demirdjian, B., Ferry, D., and Suzanne, J. (2004). Aircraft engine soot as contrail nuclei. *Geophys. Res. Lett.*, **31**, L11104.
39. Shonija, N.K., Popovicheva, O.B., Persiantseva, N.M., Savel'ev, A.M., and Starik, A.M. (2007) Hydration of aircraft engine soot particles under plume conditions: effect of sulfuric and nitric processing. *J. Geophys. Res.*, **112**, D02208. doi: 10.1029/2006JD007217
40. Fernandes, M., Skjemstad, J., Johnson, B., Wells, J., and Brooks, P. (2003) Characterization of carbonaceous combustion residues. I. Morphological, elemental and spectroscopic features. *Chemosphere*, **51**, 785–795.
41. Lyyränen, J., Jokiniemi, J., Kauppinen, E.I., and Joutsensaari, J. (1999) Aerosol characterization in medium-speed diesel engines operating with heavy fuel oils. *J. Aerosol Sci.*, **30**, 771–784.
42. Posfai, M., Simonics, R., Li, J., Hobbs, P., and Buseck, P. (2003) Individual aerosol particles from biomass burning in southern Africa: 1. Composition and size distributions of carbonaceous particles. *J. Geophys. Res.*, **108** (D13), 8483. doi: 10.1029/2002JD002291
43. Moldonova, J., Fridell, E., Popovicheva, O., Demirdjian, B., Tishkova, V., Faccinetto, A., and Focsa, C. (2009) Characterisation of particulate matter and gaseous emissions from a large ship diesel engine. *Atmos. Environ.*, **43**, 2632–2641.
44. Bockhorn, H. (1994) *Soot Formation in Combustion: Mechanisms and Models*, Springer, Berlin.
45. Marsh, H. (1989) *Introduction to Carbon Science*, Butterworths, London.
46. Mikhailov, E.F., Vlasenko, S.S., Kiselev, A.A., and Ryshkevich, T.I. (1998) Restructuring factors of soot particles. *Atmos. Ocean Phys.*, **34**, 307–317.
47. Schneider, J., Weimer, S., Drewnick, F., Borrmann, S., Helas, G., Gwaze, P., Schmid, O., Andreae, M., and Kirchner, U. (2006) Mass spectrometric analysis and aerodynamic properties of various types of combustion-related particles. *Int. J. Mass Spectrom.*, **258**, 37–49.
48. Burtscher, H. (2005) Physical characterization of particulate emissions from diesel engines: a review. *Aerosol Sci.*, **36**, 896–932.
49. Clague, A.D.Y., Donnet, J.B., Wang, T.K., and Peng, J.C. (1999) A comparison of diesel engine soot with carbon black. *Carbon*, **37**, 1553–1565.
50. Hays, M. and Vander Wal, R. (2007) Heterogeneous soot nanostructure in atmospheric and combustion source aerosols. *Energy Fuels*, **21**, 801–811.
51. Demirdjian, B., Ferry, D., Suzanne, J., Popovicheva, O.B., Persiantseva, N.M., and Shonija, N.K. (2007) Heterogeneities in the microstructure and composition of aircraft engine combustor soot: impact on the water uptake. *J. Atmos. Chem.*, **56**, 83–103.

52. Kittelson, D. (1998) Engines and nanoparticles: a review. *J. Aerosol Sci.*, **29**, 575–588.
53. Xie, Z., Blum, J., Utsunomiya, S., Ewing, R.C., Wang, X., and Sun, L. (2007) Summertime carbonaceous aerosols collected in the marine boundary layer of the Arctic Ocean. *J. Geophys. Res.*, **112**, D02306. doi: 10.1029/2006JD007247
54. Popovicheva, O.B., Persiantseva, N.M., Tishkova, V., Shonija, N.K., and Zubareva, N.A. (2008) Quantification of water uptake by soot particles. *Environ. Res. Lett.*, **3**, 025009. doi: 10.1088/1748-9326/3/2/025009
55. Grieco, W.J., Howard, J.B., Rainey, L.C., and Vander Sande, J.B. (2000) Fullerenic carbon in combustion-generated soot. *Carbon*, **38**, 597–614.
56. Sadezky, A., Muckenhuber, H., Grothe, H., Niessner, R., and Poeschl, U. (2005) Raman microspectroscopy of soot and related carbonaceous materials. Spectral analysis and structural information. *Carbon*, **43**, 1731–1742.
57. di Stasio, S. (2001) Electron microscope evidence of aggregation under three different size scales for nanoparticles in flame. *Carbon*, **39**, 109–118.
58. Kasper, A., Aufdenblatten, S., Forss, A., Mohr, M., and Burtscher, H. (2007) Particulate emissions from a low-speed marine diesel engine. *Aerosol Science Technology*, **41**, 24–32.
59. Gelencser A. (2004) *Carbonaceous Aerosols*, Springer, Berlin.
60. Pruppacher, H.K. and Klett, J.D. (1978) *Microphysics and Clouds Precipitation*, Reidel, Dordrecht.
61. Persiantseva, N.M., Popovicheva, O.B., and Shonija, N.K. (2004) Wetting and hydration of insoluble soot particles in the upper troposphere. *J. Environ. Monit.*, **6**, 939–945.
62. Gleistmann, G., and Zellner, R. (1998) A modelling study of the formation of cloud condensation nuclei in the jet regime of aircraft plume. *J.Geophys. Res.*, **103**, 19543–19556.
63. Gregg, S.J. and Sing, K.S.W. (1982) *Adsorption, Surface Area and Porosity*, 2nd edn, Academic Press, New York.
64. Chughtai, A.R., Brooks, M.E., and Smith, D.M. (1996) Hydration of black carbon. *J. Geophys. Res.*, **101**, 19505–19516.
65. Chughtai, A.R., Miller, N.J., and Smith, D.M. (1999) Carbonaceous particles hydration III. *J. Atmos. Chem.*, **34**, 259–279.
66. Popovicheva, O.B., Persiantseva, N.M., Shonija, N.K., DeMott P., Köhler, K., Petters, M., Kreidenweis, S., Tishkova, V., Demirdjian, B., and Suzanne, J. (2008) Water interaction with hydrophobic and hydrophilic soot particles. *Phys. Chem. Chem. Phys.*, **10**, 2332–2344. doi: 10.1039/b718944n
67. Crouzet, Y. and Marlow, W. (1995) Calculations of equilibrium vapor pressure of water over adhering 50–200 nm spheres. *Aerosol Sci. Technol.*, **22**, 43–56.
68. Weingartner, E., Burtscher, H., and Baltensperger, U. (1997) Hygroscopic properties of carbon and diesel soot particles. *Atmos. Environ.*, **31**, 2311–2322.
69. Popovicheva, O.B., Persiantseva, N.M., Trukhin, M.E., Shonija, N.K., Buriko, Y.Y., Starik, A.M., Demirdjian, B., Ferry, D., and Suzanne, J. (2000) Experimental characterization of aircraft combustor soot: microstructure, surface area, porosity, and water adsorption. *Phys. Chem. Chem. Phys.*, **2**, 4421–4426.
70. Alcala-Jornod, C., van der Bergh, H., and Rossi, M. (2002) Can soot particles emitted by airplane exhaust contribute to the formation of aviation contrails and cirrus clouds? *Geophys. Res. Lett.*, **29**, 1820. doi: 10.1029/2001GL014115
71. Popovicheva, O.B., Persiantseva, N.M., Trukhin, M.E., and Shonija, N.K. (2001) Water adsorption on aircraft combustor soot under plume conditions. *Atmos. Environ.*, **35**, 1673–1676.
72. Fine, P.M., Chakrabarti, B., Krudysz, M., Schauer, J.J., and Sioutas, C. (2004) Diurnal variation of individual organic compound constituents of ultrafine and accumulation mode particulate matter in the Los Angeles Basin. *Environ. Sci. Technol.*, **38**, 1296–1304.
73. Schauer, J.J., Kleeman, M.J., Cass, G., and Simoneit, B. (1999) Measurement of emissions from air pollution sources. 2. C_1 through C_{30} organic compounds

from medium duty diesel trucks. *Environ. Sci. Technol.*, **33**, 1578–1587.

74. Petters, M.D. and Kreidenweis, S.M. (2007) A single parameter representation of hygroscopic growth and cloud condensation nucleus activity. *Atmos. Chem. Phys. Discuss.*, **6**, 8435–8456.

75. Koehler, K., DeMott, P., Kreidenweis, S., Popovicheva, O., Petters, M., Carrico, C., Kireeva, E., Khokhlova, T., and Shonija, N. (2009) Cloud condensation nuclei and ice nucleation activity of hydrophobic and hydrophilic soot particles. *Phys. Chem. Chem. Phys.*, **11**, 7906–7920.

76. Popovicheva, O.B., Persiantseva, N.M., Kireeva, E.D., Shonija, N.K., Demirdjian, B., Tishkova, V., and Ferry, D. (2008) Ship particulate pollutants: characterization and interaction with environment. Proceedings of the 2nd International Conference on Harbours, Air Quality and Climate Change, 29–30 May, Rotterdam, the Netherlands, Vol. 1, pp. 8–14.

77. Aranovich, G. (1988) Principal elaboration of isotherm for polymolecular adsorption. *Russ. J. Phys. Chem.*, **92**, 3000–3008.

78. Aranovich, G. and Donohue, M. (1995) A new approach to analysis of multilayer adsorption. *J. Colloid Interface Sci.*, **173**, 515–523.

79. Dubinin, M.M. (1980) Water vapor adsorption and the microporous structures of carbonaceous adsorbents. *Carbon*, **18**, 355–364.

80. Müller, E.A., Rull, L.F., Vega, L.F., and Gubbins, K.E. (1996) Adsorption of water on activated carbon: a molecular simulation study. *J. Phys. Chem.*, **100**, 1189–1198.

81. Collignon, B., Hoang, P.N.M., Picaud, S., and Rayez, J.C. (2005) Clustering of water molecules on model soot particles: an *ab initio* study. *Comput. Lett.*, **1**, 277–287.

82. Furmaniak, S., Gauden, P., Terzyk, A.P., and Rychlicki, G. (2008) Water adsorption on carbons – Critical review of the most popular analytic approaches. *Adv. Colloid Interface Sci.*, **137**, 82–143.

83. Berezin, G.I., Vartapetian, R., Voloshuk, A.M., Petuchova, G.A., and Polakov, N.C. (1998) Model of two-stage condensation mechanism of water adsorption on nonporous carbonaceous adsorbents. *Russ. Chem. Bull.*, **47**, 1879–1884.

84. Berezin, G.I. (1995) Model of polymolecular adsorption as a process of extended liquid formation. *Russ. J. Phys. Chem.*, **69**, 304–310.

85. Paul, D.R. and Koros, W.J. (1976) Effect of partially immobilizing sorption on permeability and the diffusion time lag. *J. Polym. Sci., Polym. Phys.*, **14**, 675–677.

86. Ferry, D., Suzanne, J., Nitsche, S., Popovitcheva, O.B., Shonija, N.K. (2002) Water adsorption and dynamics on kerosene soot under atmospheric conditions. *J. Geophys. Res.*, **107**, 4734–4744.

87. Kuznetsov, B.V., Rakhmanova, T.A., Popovicheva, O.B., and Shonija, N.K. (2003) Water adsorption and energetic properties of spark discharge soot: specific features of hydrophilicity. *Aerosol Sci.*, **34**, 1465–1479.

88. Alcala-Jornod, C. and Rossi, M. (2004) Chemical kinetics of the interaction of H_2O vapor with soot in the range 190 K $\leq T \leq$ 300 K: a diffusion tube study. *J. Phys. Chem.*, **108**, 10667–10680.

6
Radioactive Aerosols – Chernobyl Nuclear Power Plant Case Study

Boris I. Ogorodnikov

6.1
Introduction

On the night of 26 April 1986, the greatest accident in nuclear-power engineering took place at the fourth power-generating unit of the Chernobyl Nuclear Power Plant (ChNPP). The active zone and the upper part of the reactor building were completely destroyed as a result of the explosion. The safety barriers and systems protecting the environment against the *radionuclides* produced in the irradiated nuclear fuel (*uranium dioxide*) over the previous two years were also destroyed.

The scientific literature contains many versions of the event: how the accident developed and how the active zone was destroyed [1]. The amount of nuclear fuel and its *fission products* released outside the power-generating unit have been discussed. However, as for any explosion, the destroyed building constructions, fragments of the reactor's active zone, and finely dispersed uranium dioxide (the "hot" fuel particles) remain as silent witnesses to the accident. Condensed micrometer and submicrometer aerosols were spread over the whole Northern Hemisphere.

All the fuel assemblies (FAs) were destroyed, and most fuel element cans (FECs) proved to be empty. Mechanical or thermal degradation of the cans resulted in the formation of fuel dust. According to expert estimates, the total amount of uranium dioxide in the form of fine dust in the upper parts of the fourth power-generating unit make up about 30 tonnes [2].

To prevent further contamination of the region around the ChNPP and the environment as a result of any migration of radionuclides, a shelter of concrete and metal construction was built on top of the ruins of the reactor building; it was brought into commission on 30 November 1986. This construction object was named the "Shelter." Observation of *radioactive aerosols* of Chernobyl genesis both in the vicinity of the ChNPP and inside the destroyed unit itself remains a priority for the provision of radiation safety and understanding of the processes going on in the "Shelter," including the evaluation of the state of the remaining nuclear fuel

Aerosols – Science and Technology. Edited by Igor Agranovski

ISBN: 978-3-527-32660-0

and the lava-like fuel-containing materials (LFCMs) formed from molten nuclear fuel and construction materials.

To reduce the irradiation of building workers, all the works were performed by a remote method, which did not allow the creation of a leakproof construction. There were very many cracks in the upper part of the construction. Besides, operational openings and hatches were made in the roof to insert gage heads into the central hall above the reactor "ruins." Finally, there is designed exhaust ventilation through the high-rise pipe HRP-2. Thus, intensive air exchange with the environment takes place in the "Shelter," and external meteorological conditions influence the state of the air (the air temperature, humidity, direction, and rate of movement, and so on) indoors.

At present, radioactive aerosols from two sources can be emitted from the "Shelter": (i) those formed during the accident and being inside the "Shelter" in the form of dust; and (ii) new ones generated in the process of physico-chemical degradation of fuel-containing materials (FCMs), including the remaining uranium dioxide. The processes of FCM degradation occur under the influence of natural (radiation fields, air humidity, temperature variation, and so on) and man-made factors (any works performed in the "Shelter"). These processes include mechanical embrittlement of FCM resulting in dust formation, radionuclide leaching from uranium dioxide matrix, and so on.

The radiological danger of aerosol products of the Chernobyl accident relates to the presence of highly toxic and long-lived fission and trans-uranium isotopes, in particular, plutonium isotopes. It is necessary to track the lowest concentrations of radioactive substances. Therefore, all the used methods of *aerosol radiometry* are based on their preliminary extraction from the air by any method followed by measurement in concentrated form. This is achieved either by forced pumping of air through filters or its expulsion through traps of different designs due to wind dynamic pressure or by gravitational settling on the surface (plates).

Pollution of air masses inside and outside the "Shelter" can be the result of a number of processes:

1) rise of dust from the soil, main traffic arteries, and vegetation in the vicinity of the ChNPP;
2) rise of dust from the surface of the "ruins" in the central hall, the reactor space, and other premises of the "Shelter";
3) dust formation in the process of construction and erection works;
4) dust formation and rise as a result of falling structural components in the "Shelter";
5) degradation of FCM as a result of radioactive processes and aging of materials;
6) leaching of radioactive substances, drying of solutions, formation and dusting of salt sediments.

The evaluation of the effect of the "Shelter" on the environment is a complex and multi-factor problem. As before, one of the main sources of potential *radioecological danger* from the destroyed fourth power generating unit (hereafter reactor 4) of ChNPP is the process of air migration of radionuclides. It is important to know the *radionuclide composition* of aerosols, their concentration, the size distribution of particles (dispersity), the place of and reasons for the generation of aerosols, their methods of transport, deposition, and dissolution in the human respiratory system, and the efficiency of use of individual and collective protective devices.

Before we start to examine the aerosols related to the "Shelter," let us explain what radioactive aerosols are and describe some of their specific properties. Radioactive aerosols are aerodisperse systems of solid or liquid particles suspended in the air or in other gaseous media and consisting wholly or partially of radioactive substances. The radioactive component can be evenly distributed within the particle volume, can be on its surface, or can occupy a certain permanent or non-permanent place inside and/or on the surface.

Radioactive aerosols are characterized by all the features typical of common aerodisperse systems. However, the presence of *ionizing radiation* imparts a number of specific properties. These include, first of all, the presence of *electrical charges* on the particles and ionization of both the material of the particles and the gaseous medium itself. Secondly, there is the possibility for spontaneous formation of nanoparticles (clusters) at radioactive decay of gaseous substances, for example, radon, thoron, xenon, krypton, and so on, as well as its daughter products. At radioactive alpha decay, for example, transuranium isotopes may generate recoil atoms. Thirdly, there is the possibility of detecting and studying aerosol particles as a result of their radioactive emission.

Radioactive aerosols are generated during various processes in the treatment of materials containing radioactive substances: fragmentation, grinding, bolting, pouring into another container, heating of solids, boiling, evaporation, pouring and bubbling of fluids, chemical interaction of substances in the gas phase, and so on. In the atomic industry, the formation of radioactive aerosols occurs mainly in the following situations: in the mechanical, metallurgical, and chemical treatment of ores (in particular, uranium ones), in the production of uranium hexafluoride, in the gas flows used to cool nuclear reactors, during radiochemical isolation of plutonium, at incineration and wet treatment of radioactive waste, and in the apparatus and plants of research laboratories. Large amounts of radioactive aerosols are generated in nuclear weapons trials, especially during explosions in the atmosphere. Accidents at nuclear power plants result in serious consequences and environmental contamination with radioactive aerosols.

All the above mechanisms of radioactive aerosol formation are special cases of dispersion and condensation processes causing the generation of common aerosols. Consequently, they have been sufficiently well investigated. However, there is a specific mechanism of generation of radioactive aerosols requiring special attention, as can be observed in the case of products of the Chernobyl disaster, that is, the appearance of radioactive recoil atoms or clusters after alpha decay in air. This mechanism was discovered at the beginning of the twentieth century soon after the discovery of radioactivity. In the case of aerosols of Chernobyl genesis, this mechanism can be observed near surfaces of lava-like materials, which appeared in 1986 in the lower part of the destroyed reactor and in the sub-reactor premises as a result of fusion and spread of the remaining nuclear fuel. Almost a quarter of a century after the accident, these lava-like materials contain, besides uranium, also alpha-emitting radioisotopes of plutonium ($^{238,239,240}Pu$), americium (^{241}Am), and curium (^{244}Cm).

A nuclear explosion in the atmosphere when the fireball does not touch the ground surface is an example of the condensation mechanism for generation of radioactive aerosols. In the epicenter of the explosion, the temperature goes up to 1000 000 °C. As the result, all nuclear reaction products and the components of mechanical constructions are immediately vaporized. In 0.01 s the temperature goes down to several thousand degrees. The formation of condensation aerosols of submicrometer size begins. If the explosion was near the ground surface, the rising and expanding fireball draws in a considerable amount of soil. This soil partially vaporizes and then condenses together with the products of the fired object. Some soil particles, for example, in the base of the "atomic cloud," become radioactive because of the deposition of explosion products on them. A number of chemical elements present in the soil become radioactive as a consequence of neutron irradiation. The study of aerosols generated in atmospheric nuclear explosions performed in 1950–1960 detected individual highly active particles named "hot" particles. Similar particles formed in the Chernobyl accident were detected both in the surroundings and at long distances, including the countries of continental Europe and Scandinavia.

Since the cessation of nuclear weapons trials in the atmosphere (the last such explosion was performed in China in 1980), radioactive products have been continually settling on the ground surface. By the end of the twentieth century, the atmosphere had practically cleared itself of the radioactive aerosols generated during nuclear trials. Against this background, radioactive aerosols formed in the Chernobyl accident and released into the *troposphere* and the *stratosphere* were distinctively detected in the Northern Hemisphere in 1986 and later. No aerosols of Chernobyl genesis have been found in the Southern Hemisphere. This is caused by the specificity of air exchange between the hemispheres of the Earth, which was revealed in the observations of radioactive products of nuclear trials already by the middle of the twentieth century.

Carriers of natural cosmogenic radionuclides occupy a special place among radioactive aerosols, which can be classified under the category of condensation ones. They are ^{3}H (tritium), ^{7}Be, ^{14}C, ^{22}Na, ^{24}Na, ^{32}P, ^{33}P, ^{35}S, and so on. They appear as a result of nuclear reactions in the upper atmospheric layers under the influence of neutrons and protons of cosmic radiation. The radioactive atoms formed settle on small atmospheric particles and then slowly get into the lower tropospheric layers. As the above-listed radionuclides have half-lives from several hours to several years, they can be used successfully to study atmospheric processes. They figure as tracers in many works on *nuclear meteorology*.

During the last decades of the twentieth century, the attention of radioecologists was drawn to radon gas and its daughter products (DPs). This was preceded by the discovery of an association between radon concentration and cancers of the lungs and upper respiratory tract in uranium miners and other underground workers. High radon concentrations can also be observed in cellars and on the first floors of dwellings and offices, where radon penetrates from the ground through cracks, poorly sealed joints, and leakage. This gas is known to appear as a result of the radioactive decay of ^{238}U. It was shown that the inert gas itself is not as dangerous as its short-lived DPs, especially alpha-emitting ones. As they appear in radon decay in the form of polonium, bismuth, and lead atoms, and have high diffusion coefficients, they deposit in the upper respiratory tract. If the air is dusty, the formed atoms settle on small atmospheric particles and penetrate into deep sections of the lungs (alveoli and primary bronchi).

The problem of radon and its DPs exists also in the "Shelter" of the ChNPP. Its sources are the huge masses of concrete and the sandy ground in which the foundations of the destroyed reactor 4 are embedded. The "Shelter" has no forced ventilation. Besides radon, the premises of the "Shelter" also emanate thoron, which is partially generated in the decay of the ^{232}U formed in the nuclear fuel of the reactor during its almost 850 days of working before the accident.

Some radioactive substances can exist in the atmosphere and in production equipment in the forms of aerosols and gases. These include mercury, polonium, iodine, ruthenium, tellurium, and so on. After the Chernobyl accident, one of the most biologically dangerous radionuclides, ^{131}I, was observed in the atmosphere in both phases for two months. The portion of radioiodine in aerosol form varied from 10% to 100%. In the gas phase, this radionuclide was present in the form of molecular iodine (I_2) and methyl iodide (CH_3I) vapors. In this connection, sorption-filtering materials entrapping both phases should be used for protection against radioiodine and for analytical purposes.

Radioactive aerosols are used for medical purposes and in scientific research, for example, in the above-mentioned nuclear meteorology. Their main advantage over other methods is the possibility of exact measurement using different radiometers. Specific methods of measurement of radioactive substances, especially alpha-emitting ones, include autoradiography and solid-state detection. One of advantages of autoradiography is the possibility of visualizing, isolating, and analyzing individual particles. This was done with aerosol samples collected at air nuclear trials for the purpose of isolation of "hot" particles and with samples collected after the Chernobyl accident.

Below are the results of observations and studies of aerosols in the "Shelter" and its vicinity performed in 1986–2008. *Petryanov filters*, which were specially developed for radioactive aerosol studies (for a detailed description of these filters, see Part 3 of this book), were used for collection and analysis of aerosol samples throughout the entire duration of this study [3, 4].

6.2
Environmental Aerosols

6.2.1
Dynamics of Release of Radioactive Aerosols from Chernobyl

Estimates of the daily release of radioactive substances (Figure 6.1) are included in "Information on the accident at the Chernobyl NPP and its consequences" [1] presented by the USSR State Committee on the Use of Atomic Energy for the Meeting of IAEA experts, which took place on 25–29 August 1986 in Vienna.

The dynamics of the release has been widely commented upon in the literature [5–9]. Here is, for example, an extract from [8]:

Radioactive substances were released into the atmosphere during the first ten days after the accident before the release was stopped. The heat of the fire increased the levels of released radioactive iodine (^{131}I and ^{133}I), a considerable portion of volatile elements from the group of metals including radioactive cesium (^{134}Cs and ^{137}Cs) as well as a somewhat smaller amount of other radionuclides that are usually present in the fuel of a reactor, which has been working for several years.

This release was not a one-time event. On the contrary, only 25% of material was released on the first day of the accident, the remaining amount was released during the subsequent nine days (Figure 6.1). The curve of the release intensity depending on the time can be divided into four areas:

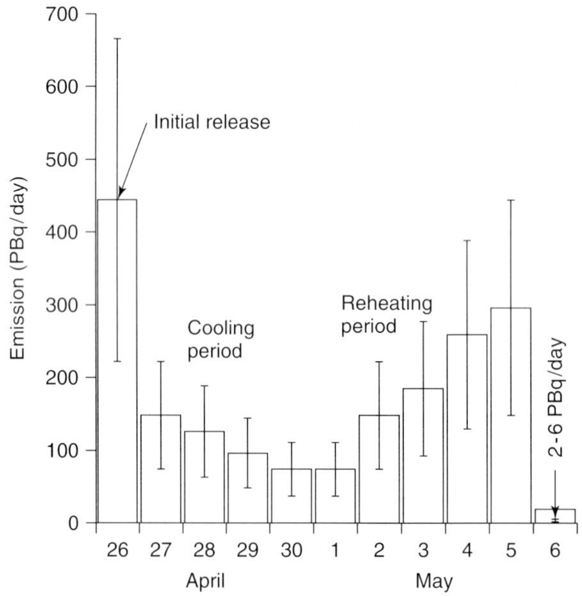

Figure 6.1 Emission of radioactive aerosols to the atmosphere during the 1986 Chernobyl disaster (24 h average with radioactive decay taken into account).

1) The first release took place on the first day of the accident. During this period, the physical release of radioactive materials was the result of explosion in the reactor and the subsequent heating as a result of the fire and the effect of the active zone.
2) During the subsequent five days, the release intensity decreased to a minimum approximately six times smaller as compared with the initial level. At this stage, the decrease in the release intensity was achieved due to measures to control graphite combustion and due to the reactor cooling. These measures, including the casting of 5000 t of boron carbide, dolomite, clay and lead from helicopters to the active zone resulted in the filtration of radioactive substances released from the active zone. At this moment, a release of fine-dispersed fuel took place immediately together with the flow of hot air and smoke from burning graphite.
3) After that there was a period of four days during which the release intensity increased again up to 70% of the initial level. First the release of volatile components, especially iodine was observed; the subsequent composition of radionuclides reminded a typical composition of spent fuel. This phenomenon was attributed to the fuel heating in the active zone to 2000 °C, which was caused by residual heat generation and isolation with materials cast down to the active zone.
4) Ten days after accident there was a sharp reduction in the release intensity to less than 1% of the initial one. Subsequently, the release intensity was constantly decreasing. The last stage, which started on May 6, was characterized by a sharp decrease in fission yield and gradual cessation of release. These events were the consequence of special measures, which resulted in binding of fission products into more stable chemical compounds [4, 10].

The dynamics of the release of radioactive substances from the destroyed reactor 4 is one of the key aspects of the acute phase of the accident. A review of the initial materials [4] used to evaluate the radioactive release in the 1986 accident report presented by Soviet experts for IAEA was published by Yu.V. Sivintsev and A.A. Khrulev (two researchers of the Kurchatov Institute – Russian National Center) in the journal *Atomic Energy* in 1995 [11].

Many authors have tried to reconstruct the *release dynamics* and to calculate the amount of released radioactive aerosols, including ^{137}Cs and especially radioiodine. Models of *atmospheric transport* that were available before the Chernobyl accident, for example, the ARAC system in the USA [12], were used, and new ones were created for different spatial and time scales. A review of the comparison of results obtained with the models used is presented in [13]. The authors of the review state that the results of reconstruction of the total amount of ^{137}Cs released with the method of solution of the inverse problem of atmospheric transport coincide quite well with each other in spite of the differences between the models. They also coincide with the results of similar evaluations performed with other methods. The differences in the reconstruction of the release dynamics are more noticeable. This is associated with the fact that it is much more difficult to calculate the spatial features of the formation of radioactive contamination fields in individual periods at the initial stage of the accident. That is why the quality of the model – that is, the detailed description of the radionuclide dispersion and deposition processes as well as the completeness and reliability of meteorological data – plays the decisive role in the reconstruction of the dynamics of nuclide activity in a given territory.

6.2.2
Transport of Radioactive Clouds in the Northern Hemisphere

The data presented in Figure 6.1 have been used in many works, for example, for the calculation of the transport of gas–aerosol discharge from the ChNPP in the Northern Hemisphere [14–16]. The heights to which they rise in the atmosphere are discussed. The results of sampling carried out from an airplane above Poland show that a portion of the radioactive aerosols was in the stratosphere [17]. As a result of the large thickness of the cloud, transport of the initial products of the explosion at different heights occurred in both the westerly and easterly directions. The monograph [6] presents the trajectories of air-mass transfer at different heights for a week after the accident.

The winds carried the torn-apart initial radioactive cloud from above the ChNPP at comparatively low height (about 3–4 km) through northwest Europe. From [6, 18–21] it follows that two or three days after the accident this part of the cloud passed above Belarus, Lithuania, Poland, the Baltic Sea, Sweden, and Finland. According to [22], air masses reached the southwestern part of Finland at about noon on 27 April 1986 (at heights of 1500 and 2000 m) and three hours later at lower heights (1000 and 1500 m). The next day, daily aerosol samplings from an airplane were started at these heights [23]. According to the data of the Finnish Center for Radiation and Nuclear Safety [24, 25], the maximal surface concentration of ^{137}Cs of about 10 Bq/m^3 was recorded on 28 April in the south of the country in Nurmiyarvi region. No other large amounts of Chernobyl aerosols coming to Finnish territory were observed.

In the report [26] it was noted that, after leaving Scandinavia, the radioactive cloud reached the North Atlantic on 2 May. At the monitoring point in Resolute in the far north of Canada, the first increase of air *activity concentration* was recorded on 2–3 May. In the sample collected here over 24 hours the aerosol concentration already exceeded the background level by five times. Some 4000 km to the southeast, on the east coast of Canada, the first noticeable increase in the air activity concentration took place in Greenwood, Halifax, Fredericton, and Digby with the filters exposed on 4–5 May. As follows from Figure 6.2, the maximal concentrations of radioactive aerosols were recorded on 10–12 May.

Other winds carried radioactive substances, which rose at the reactor explosion to heights of 5–7 km and even into the stratosphere, above Siberia. For instance, in Novosibirsk, a sharp activity increase was recorded on the night of 30 April–1 May [27]. Radioisotopes of both volatile substances (^{134}Cs, ^{137}Cs, ^{103}Ru, ^{106}Ru) and substances of low volatility (^{95}Zr, ^{95}Nb, ^{141}Ce, ^{144}Ce, ^{90}Sr) were found in the aerosols. Then these winds transported radioactive clouds above China and Japan

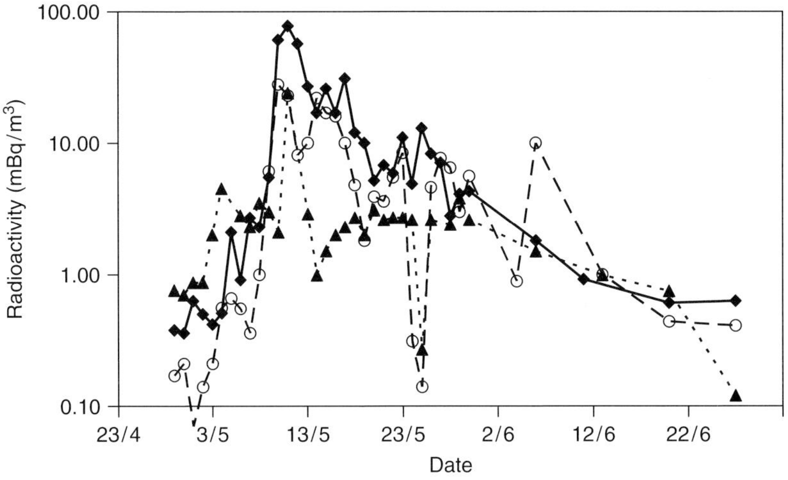

Figure 6.2 Radioactivity of beta-emitting mixtures in aerosol samples in Canada in April–June 1986. The results were obtained at Digby (open circles), Resolute (full triangles), and Edmonton (full diamonds) monitoring stations.

and brought them across the Pacific Ocean to the west coast of the USA and Canada. The authors of the report [26] think that the head of the radioactive cloud reached the west coast of Canada on 7 May. In any case, in Vancouver (the most westerly point of the Canadian monitoring network), an increased activity concentration of aerosols was observed in the sample exposed on 5–6 May, and quantitatively ^{131}I, ^{134}Cs, and ^{137}Cs were detected in the sample collected on 8–9 May.

However, in the southwestern part of Canada, that is, in Winnipeg, Saxatoon, Calgary, and Edmonton, the products of the accident were recorded for the first time somewhat earlier (4–5 May). Most probably, they were brought there through the northwestern part of the USA.

Thus, Canada turned out to be a region of the Northern Hemisphere where radioactive clouds of the first release from the destroyed reactor 4 torn apart by winds above ChNPP came together and mixed. This meeting and mixing of radioactive clouds above Canada, clouds which after the reactor explosion spread in opposite directions round the globe, can be considered one of the peculiarities of Chernobyl aerosol release. Another peculiarity consisted in the fact that the analysis of samples collected in Canada did not reveal detectable quantities of the low-volatility radionuclides ^{95}Zr, ^{95}Nb, ^{140}Ba, ^{140}La, ^{141}Ce, and ^{144}Ce, though the volatile ^{103}Ru, ^{131}I, ^{132}Te, ^{134}Cs, ^{136}Cs, and ^{137}Cs were detected. ^{131}I was in both aerosol and gaseous forms. The portion of the latter varied from 20% to 100%.

One can judge the significance of aerosol fallout from the atmosphere and its contribution to environmental pollution by the following extract from the monograph [28]: "Ingress of ^{137}Cs with waters of the Dnepr and the Danube to the Black Sea in 1986–2000 was extremely insignificant and made up approximately 1% of its atmospheric fallouts." The authors of [29–32] came to the same conclusion.

No radioactive aerosol products of the Chernobyl accident were found in the Southern Hemisphere.

6.2.3
Observation of Radioactive Aerosols above Chernobyl

Sampling of gas–aerosol radioactive substances in the vicinity of the ChNPP and along transfer routes was started the day after the accident. In [33, 34] it is reported that on the night of 27–28 April 1986 the first operative sample was collected above the ruins of reactor 4 of ChNPP from an An-24rr airplane laboratory. Then helicopters began to work together with airplanes.

However, the first (and we can say random) aerosol sample from the plume of the radioactive cloud was collected on the day of the accident (26 April) during the duty flight of an An-30rr airplane of the USSR State Hydrometeorological Committee above the western part of the European territory of the country. Having landed on one of the aerodromes, the operators discovered that the gondola with FPP-15-1.5 filtering material, which had been working during the flight, had a strong radiation background. The intensity of the exposure dose amounted to several R-units per hour. This was reported to the leadership of the USSR State Hydrometeorological Committee. The sample was promptly taken to Kiev and measured on a gamma-spectrometer. The results obtained were included in the report presented to IAEA [9].

The results of measurements of isotope composition and concentrations of radioactive aerosols collected from 28 April to 19 May 1986 with filters installed on helicopters are presented in Figure 6.3. The flights were usually performed at an altitude of about 200 m within a radius of 0.5–1 km from the ruins of the reactor building. Sampling from An-24rr airplane laboratories equipped with filtering gondolas [34–37] was performed at altitudes of 200–1200 m within a radius of 2–5 km from the ChNPP. The concentrations of radioactive aerosol products of the Chernobyl accident are presented in Figure 6.4 [35]. It can be seen that the concentrations vary considerably. In part, this is explained by differences in sampling places and times for airplanes and helicopters. Nevertheless, the comparison of Figures 6.3 and 6.4 allows us to observe the synchronism of changes in the concentrations of radioactive aerosols. High values were recorded on 28 and 29 April and on 4 and 16 May. The maximal values were observed not on 5 May, as follows from Figure 6.1, but on 4 May. On that day, the total beta activity at aircraft sampling made up about 55 000 Bq/m^3, and at helicopter sampling the total activity of nine gamma-emitting nuclides was 6900 Bq/m^3. From Figure 6.4 it is clearly seen that an increase in the concentrations of radioactive aerosols above the ruins of the reactor building was taking place with time. However, within only three months of observations, but not in the second half of May, as reported in [5], the concentrations decreased by six orders of magnitude. In Figure 6.4 we can distinguish four areas: 28 April–4 May, 5–23 May, 24 May–6 July, and 7 July–6 August.

On the whole, the first area agrees with Figure 6.1. In the second area, the maxima are observed on 7–9 May and 16–21 May. The concentration of 16 May

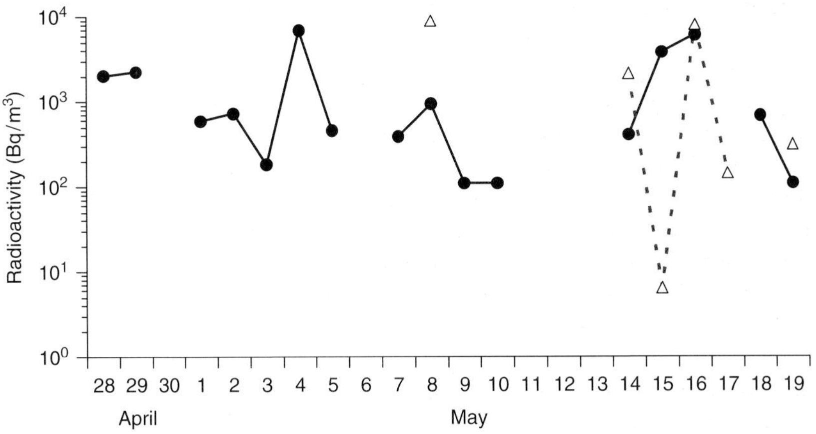

Figure 6.3 Radioactivity of nine gamma-emitting nuclides in aerosols above the ruins of the fourth block of the Chernobyl NPP in April–May 1986. Samples were obtained by using helicopter (full circles) and aircraft (open triangles).

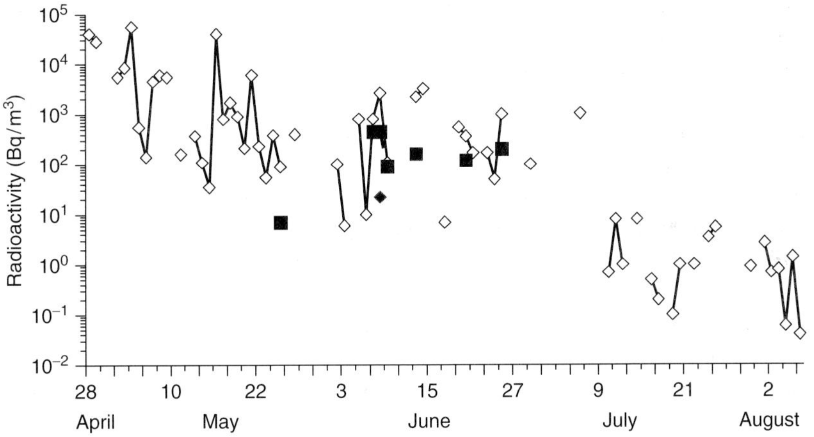

Figure 6.4 Concentration of beta-active aerosols above the Chernobyl NPP in April–August 1986 (collected from aircraft).

was conditioned mainly by refractory elements and was only ~1.5 times lower than the value obtained on 4 May. In the third area, the practically constant concentration of radionuclides remained at a level approximately 100 times lower than on 4 May. Four values exceeding 1000 Bq/m^3 were recorded in June, and one more such value was recorded early in July. The fourth (also practically horizontal area) is characterized by a sharp decrease in radioactive aerosols on average by two more orders of magnitude. In two cases, the concentration was lower than 0.1 Bq/m^3. Pronounced maxima are absent.

Thus, the results of aircraft and helicopter sampling did not confirm such a sharp decrease in the concentration radionuclide concentration of radioactive substances above the reactor as described in [4–8]. Moreover, the studies performed in the early fall of 1986 showed that the release from the destroyed reactor 4 still remained at a high level. In those experiments, in late August and early September, aerosols were collected on filters placed on a rope thrown over from the eastern to the western wall of the reactor at a height of 20–30 m from the surface of the destroyed reactor, and on 1–2 October they were collected from the arm of a construction crane at the erection of the roof of the "Shelter." It turned out that the aerosol concentrations in the flow coming out of the ruins of the reactor building made up $10^2 - 10^4$ Bq/m^3, that is, they were several orders of magnitude larger than those recorded from an airplane in July and August (see Figure 6.4). Thus, already in summer the air leaving the ruins was not warm enough to rise to the height of the patrolling airplane. It can be supposed that, since the second half of May, the aerosol concentrations recorded in aircraft samples collected even at the lowest flying height of 200 m were considerably lower than those in flows leaving the reactor – it was not possible to go down lower because of the 150 m high ventilation pipe (VP-2).

Variations of aerosol concentrations (see Figures 6.3 and 6.4) were accompanied by significant differences in their radionuclide composition. Volatile radionuclides (iodine, ruthenium, and cesium) prevailed at low activity concentrations of aerosols. These include samples collected from a helicopter on 1–3, 5, 7–10, 14, and 18 May. The activity of samples collected on 1–3 May were practically fully conditioned by ^{131}I, that of samples collected on 8 May by ^{103}Ru and ^{106}Ru, and that of samples collected on 14 May by ^{134}Cs and ^{137}Cs. Refractory radionuclides (zirconium, niobium, and cerium) prevailed in samples collected in periods of high activity concentrations of aerosols. Such samples were collected from an airplane on 28–29 April, 4, 15, and 16 May. In some cases, they contained small amounts of ruthenium and cesium radioisotopes.

In five cases, samples were collected from a helicopter and an airplane on the same day. However, on account of the differences in the heights and flight durations, the concentrations and radionuclide composition of aerosols coincided only in samples collected on 16 May. This was indicative of the variability of the discharge composition even during short time periods.

At a number of airplane probings, the gondolas were equipped with SFM-I sorption filtering materials developed at the L.Ya. Karpov Physicochemical Institute [38]. In these cases, not only aerosols but also some gaseous radioactive products, including iodine, tellurium, and ruthenium, were entrapped from the filtered air. It was revealed that ^{95}Zr, ^{95}Nb, ^{134}Cs, ^{137}Cs, ^{140}Ba, ^{140}La, ^{141}Ce, ^{144}Ce, and ^{7}Be (a radionuclide of cosmogenic origin) were practically completely entrapped in the front aerosol layer from FPP-15-1.5 material, while ^{103}Ru, ^{106}Ru, ^{131}I, and ^{132}Te were also found in the subsequent sorption layers containing activated carbon impregnated with $AgNO_3$ [39].

The detection of ruthenium and tellurium in SFM-I sorption layers was indicative of their presence in the atmosphere not only in aerosol form but also in gaseous form. The portion of ^{103}Ru in gaseous components made up 0.6–13%, that of ^{106}Ru made up 2.7–16%, and that of ^{132}Te made up 1–8%. The identity of the ^{103}Ru and ^{106}Ru distributions between sorption layers of material allowed us to conclude that they were present in the air as part of the same gaseous substances (obviously, RuO_4). Gaseous compounds of ^{132}Te were also well entrapped.

The study of radioiodine distribution over the layers of SFM-I material showed that its gaseous substances were adsorbed with more difficulty than those of ruthenium and tellurium. Obviously, ^{131}I is in the air not only in the form of molecular iodine vapor, but also as a part of organic compounds (methyl iodide, CH_3I). It should be noted that, during the sampling period above the reactor from 8 to 19 May, the portion of gaseous forms of iodine increased from 30% to 90%.

During the flight on 14 May, a three-layer filter intended for determination of the dispersion composition of aerosols was placed in one of gondolas besides the sorption filtering material [40, 41]. It was revealed that refractory radioisotopes (^{95}Zr, ^{95}Nb, ^{140}La, ^{141}Ce, and ^{144}Ce) and their compounds were concentrated on particles with active median aerodynamic diameter (AMAD) of about 0.7 μm at a standard geometric deviation $\sigma = 1.6–1.8$. Volatile radionuclides (^{131}I, ^{103}Ru, ^{106}Ru, and ^{132}Te) having gaseous compounds were associated with considerably smaller particles with AMAD of 0.3–0.4 μm at $\sigma = 2.3–2.5$. The difference in sizes of particles containing radionuclides was most probably conditioned by the physico-chemical processes taking place within the ruins of the reactor building in the remaining nuclear fuel and uranium fission products. Aerosols of detected sizes are involved in long-range atmospheric transport and can be observed at the distance of many thousands of kilometers from the place of generation.

6.2.4
Observations of Radioactive Aerosols in the Territory around Chernobyl

After the beginning of the construction of the "Shelter," the Radiometric Laboratory manned with specialists and equipped with apparatus from the V.G. Khlopin Radium Institute was entrusted with aerosol monitoring in the nearest vicinity of the ChNPP [42]. On 11 June 1986 they began daily aerosol sampling at 13 points onboard a BTR armored troop-carrier patrolling around the perimeter of the ChNPP. The air was pumped through AFA RMP-20 filters. The results of gamma-spectrometry of samples on a semiconducting detector with a Nokia LP-4900 analyzer are presented in Figure 6.5. It can be seen that the activity concentrations of the mixture of gamma-emitting nuclides were high and varied considerably; in five of 19 samples the concentrations exceeded 740 Bq/m^3, that is, the temporarily prescribed permissible norm (maximum permissible concentration–MPC). High concentrations were the consequence of construction and decontamination work near reactor 4 as well as intensive flow of traffic. By the end of June, the composition of radionuclides became stable due to the decay of iodine, lanthanum, barium, tellurium, and other isotopes. In July, a number of premises inside the plant were included in the work program. On 14 September,

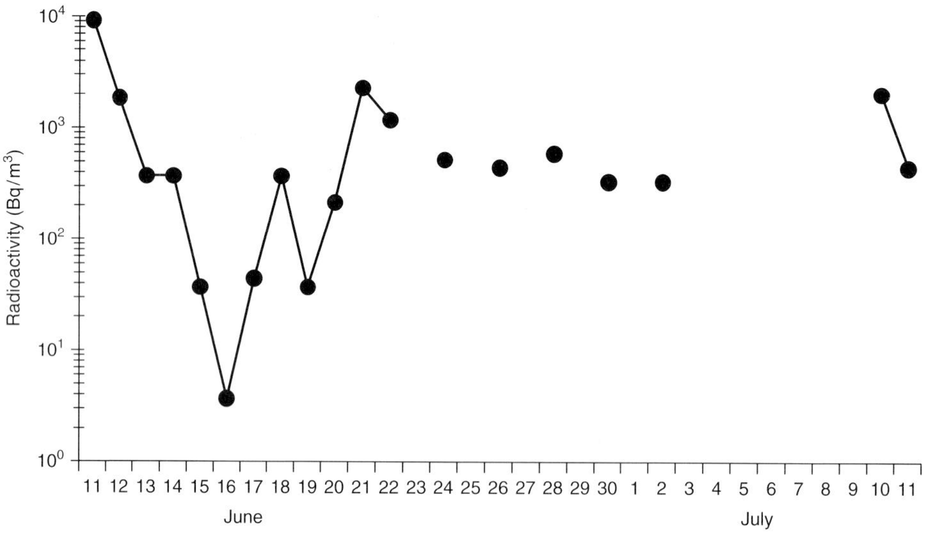

Figure 6.5 Concentration of gamma-active nuclides in aerosols at ground level of the Chernobyl NPP in June–July 1986 (samples were collected along the perimeter of the NPP).

more points, including ones in Chernobyl, were added. Figure 6.6 presents some results of this monitoring (before commissioning the "Shelter").

From Figure 6.6 it follows that, near the administrative and communal building (ACB-2) in July, August, and September, the concentrations of aerosol carriers of gamma-emitting nuclides were usually in the range of 100–1000 Bq/m^3. Such high concentrations were conditioned by the fact that the sampling point was only 300 m to the east of the place where intensive activities on erecting the northern cascade wall of the "Shelter" were going on. The pit-face made by miners who were erecting a concrete refrigerator plate under reactor 4 was nearer still. Only in October and November, when the construction works at the "Shelter" became less intensive and a large area near it was covered with "clean" soil and concreted over, did aerosol concentrations decrease to 10–100 Bq/m^3.

The sampling point near ACB-1 was approximately 1.5 km to the east of reactor 4. Though the initial explosion products were thrown away in a westerly direction, aerosol concentrations near ACB-1 and ACB-2 proved to be comparable. Most probably, this was associated with the intensive flow of traffic near ACB-1, where cars were constantly coming from regions with a high degree of pollution of soil, buildings, and machinery. Significant variations of concentrations were also likely to occur under the influence of the meteorological situation (precipitation and wind). For instance, one of the highest concentrations (4440 Bq/m^3) was recorded near ACB-2 on 21 August in a strong west wind with gusts up to 10–15 m/s.

Only in October and November, when radioactive substances that had fallen as a result of the accident were removed from rather large areas near the erected

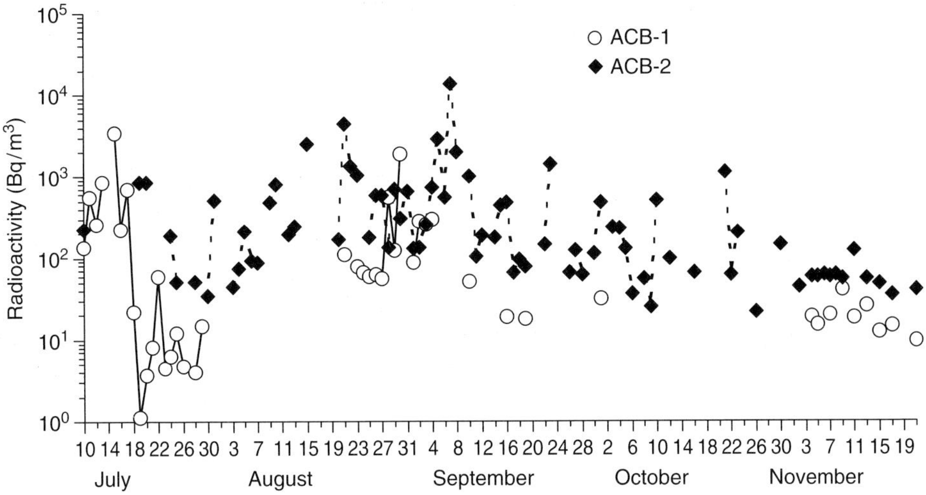

Figure 6.6 Concentration of gamma-active nuclides in aerosols at ground level of the Chernobyl NPP in June–November 1986 (samples were collected near ACB-1 and ACB-2).

"Shelter," and the areas were covered with gravel and sand and concreted over, did aerosol concentration decrease by an order of magnitude as compared with the levels observed in July to September. Rains in the fall also contributed to the reduction of dust rise.

From the end of August to the middle of September, aerosol samples near ACB-2 were also collected by specialists of the L.Ya. Karpov Physicochemical Institute [43]. They were interested in the aerosols that got into the air-raid shelter located in the basement of ACB-2. Figure 6.7 presents the results of such observations.

During the observation period, considerable variations in the concentrations were recorded at the air-raid shelter. For instance, the minimal values recorded on 20 and 25 August made up about 14 Bq/m^3, and the maximum recorded on 21 August amounted to 1110 Bq/m^3. The reason for the spike in concentration was the strong wind mentioned above. Besides the concentrations, the radionuclide composition was also unstable. There was a very high concentration of refractory elements in samples collected on 20–22, 25–26, and 30–31 August. Consequently, aerosols of fuel composition prevailed near ACB-2 on those days. The sample collected on 2 September was different. The portion of volatile substances (radioruthenium and radiocesium) in it made up about 65%. This correlated with [19–21] where it was noted that radiation of individual highly active particles collected in Sweden and Finland (thousands of kilometers away from the ChNPP) was practically completely caused by radioruthenium.

The most probable reason influencing the state of radioactive pollution of the atmosphere near ACB-2 was aerosols transported from the western part of the ChNPP area, where there are places of heavy fallout of accident products, rather

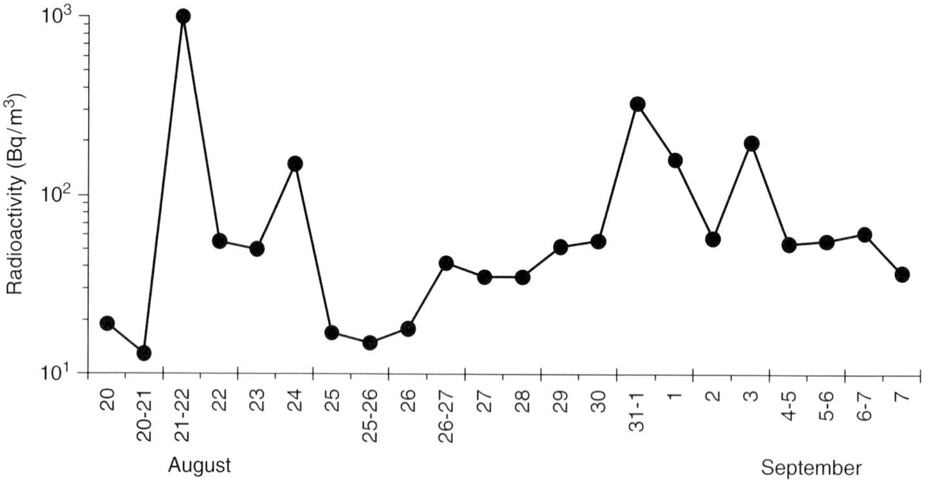

Figure 6.7 Concentration of gamma-active nuclides in aerosols at 1 m above the ground near ACB-2 in 1986.

than aerosols coming directly from the ruins of the reactor building. This confirms the relation of the aerosol concentrations with the wind rose. The highest aerosol concentrations were observed for west winds recorded on 21, 22, 24, and 31 August as well as on 3 September. The stronger the wind (for example, on 21 August its gusts achieved 15 m/s), the higher the concentration. The lowest concentrations of radioactive aerosols were observed in east winds recorded twice (on 20 and 25 August).

Besides the specialists of the V.G. Khlopin Radium Institute and the L.Ya. Karpov Physicochemical Institute, in the fall of 1986 regular aerosol sampling was performed by employees of the F.E. Dzerzhinsky All-Union Heat Engineering Institute. Their four control points were located at the plant periphery at a height of 1 m from the ground; three of these were in the places of the air intakes for input ventilation in three units of the first line of the ChNPP [44]. Figure 6.8 presents the results obtained at surface level. Aerosol concentrations varied from 1 to 220 Bq/m^3 and coincided with the data obtained near ACB-1 (see Figure 6.6) in order of magnitude. On the whole, as noted previously (see Figure 6.6), in the fall of 1986, the concentration of radioactive aerosols in the vicinity of the ChNPP distinctly decreased.

Besides AFA filters, a cascade impactor was used for sampling on 9–26 July. The results revealed the existence of bimodal size distribution of particles around the perimeter of the main building of the ChNPP first line. The portion of finely dispersed fraction (the activity of which was 90% conditioned by ^{103}Ru and ^{106}Ru) made up on average 8% of the total activity of gamma-emitting nuclides. The AMAD of this fraction was about 0.7 μm at $\sigma = 2–3$. The rest of the activity fell into the fraction of coarsely dispersed aerosols with AMAD of 8–12 μm at the same σ [44].

After the commissioning of the "Shelter," the arrangement of the radioactive aerosol control system on its territory took about a year. When, in 1988, well-boring was started to seek FCMs in reactor 4, a team of specialists from the Complex Expedition at the I.V. Kurchatov Institute of Atomic Energy was entrusted with the control of the aerosol situation inside the "Shelter" and its surroundings. Four filtering plants (FPs) were already functioning within a radius of 300 m from the reactor by that time. They provided round-the-clock air pumping through FPP-15-1.5 material at a flow rate of 400–500 m^3/h. The filter exposure lasted five days. The results obtained in the fall of 1988 are presented in Figure 6.9.

As can be seen in Figure 6.9, the highest values amounting to 100 Bq/m^3 were observed in the southern part of the territory. For all three plants the difference between the maximal and minimal concentrations made up 1–2 orders of magnitude. Though the distance between the plants was only 200–300 m, the

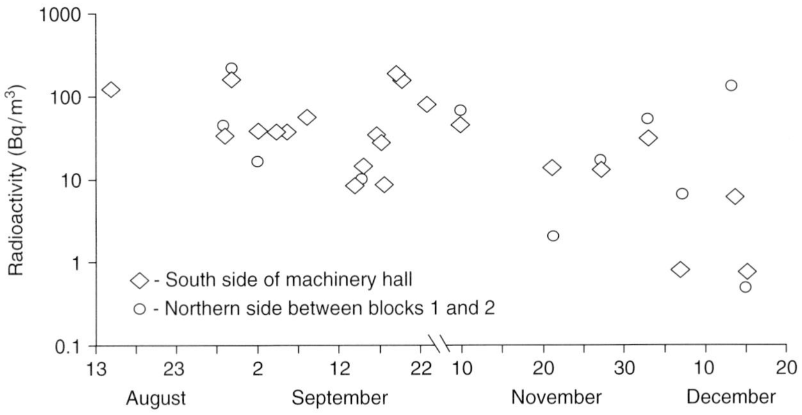

Figure 6.8 Concentration of gamma-active nuclides in aerosols collected near the first line of the ChNPP in August–December 1986.

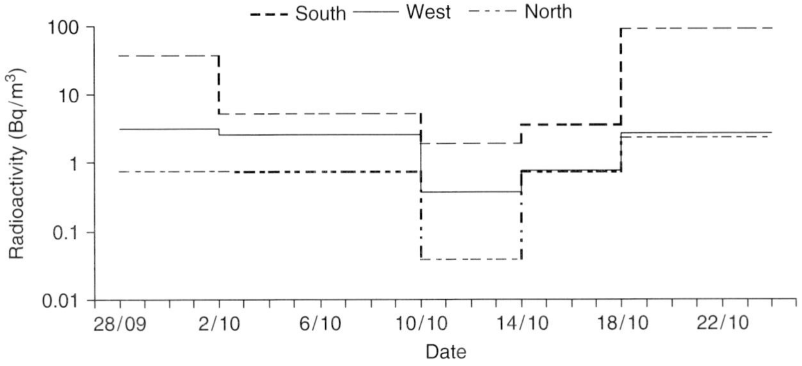

Figure 6.9 Integrated concentration of gamma-active nuclides in aerosols collected in various directions in the vicinity of the "Shelter" in 1988.

concentrations at simultaneous samplings practically did not coincide. From this, it follows that the aerosol situation in the places where the plants are located was determined by the effects of local sources rather than by regional air masses, probably like in other points near the "Shelter." Such a situation was still observed in 1989–1990.

Three years after the accident, when radioactive products with half-lives of less than 100 days had practically disappeared, attention was mainly focused on the content of alpha-emitting aerosols, as they began to determine the dose due to ingress of substances by inhalation, and ^{137}Cs, which made a major contribution to external irradiation.

Summarized data on the concentrations of ^{137}Cs and ^{241}Am aerosols in the local zone collected during 15 years in the "Areal" laboratory created at the Interdisciplinary Science and Technology Center "Shelter" of the National Academy of Sciences of the Ukraine are presented in Figure 6.10. Practically all the samples were obtained at 15-day exposure of FPP-15-1.5 filters. Considerable (up to three orders of magnitude) differences in the minimal and maximal values of both ^{137}Cs and ^{241}Am during comparatively short time periods are observed. The greatest variations are recorded for samples collected using the plant installed in the southern part of the territory. The reason is probably associated with the higher concentrations of uranium, and consequently its fission products, found here after the accident. Nevertheless, in Figure 6.10 it can be seen that, on the whole, ^{137}Cs and ^{241}Am concentrations in the places where the northern, northwestern, and southern plants are installed are very close to each other. One other important conclusion follows from the graphs in Figure 6.10: since December 1992, the concentration of ^{137}Cs aerosols decreased within seven years by approximately five times. This value is much larger than that achieved due to radioactive decay (the half-life of ^{137}Cs is $T_{1/2} = 30.2$ years).

A still sharper decrease in ^{137}Cs concentration in the air during the first years after the accident was reported in [45, 46]. Samplings in Pripyat City, situated at a distance of 5 km to the northwest of the "Shelter," over a period of five years (from July 1987) revealed that the concentration of aerosol carriers of ^{137}Cs had decreased by approximately 20 times. During subsequent years, this process slowed down and even stabilized. The authors of [46] represented the concentrations of ^{137}Cs (μBq/m^3) by the equation

$$C_0(t) = 6725 \exp(-t/30) + 346$$

where t is the number of years after commencement of observations. They predicted that the persistence of such regularity will be valid till 2009.

The reason for the sharp (during the first years after the accident) and then delayed decrease in the concentration of aerosols of ^{137}Cs and other radionuclides (in particular, plutonium isotopes) in the air is probably associated with natural and man-made factors: embedding of accident products in the soil, decontamination of the region around the "Shelter" and planting of greenery on it, dust control at construction and erection works, and so on. At the same time, a more rapid decrease in the concentration of ^{137}Cs aerosols in the air than that calculated for

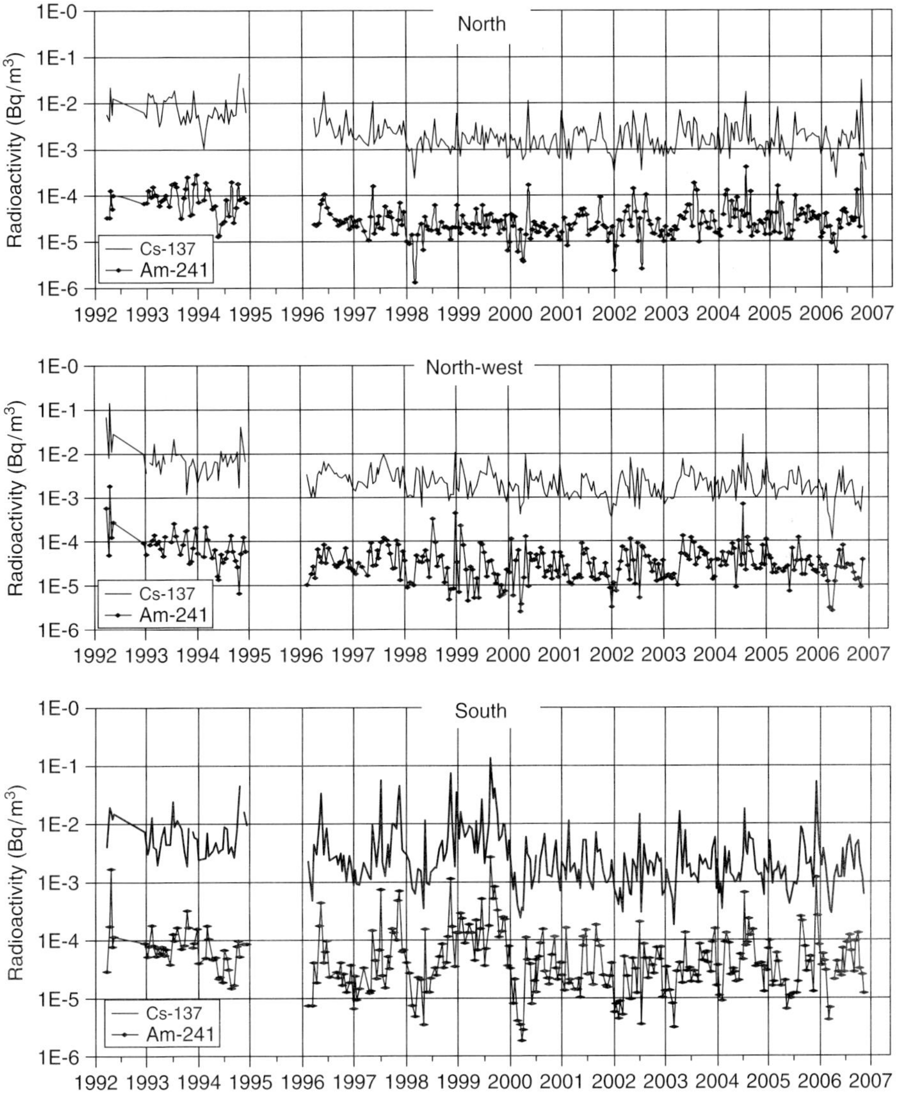

Figure 6.10 Radioactivity of aerosol particles containing ^{137}Cs and ^{241}Am obtained in the vicinity of the "Shelter" (north, northwest, and south) in 1992–2007.

$T_{1/2} = 30.2$ years indicates that the contribution of discharge from the "Shelter" to the current aerosol situation is insignificant.

Though the nuclear physical origins of ^{137}Cs and ^{241}Am are different – the first one being the product of uranium fission, and the second one the product of beta-decay of ^{241}Pu ($T_{1/2} = 15.2$ years) produced in the fuel – the dynamics of their concentrations in 1992–2007 in the air in the region of the "Shelter" were practically identical (see Figure 6.10). ^{241}Am concentrations in the range of 0.01–0.1 mBq/m^3 prevailed among almost 900 samples. The minimal concentrations were usually not lower than 0.003 mBq/m^3, and the maximal ones only in several cases approximated to or somewhat exceeded 1 mBq/m^3. The highest concentrations were more often recorded near the southern FP-3. The three times that activity concentrations of ^{241}Am amounted to 1 mBq/m^3, it was associated with construction works. For example, in the late spring–early summer of 2000, intensive excavation works were going on at the site of the construction of a spent nuclear fuel repository (SNFR-2) located 2 km to the southeast of the "Shelter." In August–September 2006, a path was laid (scraping and loading of soil, delivery, and leveling of gravel) onto the berm near FP-3. It should be noted that these roadworks caused only an increase in ^{137}Cs and ^{241}Am concentrations in samples from the southern plant. No increased radionuclide concentrations were recorded in samples collected simultaneously at the northern and northwestern plants (see Figure 6.10).

The radionuclide ratio is known to be very important for the identification of radioactive aerosols and the detection of their sources of origin. From Figure 6.10 it can be easily estimated that the ^{137}Cs/^{241}Am ratio was usually in the range of 50–70. These values are typical of nuclear fuel with burnup of 11 MWt day/kg of uranium [47]. Thus, the aerosol situation around the local zone of the "Shelter" is determined by the so-called radioactive fuel particles. It is these particles that fell near reactor 4 after the accident on 26 April 1986. The high ^{137}Cs/^{241}Am ratios (120–190) in December 2000 and January 2001 are probably associated more with inaccuracy of the sample measurements on account of the very small concentrations of radioactive substances observed during the winter season, when the rise of dust from the ground was complicated, than with the ingress of so-called condensation aerosols (see Figure 6.10).

Soon after the commissioning of the "Shelter," the specialists of the "Kombinat" production association started to control the state of the air medium around it at distances above 0.5 km [48]. VP-2 was chosen as the zone center. Within a radius up to 1 km, samples were collected in 11 points, and at distances of 1–3 km in 13 points. Figure 6.11 presents summarized aerosol concentrations of the gamma-emitting nuclides from March to December 1987. It can be seen that the concentrations of radioactive aerosols in both zones were decreasing synchronously. They were usually five times lower at 1–3 km distance than at 0–1 km distance. This is explained by a decrease in both the density of local contamination with radionuclides that had fallen after the accident and the intensity of dust rise due to human activities and strong winds. From the approximating dotted straight lines, it follows that the period of the activity decrease by half was about 22 days. As at the beginning of these observations all radionuclides of accident origin with half-lives less than 30 days no longer existed, the high rate of activity decrease can be explained only by the effect of natural and anthropogenic factors: the decontamination of the locality, roads, constructions, as well as embedding of the fallen substances into soil. Besides, the intensity of construction works including the flow of traffic also decreased.

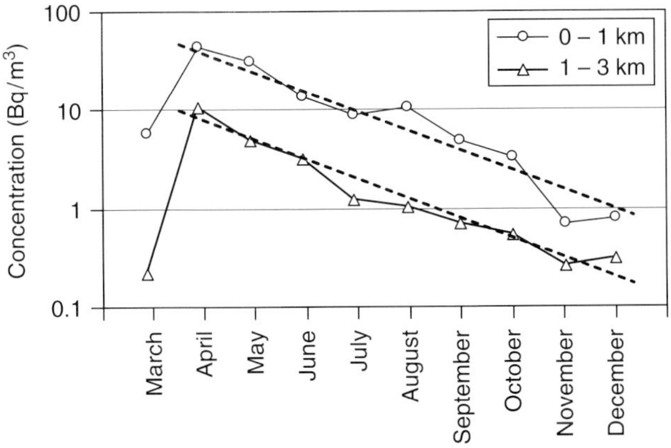

Figure 6.11 Concentration of gamma-active nuclides in aerosols collected at distances of 0–1 and 1–3 km from the fourth block of the ChNPP in March–December 1987.

The Automated Radiation Monitoring System began to function in the 30 km Chernobyl exclusion zone in 1988. It included about 30 stationary points of round-the-clock aerosol sampling using FPP-15-1.5 materials. Five points of this type systematically collecting aerosols are located within the radius of 0.5–3.0 km from the "Shelter" [49, 50]. The specialists of the Exclusion Zone Radioecological Monitoring Center work at these points. Their data are an important supplement to observations carried out in the surrounds of the "Shelter."

At the control point under consideration, the aerosol situation largely depends on the works performed near the FPs and the weather conditions, in particular the *wind speed*. For instance, excavation and construction works in the vicinity of the "Shelter" in the spring of 2000 were the source of radionuclide-contaminated dust, which was recorded practically simultaneously at sampling on 4–18 May in the southern part of the "Shelter" region and on 7–12 May at the points of the outdoor switchgear and Petroleum Storage Depot. At this time, the ^{137}Cs concentration (135 mBq/m^3) in the sample from near the "Shelter" proved to be maximal during the 10-year observation period starting from 1993, and was also maximal at the outdoor switchgear (16 mBq/m^3) and the Petroleum Storage Depot (2.5 mBq/m^3) [51].

In the second half of 2002, radioactive aerosol monitoring with a "Typhoon" aspirator was commenced near the office building standing 2 km to the west of the "Shelter" [46]. The high efficiency of the aspirator (about 4000 m^3/h) allowed representative samples to be obtained within short time periods, for example, a day. Data on ^{137}Cs concentration are given in Figure 6.12. It can be seen that most often the concentrations were within the range of 0.1–0.3 mBq/m^3, that is, an order of magnitude lower than in the vicinity of the "Shelter." During the period under consideration, sharp increases in radioactive aerosol concentration caused by natural factors were observed twice: early in September as a consequence of forest fires, and early in December on account of strong winds. These situations will be described below in more detail. Here it should also be noted that the

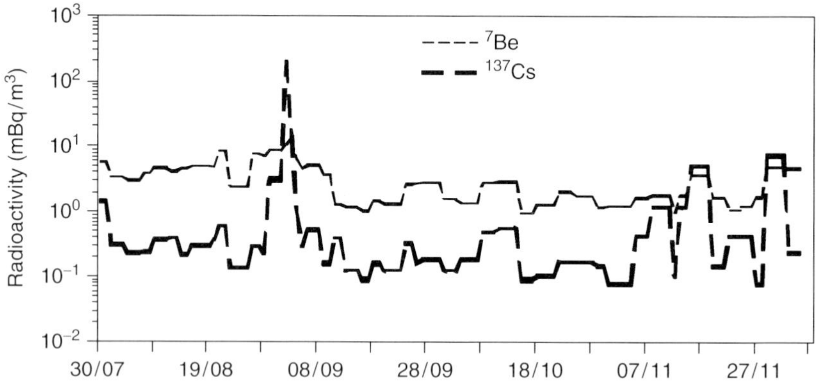

Figure 6.12 Radioactivity of ^{7}Be and ^{137}Cs containing aerosols at ground level 2 km from the "Shelter" (near the administrative building) from 30 July to 5 December 2002.

concentrations of cosmogenic ^{7}Be – which does not bear any relation to the products of the Chernobyl accident and is a peculiar marker of air masses coming from the upper troposphere and from the stratosphere – were high in August and in the first half of September and then decreased by 4–5 times.

The influence of high wind speeds is manifest in an increase in the number and variation of sizes of particles, which start to take off from the underlying surfaces (ground, roads, roofs, leaves of trees and bushes, grass, and so on). Naturally, the higher the density of local contamination with the products of the accident and the subsequent ingress of radionuclides of Chernobyl genesis as a result of human activities, the larger will be the concentrations of radioactive aerosols in the air.

The effect of high wind speeds can be traced by the example of the situation in March 2003, which influenced aerosol samples collected by stationary plants in the region of the "Shelter" as well as those located in the vicinity of the ChNPP [52]. According to the data of the "Chernobyl" weather station, the month's highest wind speeds were observed on 20–22 March: average wind speeds up to 5 m/s and gusts up to 14 m/s. The mean daily temperatures of the air, which had been above freezing since 11 March, went down below zero (to $-5\,^{\circ}C$), and then went up again to 2–6 °C after 24 March. Snow, which fell on 20–21 March, reduced dust rise from the soil and roads to a certain extent.

High concentrations of ^{137}Cs aerosols were recorded around the "Shelter" at exposure of the filters from 17 March to 2 April (Figure 6.13). The concentration increased by an order of magnitude as compared with the average annual value in the southern part of the territory. A similar situation was observed at three aerosol monitoring points located in the vicinity of the ChNPP during the sampling period from 18 to 24 March: ^{137}Cs concentrations at the points of the outdoor switchgear, Petroleum Storage Depot, and bored piles increased by an order of magnitude as compared with the average annual level. It should be noted that, at the Petroleum Storage Depot and bored piles, the increase in aerosol concentration could not be

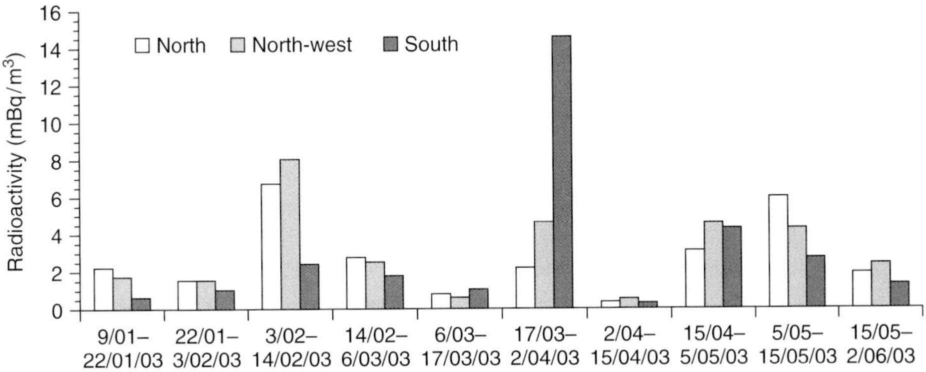

Figure 6.13 Concentration of ^{137}Cs in aerosols at ground level obtained in the vicinity of the "Shelter" in January–May 2003.

the consequence of carryover of radioactive substances through the VP-2 pipe and leakages (cracks) of the "Shelter," because a stable northwest wind was observed on 17–22 March. For the Petroleum Storage Depot this is important as the point is located 2 km to the northwest of the "Shelter." The bored piles are located outside the plume of aerosol carryover, 3 km to the east of the ChNPP.

Observations at the "Chernobyl" weather station over many years show that dust storms can be observed in the Chernobyl exclusion zone, with gusts stronger than 15–18 m/s. One of them occurred a year after the accident. Gusts of 25 m/s were recorded by the "Chernobyl" weather station on 18 April 1987. The probability of occurrence of dust storms in the Kiev region is from 2.1% to 31.7% in different months. The maximal probability falls at July.

A synoptic situation, which also caused a dust storm, emerged in the first 10 days of September 1992 in the northern part of the Ukraine, in the Bryansk region of Russia, and in the southeastern part of Belarus. These regions are among areas highly contaminated with the products of the Chernobyl accident, which fell down from radioactive clouds in April–May 1986. The dust storm was caused by a powerful cyclone, which arose in the Balkans on 4 September and was above the Baltic Sea on the coast of Lithuania on 8 September. Thanks to scheduled and unscheduled aerosol samplings performed on these days at a number of weather stations in the Ukraine and Belarus, as well as at monitoring points in the Chernobyl exclusion zone, it became possible to obtain unique data on the composition of dust raised into the air by strong winds.

A considerable increase in the concentrations of atmospheric dust and radioactive products of the accident was recorded in all the above regions. In the Chernobyl exclusion zone, the duration of exposure of FPP-15-1.5 filters was reduced from 5–7 days to 2–23 hours. It was revealed that the concentrations of the accident products increased during the storm by 1–2 orders of magnitude. Aerosol sampling in Vilnius (500 km to the northwest of the ChNPP) revealed a 100-fold increase in ^{137}Cs concentration. The calculation of the trajectories of air-mass transfer on

6–7 September showed that this ^{137}Cs could have come to the environs of the Lithuanian capital at altitudes of 1.5 km from the Kiev region (Ukraine) and the Gomel region (Belarus).

The evaluation showed that, on 6–7 September 1992, when average wind speeds of 8–12 m/s with gusts up to 25 m/s were observed in the Chernobyl exclusion zone for 30 hours, the amount of the ^{137}Cs radioisotope that could rise into the atmosphere from territories with a density of contamination of this isotope of 3.7×10^6 Bq/m^2 could exceed by four times the admissible monthly release of a normally functioning 6 GWt thermonuclear power plant.

It was mentioned above that fires on areas contaminated with the products of the Chernobyl accident could result in a sharp increase in the concentration of radioactive substances in the air. From 30 August to 4 September 2002 near the administrative building of "Shelter" the smell of smoke was in the air, and abnormally high ^{137}Cs concentrations were recorded [53].

As follows from Figure 6.12, the maximal value of about 170 mBq/m^3 fell on 2–3 September. At the same time, increased concentrations of ^{134}Cs, ^{90}Sr, ^{241}Am, and plutonium isotopes were recorded. On those days, the average concentration of ^{137}Cs in the air increased by three orders of magnitude. The $^{137}Cs/^{90}Sr$ ratio was about 30, which is approximately 20 times higher than that in nuclear fuel at the moment of the accident at reactor 4 [47]. Such an increase in cesium concentration in aerosols had already been observed after forest fires in regions contaminated with the products of the Chernobyl accident [54–56].

Space photographs were used to detect the sources of smoke. The fires were detected using an infrared imager. Not less than 45 forest fires occupying an area of 0.3–1.2 km^2 were recorded over an area of 200×200 km^2. All of them were in the eastern sector relative to the ChNPP near the border between the Ukraine and Belarus, half of them being at the distance of only 20–40 km from the station. The weather favored both the development of the fires and the transport of combustion products toward the ChNPP. According to the data of the "Chernobyl" weather station, from 29 August to 5 September 2002 the night temperature was 11–15 °C, and the day temperature went up to 30 °C. Air pressure was in the range of 752–758 mmHg. Wind speed varied from 1 to 2 m/s. Sometimes calm was observed at night, and only in the daytime on 30 August did the wind speed amount to 3 m/s. Maximal gusts up to 8 m/s were recorded. In this period from 12 to 6 p.m. the wind had a southern direction. On all the other days, air masses were coming only from the southern and eastern quarters, which provided the transport of combustion products to the vicinity of the ChNPP.

Thus, the analysis of space photographs and the meteorological situation in late August–early September 2002 showed that the smoke seen in the area around the "Shelter" and its administrative building as well as the 100-fold increase in the concentrations of ^{137}Cs and other radionuclides were associated with forest fires in the area between the Dnepr and Pripyat rivers near the northeastern boundary of the Chernobyl exclusion zone, where there are vast territories contaminated with radioactive products of the accident.

The works [57–59] carried out on experimental sites in the Chernobyl exclusion zone by specialists from the Ukrainian Institute of Agricultural Radiology showed that high concentrations of radioactive products of the Chernobyl accident are generated not only in forest fires but also in grassland fires. The formed smoke particles of micrometer sizes containing radionuclides are involved in long-range atmospheric transport.

6.2.5
Dispersity of Aerosol Carriers of Radionuclides

Undoubtedly, the results of the measurement of the disperse composition of radioactive aerosols are interesting from the point of view of radiation safety. Dispersity is the basic parameter determining *aerosol deposition* in respiratory organs. The transport of particles in the atmosphere and working areas as well as the functioning of treatment plants, analysis, and individual protection devices depend on their sizes.

After the accident, the first samples for evaluation of the dispersion composition of radioactive aerosols were obtained by specialists of the USSR Ministry of Defense. In [60] there are data on the distribution of radioactive aerosol at a height of 2 m over the territory of the ChNPP on 12 May 1986. Exact sampling points and techniques are not reported. Only the distribution of particles in 1 cm^3 of air in the range from 0 to 1.6 μm in 0.2 μm steps is presented. Data processing showed that they are well approximated by the log-normal distribution with median aerodynamic diameter of 0.62 μm and $\sigma = 1.5$.

Sampling was performed above the ruins of reactor 4 from an An-24rr airplane at an airspeed of 350–400 km/h, at a height of 300 m above the ground, on two days [36]. In one of the two gondolas installed on the fuselage, there was a pack of three-layer Petryanov filters with an area of 1 m^2 through which air passed at a rate of 0.9–1.2 m/s. The flight over, each of the pack layers was measured on a gamma-spectrometer. On 14 May 1986, it was found that the refractory radionuclides ^{95}Zr, ^{95}Nb, ^{140}La, ^{141}Ce, and ^{144}Ce had AMAD of about 0.7 μm at $\sigma = 1.6$–1.8. These results were practically identical to those recorded on 12 May. As for the volatile radionuclides ^{103}Ru, ^{106}Ru, ^{131}I, and ^{132}Te, they were bound with considerably smaller particles whose AMAD was in the range of 0.3–0.4 μm at $\sigma = 2.3$–2.5. This difference was probably caused by the high temperatures in the ruins of the reactor building and the physico-chemical processes taking place there.

The studies performed on 9–26 July 1986 by the specialists of the F.E. Dzerzhinsky All-Union Heat Engineering Institute in the area of the first line of the ChNPP were mentioned above. They used a cascade impactor to reveal a bimodal size distribution of particles. Approximately 8% of the total gamma-activity of the samples fell in the finely dispersed fraction with AMAD of about 0.7 μm at $\sigma = 2$–3; 90% of the activity was determined by ^{103}Ru and ^{106}Ru. The coarsely dispersed fraction had AMAD of 8–12 μm. Most probably, the generation of large particles occurred as a result of construction works and traffic flow.

On 10 September 1986, before the erection of the roof of the "Shelter" was started, a sample was collected on a pack of three-layer Petryanov filters located approximately 20 m from the surface of the ruins of the reactor building [61]. Measurements and calculations showed that the AMADs of the particle carriers of ^{141}Ce, ^{144}Ce, ^{134}Cs, ^{137}Cs, ^{95}Zr, and ^{95}Nb were equal and in the range 0.98–1.14 μm. Obviously, each aerosol particle contained all the isotopes in constant ratio. Only the AMAD of the particle carriers of ^{103}Ru and ^{106}Ru was somewhat smaller (0.74–0.92 μm).

The specificity of the particle carriers of radioruthenium was clearly seen from the results of observations performed in May, July, and September 1986 in different places: they systematically had smaller sizes. This was obviously associated with the presence of the volatile compound RuO_4, which in the gaseous state deposited on small atmospheric particles having sizes of 0.1–0.3 μm.

After the arrangement of radiation observation points in the Chernobyl exclusion zone, packs of three-layer Petryanov filters were regularly used to determine the dispersion composition of radioactive aerosols. Filters with an area of 0.3 m^2 were continuously exposed for 5–7 days at a flow rate of about 1 m/s. Air pumping over, each layer was measured on a gamma-spectrometer and transferred for radiochemical analysis. Not only the isotope products of the Chernobyl accident, but also the DPs radon and thoron, as well as cosmogenic ^{7}Be, were determined. The results of measurements of particle sizes collected at the ChNPP are shown in Figure 6.14 [62]. For almost 20 years, the sizes of the carriers of the radionuclide products of the accident remained stable: their AMAD was in the range of 3–8 μm. No separation of radionuclides (^{134}Cs, ^{137}Cs, ^{144}Ce, and ^{239}Pu) by particle size was observed. At the same time, ^{212}Pb (the DP of thoron) and ^{7}Be were on considerably smaller particles with AMAD

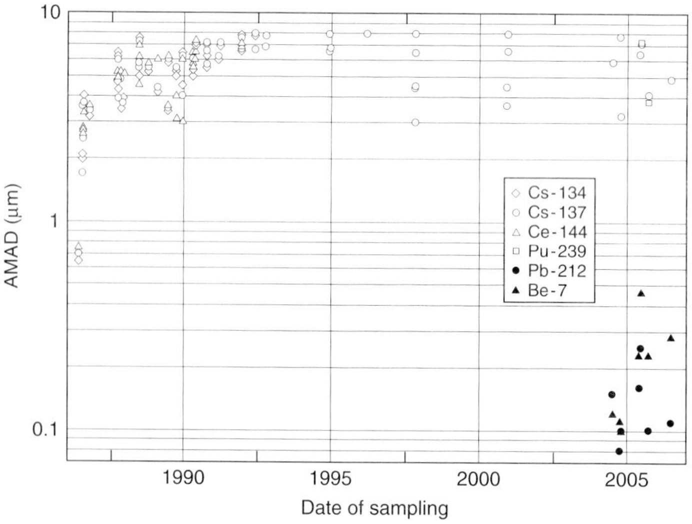

Figure 6.14 Size of radioactive particles (^{134}Cs, ^{137}Cs, ^{144}Ce, and ^{239}Pu) and natural nuclides (^{212}Pb and ^{7}Be) collected at the ChNPP in 1986–2006.

0.1–0.3 μm. This is quite natural, as ^{212}Pb and ^{7}Be atoms on emergence deposit on small atmospheric particles, and the radionuclide products of the Chernobyl accident are associated with dispersion particles formed during mechanical processes and the wind rise of dust. It should be added that in 1988–1991, when units 1–3 of the ChNPP were working, ^{131}I was found in some samples. Its carriers were particles with AMAD of 0.3–1 μm [63]. Besides, radioiodine was present in gaseous form.

6.3
Aerosols inside the Vicinity of the "Shelter" Building

6.3.1
Devices and Methods to Control Radioactive Aerosols in the "Shelter"

The studies and measurements of radioactive aerosols inside the "Shelter" were started soon after its commissioning (30 November 1986). They took on special significance in 1988–1991 after drilling activity to seek nuclear fuel remaining in the ruins of the reactor building.

All scheduled samplings were carried out with analytical filters AFA RMP-20 or AFA RSP-20 made of Petryanov fibrous polymeric materials. The use of RSP filters is preferable, as their main filtering layer is made of ultrathin perchlorovinyl fibers with a diameter of about 0.5 μm. This provides for entrapment of aerosols in the front layer and allows measurement of radioactivity without the need for a correction due to the absorption of alpha-particles by fibers. The filters are made in the form of disks with an area of 20 cm^2. The linear flow rate varied from 50 to 150 cm/s. AFA filters have high efficiencies of aerosol entrapment for a broad range of particle sizes and flow rates [3]. The scheduled aerosol sampling in the "Shelter" is carried out using portable blowers powered from the electricity network or accumulators.

In premises where personnel stay all the time, aerosol samples are collected once a day, and in the "Shelter" once a week. However, when repair, construction or decontamination works as well as special studies are conducted, both the number of points and the sampling periodicity are increased. Statistical data indicate that 1200 samples were obtained in 1992. Later on, their number continually increased. The maximum (20 800 samples) was achieved in 2000–2003. Approximately 10 000 samples were collected in 2006 and 2007. The decrease is associated with the rearrangement of the Radiation Monitoring System at the "Shelter" and the introduction of individual aerosol samplers, the filters or impactor heads of which are located in the workers' respiratory areas [64]. On average, such devices collect about 1000 samples annually.

6.3.2
Control of Discharge from the "Shelter"

Control of discharge from the "Shelter" is performed in the "Bypass" system, where aerosols from the former central hall of reactor 4 come through an air-duct.

They then go either to the VP-2 or to the filtering station. From the "Bypass," an aerosol sampling line is laid (a 35 m long tube with a diameter of 15 mm) to the radiometric plant RKS2-03 No. 9 "Kalina" and a cartridge with an AFA filter. Automatic aerosol sampling on a filtering tape made of Petryanov material, the measurement of activity, and transfer of the results to the duty operator control desk are performed in "Kalina." The tape is changed every 6, 12, or 24 hours. The AFA filter may be removed and measured on the instruction of the duty operator at any time. The results obtained form the basis of the calculation of the total discharge from the "Shelter" (daily, monthly, and annual).

Air release from reactor 4 through the "Bypass" occurs due to the natural draft in the VP-2 pipe. Its top is at a height of 150 m. If the release of the mixture of long-lived beta-emitting nuclides exceeds 26 MBq/day, it can be decided to close the shutter of the "Bypass" and to switch the fans for air transport to the filtering station for purification and then to VP-2. However, in the 23 years since the accident, the shutter in the "Bypass" has not been closed. This is indicative of the normal aerosol situation in the "Shelter."

The problem of the transport of radioactive aerosols from the premises of the "Shelter" that houses the remaining nuclear fuel and its fission products has been a subject that has attracted the attention of scientists and specialists of the Radiation Safety Service for many years. The first studies of air flows in the reactor cavity using helium labeling were performed in 1988–1989 by researchers of the V.G. Khlopin Radium Institute. The results showed that air flows often change their direction and speed. In 1990, the work was continued, using not only helium but also molecular tritium, ^{14}C (as a part of methane), and ^{85}Kr. After that, about 30 points were selected, where the temperature, relative humidity, speed, and direction of air flows are measured every week.

The release of radioactive aerosols from the "Shelter" occurs not only through the "Bypass" to the VP-2 pipe, but also through numerous cracks in the external constructions. Their total area in the late 1980s was estimated at 1200 m^2. Later on, after repair and stabilization work, most cracks were stopped up. By the end of the 1990s, the area of holes had been stabilized at a level of 120 m^2. Air gets into the "Shelter" and leaves it through cracks. Much depends on the air temperature and pressure, the wind direction and velocity in the environment, the season, and a number of other factors. Observations showed that air flows in cracks are unstable. The measuring equipment should be placed in all the main openings for exact measurements of the carryover of radioactive aerosols. However, this is not feasible. Besides, the complex radiation situation impedes their maintenance.

6.3.3 Well-Boring in Search of Remaining Nuclear Fuel

Examination of the central part of the "Shelter" with well-bores began in 1988. The main task was to seek the remaining nuclear fuel. About 40 horizontal and inclined wells, usually with a diameter of 172 mm, were examined during three years. The length of some of them amounted to 20 m.

Though boring was accompanied by washing, the pollution level of the air medium in the working and auxiliary premises was rather high. Besides dust escaping at drifting of concrete walls and floors, sometimes disintegrated material was spilled onto the floor because of inaccurate extraction of the core and the boring tool. About 1000 people were carrying out boring at the most intensive stage. Dust control using localizing and accumulating solutions was performed to reduce aerosol pollution of the air.

The results of measuring the samples collected in August 1988 in rooms 207/5 and 427/2 serve as an example of the sharp increase in the concentration of radioactive aerosols at boring. When the borers were switched off, the concentrations of alpha-emitting aerosols in these rooms were 0.037 and 0.081 Bq/m^3, respectively. In the process of boring, the pollution level increased to 0.55 and 1.92 Bq/m^3, that is, by 15–20 times. At the same time, in the turbine island of the third unit, which is not connected with the "Shelter," the concentration was 0.0022 Bq/m^3.

Systematic aerosol sampling in the above rooms during the work conducted in 1988–1991 (Figure 6.15) provided important information. When the borers were switched off, the concentrations of the mixture of alpha-emitting aerosols were usually at the level of 0.03 Bq/m^3. On resuming the boring process, the amount of radioactive substances in the air sharply increased by one or two orders of magnitude. In some cases, the concentrations increased by three or even four orders of magnitude. In room 427/2, a concentration peak of 150 Bq/m^3 was recorded on 22 November 1988. Something like that occurred three years later, on 10 October 1991 in room 207/5, when the concentration increased rapidly to 90 Bq/m^3, and on the next day decreased by three orders of magnitude.

Summarized results of average annual concentrations of aerosols of alpha-emitting nuclides in 1989–1991 in seven main premises where boring was

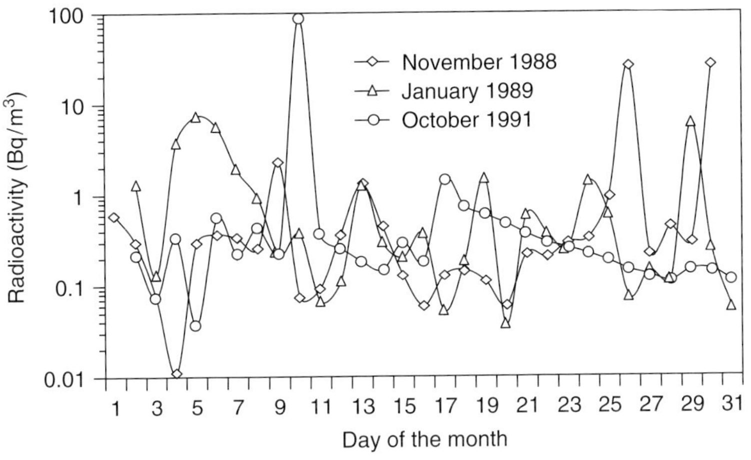

Figure 6.15 Concentration of aerosols containing alpha-emitting nuclides during well-boring activities.

Table 6.1 Average annual concentration of long-lived alpha-emitting aerosols in various areas of the "Shelter" (Bq/m^3) in 1989–1991.

Room number	1989	1990	1991 (first half)
207/4	0.53	0.35	0.093
207/5	0.32	0.36	0.11
208/10	0.32	–	0.15
318	0.74	0.21	0.1
427/2	0.92	0.15	0.093
515	0.20	0.26	0.067
605	0.21	–	–

performed are presented in Table 6.1. As noted in [65], only the days when active boring was going on were taken into account in the calculations. Figure 6.15 shows that the bursts of aerosol concentrations were not very long. However, they influenced average daily and even average annual values.

In May 1989, on completion of boring, it was necessary to remove a metal platform in room 208/10. By this time, a large amount of dust had accumulated on the floor, the walls, and the equipment. That is why aerosols rising in the air during metal cutting were supplemented with dust that appeared as a result of the movement of people and use of instruments. The dynamics of the concentrations of alpha-emitting aerosols are shown in Figure 6.16, from which it can be seen that the start of work was accompanied by a sharp increase in the concentration of radioactive aerosols in the air. On the first and second days of the work, the concentrations increased by two orders of magnitude from the background level of about 0.1 Bq/m^3. However, already 1 h after the completion of the work, air pollution considerably decreased, and within 8 h it practically reverted to the initial value.

After the central part of the "Shelter" including the reactor cavity had been examined using the wells, some wells were plugged and diagnostic equipment was placed in the others. On completion of intensive boring in 1992, the concentration of radioactive aerosols in decontaminated premises of the "Shelter" went on decreasing and in three years did not exceed 0.08 Bq/m^3 for alpha-emitting and 37 Bq/m^3 for beta-emitting substances.

6.3.4 Clearance of the Turbine Island of the Fourth Power Generating Unit

In the fall of 1988, the clearance of the turbine island of reactor 4 was started; in the process of construction of the "Shelter" it was separated from the turbine island of the third unit with a metal wall. Among operations with intense dust formation, we can distinguish removal of the roof accompanied by falling of building material fragments and clearance of the island by a ladle robot. The aerosol concentration

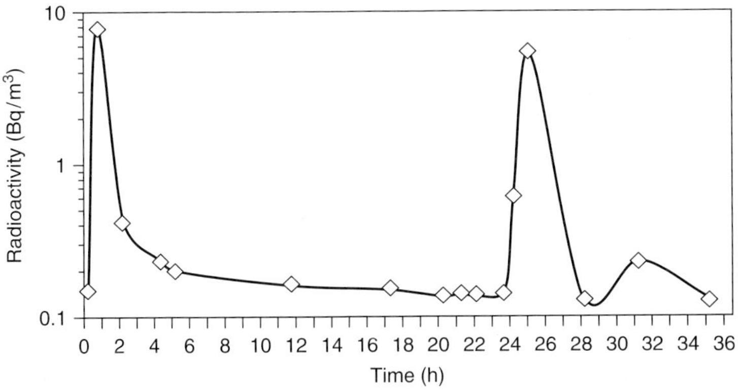

Figure 6.16 Concentration of aerosols containing alpha-emitting nuclides during metal-cutting activities in room 208/10 in the "Shelter" in May 1989.

in the air sometimes changed sharply within a short time. From 30 April to 25 May, operative control of aerosol concentration with simultaneous measurement of radioactivity of individual fractions was performed by specialists from the A.A. Bochvar All-Union Institute of Inorganic Materials [66]. A system consisting of a laser aerosol spectrometer (for optical determination of the size distribution of aerosols) and an impactor (to measure both the dispersity and radionuclide composition of individual fractions) was created. The results of observations showed that after carrying out the work the total aerosol concentration was increased from 3.6×10^4 to 20×10^4 particles per liter. Up to 70 particles per liter with the diameter of more than 7 μm appeared in the air, while there had been practically no such particles before the work started. On the completion of the work, the concentration of large aerosols (diameter of 1–10 μm) quickly (in about an hour) decreased by 5–10 times. The concentration of submicrometer particles (diameter less than 1 μm) decreased much more slowly – by two times within 2–3 h.

Dust control and treatment of surfaces with decontaminating solutions were performed to improve the aerosol and radiation situation in the turbine island. After that, the reduction of the number of particles larger than 5 μm and an increase in the concentration of fine aerosols up to 2×10^5 particles per liter were observed.

6.3.5
Strengthening of the Seats of Beams on the Roof of the "Shelter"

In the fall of 1999, the bases under the beams B1/B2 were strengthened at the western wall of reactor 4 to prevent the collapse of the roof of the "Shelter." These beams consist of double-T iron constructions B1 and B2; they are 3.2 m high and about 40 m long. On the beams, there is a roll of 27 tubes with 2 m diameter serving as the floor above the ruins of the central hall [67].

The specialists of the "Shelter" were controlling radiation, including the aerosol situation, on carrying out stabilizing work. Aerosols samples were collected during the work on AFA RSP-20 filters with "Typhoon" portable blowers at two monitoring points (MP1 and MP2). Measurements of alpha- and beta-emitting nuclides were performed with a radiometer; the concentration of ^{137}Cs was calculated based on the results of gamma-spectrometry of samples.

In October to December 1999, preparatory operations were performed: cabling, lighting equipment assembly, transport and installation of welding equipment, and fastening of lead sheets to reduce irradiation of personnel. In addition, metal cutting and welding, clearing and slotting of concrete constructions with a perforator, and underpouring were carried out. Naturally, all the works were accompanied by dust rise and aerosol generation.

In the month and a half from 27 October to 13 December 1999, the unique data presented in Figure 6.17 were obtained regarding control of the radiation situation. The highest concentrations of ^{137}Cs (above 1000 Bq/m^3) were recorded during welding. For samples obtained near the bases of beams B1/B2, the aerosol composition depended on at least three factors: dust generation during working

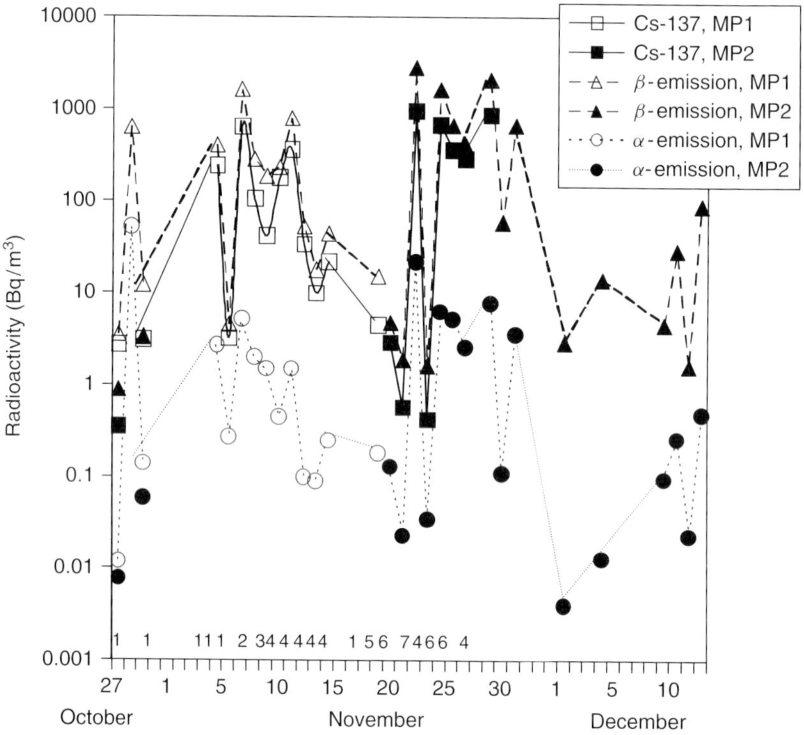

Figure 6.17 Concentration of radioactive aerosols carrying alpha- and beta-emitting nuclides, ^{137}Cs at two locations (MP1 and MP2) of the "Shelter" in 1999. Types of work are shown along the *x*-axis: 1, preparation; 2, mounting of welding equipment; 3, covering by lead sheets; 4, welding; 5, metal cutting; 6, hammering; and 7, mounting lights.

operation, air intake from the central hall of the "Shelter," and air inflow from the free air. Naturally, under these conditions the ratios of radionuclides, in particular, ^{137}Cs and alpha-emitting nuclides, significantly changed, which can be seen in Figure 6.17. The aerosol situation during the work directly or indirectly also influenced other premises of the "Shelter."

6.3.6
Aerosols Generated during Fires in the "Shelter"

Seven fires occurred on completion of the acute phase of the Chernobyl accident at the former reactor 4. The most dangerous ones were at the end of May 1986 in rooms 402/3 and 403/3, and in January 1993 in room 805/3 [68, 69]. Ignition in room 805/3 (2880 m^3 – air-duct of exhaust ventilation) caused complete burning of 1 m^3 of sleepers and 20 m of cables (the mass of burned material was 3500 kg). The average rate of flame spread over the cable surface was 0.5 m/min. The temperature in the fire zone exceeded 800 °C. Toxic and poisonous combustion products were released into the ambient air of the "Shelter."

The release of radioactive smoke from the "Shelter" occurred both through cracks in external constructions and through the pipe VP-2. The results of gamma-spectrometry on the collecting filter working in the system for control of air passing through "Bypass" showed that there was a sharp increase of radioactive aerosol concentration on 14 January (Figure 6.18). While on the preceding and subsequent days of January the daily release of the mixture of gamma-emitting nuclides varied in the range of 0.37–3.7 MBq/day, on the day of the fire it increased up to 33 MBq/day. As the air flow in the "Bypass" made up 25×10^3 m^3/h, the average daily concentration was 55 Bq/m^3. To calculate the average concentration

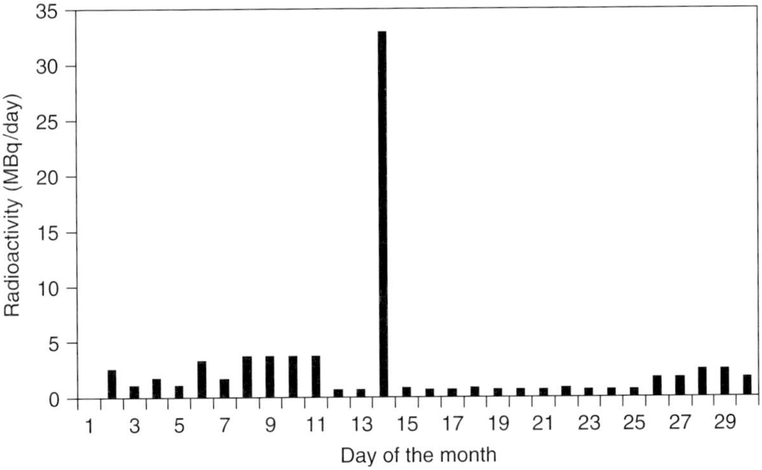

Figure 6.18 Daily emission of gamma-emitting aerosols for the "Shelter" in January 1993 (the fire event occurred on 14 January).

for the actual time of the fire, which lasted for about 6 h, this value should be increased by four times, that is, to 220 Bq/m^3. This is two orders of magnitude larger than the concentration of aerosols released on ordinary days.

The danger of fire from the point of view of the formation and carryover of radioactive products is also associated with the appearance of powerful convection currents. The motions of air masses that exceed the routine flows in their intensity and direction can emerge inside the "Shelter." In addition, high temperatures can cause crumbling of different materials and the formation of *erosive aerosols*.

6.3.7 Dust Control System

The reduction of dust rise has always been a primary task when carrying out technological and research works in and around the "Shelter." A complex series of activities involving the application of special polymeric materials on the building constructions and equipment was started after the commissioning of the "Shelter" to fix radioactive substances onto the surfaces. At first, manual sprayers were used for this purpose. The efficiency of the localizing action of the coverings was evaluated by taking smears from representative samples before and after film application. The determined activity decreased by 2–3 orders of magnitude.

A stationary dust control system intended for the application of coverings directly on the surface of the ruins in the former central hall of reactor 4 was created at the end of 1989. The materials were sprayed using 14 injectors inserted into the airspace above the ruins of the reactor building through roof hatches of the "Shelter." However, spraying through the roof hatches only allowed less than a half of the roof space to be treated [70]. To increase the sprayed area, 35 additional injectors were installed in 2003. The aerosol concentration in the space of the former central hall increased more than once on testing the new system. This was caused by the fact that, before getting into the injectors, the solution forced air out of the collectors. Reaching the dry surface of the ruins, the compressed gas flows raised dust. The first drops hitting the surface and breaking into pieces also caused an inertial rise of dust particles. In 2004–2005, an 80–100 μm thick polymeric film was created on the surface of the ruins of the reactor building [70]. This considerably reduced the dust rise at the time of spraying solutions through injectors [71]. The system was put into commission in 2006. In accordance with the regulations, the film is replaced once a year.

6.3.8 Control of the Release of Radioactive Aerosols through the "Bypass" System

After the commissioning of the "Shelter" in November 1986, the specialists of the ChNPP Radiation Safety Service began to carry out scheduled observations of the release of radioactive aerosols from it using the "Bypass." Aerosols were collected through 35 m pipes with the diameter of 15 mm onto LFS-2 filter tape located in the "Kalina" radiometer, and on AFA RMP-20 or AFA RSP-20 filters. The

Figure 6.19 Mounting of the aerosol sampling pump in the "Bypass."

imperfections in this system are discussed in [72]. It was shown that, depending on the dispersity of aerosols getting into the "Bypass," the concentrations can be reduced down to 100%, especially for particles larger than 3 μm.

To get rid of uncertainties associated with aerosol deposition in the tubes, it was decided to perform aerosol sampling with a blower located immediately in the "Bypass" air flow (Figure 6.19). The specialists of the Institute for Safety Problems of NPP started such control in 2002. The blower with a pack of three-layer Petryanov filters sucked 10–12 m^3 of air within 2 h at a rate of about 0.8 m/s. On completion of the session, each filter layer was measured on radiometers, and the radionuclide composition, concentration, and dispersity of aerosols were determined. To provide isokineticity of aerosol sampling, a conical nozzle directed toward the flow was attached before the filter pack.

In 2002–2008, about 300 aerosol samples were collected in the "Bypass." Figure 6.20 presents data on the activity concentration of the mixture of beta-emitting nuclide products of the accident ($\sum \beta$) from September 2003 to December 2004. As follows from Figure 6.20, the $\sum \beta$ values were mainly in the

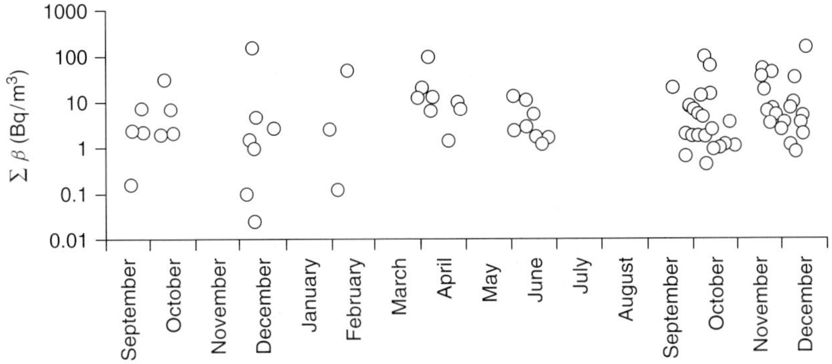

Figure 6.20 Concentration of radioactive aerosol containing a mixture of beta-emitting nuclides measured from September 2003 to December 2004 in the "Bypass" of the "Shelter."

range of 1–10 Bq/m^3. About 30% of the activity was due to ^{137}Cs. The highest concentrations were conditioned by strong winds. For instance, on 8 December 2003 when $\sum \beta = 165$ Bq/m^3, the gusts amounted to 12–13 m/s. In four months, on 5 April 2004, $\sum \beta = 110$ Bq/m^3 was recorded. This was preceded by three days (1–3 April) with gusts of 10–11 m/s and one day (4 April) with the gusts of 8 m/s.

A considerable increase in the concentration of radioactive aerosol occurred after intensive technological activities within the premises of the "Shelter," for example, as noted above, at spraying of polymeric solutions through injectors located under the light roof of the "Shelter." For instance, on 11 February 2004, immediately on completion of the injectors' work, the average $\sum \beta$ within 2 h of sampling came to 164 Bq/m^3. Similar data were obtained during the subsequent years both at switching on injectors and for other works. For example, in September 2005 and in February 2006 on carrying out stabilizing measures in the southern part of the "Shelter," including the welding and slotting works, not only were $\sum \beta$ values amounting to 500–1000 Bq/m^3 observed but also aerosols with unusual staining (black and red).

Figure 6.21 presents $\sum \beta$ values for 2008 when the technological activities conducted in the premises of the "Shelter" were insignificant, most of the stabilizing works having been completed in 2006–2007. The comparison of the data in Figures 6.20 and 6.21 shows that the $\sum \beta$ values again were mainly within the range of 1–10 Bq/m^3: $\sum \beta$ exceeded 10 Bq/m^3 in only nine out of 72 samples collected in 2008. A strong wind was observed in the environment in all these cases either during sampling or a few hours before it. For instance, the largest value of $\sum \beta = 100$ Bq/m^3 was recorded on 8 April when, according to the data of the "Chernobyl" weather station, the gusts amounted to 14 m/s. Half a year later (22 September), after gusts of 10 m/s, $\sum \beta$ made up 34 Bq/m^3.

The effect of the wind on the aerosol situation inside the "Shelter" is one of the properties of this construction. It is conditioned by the presence of cracks and operational openings in its external construction. The effect of meteorological conditions on aerosol behavior in the "Shelter" is discussed in more detail in [73].

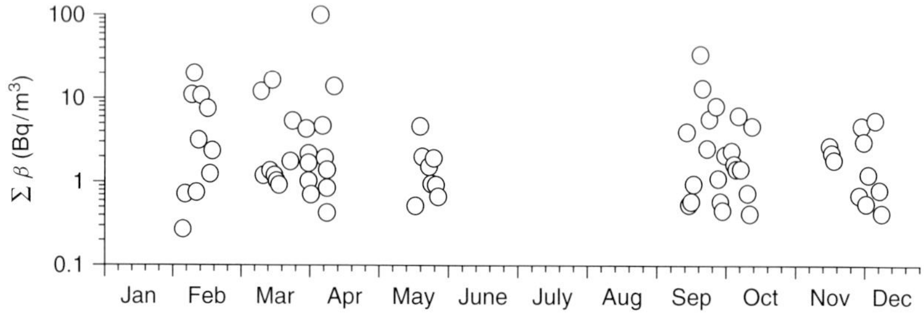

Figure 6.21 Concentration of radioactive aerosol containing a mixture of beta-emitting nuclides measured in 2008 in the "Bypass" of the "Shelter." The results were obtained for the time intervals when no technological procedures were being performed.

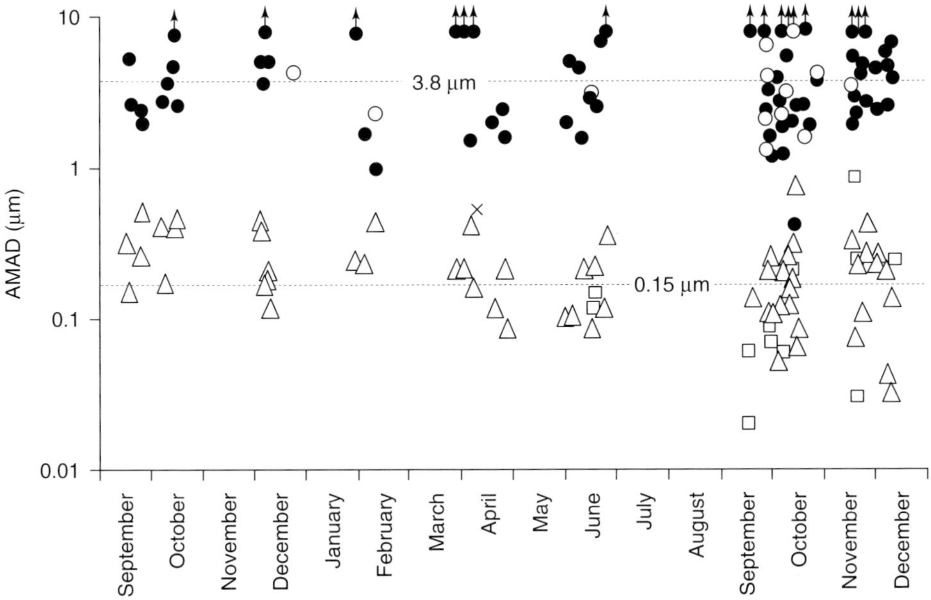

Figure 6.22 Size of radioactive aerosols measured in 2003–2004 in the "Bypass" of the "Shelter." *Key*: beta-emitting aerosols aerosolized naturally by the wind (full circles) and during dust suppression activities (open circles); and DPs of radon and thoron containing aerosols aerosolized naturally by the wind (open triangles) and during dust suppression activities (open squares); AMAD larger than 8 µm (up-arrows).

In particular, it is noted that, in low winds and fog, the concentrations of aerosol products of the Chernobyl accident are usually low.

Processing of the results of measurements of the "Bypass" samples showed that aerosol carriers of radionuclide products of the Chernobyl accident usually had AMAD within the range of 1–10 µm at σ values from 1.1 to 3.5. Data for 2003–2004 are presented in Figure 6.22. For 80 selected samples, the average value was AMAD = 3.8 µm, and 67% of the samples had AMAD from 1.7 to 8.5 µm. Similar data were obtained in 2005–2008. From this it follows that carriers of radionuclides of Chernobyl genesis usually have a dispersion origin. No fractionation of radionuclides (^{90}Sr, ^{137}Cs, and ^{241}Am) by particles of different sizes was observed.

6.3.9
Radon, Thoron and their Daughter Products in the "Shelter"

The DPs of radon (^{222}Rn) and thoron (^{220}Rn) take a special place among the radioactive aerosols that are present in the "Shelter." They influence the radiation situation in the "Shelter" and the detection of radioactive aerosols of Chernobyl genesis. The two gases emanate from concrete constructions of both the former reactor 4 and new elements of the "Shelter" (the cascade and separating walls, materials thrown onto the reactor from helicopters, and so on) containing the natural radionuclides ^{226}Ra and ^{232}Th. Radon and thoron also come from the soil (mainly sand) in which the foundations and lower levels of the ChNPP are embedded. A certain amount of thoron can appear from the irradiated fuel remaining in the ruins of reactor 4 as a result of the decay of ^{232}U formed during two years of reactor functioning before the accident. A chain of three successive alpha-decays of ^{232}U results in the appearance of thoron-generating ^{216}Po and ^{212}Pb. As the half-life of ^{232}U is 72 years, thoron will be generated from it for a long time.

If air containing radon and thoron is inhaled, their DPs present the greatest danger. First, some of them emit alpha-particles. Second, being on submicrometer aerosols, they penetrate into the lower sections of lungs, the bronchi and alveoli.

For radioactive aerosol monitoring in the "Shelter," the DPs of radon and thoron are an interfering factor, because they complicate the radiometry of the samples. In 1987, this was noted by the specialists of the All-Union Instrument-Making Institute on the creation of the system for radiation control and diagnostics at the "Shelter" [74]. According to their data, the activity concentration of aerosols of DPs of natural radioactive gases in controlled premises on average made up 150 Bq/m^3. That is why, in determining the concentration of aerosol products of the Chernobyl accident, filters with collected samples should be kept for about 6 h so that the content of radon DPs has decreased by approximately 1000 times and for about 4 days for the same decrease in the amount of thoron DPs. However, there is also a positive thing: radon and thoron DPs present an original label of submicrometer aerosols. They can be used, for example, to evaluate the efficiency of work FPs and respirators.

For the first time, long-term observations of radon and thoron DPs were conducted in the "Shelter" in December 2000 [75]. Systematic samplings from the ventilation flow coming to the "Bypass" were started in 2002. Radon and thoron DPs were entrapped simultaneously with aerosol products of the Chernobyl accident in a pack of three-layer Petryanov filters. The activity concentrations of ^{212}Pb in 2002–2008 were within the range of 0.5–9 Bq/m^3 (see Figure 6.23). This is 50–100 times higher than estimates obtained in simultaneous samplings performed according to the same technique in March–May 2007 in the vicinity of the "Shelter" [76]. Thus, the thoron source was inside the "Shelter." Consequently, ^{212}Pb was also generated inside the "Shelter" and did not come from outside with

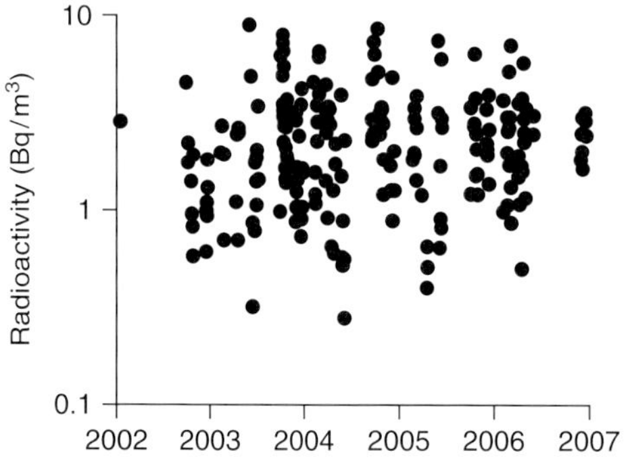

Figure 6.23 Concentration of radioactive aerosol containing ^{212}Pb in the "Bypass" of the "Shelter."

external air. This is confirmed by the data obtained in February–March 2008 in rooms 207/4 and 318/2 of the "Shelter" by the employees of the Department of Radiation Technologies, Materials Science and Environmental Research of the Institute for Safety Problems of NPP. They revealed that ^{212}Pb concentrations made up about 2 Bq/m^3 [77].

Aerosol particles with AMAD of 0.05–0.4 μm (see Figure 6.22) are carriers of radon and thoron DPs both in the "Bypass" and in the "Shelter" [62, 76]. In accordance with Figure 6.22, the average value is AMAD = 0.15 μm, which is typical of carriers of radon and thoron DPs in other regions both outdoors and indoors [78]. This is associated with the fact that carriers of radon and thoron DPs usually have a condensation origin.

The studies in some parts of the "Shelter" showed that radon and thoron DPs, like the parent substances themselves, can accumulate when there is no longer an air draft through the "Bypass." This usually occurs in the second half of spring when in the day time the air temperature in the vicinity of the ChNPP amounts to 30 °C, and non-heated premises inside the "Shelter" still remain cold after winter. Such situations were recorded in room 207/5 during the last 10 days of May 2003 and on 23–24 May 2007. In the first case, ^{212}Pb concentration amounted to 15 Bq/m^3, and in the second to 12–13 Bq/m^3 [76].

The detection of radon concentration exceeding 100 Bq/m^3 and thoron concentration higher than 6 Bq/m^3 in premises of the "Shelter" is a negative factor, which has not been taken into account previously in monitoring the radiation situation. Owing to inhalation of aerosol carriers of the DPs of these noble gases, the average annual radiation dose for the personnel of the "Shelter" can amount to tens of percent of the maximal equivalent dose of 20 mSv/year.

References

1. Abagyan, A.A., Asmolov, V.G., Gus'kova, A.K. *et al.* (1986) Information on the accident at the Chernobyl NPP and its consequences, prepared for IAEA. *At. Energy*, **61** (5), 301–320 (in Russian).
2. Borovoy, A.A., Bogatov, S.A., and Pazukhin, E.M. (1999) Sarcofagus: current state and environmental impact. *Radiochemistry*, **41** (4), 390–401.
3. Filatov, Yu., Budyka, A., and Kirichenko, V. (2007) *Electrospinning of Micro- and Nanofibers and their Application in Filtration and Separation Processes*, Begell House, New York.
4. Petryanov, I.V., Kozlov, V.I., Basmanov, P.I., and Ogorodnikov, B.I. (1968) *PF Fibrous Filtering Materials*, Znaniye, Moscow.
5. Kupny, V.I. (1996) The object "Shelter": yesterday, today, tomorrow in *The object "Shelter"–10 years. Basic results of scientific research*. Chernobyl, pp. 57–77 (in Russian).
6. Izrael, Yu.A., Vakulovsky, S.M., Vetrov, V.A. *et al.* (1990) *Chernobyl: Radioactive Contamination of Natural Media*, Hydrometeoizdat, Moscow (in Russian).
7. International Advisory Group on Nuclear Safety (1988) Final Report of the Meeting on Examination of Causes and Consequences of the Chernobyl Accident. Report of the International Advisory Group on Nuclear Safety. Series of Publications on Biosafety, No. 75, INSAG-1. IAEA, Vienna (in Russian).
8. The International Chernobyl Project (1991) Assessment of Radiological Consequences and Evaluation of Protective Measures. Technical report, Report by

an International Advisory Committee, IAEA, Vienna. ISBN 92-0-129191-4.
9. USSR State Committee on the Utilization of Atomic Energy (1986) The accident at the Chernobyl NPP and its consequences. IAEA Post Review Meeting, 25–29 August, Vienna.
10. United Nations (1988) Sours, Effects and Risks of Ionizing Radiation (Report to the General Assembly). Scientific Committee on the Effects of Atomic Radiation (UNSCEAR), UN, New York.
11. Sivintsev, Yu.V. and Khrulev, A.A. (1995) The evaluation of radioactive discharge at the 1986 accident at the fourth power generating unit of the Chernobyl NPP. *At. Energy*, **78** (6), 403–417 (in Russian).
12. Gudiksen, P.H. *et al.* (1989) Chernobyl source term, atmospheric dispersion, and dose estimation. *Health Phys.*, **57**, 697–706.
13. Talerko, N.N. and Garger, E.K. (2006) The evaluation of the primary discharge from the destroyed unit of the Chernobyl NPP by modeling the atmospheric transport process. *Prob. Nucl. Power Plants Safe and of Chornobyl*, **5**, 80–90 (in Russian).
14. Apsimon, H. *et al.* (1989) Analysis of the dispersal and deposition of radionuclides from Chernobyl across Europe. *Proc. R. Soc. London*, **425** (1869), 245–483.
15. Commissariat a l'Energie Atomique (1986) L'Accident de Tchernobyl, Report IPSN 2/86, Rev. 3, CEA, Paris.
16. Persson, C. *et al.* (1987) The Chernobyl accident. A meteorological analysis of how radionuclides reached and were deposited in Sweden. *Ambio*, **16** (1), 20–31.
17. Kownacka, L. and Jaworwski, Z. (1987) Vertical distribution of ^{131}I and radiocesium in the atmosphere over Poland after Chernobyl accident. *Acta Geophys. Polon.*, **35** (1), 101–109.
18. Mastauskas, A., Nedveckaite, T., and Filistovic, V. (1997) Consequences of the Chernobyl accident in Lithuania. Poster presentation at One Decade after Chernobyl: Summing up the Consequences of the Accident. International Conference, 8–12 April 1996, Vienna, IAEA, vol. 2, pp. 23–29.
19. Finnish Centre for Radiation and Nuclear Safety, Finland (1986) Interim Report on Fallout Situation in Finland from April 26 to May 4 1986, May 1986, STUK-B-VALO 44.
20. Devell, L., Tovedal, H., Bergstrom, U. *et al.* (1986) Initial observations of fallout from the reactor accident in Chernobyl. *Nature*, **321**, 192–193.
21. Devell L., Tovedal H., Bergstrom U. *et al.* (1986) Initial observations of fallout at Studsvik from the reactor accident at Chernobyl. Report No. NP-86/56, Studsvik Energiteknik AB, Sweden.
22. Valkama, I., Salonoja, M., Toivonen, H. *et al.* (1995) *Transport of Radioactive Gases and Particles from the Chernobyl Accident. Environmental Impact of Radioactive Releases*, IAEA-SM-339/69, International Atomic Energy Agency, Vienna, pp. 57–68.
23. Sinkko, K., Aaltonen, H., Mustonen, R. *et al.* (1987) Airborne Radioactivity in Finland after the Chernobyl Accident in 1986, Internal Report, STUK-A56. Finnish Centre for Radiation and Nuclear Safety, Helsinki.
24. Finnish Centre for Radiation and Nuclear Safety (1986) Interim Report. Radiation Situation in Finland 26.4–4.5.1986, STUK-B-VALO-44. Finnish Centre for Radiation and Nuclear Safety, Helsinki.
25. Finnish Centre for Radiation and Nuclear Safety (1986) Second Interim Report. Radiation Situation in Finland 5–16.5.1986, STUK-B-VALO-45. Finnish Centre for Radiation and Nuclear Safety, Helsinki.
26. Health and Welfare Canada (1987) Environmental Radioactivity in Canada–1986. Radiological Monitoring Annual Report, No. 87-EHD-136. Health and Welfare Canada, Ottawa.
27. Selegey, V.V. (1997) Radioactive contamination of Novosibirsk – the past and the future. *Novosibirsk*, 162 (in Russian).
28. Polikarpov, G.G., Egorov, V.N., Gulin, S.B. *et al.* (2008)in *Radioecological Response of the Black Sea to the Chernobyl Accident* (eds G.G. Polikarpov and V.N. Egorov), ECOSEA-Hydrophysics, Sevastopol (in Russian).

29. International Atomic Energy Agency (2006) Environmental Consequences of the Chernobyl Accident and their Remediation: Twenty Years of Experience. Report of the UN Chernobyl Forum Expert Group "Environment". Radiological Assessment Report, IAEA, Vienna.

30. Anspaugh, L.R. (2008) Environmental consequences of the Chernobyl accident and their remediation: 20 years of experience Chernobyl: looking back to go forward. Proceedings of an International Conference on Chernobyl: Looking Back to Go Forward, 6–7 September, 2005, Vienna, pp. 47–76.

31. International Atomic Energy Agency (2006) *Radiation Conditions of the Dnieper River Basin: Assessment by the IAEA Project Team and Recommendations for Strategic Action Plan*, IAEA, Vienna.

32. Ukrainian Hydro-Meteorological Institute (2004) Database of Radiation Measurements of Aquatic Samples. Ukrainian Hydro-Meteorological Institute, Kiev.

33. Matushchenko, A.M. (1996) *In the Air – Radiation Exploration Aircrafts. Chernobyl: Disaster. Feat. Lessons and Conclusions*, Inter-Vesy, Moscow, pp. 436–456 (in Russian).

34. Mezelev, L.M. (2002) *They Were the First*, Part 2, CNII Atominform, Moscow (in Russian).

35. Dobrynin, Yu.L. and Khramtsov, P.B. (1993) Date verification methodology and new data for Chernobyl source term. *Radiat. Prot. Dosim.*, **50** (2–4), 307–310.

36. Borisov, N.B., Verbov, V.V., Kaurov, G.A. *et al.* (1992) The composition and concentration of gas-aerosol radioactive substances above the ruins of the fourth unit of the Chernobyl NPP and in the remote zone in May 1986. *Environment Control, Problems of Ecology and Product Inspection*, Vol. 1, Research Institute for Technical and Economic Information in Chemical Industry, Moscow, pp. 11–17 (in Russian).

37. Ogorodnikov, B.I. (1998) A trap for radioactive aerosols ("Petryanov filters"). *Energ. Econ. Technol. Ecol.*, **8**, 34–39 (in Russian).

38. Borisov, N.B. (1978) *New Sorption-Filtrating Materials for Analysis of Aerosols and Vapors. Isotopes in the USSR*, vol. 52/53, Atomizdat, Moscow, pp. 66–67 (in Russian).

39. Borisov, N.B., Ogorodnikov, B.I., Kachanova, N.I. *et al.* (1992) Observation of gas-aerosol components of radioiodine and radioruthenium during the first weeks after the accident at the Chernobyl NPP. *Environment Control, Problems of Ecology and Product Inspection*, Vol. 1, Research Institute for Technical and Economic Information in Chemical Industry, Moscow, pp. 17–24 (in Russian).

40. Budyka, A.K., Ogorodnikov, B.I., and Skitovich, V.I. (1998) Evaluation of the aerosol particle size by means of three filters. PARTEC 98, 7th European Symposium on Particle Characterisation, 10–12 March, 1998, Nürnberg, Germany, Preprint III, pp. 1239–1245.

41. Budyka, A.K., Ogorodnikov, B.I., and Skitovich, V.I. (1993) Filter pack technique for determination of aerosol particle sizes. *J. Aerosol Sci.*, **24** (Suppl. 1), S205–S206.

42. Belovodsky, L.E. and Panfilov, A.P. (1997) Ensuring radiation safety during construction of the facility "Ukrytie" and restoration of unit 3 of the Chernobyl nuclear power station. Poster presentation at One Decade after Chernobyl: Summing up the Consequences of the Accident. International Conference, 8–12 April, 1996, Vienna, IAEA, vol. 2, pp. 574–590.

43. Ogorodnikov, B.I., Budyka, A.K., Skitovich, V.I., and Todoseichuk, S.P. (1988) The study of efficiency of the air supply system of civil defense shelters at the Chernobyl NPP. The 1st Scientific–Technical Seminar on the Basic Results of Elimination of the Consequences of the Accident at the Chernobyl NPP, April 1988, Chernobyl, Collected Abstracts, p. 13 (in Russian).

44. Teplov, P.L., Shaposhnikov, B.G., Groshev, I.M. *et al.* (1989) The study of the radiation situation at the Chernobyl NPP after the accident. Report to the 1st All-Union Scientific–Technical Meeting on the Results of Elimination of the Consequences of the Accident at the Chernobyl NPP, Chernobyl, vol. 1, pp. 40–133 (in Russian).

45. Geras'ko, V.N., Klyuchnikov, A.A., Korneev, A.A. *et al.* (1997) *The Object "Shelter". History, State and Prospects*, Intergrafik, Kiev (in Russian).

46. Garger, E.K. and Kuz'menko, Yu.I. (2007) The prediction of activity concentration of ^{137}Cs in the surface atmospheric layer of the 30 km zone of the Chernobyl NPP. *Prob. Nucl. Power Plants Safe and of Chornobyl*, **7**, 96–102 (in Ukrainian).

47. Begichev, S.N., Borovoy, A.A., Stroganov, A.A. *et al.* (1990) Reactor Fuel of the Fourth Unit of the Chernobyl NPP, IAE 5268/3, I.V. Kurchatov Institute of Atomic Energy, Moscow (in Russian).

48. Komarov, V.I., Malkov, V.L., Smirnov, N.V. *et al.* (1997) The concentration dynamics of radioactive atmospheric aerosols in the area of the Chernobyl NPP. Report to the 1st All-Union Scientific–Technical Meeting on the Results of Elimination of the Consequences of the Accident at the Chernobyl NPP, Chernobyl, Vol. 5, pp. 115–121 (in Russian).

49. Derevets', V.V., Kireev, S.I., Obrizan, S.M. *et al.* (2001) Radioactive state of the exclusion zone. 15 years after the accident. *Bull. Ecol. State Exclusion Zone*, **17**, 5–19 (in Ukrainian).

50. NPO (1994) Radioactive State of the Environment and Radiation Safety in the Exclusion Zone of the Chernobyl NPP. Report NPO "Pripyat" No. 1002 – Chernobyl (in Russian).

51. Ogorodnikov, B.I. (2002) The origin and components of radioactive aerosols at the industrial site of the object "Shelter" of the Chernobyl NPP. *At. Energy*, **93** (5), 375–383 (in Russian).

52. Ogorodnikov, B.I., Pavlyuchenko, N.I., Khan, V.E. and Krasnov, V.A. (2004) The relation between the concentrations of radioactive aerosols in the "Bypass" of the object "Shelter" and weather conditions. *Prob. Chernobyl*, **15**, 14–23 (in Russian).

53. Garger, E.K., Kashpur, V.A., Skoryak, G.G. *et al.* (2004) Aerosol radioactivity and disperse structure at the Chernobyl NPP site during the period of forest fires. *Agroecol. J.*, **3**, 6–12.

54. Budyka, A.K. and Ogorodnikov, B.I. (1995) Radioactive aerosols at fires on territories contaminated with products of the Chernobyl accident. *Radiat. Biol. Radioecol.*, **35** (1), 102–109 (in Russian).

55. Ogorodnikov, B.I., Miroshnichenko, A.V., and Sharapov, A.G. (2000) Behavior of ^{137}Cs of Chernobyl genesis in aerosols and soil at temperatures up to 900 °C. International Conference on Radioactivity at Nuclear Explosions and Accidents, St Petersburg, April 24–26, 2000, vol. 1, Hydrometeoizdat, Moscow, pp. 720–726 (in Russian).

56. Pazukhin, E.M., Borovoy, A.A., and Ogorodnikov, B.I. (2004) A forest fire as a factor of environmental redistribution of radionuclides of Chernobyl genesis. *Radiochemistry*, **46** (1), 93–96 (in Russian).

57. Kashparov, V.A., Lundin, S.M., Kadygrib, A.M. *et al.* (2000) Forest fires in the territory contaminated as a result of the Chernobyl accident: radioactive aerosol resuspension and exposure of fire-fighters. *J. Environ. Radioact.*, **51**, 281–298.

58. Yoschenko, V.I., Kashparov, V.A., Protsak, V.P. *et al.* (2006) Resuspension and redistribution of radionuclides during grassland and forest fires in the Chernobyl exclusion zone: Part 1. Fire experiments. *J. Environ. Radioact.*, **86**, 143–163.

59. Yoschenko, V.I., Kashparov, V.A., Levtchuk, S.E. *et al.* Resuspension and redistribution of radionuclides during grassland and forest fires in the Chernobyl exclusion zone: Part II. Modeling the transport process. *J. Environ. Radioact.*, **87**, 260–278.

60. Vladimirov, V.A. and Malyshev, V.P. (2001) The results of overcoming the consequences of the Chernobyl disaster, in *Chernobyl. Duty and Courage* (ed. A.A. Dyachenko), Voenizdat, Moscow (in Russian).

61. Budyka, A.K., Ogorodnikov, B.I., and Skitovich, V.I. (1990) *Fission Product*

Transport Processes in Reactor Accidents, Hemisphere, New York, pp. 779–787.

62. Budyka, A.K. and Ogorodnikov, B.I. (2008) Radioactive aerosols of Chernobyl origin in 1986–2007. European Aerosol Conference – 2008, Thessaloniki, Abstract T07A020P.
63. Budyka, A.K. and Ogorodnikov, B.I. (1999) Radioactive aerosols generated by Chernobyl. *Russian J. Phys. Chem.*, **73** (2), 310–319.
64. Aryasov, P., Nechaev, S., Dmitrienko, A., and Konstantinenko, S. (2009) Radioactive aerosol monitoring program with personal impactors usage during the work at the Object Shelter. *Nucl. Radiat. Safe.*, **12** (1), 49–54 (in Ukrainian).
65. The Shelter's Current Safety Analysis and Situation Development Forecasts (Updated Version) (1998). Executive in charge A. A. Borovoy. Tacis services DG IA. European Commission. Brussels. 103 p.
66. Poluektov, P.P., Kolomeitsev, G.Yu., and Timonin, V.V. (1994) Express control of aerosols radioactivity at restoration activities at the Chernobyl NPP. *At. Energy*, **76** (5), 435–441 (in Russian).
67. Nemchinov, Yu.I., Krivosheev, P.I., Sidorenko, M.V. *et al.* (2006) *From the Shelter to Confinement of the Fourth Unit of the Chernobyl NPP. Construction Aspects* (eds P.I. Krivosheev *et al.*), Logos, Kiev (in Russian).
68. Azarov, S.I., Tokarevsky, V.V. (1997) Evaluation of radiation state at fire on object "Shelter". *At. Energy*, **82** (3), 235–237 (in Russian).
69. Azarov, S.I. (2001) The analysis of damaging factors at fires in the object "Shelter". *At. Energy*, **90** (4), 296–304.
70. Klyuchnikov, A.A., Krasnov, V.A., Rud'ko, V.M., and Shcherbin, V.N. (2006) The object "Shelter": 1986–2006, Institute for Safety Problems of NPP NAS of Ukraine, Chernobyl (in Russian).
71. Ogorodnikov, B.I., Khan, V.E., Krasnov, V.A., and Kovalchuk, V.P. (2008) Monitoring of aerosol situation in the object "Shelter" at the MSDS session in April 2008. *Prob. Nucl. Power Plants Safe and of Chornobyl*, **10**, 133–135 (in Ukrainian).
72. Dovyd'kov, S.A. and Ogorodnikov, B.I. (2007) The evaluation of the deposition of radioactive aerosols in the sampling channel of the "Bypass" system of the object "Shelter". *Prob. Nucl. Power Plants Safe and of Chornobyl*, **7**, 110–115 (in Ukrainian).
73. Ogorodnikov, B.I. and Skorbun, A.D. (2007) The relation between radioactive aerosols released from the object "Shelter" and the meteorological situation. The effect of wind velocity. *Prob. Nucl. Power Plants Safe and of Chornobyl*, **7**, 103–109 (in Ukrainian).
74. Zalmanzon, Y.E. and Fertman D.E. (1989) The evaluation of the possibility of automated control of contamination of the object "Shelter" air with radioactive aerosols. Report to the 1st Scientific–Technical Meeting on the Results of Elimination of the Consequences of the Accident at the Chernobyl NPP, Chernobyl, vol. 7, pp. 61–65 (in Russian).
75. Ogorodnikov, B.I. and Budyka, A.K. (2001) Monitoring of radioactive aerosols in the object "Shelter". *At. Energy*, **91** (6), 471–475.
76. Ogorodnikov, B.I., Budyka, A.K., Khan, V.E. *et al.* (2008) Aerosols-carriers of ^{212}Pb in the object "Shelter". *Prob. Nucl. Power Plants Safe and of Chornobyl*, **9**, 54–65 (in Ukrainian).
77. Badovsky, V.P., Klyuchnikov, A.A., Kravchuk, T.A., Melenevsky, A.E., and Shcherbin, V.N. (2008) Monitoring of aerosol situation in some sub-reactor premises of the object "Shelter". *Prob. Nucl. Power Plants Safe and of Chornobyl*, **10**, 99–110 (in Ukrainian).
78. Postendörfer, J. (1994) Properties and behaviour of radon and their decay products in air. *J. Aerosol Sci.*, **25**, 219–263.

Part II
Aerosol Measurement and Characterization

Aerosols – Science and Technology. Edited by Igor Agranovski

ISBN: 978-3-527-32660-0

7
Applications of Optical Methods for Micrometer and Submicrometer Particle Measurements

Aladár Czitrovszky

In memory of my father – Dr. Aladár Czitrovszky (physician)

7.1
Introduction

It was 1914 when Richard Zsigmondy (winner of Nobel Prize in 1925) developed an instrument for observation of the light scattered by small particles – the ultramicroscope (see Figure 7.1) [1]. Since then, the study and application of this instrument have become important in colloid chemistry, hydrosols, toxicology, and so on where micrometer and submicrometer particles play a substantial role. A number of other scientists in the middle of the twentieth century (see the next paragraph), working in optics and optical engineering, also directly or indirectly influenced the field of *particle measurements* and created something enduring for this science.

This knowledge gained in the optical investigation of aerosols and *optical engineering*, firmly rooted in the past, was developed intensively in the last few decades. The study of *light scattering* and its applications – the development of light scattering, extinction, and other *optical measurement* methods, and instrumentation, which was accelerated especially after the development of the laser by Charles Townes, Alexander Prochorov, and Nikolay Basov (winners of the Nobel Prize in 1964) – is now the field of activity of hundreds of research centers worldwide.

The common feature of these methods is the possibility to perform *non-contact, real-time*, mainly *in situ*, measurements with a short sampling time and high accuracy. As the medium holding the information is the light itself, in most cases the sampling can be done at a distance from the sample. The next advantage of optical methods is the possibility to determine simultaneously a number of different parameters – concentration, size distribution, *complex refractive index* (real and imaginary part), density, and so on.

The aims of this chapter are (i) to give a short overview of existing light scattering methods and theories, (ii) to review the state of the art of their application to particle measurements in the micrometer and submicrometer range, (iii) to demonstrate

Aerosols – Science and Technology. Edited by Igor Agranovski

ISBN: 978-3-527-32660-0

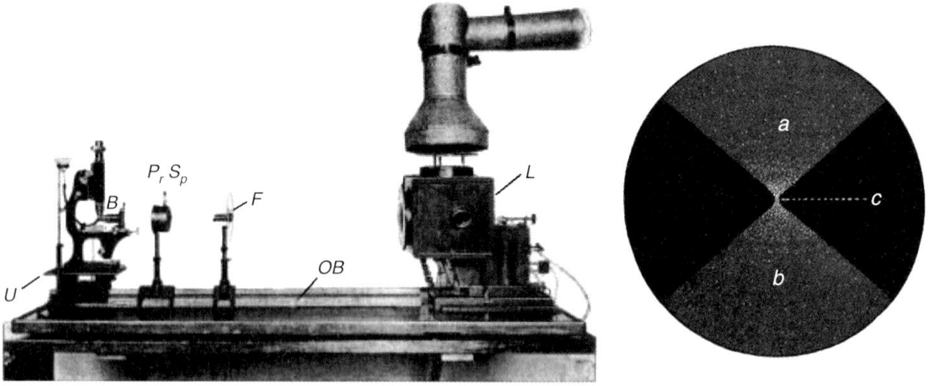

Figure 7.1 The ultramicroscope of Richard Zsigmondy and the dark-field image. From Zsigmondy and Bachmann [1].

some examples of the development of particle measurement instrumentation, emphasizing the development of new airborne particle counters and sizers, and new methods for the determination of their optical properties, electrical charge, density, and so on, (iv) to give a brief comparison of different instruments, and (v) to present a short analysis of possible further applications.

After outlining the history of light scattering and introducing the background of two light scattering theories in the next two sections, we present the classification of the optical instruments for particle measurements. After that, the development of airborne and liquid-borne particle counters and sizers are described. The next section contains new methods used to characterize the electrical charge, density, and complex refractive index of aerosol particles. The operational principle and novelties of the instruments are briefly described, as well as their advantages and limitations. Further, finally, we compare the technical parameters of several commercially available instruments with newly developed ones and analyze their further applications.

We hope that this chapter will help to better understand the optical methods applied for the characterization of aerosols.

7.2
Optical Methods in Particle Measurements

Light is scattered because there is a perturbation of the *refractive index*, Δn_i, in the particular medium through which the light is traveling. Such perturbation may be caused by, for example: isobaric fluctuation; adiabatic fluctuation, produced by pressure, which propagates via an elastic wave (for example, an acoustic wave); fluctuation of anisotropy; fluctuation of concentration; and change of refractive index caused by particles suspended in the air or liquid having optical properties different from those of the surrounding medium. Therefore, if a light beam is passing through such an airborne or liquid-borne suspension, it will be scattered in all directions. When studying the light scattered by particles suspended in air or in a pure transparent liquid, all these kinds of Δn_i caused by fluctuation of the molecules – except the last one corresponding to suspended particles – can usually be neglected. This type of light scattering was first observed by Arago [2] and Govi [3], and experimentally investigated by Tyndall [4] for particles smaller than the wavelength of the incident white light.

Tyndall discovered that the scattered light has a bluish coloration, while the transmitted light has a red coloration, and observed the polarization of the light scattered at 90°. The light *scattering intensity* and its angular dependence (light scattering indicatrix) can be determined by the optical parameters of the medium and by the optical and geometrical parameters of the particles. The first light scattering theory was described by Rayleigh between 1871 and 1899, which until now has been found to be applicable for particles that are small with respect to the wavelength (diameter $d < 0.05$ μm for $\lambda = 0.5$ μm) [5–8]. For these particles the Rayleigh scattering law predicts that the total scattering is proportional to d^6 and inversely proportional to λ^4.

When the dimensions of the scattering particles are comparable to the wavelength, the analysis of *Mie scattering theory* is required. Mie developed his more complicated light scattering theory in 1908 [9]. In this theory, the *scattering efficiency* exhibits rapid fluctuation with wavelength. The scattering, extinction, and absorption efficiencies are defined, respectively, as the ratio of the scattering, extinction, and absorption cross-sections to the geometric cross-sectional area. When the scattering particles are much larger than the wavelength, the scattering cross-section becomes independent of wavelength and equal to twice the cross-sectional area of the particles. The dependence upon particle diameter becomes weaker with increasing particle diameter: for 0.3 μm $< d <$ 0.6 μm, the scattering is proportional to d^2. The scattered radiation in Mie theory is polarization dependent and the normalized amplitudes of the flux scattered through a certain angle can be described by two factors that depend on the optical and geometrical properties of the particle differing from the optical properties of the surrounding medium and on the angular dependence of the scattered amplitude [9].

More complicated light scattering theories, taking into account also the *shape factor*, homogeneity, and other parameters of the particles, were developed much later by van de Hulst, Kerker, Hodkinson, and others [10–13]. Light scattering corresponding to the centers that are caused by fluctuations of the molecules was intensively developed after establishing the thermodynamic theory of the fluctuations by Smoluchowski [14] and Einstein [15] between 1905 and 1908. The *molecular light scattering* was developed on the basis of the theory of the fluctuations by Cabannes [16], Strutt (the son of Lord Rayleigh) [17], and Wood [18]. Based on the theory of molecular scattering, the molecular-weight and size distributions of polymers, electrolytic solutions, and so on were determined by Putzeys and Brosteaux [19] and Debye [20], and phase transitions were studied by Ginzburg [21].

Until now we have supposed that the perturbations of the refractive index, Δn_i, caused by fluctuations of the molecules are frozen in space. In real cases

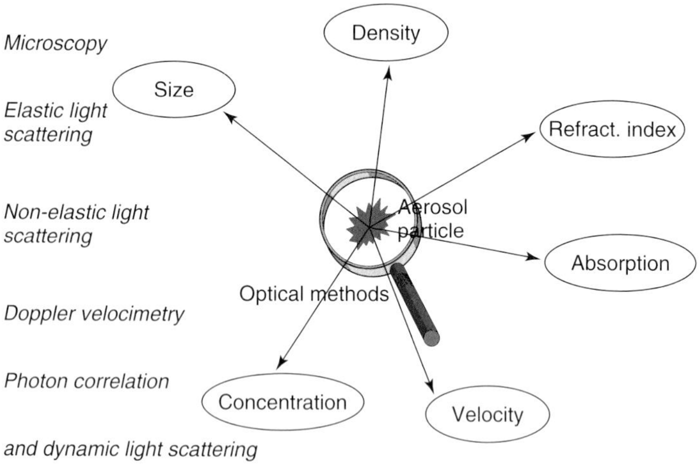

Figure 7.2 Optical methods in aerosol measurements.

these perturbations are permanently moving, which changes the spectrum of the scattered light. This phenomenon was first observed by Brillouin [22]. He observed that in the spectrum of monochromatic Rayleigh-scattered light there also exist two additional lines [22]. This phenomenon – named *elastic scattering* – was observed also by Mandelstam [23], and later by Raman [24], Landsberg [25], Cabannes [26–28], and Fabelinski [29–31].

Mie scattering theory can be used in particle measurements for the determination of particle size and concentration, which is described in [32–48]. Based on previous light scattering experiments [49–51], the development of airborne particle counters and sizers started in the middle of the twentieth century and accelerated in the 1960s after the invention of lasers. The *particle size distribution* in *single-particle counters* is determined by comparing the detected pulse heights of the optical signals that correspond to the single-particle flow through a small illuminated zone with a standard calibration curve obtained from a set of uniform particles of known diameter. The density of the particles can be calculated from the equations of motion in the case of accelerating particles. The electrical charge can be measured by analyzing the trajectory of particles moving in an electric field. The refractive index can be determined from forward- and back-scattering at different wavelengths. The velocity of the particles can be measured using laser Doppler methods. These methods are summarized in Figure 7.2 and described in the following sections.

7.3
Short Overview of Light Scattering Theories

Light scattering can be described as an interaction of light with matter, which leads to a change in the direction, intensity, sometimes the polarization, and – in the case of non-elastic scattering – the length of the incident light wave. The incident light interacting with an optical heterogeneity center of the medium – a local change of refractive index (for example, due to particles) – is deflected from its incident direction, so scattering occurs as an effect of heterogeneity in the optical properties of the medium. Scattering is usually accompanied by absorption, which leads to the attenuation of the light energy propagating in the initial direction and transformation to some other form of energy. As described in the previous section, many authors have studied theoretically and experimentally the phenomena of light scattering and absorption during the interaction of light with various media, so for a detailed description the interested reader can consult the previously mentioned references. In this section we present two basic approaches: (i) for single particles (spheres) much smaller than the wavelength of the incident light, and (ii) for particles comparable to or bigger than the wavelength of the incident light. As the relations between the scattering intensity and the properties of the particles in these two cases are different, the *scattering coefficient* β_P can be described by the sum

$$\beta_P = \beta_M + \beta_R \tag{7.1}$$

where β_M corresponds to the particles comparable to or bigger than the wavelength of the incident light (Mie scattering), and β_R corresponds to the particles smaller than the wavelength of the incident light (Rayleigh scattering). Both components depend on the optical and geometrical parameters of the particles, scattering angle, concentration, and wavelength of the incident light.

For small spherical particles ($a < \lambda$), when the incident light is linearly polarized and the concentration of the particles is not too high (multiple scattering can be neglected) – in the case of Rayleigh scattering – the scattering coefficient will be [36]

$$\beta_R = \frac{8\pi^4 a^6 N}{\lambda^4} \frac{n^2 - 1}{n^2 + 1} \left(1 + \cos^2 \Theta\right) \tag{7.2}$$

where λ is the wavelength of the light, a is the radius, N the concentration and n the refractive index of the particle, and Θ is the scattering angle. As we can see, in this case the scattering intensity is proportional to the sixth power of the radius of the particle and inversely proportional to the fourth power of the wavelength. Molecular light scattering in the atmosphere (arising from optical inhomogeneity caused by the fluctuation of the refractive index due to stochastic motion of the molecules) can also be described by this type of scattering, where the shorter wavelengths are scattered much more strongly (explaining the blue color of the sky). The calculations show that β_R depends on the wavelength: in the range $0.6 < \lambda < 1.0\ \mu m$, β_R varies by one order of magnitude; and in the range $1.0 < \lambda < 10\ \mu m$, by nearly four orders of magnitude.

When the size of the particles is neither very large nor very small compared to the wavelength of light, Mie scattering theory must be used to obtain an accurate prediction of the scattering. This theory represents an exact solution of Maxwell's equation for scattering spheres with complex refractive index. We present here only the main consideration and final result of this theory – the complete derivation can be found in the monograph of van de Hulst [10]. In the case of Mie scattering the scattered light is polarization dependent and we denote by $S_1(\Theta)$ and $S_2(\Theta)$

the normalized amplitudes of the flux scattered through angle Θ; the subscripts 1 and 2 refer to flux polarized normal to and parallel to the scattering plane, respectively. The normalization is made relative to the amplitude incident on the particle (sphere) cross-section.

According to Mie scattering theory

$$S_1(\theta) = \sum_{m=1}^{\infty} \frac{2m+1}{m(m+1)} \left[a_m \pi_m(\cos\theta) + b_m \tau_m(\cos\theta)\right]$$
$$S_2(\theta) = \sum_{m=1}^{\infty} \frac{2m+1}{m(m+1)} \left[b_m \pi_m(\cos\theta) + a_m \tau_m(\cos\theta)\right] \tag{7.3}$$

where

$$\pi_m(\cos\theta) = \frac{1}{\sin\theta} P_m^1(\cos\theta)$$
$$\tau_m(\cos\theta) = \frac{\mathrm{d}}{\mathrm{d}\theta} P_m^1(\cos\theta)$$

P_m^1 is an associated Legendre polynomial, and the coefficients a_m and b_m are given by

$$a_m = \frac{\psi'_m(\hat{n}x)\psi_m(x) - \hat{n}\psi_m(\hat{n}x)\psi'_m(x)}{\psi'_m(\hat{n}x)\xi_m(x) - \hat{n}\psi_m(\hat{n}x)\xi'_m(x)}$$
$$b_m = \frac{\hat{n}\psi'_m(\hat{n}x)\psi_m(x) - \psi_m(\hat{n}x)\psi'_m(x)}{\hat{n}\psi'_m(\hat{n}x)\xi_m(x) - \psi_m(\hat{n}x)\xi'_m(x)} \tag{7.4}$$

where

$$\psi_m(x) = \sqrt{\pi x/2}\, J_{m+1/2}(x),$$
$$\xi_m(x) = \sqrt{\pi x/2}\left[J_{m+1/2}(x) + (-1)^m \mathrm{i} J_{-m-1/2}(x)\right]$$
$$x = ka = \frac{2\pi a}{\lambda}$$

In these equations, a_m and b_m are coefficients characterizing the geometrical and optical parameters of the scattering particle, π_m and τ_m describe the angular dependence of the scattered light, ξ_m and ψ_m are Riccati–Bessel functions, J_m are Bessel functions, a is the radius of the sphere (particle), and x is the size parameter–the ratio of the sphere circumference to the wavelength. As we can see, in this case the scattering coefficient consists of two components – one is connected with the geometrical and optical parameters of the particles, and the other describes the angular distribution of the scattered light (see Figure 7.3).

In the case of a *polydisperse mixture* of particles, the scattering on bigger particles (Mie scattering) is dominant. In this case, the dependence on the wavelength is not as strong as in the case of smaller particles.

As can be seen from the Mie scattering theory, the *scattered intensity* also depends on the complex refractive index. This dependence is especially important in optical measurement methods of aerosols such as optical particle sizing, because usually we try to determine the size distribution of a mixture containing particles with different refractive index. The intensity dependence is shown in Figure 7.4, for

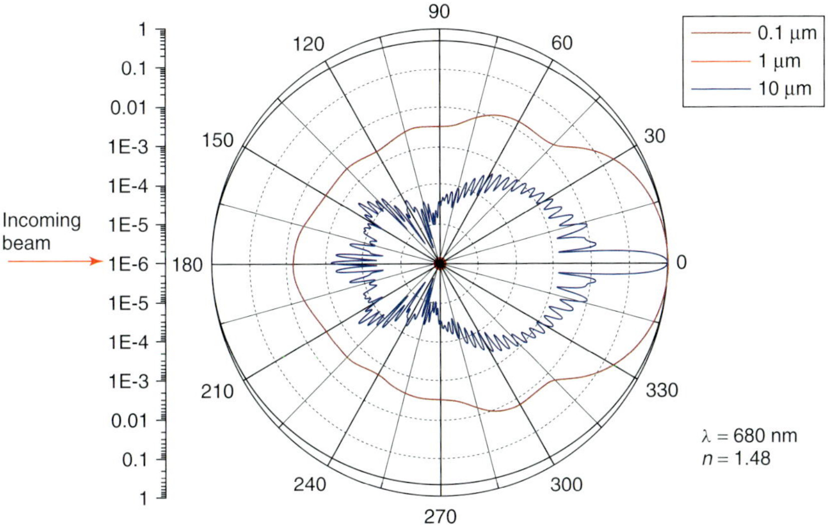

Figure 7.3 Angular distribution of the scattered intensity for different sizes. The curves are normalized to the maximal value for forward scattering. We note that the maximal scattered intensity for a 10 μm particle is more than six orders of magnitude larger than for a 0.1 μm particle.

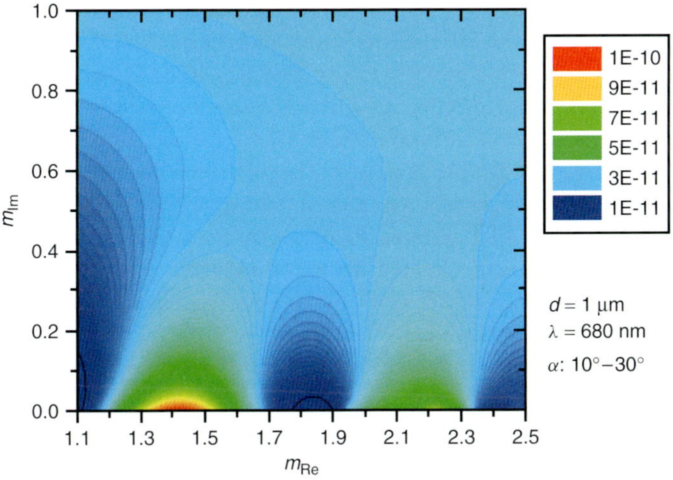

Figure 7.4 Intensity distribution of the scattered light calculated using Mie theory of scattering for 1 μm particles versus the complex refractive index. The color scale indicates the forward scattered intensity in the integration range $10^\circ - 30^\circ$.

unpolarized light in the case of 1 μm particles. In both the previous theories we suppose that the particles are frozen in the medium in which they are located.

If the incident light is coherent and monochromatic (for example, a laser beam), with an appropriate detector working in photon counting regime, it is possible to observe time-dependent fluctuations in the scattered intensity (*dynamic light scattering*). The fluctuations arise from the fact that the particles are small enough to undergo random thermal motion (*Brownian motion*) and the distance between them is therefore permanently varying. Constructive and destructive interference of light scattered by neighboring particles gives rise to the *intensity fluctuation* at the detector, which, as it arises from particle motion, contains information about this motion. Analysis of the time dependence of the intensity fluctuation can therefore yield the diffusion coefficient of the particles, from which, via the *Stokes–Einstein equation*, knowing the viscosity of the medium, the *hydrodynamic radius* or diameter of the particles can be calculated. With Fabelinski, we measured such time dependence of the intensity fluctuation using *photon correlation* techniques and special low-noise photon counting photomultipliers [49, 44, 51]. Using this equipment we measured the intensity *autocorrelation function*, which can be described as the ensemble average of the product of the signal with a delayed version of itself as a function of the delay time. The signal in this case is the number of photons counted in one sampling interval. At short delay times, correlation is high, and, over time, as particles diffuse, the correlation diminishes to zero, and the exponential decay of the correlation function is characteristic of the diffusion coefficient of the particles. Data are typically collected over a delay range of 100 ns to several seconds, depending upon the particle size and viscosity of the medium. Analysis of the autocorrelation function in terms of particle size distribution is done by numerically fitting the data with calculations based on assumed distributions. A truly monodisperse sample would give rise to a single-exponential decay, to which fitting a calculated particle size distribution is relatively straightforward. In practice, polydisperse samples give rise to a series of exponentials, and several quite complex schemes have been devised for the fitting process. One of the methods most widely used today is known as non-negatively constrained least squares (NNLS). Several correlator software programs – for example, the Brookhaven correlator software – include this along with several other approaches to the problem.

Particle size distributions can be calculated either by assuming some standard form such as log-normal or without any such assumption. In the latter case, it becomes possible, within certain limitations, to characterize multimodal or skewed distributions. The size range for which dynamic light scattering is appropriate is typically submicrometer, with some capability to deal with particles up to a few micrometers in diameter. The lower limit of particle size depends on the scattering properties of the particles concerned (relative refractive index of particle and medium), the incident light intensity (laser power and wavelength), and the detector–optics configuration.

Dynamic light scattering – also known as quasi-elastic light scattering (QELS) and photon correlation spectroscopy (PCS) – is particularly suited to determining small changes in mean diameter such as those due to adsorbed layers on the particle surface or slight variations in manufacturing processes.

7.4 Classification of Optical Instruments for Particle Measurements

Optical particle measuring instruments may be further divided according to whether the sensing zone contains one particle or numerous particles at a given time. Multi-particle instruments will be considered first, then single-particle light scattering direct-readout instruments.

7.4.1 Multi-Particle Instruments

Light scattering photometers are multi-particle sensing zone instruments, in which light scattered from particles in the sensing zone falls on a detector off the optical axis. As the number of particles increases, the light reaching the detector increases. The angular pattern of scattering from a sphere is a complicated function of particle diameter, refractive index, and wavelength. Forward-scattering photometers, which employ a laser light source and optics similar to dark-field microscopy, have been commercially produced. A narrow cone of light converges on the aerosol cloud, but it is prevented from falling directly on the photodetector by a dark stop; only light scattered in the near forward direction falls on the detector. The readout of these instruments is in mass or number concentration, but the calibration may change with the composition and size distribution of the particles to be measured [52].

In *integrating nephelometers* the particles are illuminated in a long sensing volume and scattered light reaches the detector at angle from about 8 to 170° off-axis. This simplifies the complex angular scattering relationship by summing the scattering over nearly the entire range of angles. In some cases, the scattering was shown to be well correlated with the atmospheric mass concentration. Some caution must be exercised when using nephelometers in an environment with sooty particles, since the scattering will be attenuated because of light absorption. In this case the particle concentration will be lower than expected [52, 53].

A multi-particle, light scattering instrument that employs a long-pathlength back-scattering light collection is *light detection and ranging* (*LIDAR*) [54]. A powerful pulsed laser is used, and the temporal analysis of back-scattered light indicates the spatial distribution of particles. This type of instrument has been used to map smoke plume opacity in the vicinity of the stack. Unless the size distribution and composition of the particles are known, only a qualitative comparison of aerosol concentration at different locations can be made.

Another method that is similar to scattering is the *laser diffraction method.* This is a multi-particle method that can be used only for size distribution measurement of aerosol or hydrosol particles; the concentration of the particles cannot be determined [55]. The principle of the measurement is based on the observation of the diffraction parameters of laser light on an ensemble of polydisperse particles.

The *aethalometer* is used to measure suspended carbonaceous particulates, including aerosol black carbon (BC) or elemental carbon (EC). This technology uses continuous filtration and optical transmission to measure the concentration of BC in almost real time. Aethalometers are designed for a number of specific applications and measure the absorption at different wavelengths. Magee Scientific Inc. produces a dual-wavelength (880 and 370 nm) version for measurement of aromatic organic species, that is, tobacco smoke, fresh diesel exhaust, and wood smoke, and seven-wavelength aethalometers working in the 370–950 nm range for the study of atmospheric optical properties, radioactive transfer, and so on.

7.4.2
Single-Particle Instruments

Light scattering particle counters or optical particle counters (OPCs) or spectrometers employ a small sensing volume, either by a focused incandescent lamp or a laser source. In such instruments, it is important to avoid coincidence errors resulting from more than one particle in the sensing volume. The instrument manufacturer specifies the maximum number concentration that can be handled. Generally, commercial instruments handle a concentration of up to $\sim 10^6$ particles per liter. Beyond this concentration limit, sample dilution is usually used, which decreases the accuracy of the determination of the concentration. The range of particle diameters that single-particle instruments are able to handle is ~ 0.3–$10\ \mu m$. OPCs have found wide use, first in clean-room monitoring and more recently in community air pollution and industrial hygiene studies. A number of laboratory instruments employing single-particle scattering have been constructed. A critical review of such instruments is given by Chigier and Stewart [56]. The principle employed in these instruments will likely be used in the next generation of commercial particle counters.

Condensation nuclei counters or condensation particle counters (CPCs) are used to measure the total number concentration of airborne particles much smaller than 0.5 μm, which cannot be detected directly by light scattering. These instruments use the principle of adiabatic expansion or cooling in a vapor-saturated chamber. Vapor condenses upon nuclei and the particles grow to a detectable final diameter. They are then counted by light scattering. The instrument measures total nuclei number concentration, since the final particle size is relatively independent of the number of nuclei present. Three types of instruments are currently in use. The first is a manual type in which a single expansion is performed in a water vapor-saturated chamber. In the second type, the expansions are performed cyclically, two per second, in a smaller water vapor-saturated chamber. A third type has been developed in which the particles are passed continuously through an alcohol chamber and are grown by cooling [57].

In the above instruments, the light interaction with a particle (or particles) is measured to obtain information directly. The particle motion is not important as long as it remains for an appropriate time in the sensing volume. Several instruments have been developed in which optical detection is used to infer particle motion.

In *aerodynamic particle counters* the aerodynamic diameter of a particle can be measured by determining the particle velocity in an accelerating flow by measuring the time of flight of the particle between two spots in the acceleration region after aerodynamic focusing by an orifice nozzle. Commercially available instruments have been designed for aerosols in the range of 0.5–10 μm. The advantage of this method is that it is independent of the optical properties of the particles as long as the scattered light can be detected [58–67].

The study of light scattering and the development of light scattering instrumentation, especially particle counters and sizers, in the past two or three decades were motivated by the widening of the market for clean room technologies. At the end of the 1960s and beginning of the 1970s, several big firms for the production of OPCs and sizers were established, mainly in the USA (TSI, Hiac-Royco, Climet, Met-One, Amherst, Particle Measuring Systems), in England (Malvern), and in Germany (Polytec, Topas, and others).

At the end of the 1980s the market for light scattering instrumentation widened as well as the number of companies. At the same time the number of technologies that needed clean room monitoring instrumentation increased – now clean room technology is involved not only in microelectronics, but also in pharmacology, in medicine, in the photochemical industry, in several branches of the packaging industry, in the paint industry, in the production of precise mechanics and optics, in the production of filters, and so on. In connection with the firms, the production of clean room monitoring equipment has been increased up to about 30 firms, but the influence of the old firms having a wide service network is still strong.

Another branch of application of particle counters and sizers based on light scattering is the field of high particle concentration – for example, industrial air quality control, environmental monitoring, toxicology, aerosol testing, health control, and so on. For these applications, a series of instruments were developed by Amherst, TSI, Topas, Grimm, Technoorg, and so on.

A detailed analysis of more than 20 instruments produced by 15 manufacturers is presented in Section 7.8, where the main parameters and a comparison of the different commercially available instruments are described.

In the next sections we will describe some of the airborne and liquid-borne particle counters developed by various authors and groups.

7.5 Development of Airborne and Liquid-borne Particle Counters and Sizers

The airborne and liquid-borne particle counters are based on the same principle: counting and sizing of particles (one-by-one) when they pass through the illuminated sensing volume. For determination of the concentration, the flow rate (the volume to be tested) is also measured simultaneously. The main difference between airborne and liquid-borne particle counters is the handling of the flow. At the same time, the relative refractive index of particles in liquids is smaller than in the air or gases, so the lower size limit is usually higher.

The different light detection geometries applied in different particle counters yields a wide range of instrument designs and constructions. As the dependence of the scattered intensity on the refractive index of the particles is less pronounced around 90° scattering angle, in most airborne particle counters perpendicular scattering geometry is implemented.

In several particle counters – for example, in the Dual Wavelength Optical Particle Spectrometer (DWOPS) described in Section 7.7 – simultaneous measurement of some other parameters is also possible (in DWOPS, the complex refractive index). A special sampling and detection geometry providing the measurement of very high concentrations is used in particle counters made by Topas. Several particle counters have some additional modular units to measure specific aerosols – for example, the particle counters of Grimm have a modular unit for measurement of polyaromatic hydrocarbons or nanoparticles, which can be attached to different Grimm instruments.

Special types of liquid-borne particle counters have been developed for pharmacological and hematological applications. These instruments usually are designed for measurement of special particles – for example, the concentration of erythrocytes and leucocytes in blood (Diatron Ltd), for measurement of aerosol drugs, and so on.

7.5.1
Development of Airborne Particle Counters

The principle of airborne particle counters is shown in Figure 7.5. The aerosol to be tested is passed through an illuminated zone and scatters the light of a laser beam in all directions. Part of this scattered light in a certain acceptance angle is collected by the optical system of the detector. The amplitude of the photoelectric signal generated on the detector by the particle crossing the illumination zone is compared with the standard calibration curve obtained from a set of known particles.

Calibration curves, calculated using Mie scattering theory for different scattering and integration angles, are shown in Figure 7.6. As we can see, especially in the case of forward and back-scattering, the relation between size and scattered intensity in certain size ranges is not uniform: the long-period oscillations for back-scattering are pronounced in the 0.4–1 μm range, and for forward scattering in the 1–3 μm range. In the case of perpendicular scattering, shorter oscillations are pronounced in the 2–5 μm range, but if we measure the size distribution, this geometry is even better. In this case, the dependence on the refractive index of the particles is also less pronounced.

For further considerations we have also plotted the calculations of the combination of forward and back-scattering, which we applied on our new aerosol analyzer, described in Section 7.7. This device is capable of measuring not only the size

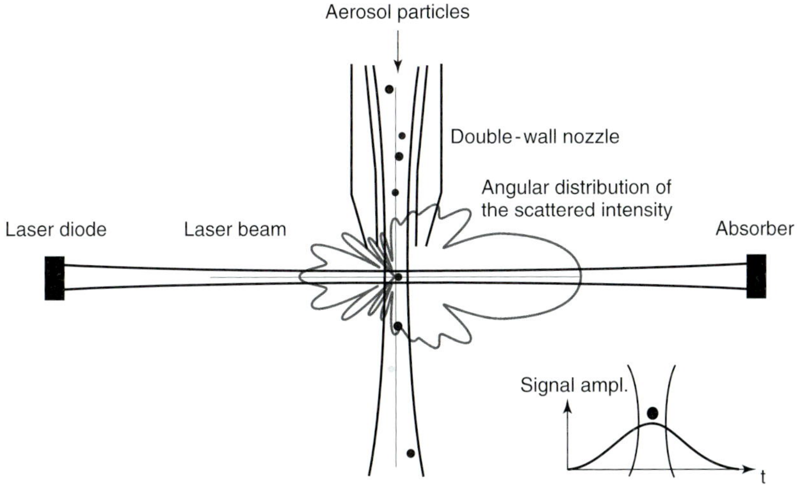

Figure 7.5 The principle of airborne particle counters. The particle crossing the illumination zone produces a photoelectric impulse, which is connected with the size of the particle to be measured.

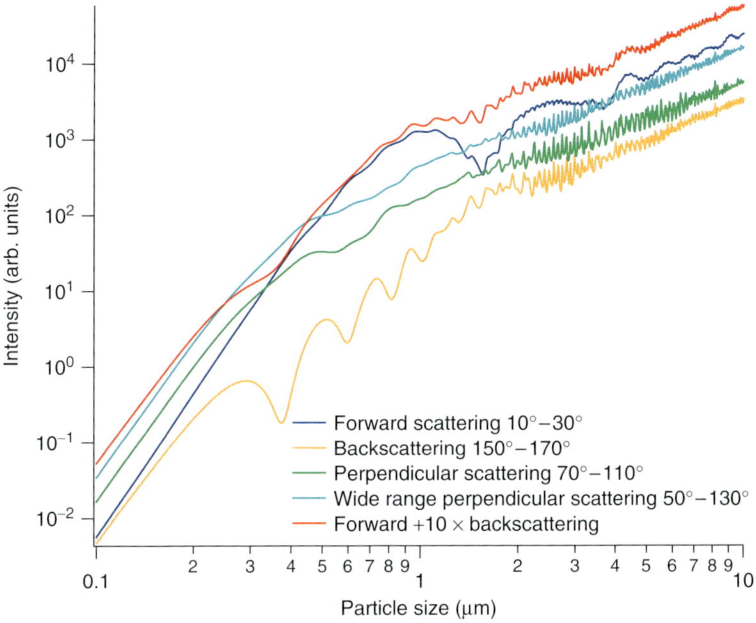

Figure 7.6 Calculated calibration curves for different scattering angles and integration ranges for polystyrene latex.

distribution and concentration but also the complex refractive index of aerosol particles.

Previous studies of the measurement of particle size distribution in aerodynamic particle sizers [59, 60, 68] showed that devices operating by means of measuring the velocity of each individual particle in the incoming aerosol stream accelerated through an orifice nozzle underestimate the aerodynamic diameter by an average of 25% [61, 62], even in the case of regular-shaped non-spherical particles. The reason for such an underestimation is that these devices work under ultra-Stokesian flow conditions (within the measurement zone, particles having an aerodynamic diameter greater than 1 μm have a Reynolds number greater than 1 [63]). Similar effects are also obtained when measuring liquid droplets larger than a few micrometers, since liquid spheres are distorted when passing through the region of strong acceleration in the tapered nozzle [64, 65]. According to [66, 69], the measurement results depend on the particle density even in the case of spherical particles. The same devices overestimate the size of particles that have a density appreciably greater than unity. Because of these, several manufacturers have introduced an option in the operating software to correct for deviations in particle density [67]. An additional problem is that calibration curves obtained at reduced ambient pressure are different from the manufacturer's data, indicating that recalibration of the device is required if other than standard operating conditions occur [70].

Here we describe the design of a light scattering airborne particle counter that differs from the well-known OPCs, which have forward scattering geometry and a lower limit of registration of 0.5 μm [71], from particle counters that have a lower limit of registration of 0.1–0.3 μm, which have the great disadvantage of placing the sampling chamber inside the laser resonator and passing the particles through the open cavity laser [72–74], and from particle counters in which the scattered light is collected by two elliptical mirrors [75].

In this device, illumination and light detection are performed outside the laser resonator and the permanent stabilization of the photodetector is combined with the aerodynamic focusing of the aerosol stream to be tested. The size detection limit of the system is not altered, and it remains low (~0.2 μm).

The optical scheme of our instrument is shown in Figure 7.7. The light from a laser falls on the L_1–L_4 focusing lens system and the diaphragm D_1. The appropriate diameter and position of the light spot inside the hermetic optical chamber is achieved by fine adjustment of lens L_3. The air to be tested is forced through a vertical nozzle, which is located at point P perpendicularly to the plane of the figure. The curvature of mirror M_1 is exactly matched to the beam divergence, so the beam is reflected into itself. Thus the light scattered opposite to the direction of detection is reflected back towards the detector. By this arrangement the collected light intensity is increased approximately two times. Moreover, in this way, the optical system can easily be adjusted. Diaphragm D_2 is very important, as it reduces the amount of light scattered from the optical elements and mechanical parts of the system.

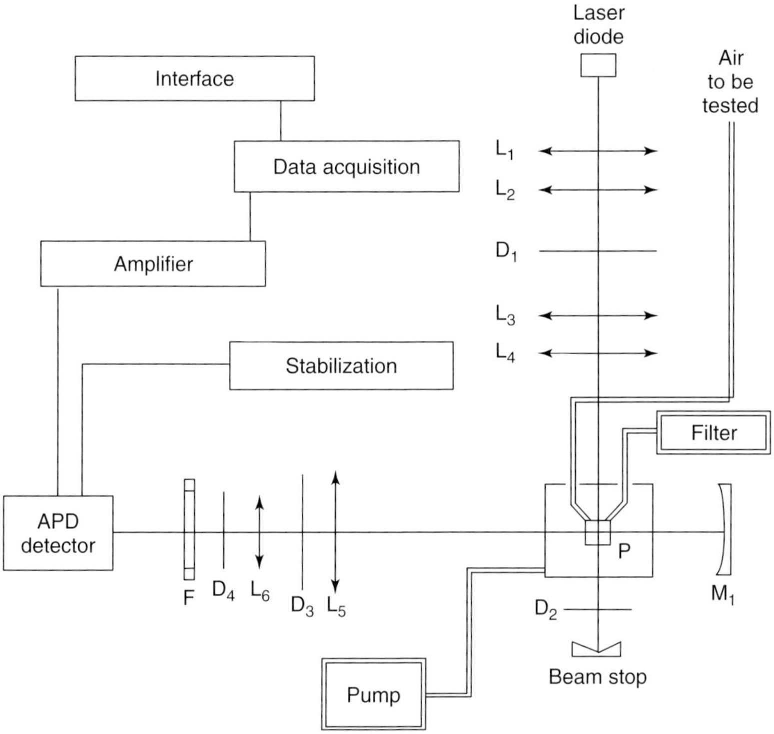

Figure 7.7 The optical scheme of the airborne particle counter APC-03-2C.

The light collecting optics consists of lenses L_5 and L_6 and diaphragms D_3 and D_4. The measurement volume, with dimensions approximately $400 \times 500 \times 500\,\mu m^3$, is seen at 60° by the collecting optics. This relatively large collecting angle has the benefit of integrating the intensity oscillations of the light scattered from large particles, which is especially important in the case of 90° scattering geometry. Perfect alignment of mirror M_1 was achieved by monitoring the signal of the photomultiplier when single particles were passing through the measurement volume.

The air to be tested is forced through the optical chamber by a high-capacity push–pull diaphragm pump. Aerodynamic focusing of the sample air is performed by a double-wall confocal nozzle. The air enters the inner inlet of the nozzle; the clean air flow (after passing a zero-count filter) exits via the external inlet surrounding the air to be tested and accelerates it. Within the external inlet, the inner one can be hermetically adjusted so that it is vertically aligned with the former. After choosing the appropriate pressure conditions and the aerodynamic focusing parameters, the location and diameter of the waist of the sample air stream can be matched to the optical beam.

In the present system, the photomultiplier is permanently stabilized to give a standard output signal in response to a standard optical input signal. The standard optical input signal is produced by stabilized light-emitting diode impulses applied to the photomultiplier. The frequency of this signal is 70 Hz, and its amplitude is equivalent to the light scattered from the largest particles. This frequency is subtracted from the counted number in the appropriate size range. A stabilizer is built into the system to compensate for fluctuations of the photomultiplier by controlling its anode–cathode voltage through a feedback loop. The light scattered from the measured particles is filtered by a narrow-band filter F, which is adjusted to the wavelength of the laser light.

The size ranges were chosen in accordance with the standards for particle size distribution measurements. The lower limit of the first size range is 0.3 μm; however, the size detection limit is 0.2 μm. The computer-controlled data evaluation system takes care of the setting of the measurement parameters, the control of the measurement, and the display of the measured data in several modes. The connection between the optomechanical part and the controlling PC computer is realized by the use of a standard RS-232 or USB line with 9600 baud, so any IBM-compatible PC could be used. With the PC, we can set the number of measuring cycles, the duration of one cycle, the alarm levels for each size range, and so on.

The calibration measurements were carried out by the use of a PG 100 Pacific Instruments aerosol generator with polystyrene latex from Particle Measuring Systems Inc. and Dow Chemical Inc. These calibration particles have a low standard deviation (from 0.7% to a small percentage) of diameter [76]. In the first step of the calibration for better resolution, the pulse height distribution of the photomultiplier impulses corresponding to a certain size of calibration particles was analyzed by a Norland 5500 multichannel analyzer [77]. The typical distribution for the 1.1 μm monodisperse calibration latex shows two maxima. Such effects were also observed by Pinnick in Knolleberg-type light scattering counters [78]. This phenomenon is associated with coagulation [79], so the second maximum corresponds to the double bunching of the aerosol particles.

In this device the main source of the finite linewidth of the size resolution is the non-uniformity of the illuminating light intensity in the optical chamber. The best approximation for this is the Gaussian shape. The light detection statistics of the photomultiplier in this case play a secondary role. If we take the rule that the integrated scattered light intensity is proportional to the surface of the particle [80, 81], then from this distribution we can assume that the full width at half-maximum of the distribution equals one-quarter of the particle diameter. Owing to this the present number of size limits (0.3, 0.5, 1.0, 3, 5, 10 μm) could be at least doubled.

The main cause limiting the concentration of the measured aerosol is particle bunching in the measurement volume. This leads to false particle size and number determination. Assuming Poisson statistics for the interval between the arrival times of two particles, a simple calculation leads to the conclusion that, for a 92% certainty for two particles not to fall in the illuminated area simultaneously, the time interval between them must be 10 times more. However, this requires that the mean interval between the arrival times should be greater than 25 μs. The flow rate of the present instrument is around 3 l/min. This means that particle concentrations up to 0.8×10^7 particles per liter can be measured

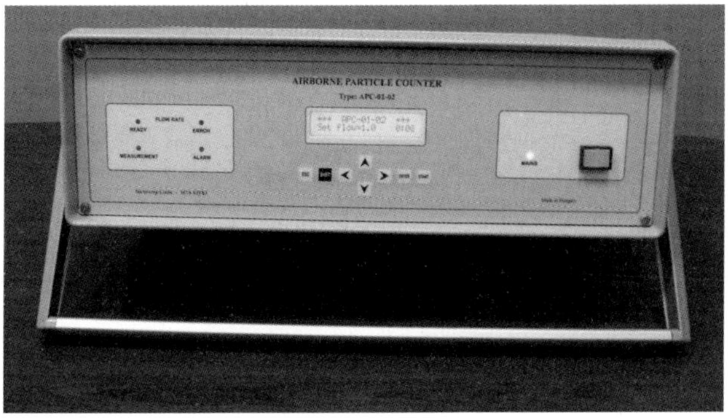

Figure 7.8 Airborne particle counters APC-03-2C and APC-01-02 produced by Technoorg, Hungary.

with better than 10% accuracy. This relatively large concentration limit offers applications not only in clean room monitoring, but also in highly contaminated environments. It is necessary to mention, however, that, as is usual in optical particle size measurements based on scattered light analysis, our measurements of size distribution also contain errors connected with the shape and refractive index inequalities of particles. Errors of this type are analyzed in [82].

Figure 7.8 shows photographs of the airborne particle counter described and a similar one (both manufactured by Technoorg-Linda Ltd). In the past few years, this device has been applied for clean room monitoring measurements [83–85], tests of various types of air filtering equipment (laminar boxes, climatic, and air cleaning systems in operating theaters, hi-tech laboratories), checking the particle size distribution and concentration in inhalation chambers for toxicological experiments (studying the toxicity of agricultural chemicals), monitoring the aerosols and air in various stages of the pharmacological processing of medicines, and the measurement of urban aerosols within the city of Budapest. This device was applied also for monitoring the air quality in industrial works in Hungary – Forte Photochemical Works (Vác), Chinoin Pharmaceutical Factory (Budapest), Parma Pharmaceutical Factory, and Viscosa Works (Nyergesújfalu) – and in the operation theaters of several hospitals [86–98].

7.5.2
Development of Liquid-borne Particle Counters

The LPC-1-200 liquid-borne particle counter is designed to measure the size distribution and concentration of particles suspended in various liquids whose viscosity is between about 0 and 100 P. This particle counter can be used to test the purity of parent solutions in various fields: pharmacology, hematology, biology, toxicology, health care, semiconductor process liquids, hydraulic fluids, environmental control, photochemical industry, refining industry, control of screening, and cleaning efficiency, and so on [99–101].

The optical scheme of the measurement chamber is shown in Figure 7.9. The principle of the measurement of the size distribution and the concentration of the particles suspended in liquids is light scattering or extinction (the theoretical approach is based on Mie scattering theory). However, as the velocity distribution of liquid-borne particles in a laminar fluid flow is sufficiently uniform, in order to determine the size of the particles to be tested (in a certain size range, and for a narrow slab-like region of uniform illumination), beside the pulse height measurement, the pulse duration of the optical signal (Figure 7.10) – which corresponds to the single-particle flow through the illuminated zone – can be used. The benefit of this method is that, in the case of measurement of the optical signal from the extinction of the incident light by the particles to be tested, the size distribution obtained in such a way is less dependent on the refractive index of the particles to be tested than in the case of the 90° scattering geometry used in the airborne particle counter. Moreover, if the size range is higher than several micrometers, because of the good resolution of the time interval measurement in the microsecond or millisecond region, the method can have a higher resolution than that based on commonly used pulse height measurements.

Figure 7.11 shows photographs of the liquid-borne particle counter described and a similar one (both manufactured by Technoorg-Linda Ltd). Integrated into the LPC-1-200 are methods based on pulse height and pulse duration measurements using a signal evaluation system, which consists of a 64-channel amplitude analyzer with a logarithmic preamplifier for pulse height analysis, and a higher-resolution system that consists of a 128-channel amplitude analyzer for pulse duration measurement.

The liquid to be tested is pumped through a small illuminated zone by a high-capacity peristaltic pump with a computer-controlled step motor drive. The rotation frequency of the pump can be adjusted over a wide range depending on

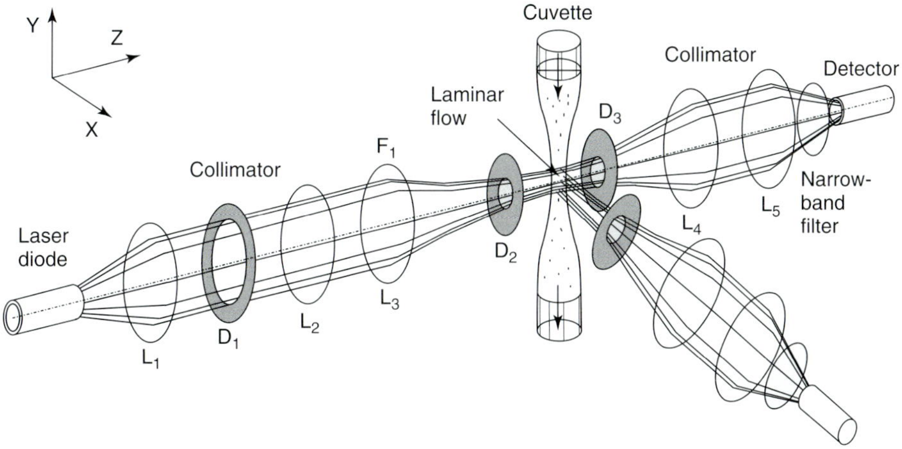

Figure 7.9 The optical scheme of the liquid-borne particle counter.

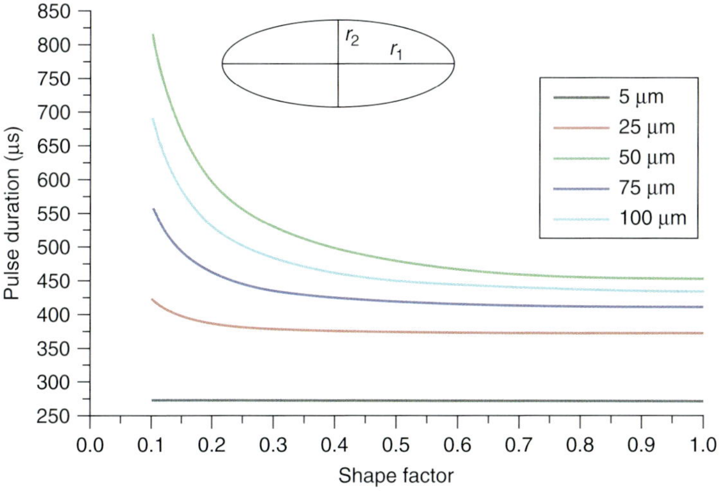

Figure 7.10 Liquid-borne particle counters of Technoorg, Hungary.

the viscosity and volume of the liquid to be tested. The controlled flow rate of the liquid is also measured, so not only the size distribution but also the concentration of the single particles can be determined. The light intensity of the illumination at the beginning of the measurement is automatically fitted to the transparency of the given liquid and controlled during the measurement. The flow rate is also continuously controlled and stabilized at the preset value.

The calibration of OPCs plays an increasingly important part in confirming the validity of measurements of suspended particle size distribution and concentration, since calibration depends not only on the standard size deviation of the calibration particles, but also on their optical properties. Calibration of airborne

Figure 7.11 The dependence of the pulse duration on the shape factor of the particles [102].

and liquid-borne particle counters usually utilizes monodisperse or polydisperse spherical polystyrene latex with known size distribution, known refractive index, and appearance verified by electron microscopy [100, 101]. These particles are quite suitable for calibration of airborne particle counters, but there are three additional problems for liquid-borne particle counters operating in the size range higher than 100 μm. In liquids, the coagulation of the polystyrene particles not only into doublets, but also into triplets and multiplets is likely to be more significant; and the upper size limit may exceed the dimensions of commercially available polystyrene latex. The third problem is that the standard deviation of the bigger polystyrene particles is much larger than that of the smaller ones. In some cases calibration in this range is carried out by approximating the calibration curve, but obviously this is not so accurate. Thus, a new type of calibration particle is needed. We used metallic spheres produced by ultrasonic evaporation from a liquid metallic melt, which has a low coagulation factor. The standard size deviation of such particles is several per cent. To prevent the sedimentation of the particles during the calibration, we performed permanent mixing and high velocity of sampling of the calibration liquid.

The measurement of the size distribution up to such a size region is important, for example, in testing pharmacological ingredients obtained by grinding some natural solid substance emulsified into a parent liquid. A computer-controlled data evaluation system takes care of the setting of the measurement parameters, the control of the measurement, the display of the measured data, and their storage. With the PC, we can set the number of measuring cycles, the duration of one cycle, the alarm levels for each size range, and so on.

7.6
New Methods Used to Characterize the Electrical Charge and Density of the Particles

After the development of different airborne and liquid-borne particle counters, the possibility of extending the laser method of aerosol sizing to obtain particle size and charge distribution was proposed [103]. This possibility can be realized by using an electric field perpendicular both to the aerosol jet and to the laser beam. Deflection of charged particles in the electric field results in the change of intensities of the scattered light pulses because of the inhomogeneous transverse *intensity distribution* of the laser beam. Aerosol particle distribution with respect to the *scattering amplitudes* is measured at several values of the field strength. The set of distributions thus obtained can be transformed to a two-dimensional size and charge distribution. Suppose a focused air jet containing aerosol particles is crossing a laser beam. If a voltage is applied to the capacitor plates, the charged aerosol particles will be deflected from their original trajectory (Figure 7.12).

The amplitude of the scattered light pulse detected in the experiment produced by a particle crossing the laser beam depends on the value of the deflection. For simplicity, we assume that, in the absence of an electric field, the aerosol jet crosses the laser beam axis. A particle of size a deflected from its original trajectory by distance l (*depending on* E, q *and* a) gives rise to a light pulse with intensity depending on the electrical field strength inside the capacitor, the radius of the particle a, and their charge q. Analyzing aerosol samples at different values of E_i, $i = 1, 2, 3, \ldots, i_{max}$, one can determine the number of particles giving rise to scattered light intensity Y_i at a fixed value of E_i. In other words, one can determine $N_{E_i}(Y_i)$, the distribution function of the aerosol particles with respect to Y_i. If a certain range of values ΔE_i from $E_i - \mathrm{d}E_i/2$ to $E_i + \mathrm{d}E_i/2$ can be put in correspondence to each dependence $N_{E_i}(Y_i)$, then the function $N_{E_i}(Y_i)$ can be transformed to the normalized two-dimensional distribution of aerosol particles with respect to E_i and Y_i. The problem is to derive the aerosol particle charge and size distributions from the dependence $N^2_{E_i}(Y_i)$. This derivation can be done by using one-dimensional distribution functions [102]. The accuracy of calculation of the distribution functions is determined by the ratio of the width l of the air jet containing aerosols to the characteristic laser beam radius (ideally, one should use

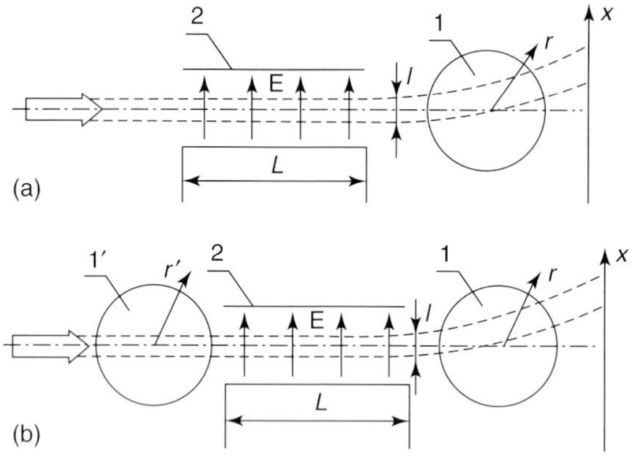

Figure 7.12 The scheme of the measurement of the electrical charge on aerosol particles.

an air jet as thin as possible). At $l/a < 0.2$ the accuracy of the method proposed is better than that of the traditional ones, based on the light scattering amplitude [102].

However, this possibility can be realized only for stationary (during the time of measurement) states of the aerosol. One can estimate this drawback, provided the range of charge distribution is not too broad, so that it is sufficient to measure the distribution of the aerosol particles with respect to scattering amplitudes at a fixed value of the field. In this case it is possible to make two measurements simultaneously by illuminating the aerosol jet at two different points [102]. In the first laser spot, processing the scattered signal as traditionally for laser sizers is used to determine the particle size distribution. Thus, after passing through the electric field of the capacitor, particles arrive at the second laser spot, where the second measurement is made simultaneously, which allows determination of the charge distribution.

Thus, the use of an electric field to deflect the charged particles in the transverse direction with respect to the laser beam in the aerosol particle sizer allows one to determine the two-dimensional size and charge distribution functions of small-sized aerosol particles. Experiments that have been performed show the applicability of this method.

The determination of the density of the aerosol particles is based on the measurement of the time of flight of the particle between two illuminated spots in the acceleration region of the aerodynamic focusing after an orifice nozzle (Figure 7.13). The motion of a particle with initial velocity v can be described by the *Stokes equation*

$$m\frac{\Delta v}{\Delta t} = 6\pi \mu a \overline{v}$$

where m is the mass of the particle, $A = \Delta v/\Delta t$ is the acceleration, μ is the dynamic viscosity of the air (or gas), a is the radius of the particle, $\overline{v} = (v_0 + v_1)/2$ is the average velocity, and Δt is the time between the pulses. When the distance between the two illuminated spots is small, the time to cover this distance with acceleration

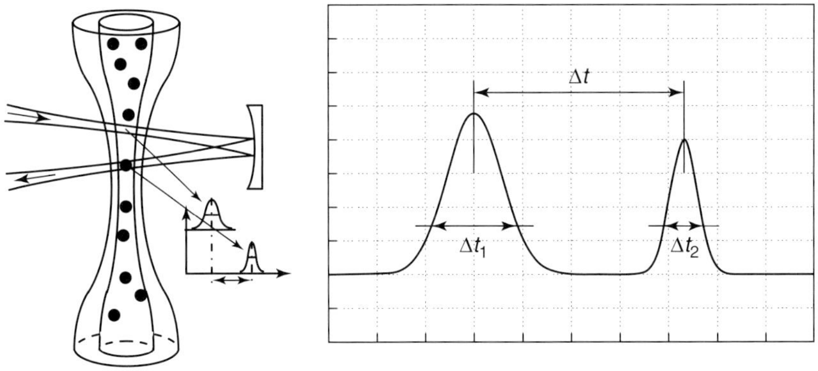

Figure 7.13 The principle of the estimation of the density of aerosol particles.

a (when ν is changing from ν_0 to ν_1) is Δt, and then the previous equation can be described as follows:

$$m(v_0 - v_1)/\Delta t = 6\pi \mu a(v_0 + v_1)/2$$

Taking into account that the mass of a spherical particle can be written as $m = \frac{4}{3}a^3\pi\rho$, the density ρ can be written as:

$$\rho = \frac{9\mu\overline{v}\Delta t}{2a^2\Delta v}$$

Since μ is known, a can be determined from the size measurement, and Δt and Δv can be defined from the experiment, the density of the particles can be measured.

7.7
Aerosol Analyzers for Measurement of the Complex Refractive Index of Aerosol Particles

As we can see from the previous sections, the scattered intensity from a particle depends not only on its size but also on the complex refractive index. When we measure scattering at different angles with different integration ranges using appropriate software based on generalized light scattering theory, the refractive index of the particle to be measured can also be retrieved. More comprehensive information could be achieved using two different wavelengths and a combination of forward and back-scattering for each detected particle. This principle was recently introduced for simultaneous measurement of the size and the complex refractive index of aerosol particles [103–107].

The method is based on a dual-wavelength illumination system where the scattered light is collected over four angular ranges, in forward and backward scattering directions compared to the two illumination laser light beams (Figure 7.14). The measured and digitized quartet is then compared to a pre-computed table calculated using the Mie scattering theory. The rows of the table contain the size, complex

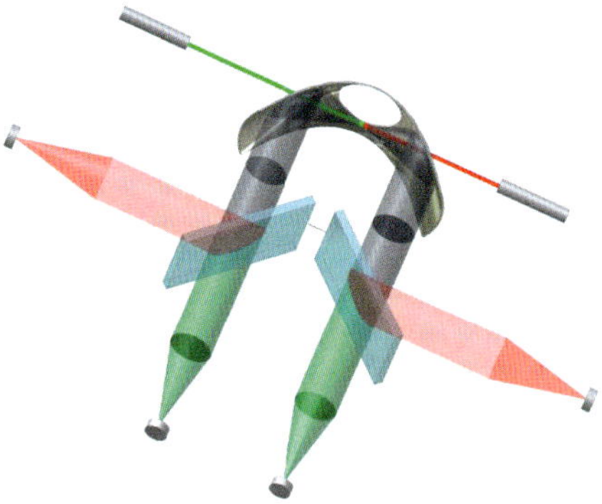

Figure 7.14 The scheme of the Dual Wavelength Optical Particle Spectrometer (DWOPS).

refractive index, and the corresponding four scattered signals from the four angular ranges (Figure 7.15). The four measured scattered intensity amplitudes – the responses of the detectors – are digitized and mapped according to the procedure demonstrated in Figure 7.15 (step 1). Then a search is performed (steps 2 and 3) to find the best-fitting row in the evaluation table by using the developed software. The measured size and complex refractive index are finally acquired from the appropriate row of the evaluation table (step 4). Figure 7.16 shows a photograph of the Dual Wavelength Optical Particle Spectrometer described (manufactured by Envi-Tech Ltd).

The resolution and the accuracy of the method depend on many parameters, for example, the instabilities in the sample flow system, uncertainties in the illumination and detection electronics, the applied analog-to-digital converter, and also the dimensions of the evaluation table. A comprehensive numerical and experimental study showed the feasibility of the method [108].

The calibration of the instrument was performed using laboratory particle standards. Initially, scale factors were determined for each detector to link the scattered intensities calculated using the Mie theory with the detector responses.

To evaluate the performance of the method, particles were generated from a suspension using pneumatic atomization and electrostatic classification were used to obtain monodisperse distributions. These particles were then introduced into the instrument's measuring volume after aerodynamic focusing. The four signals from the four detectors were obtained by a four-channel peak detection system, which utilizes certain logic to minimize the effects of possible crosstalk between channels. The detected signals were digitized and, using the scale factors, a search was performed in the evaluation table to obtain the particle size and complex refractive index.

Laboratory standard particles with different sizes from aqueous suspension, diethylhexyl sebacate (DEHS) particles generated from an isopropyl alcohol–DEHS solution, paraffin oil, and carbon-like absorbing particles generated from black ink diluted in pure distilled water were used for the test measurements.

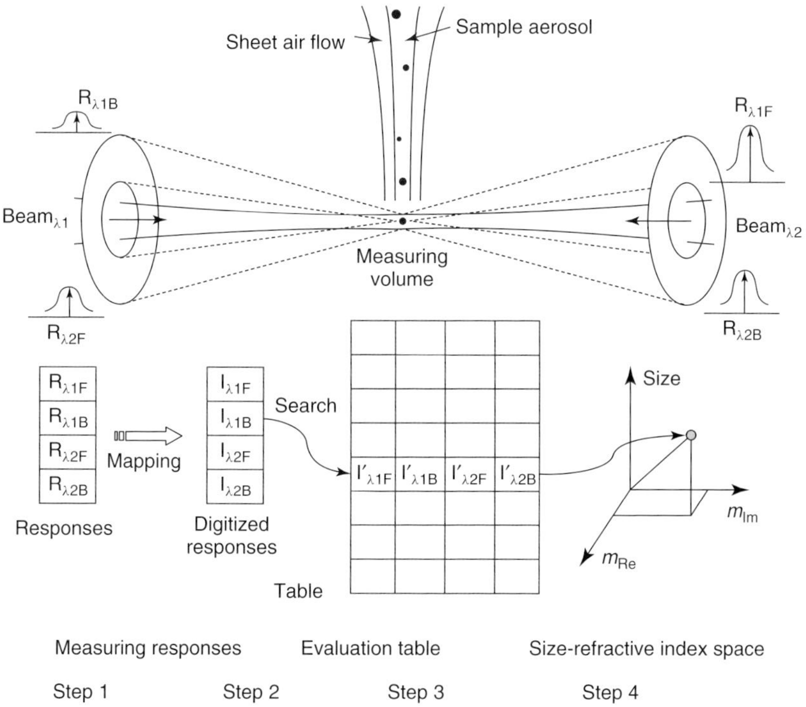

Figure 7.15 The data evaluation scheme of DWOPS.

Figure 7.16 The Dual Wavelength Optical Particle Spectrometer (Analyzer) developed by Envi-Tech Ltd, Hungary.

7.8
Comparison of Commercially Available Instruments and Analysis of the Trends of Further Developments

In this section, we present the parameters of the commercially available instruments – single-particle light scattering direct-readout counters and sizers for airborne and liquid-borne particle measurement (a method for determination of the aerosol parameters in case of multi-particle sensing where multiple scattering takes place is described in [109]).

Depending on the construction and application possibilities, the particle counters are classified as portable airborne particle counters, remote particle counters, manifold particle counters, and handheld particle counters. Table 7.1 provides a summarized review and very brief description of the main features and capabilities the instruments of the main manufacturers. The manufacturers are listed in [110–119].

7.8.1 Portable Particle Counters

Portable particle counters have an entire fully functioning air monitoring system built into one compact unit. Extremely versatile, mobile, and with a high air flow sample rate, they can be used cost effectively for contamination control via clean room certification or monitoring. These counters generally have built-in printers, displays, alarms, and input–output ports such as serial, RJ-45 or USB for computer connectivity. Some portable particle counters also have other features, such as temperature, relative humidity, and pressure differential and air velocity. Data from these sensors can be downloaded into a computer as well. Most portable particle counters from different manufacturers have been made to do the same job, but come in a variety of packages.

7.8.2 Remote Particle Counters

These are small particle counters that are used in a fixed location typically inside a clean room to continuously monitor particle levels 24 hours per day, seven days a week. These smaller counters typically do not have a local display and are connected to a network of other particle counters and other types of sensors to monitor the overall clean-room performance. This network of sensors is typically connected to a facility monitoring system, data acquisition system or programmable logic controller. This computer-based system can integrate into a database and/or an alarming/alerting system, and may have e-mail capability to notify facility or process personnel when conditions inside the clean room have exceeded predetermined environmental limits. Remote particle counters are available in several different configurations, from single channel to models that detect up to eight channels simultaneously; they can have a particle size detection range from 0.1 to 100 μm, and may feature one of a variety of output options, including 4–20 mA, RS-485 Modbus, ethernet, and pulse output.

Table 7.1 List of different types of optical instruments for measurement of aerosol particles.

Manufacturer	Model	Name	Light source	Detection angle (°)	Sample flow	Size range (μm)
TSI	3340	Laser aerosol spectrometer	Laser	–	0.1 l/min	0.09 to 7.5
TSI	3321	Aerodynamic particle sizer, APS	Laser	–	–	0.5 to 20 (0.37 to 20)
TSI	3785	Water-based condensation particle counter	Laser	–	–	0.005 to 3.0
TSI	3776	Ultrafine condensation particle counter, UCPC	Laser	–	$50\,cm^3$/min	0.0025 to 0.020
CCSTech	Solair 3100Rx	Airborne particle counter	Laser diode	90	1.0 CFM = 28 l/min	0.3 to 25
CCSTech	Solair 5100Rx	Airborne particle counter	Laser diode	90	1.0 CFM = 28 l/min	0.5 to 25
Climet	CI 150t	Airborne particle counter	Laser diode	–	1.0 CFM	0.3 to 5
Climet	CI-226	Remote airborne particle sensor	Halogen lamp	15–105	7–28 l/min	0.3 to 20
Climet	CI-6400	Airborne particle counter	He–Ne laser	45–135	1.0 CFM = 2.8 l/min	0.1 to 0.5
Climet	CI-8060	Airborne particle counter	Halogen lamp	15–150	1.0 CFM	0.3 to 10
Climet	CI 453	Airborne particle counter				
Faley Instrum. Corp.	Status 2100	Airborne particle counter	Halogen lamp	70	0.1 CFM	0.5 to 5
GAL, Galai, Brinkmann Instrum.	CIS-1	Airborne particle size analyzer	He–Ne laser	–	–	0.7 to 150
Grimm Aerosol Technik	1.109	Portable aerosol spectrometer	Laser diode	60–120	1.2 l/min	0.25 to 32
Grimm Aerosol Technik	5.401, 5.403	Condensation particle counter	Laser diode	~90	0.3 l/min	0.0045 to 3.0
Grimm Aerosol Technik	EDM 107	Environmental dust monitor	Laser diode	60–120	1.2 l/min	0.25 to 32

(continued overleaf)

Table 7.1 *(continued)*

Manufacturer	Model	Name	Light source	Detection angle (°)	Sample flow	Size range (μm)
Malvern	3600E	Particle sizer	He–Ne laser	Forward	–	0.5 to 560
MIE, Monitoring Instrum. for Environment	FAM-1	Fibrous aerosol monitor	He–Ne laser	Forward	2 l/min	2 to 200
PMS, Particle Measuring Systems	FSSP	Forward scattering spectrometer probe	He–Ne laser	Forward	1.0 CFM	0.3 to 10
PMS, Particle Measuring Systems	ASASP	Active scattering spectrometer probe	He–Ne laser	35–120	0.12 l/min	0.15 to 3
PMS, Particle Measuring Systems	LPC	Passive laser cavity particle counter	He–Ne laser	Forward	0.1 CFM	0.1 to 5
TSI	3755	Laser particle counter	Laser diode	Forward	0.1 CFM	0.5 to 20
Amherst Process Instrum.	MACH 2	High-resolution particle size analyzer	He–Ne laser	Forward	~0.1 l/min	0.2 to 150
PPM	PCAM	Portable continuous aerosol monitor	IR source	Forward	1 l/min	0.2 to 20
WYT	DAWNA	Aerosol particle analyzer	He–Cd laser	–	Variable	0.5 to 5
POI, Polytec Optronics Inc.	HC 15	Particle size analyzer	Halogen lamp	90	Variable	0.4 to 100
Technoorg-Linda	APC-03-2A	Airborne particle counter	He–Ne laser	65–115	1.0 CFM	0.3 to 10
Technoorg-Linda	LQB-1-200	Liquid-borne particle counter	Laser diode	Forward	Variable	1 to 200
MetOne	237A/B	Portable airborne particle counter	Laser diode	90	0.1 CFM	0.3 to >5
Topas	LAP	Airborne particle counter	He–Ne laser	Near forward	Up to 3 l/min	0.3 to 20

7.8.3
Multi-Particle Counters

These are modified portable aerosol particle counters that have been attached to a sequencing sampling system. The sequencing sampling system allows for one particle counter to sample multiple locations, via a series of tubes drawing air from up to 32 locations inside a clean room. They are typically less expensive than utilizing remote particle counters; however, each tube is monitored in sequence.

7.8.4
Handheld Particle Counters

These are small self-contained particle counters that are easily transported and used. Though the flow rates of $0.2\,m^3/h$ (1/10th of a cubic foot per minute) are lower than for larger portables with $2\,m^3/h$ (one cubic foot per minute), handhelds are useful for most of the same applications. However, longer sampling times may be required when doing clean-room certification and testing. Most handheld particle counters have direct mount isokinetic sampling probes.

Further development of the instruments can be performed by widening the size ranges, and combining particle counting and sizing with other facilities – for example, determination of the electrical charge, refractive index, density of the particles, and so on.

7.9
Conclusions

Light scattering and its applications in different fields of particle measurement – such as the study of the kinetic stability and aggregation behavior of colloid dispersions and suspensions, the investigation of macromolecules, polymers, and micellar systems, the coagulation kinetics of colloid particles, mining engineering, material processing, analysis of geological ingredients and liquid suspensions, and studying the kinetics and dynamics of aerosols, and so on – is the field of activity of hundreds of research centers all over the world.

A number of research firms and institutions have developed a series of airborne and liquid-borne particle counters and sizers with competitive technical specifications. These devices have successful applications in clean-room monitoring, in laminar box testing, in environmental air pollution monitoring, in medicine, in toxicology, in pharmacology, in chemistry, in filter testing, in industrial environment monitoring, in nuclear air ingress experiments, in hematology, and so on [120–134].

Besides demonstrating the applicability for these purposes, the obtained new results also serve for the possible judgment of future applications and give information on the concentration, size distribution, density, refractive index, absorption, extinction, and shape factor of airborne particles in various places under various conditions. The short sampling time of the optical methods can serve for studying the dynamics of aerosols.

About 30 big companies all over the world are involved in the production of laser light scattering instrumentation, and they are developing and widening the capability of the new devices and spreading their fields of application.

References

1. Zsigmondy, R. and Bachmann, W. (1914) Handhabung des Immersionultramikroskops. *Kolloid Z.*, **XIV** (6), 18–30.
2. Arago, F. (1858) *Oeuvres*, vol. 7, Paris, p. 10.
3. Govi, G. (1860) *C. R. Acad. Sci., Paris*, **51**, 360, 699.
4. (a) Tyndall, J. (1869) On the blue colour of the sky, the polarization of skylight, and on the polarization of light by cloudy matter generally. *Philos. Mag.*, **37**, 384; (b) Tyndall, J. (1869) On the blue colour of the sky, the polarization of skylight, and on the polarization of light by cloudy matter generally. *Proc. R. Soc.*, **17**, 223.
5. (a) Rayleigh, Lord (1899) On the light from the sky, its polarization and colour, *Scientific Papers*, vol. I, University Press, Cambridge, pp. 87–110; (b) Reprinted in Rayleigh, Lord (1964) *Scientific Papers*, vol. **1**, Dover, New York.
6. Rayleigh, Lord (1899) On the electromagnetic theory of light, *Scientific Papers*, vol. I, University Press, Cambridge, pp. 518–536.
7. Rayleigh, Lord (1903) On the transmission of light through an atmosphere containing small particles in suspension, and on the origin of the blue of the sky, *Scientific Papers*, vol. IV, University Press, Cambridge, pp. 397–405.
8. (a) Rayleigh, Lord (1871) On the scattering of light by small particles. *Philos. Mag.*, **41**, 447–454; Reprinted in (b) Rayleigh, Lord (1964) *Scientific Papers*, vol. 2, Dover, New York.
9. Mie, G.A. (1908) A contribution to the optics of turbid media. *Ann. Phys. (Leipzig)*, **25** (4), 377–445.
10. van de Hulst, H.C. (1957) *Light Scattering by Small Particles*, John Wiley & Sons, Inc., New York.
11. Kerker, M. *et al.* (1966) Color effects in the scattering of white light by micron and submicron spheres. *J. Opt. Soc. Am.*, **56**, 1248–1258.
12. Kerker, M. (1969) *The Scattering of Light and Other Electromagnetic Radiation*, Academic Press, New York.
13. Hodkinson, J.R. (1966) The optical measurement of aerosols, in *Aerosol Science* (ed. C.N. Davies), Academic Press, New York, pp. 287–357.
14. Smoluchowski, M. (1908) Molekular-kinetische Theorie der Opaleszenz von Gasen in kritischen Zustande, sowie einiger Verwendter Erscheinungen. *Ann. Phys.*, **25**, 205.
15. Einstein, A. (1910) Theorie der Opaleszenz von homogenen Flüssigkeiten und Flüssigkeitsgemischen in der Nahe der kritischen Zustandes. *Ann. Phys.*, **33**, 1275.
16. Cabannes, J. (1915) Sur la diffusion de la lumière par l'air, *C. R. Acad. Sci., Paris*, **160**, 62.
17. (a) Strutt, R.J. (1918) The light scattering by gases, its polarization and intensity. *Proc. R. Soc.*, **94**, 453; (b) Strutt, R.J. (1918) The light scattering by gases, its polarization and intensity. *Proc. R. Soc.*, **95**, 155.
18. Wood, R.W. (1920) Light scattering by air and the blue colour of the sky. *Philos. Mag.*, **39**, 423.
19. Putzeys, P. and Brosteaux, J. (1935) *Trans. Faraday Soc.*, **31**, 1314.
20. Debye, P. (1944) *J. Appl. Phys.*, **15**, 338.
21. Ginzburg, V.L. (1955) O rasseanii sveta v blizi tochek fazovoho perehoda

vtoroho roda. *Dokl. Akad. Nauk SSSR*, **105**, 240 (in Russian).

22. Brillouin, L. (1922) Diffusion de la lumière et des rayonnes X par un corps transparent homogène; influence de l'agitation thermique. *Ann. Phys. (Paris)*, **17**, 88.
23. Mandelstam, L.I. (1926) K voprosu o rasseianii sveta neodnorodnoi sredoi. *Zs. RFHO*, **58**, 381
24. Raman, C.V. (1928) A new type of secondary radiation. *Nature*, **121**, 3048.
25. Landsberg, G.S. (1927) Molekuliarnoie raseianie sveta v tviordom tyele, I. Rasseianie sveta v kistalicheskom kvarce i evo temperaturnaia zavisimosty. *Z. Phys.*, **43**, 733.
26. Cabrannes, J. and Daure, P. (1928) Analyse spectroscopique de la lumière obtenue par diffusion moléculaire d'une radiation monochromatique au sein d'un fluide. *C. R. Acad. Sci., Paris*, **186**, 1533.
27. Cabannes, J. (1927) La diffusion de la lumière dans les liquides. Controle expérimental des formules théoretiques. *J. Phys. Radium*, **8**, 321.
28. Cabannes, J. (1929) *La Diffusion Moléculaire de la Lumière*, Presse Université France, Paris.
29. Fabelinski, I.L. (1957) Some questions of light scattering in liquids. *Uspekhi Fiz. Nauk*, **63**, 355 (in Russian).
30. Fabelinski, I.L. (1958) Molecular light scattering in liquids. *Trudi Fizicheskovo Instituta Akademii Nauk*, **9**, 181 (in Russian).
31. Fabelinski, I.L. (1965) Molekuliarnie rasseanie sveta. Chapter IV, *Izd. Nauka, Moscow*, (in Russian), pp. 220–233.
32. Stratton, J.A. (1941) *Electromagnetic Theory*, McGraw-Hill, New York.
33. Rosenberg, G.V. (1960) Rassejanie sveta v zemnoi atmosfere. *Uspekhi Fiz. Nauk*, **71** (2), 173–213 (in Russian).
34. McCartney, E.J. (1977) *Optics of the Atmosphere, Scattering by Molecules and Particles*, John Wiley & Sons, Inc., New York.
35. Penndorf, R. (1962) Angular Mie scattering. *J. Opt. Soc. Am.*, **52**, 402–408.
36. Levi, L. (1968) *Applied Optics*, John Wiley & Sons, Inc., New York.
37. Driscoll., W.G. and Vaughan, W. (eds) (1978) *Handbook of Optics*, McGraw-Hill, New York.
38. Boren, C.F. and Huffman, D.R. (1983) *Absorbing and Scattering of Light by Small Particles*, John Wiley & Sons, Inc., New York.
39. Shifrin, K.S. (1951) Rasseianie sveta v mutnoi srede, Moscow-Leningrad (now St. Petersburg) (in Russian).
40. *Fizicheskaja Enciklopedia*, Izd. Sovietskaia Enciklapedia, vol. 3, Moscow (1991) (in Russian).
41. Lowan, A.N. (1949) *Tables of Scattering Functions for Spherical Particles*, Applied Mathematics Series, vol. 4, National Bureau of Standards, Washington, DC.
42. Cumprecht, R.O. *et al.* (1952) Angular distribution of intensity of light scattered by large droplets of water. *J. Opt. Soc. Am.*, **42**, 226–231.
43. Penndorf, R. and Goldberg, B. (1956) New tables of Mie scattering functions for spherical particles, Part 1-5, Amplitude functions *a* and *b*. Report AFCRCTR-56-204, AFCRL, Bedford, MA. AD 98767-AD 98771, NTIS, Springfield, VA.
44. Pangonis, W.J. and Heller, W. (1957) *Angular Scattering Functions of Spherical Particles*, Wayne State University Press, Detroit.
45. Giese, R.H. *et al.* (1961) Tables related to scattering functions and scattering cross sections of particles according to Mie theory. *Abh. Deut. Akad. Wiss., Berlin, Kl. Math. Phys. Tech.*, **6**.
46. Dahneke, B.E. (ed.) (1983) *Measurement of Suspended Particles by Quasi-Elastic Light Scattering*, John Wiley & Sons, Inc., New York.
47. Allen, P.W. (ed.) (1959) *Techniques of Polymer Characterization*, Butterworths Scientific, London.
48. Penndorf, R. (1957) Tables of the refractive index for standard air and Rayleigh scattering coefficient for the spherical region between 0.2 and 20 mm and their applications in atmospheric optics. *J. Opt. Soc. Am.*, **47**, 176–182.
49. Zaitsev, V.P., Krivokhizha, S.V., Fabelinski, I.L., Czitrovszky, A., Chaikov, L.L., and Jani, P. (1986)

Correlation radius near the critical points of guaiacol–glycerin solution. *JETP Lett.*, **43**, 112–116.
50. Krivokhizha, S.V., Lugovaia, O., Fabelinski, I.L., Chaikov, L.L., Czitrovszky, A., and Jani, P. (1993) Temperature dependence of the correlation radius of the fluctuation of the concentration of guaiacol–glycerin solution. *JETP*, **103**, 115–124.
51. Chaikov, L.L., Fabelinski, I.L., Krivikhizha, S.V., Lugovaia, O., Czitrovszky, A., and Jani, P. (1994) Light scattering spectrum in the region of the upper, lower and double critical point in guaiacol–glycerin solution. *J. Raman Spectrosc.*, **25** (7–8), 463–468.
52. Waggoner, A.P. and Charlson, R.J. (1976) Measurement of aerosol optical parameters, in *Fine Particles* (ed. B.Y.H. Liu), Academic Press, New York, pp. 511–534.
53. Lillenfeld, P. (1982) in *Aerosols in the Mining and Industrial Work Environments* (eds V.A. Marple and Y.H. Liu), Ann Arbor Science Publishers, Ann Arbor, MI, pp. 879–895.
54. Cook, C.S. *et al.* (1972) Remote measurement of smokeplume transmittance using LIDAR. *Appl. Opt.*, **11**, 1742–1848.
55. Barth, H.G. (ed.) (1984) *Modern Methods of Particle Size Analysis*, John Wiley & Sons, Inc., New York, pp. 137–157.
56. Chigier, N. and Stewart, G. (1984) Particle sizing and spray analysis. *Opt. Eng.*, **23**, 554.
57. Bricard, J. *et al.* (1976) Detection of ultra fine particles by means of a continuous flux condensation nuclei counter, in *Fine Particles* (ed. B.Y.H. Liu), Academic Press, New York, pp. 565–580.
58. Wilson, J.C. and Liu, B.Y.H. (1980) Aerodynamic particle size measurement by laser Doppler velocimetry. *J. Aerosol Sci.*, **II**, 139.
59. Remiraz, R.J. Agarval, J.K., Quant, F.R., and Sem, G.J. (1983) in *Aerosols in the Mining and Industrial Work Environments*, vol. 3, *Instrumentation* (eds V.A. Marple and Y.H. Liu), Ann Arbor Science Publishers, Collingwood, pp. 879–895.
60. Baron, P.A. (1983) in *Aerosols in the Mining and Industrial Work Environments*, vol. 3, *Instrumentation* (eds V.A. Marple and Y.H. Liu), Ann Arbor Science Publishers, Collingwood, pp. 861–877.
61. Marshal, I.A., Mitchell, J.P., and Griffiths, W.D. (1991) The behavior of regular-shaped non-spherical particles in a TSI aerodynamic particle sizer. *J. Aerosol Sci.*, **22** (1), 73–89.
62. Marshal, I.A. and Mitchell, J.P. (1991) The behavior of non-spherical particles in a Malvern aerosizer. API, Progress Report, AEA Technology Report AEA RS 5167, March–August.
63. Baron, P.A. (1986) Calibration and use of the aerodynamic particle sizer APS 3300. *Aerosol Sci. Technol.*, **5**, 55–67.
64. Marshal, I.A. and Mitchell, J.P. (1992) The behaviour of spheroidal particles in time-of-flight aerodynamic particle sizers. *J. Aerosol Sci.*, **23** (Suppl. 1), 297–300.
65. Griffiths, W.D., Iles, P.J., and Vaughan, N.-P. (1986) The behavior of liquid droplet aerosol in an APC 3300. *J. Aerosol Sci.*, **17** (6), 921–930.
66. Wang, H.C. and John, W. (1987) Particle density correction for aerodynamic particle sizer. *Aerosol Sci. Technol.*, **6**, 191–198.
67. Wang, H.C. and John, W. (1989) A simple iteration procedure to correct for the density effect in aerodynamic particle sizer. *Aerosol Sci. Technol.*, **10**, 501–505.
68. Griffiths, W.D., Iles, P.J., and Vaughan, N.P. (1986) Calibration of the APS 33 aerodynamic particle sizer and its usage. *TSI J. Part. Instrum.*, **1** (1), 3–9.
69. Ananth, G. and Wilson, J.C. (1988) Theoretical analysis of the performance of the TSI aerodynamic particle sizer. *Aerosol Sci. Technol.*, **9**, 189–199.
70. Chen, B.T., Cheng, Y.S., and Yeh, H.C. (1985) Performance of a TSI aerodynamic particle sizer. *Aerosol Sci. Technol.*, **4**, 89–97.
71. TSI Incorporated, Particle Instruments Group (1989, 1990) Submicrometer sizer, particle technology, St Paul, MN.
72. Herring, S.V. (1989) *Air Sampling Instruments for Evaluation of Atmospheric*

Contaminations, American Conference of Governmental Industrial Hygienists, Cincinnati, OH, pp. 489–500, 503–507.
73. Barth, H.G. and Sun, S.T. (1985) Particle size analysis. *Anal. Chem.*, **57** (5), 151R–175R.
74. Belyaev, S.P., Nikiforova, N.K., Smirnov, V.V., and Shelchkov, G.I. (1981) *Opto-electronical Methods for Aerosol Investigation*, Energoisdat, Moscow (in Russian).
75. Liberman, A. and Bates, A.D. (1983) in *Aerosols in the Mining and Industrial Work Environments*, vol. 3, *Instrumentation* (eds V.A. Marple and Y.H. Liu), Ann Arbor Science Publishers, Collingwood, pp. 811–824.
76. Mitchell, J.P. (1992) Aerosol standards. *J. Aerosol Sci.*, **23** (Suppl. 1), 289–292.
77. Czitrovszky, A., Jani, P., and Ferrari, A. (1990) Design and development of a laser airborne particle counter. Proceedings of IMECO Symposium on Measurement and Inspection in Industry, Balatonfüred, Hungary, pp. 335–342.
78. Pinnick, R.G. and Rosen, J.M. (1979) Response of Knollenberg light scattering counters to non-spherical doublet polystyrene latex aerosols. *J. Aerosol Sci.*, **10**, 533–538.
79. Green, H.L. and Lane, W.R. (eds) (1964) in *Particle Clouds: Dust, Smokes and Mists*, E. and F.N. Spon, London, pp. 138–170.
80. Hinds, W.C. (1982) *Aerosol Technology*, John Wiley & Sons, Inc., New York, pp. 337–340.
81. Gebbhard, V.J., Blankenberg, P., Bormann, S., and Roth, C. (1983) Vergleichsmessung an optischen Partikelzahlern. *Staub, Reinhalt, Luft*, **43** (II), 439–447.
82. Reist, P.C. (1984) *Introduction to Aerosol Science*, Macmillan, London.
83. ANSI (1976) Federal Standard 209B, *Clean Rooms*. American National Standards Institute, New York.
84. Cooper, D.W. (1986) Rationale for proposed revision to Federal Standard 209B. *J. Environ. Sci.*, **3–4**, 25–29.
85. IES (1984) Recommended Particle for Testing Clean Rooms, IES-CC-006-84T, Institute of Environmental Science, Chicago, IL.
86. Czitrovszky, A. and Jani, P. (1993) New design for a light scattering airborne particle counter and its application. *Opt. Eng.*, **32** (10), 2557–2562.
87. Czitrovszky, A. and Jani, P. (1993) Design and application of a light scattering airborne particle counter developed in KFKI. *J. Aerosol Sci.*, **24** (Suppl. 1), S227–S228.
88. Czitrovszky, A. and Jani, P. (1993) Performance of the new type of general purpose airborne particle counter. *Proc. SPIE*, **1983**, 998–999.
89. Czitrovszky, A., Jani, P., Poluektov, P.P., Kolomeitsev, G.Yu., and Yefankin, V.G. (1993) A new method for laser analysis of aerosol particles to determine the distribution of electric charge, size, density and concentration. *Proc. SPIE*, **1983**, 969–970.
90. Czitrovszky, A. and Jani, P. (1994) Efficiency of sampling by the APC-03-2 single particle counter. *J. Aerosol Sci.*, **25** (Suppl. 1), 465–466.
91. Czitrovszky, A., Frecska, J., Matus, L., and Jani P. (1994) Investigation of aerosol released from heated LWR fuel rod and its properties. Proceedings of the IVth International Aerosol Conference, Los Angeles, CA, pp. 784–785.
92. Czitrovszky, A. and Jani, P. (1995) Application examples of APC-03-2 and APC-03-2A airborne particle counters in a highly contaminated environment. *J. Aerosol Sci.*, **26** (Suppl. 1), 793–794.
93. Czitrovszky, A., Frecska, J., Jani, P., Matus, L., and Nagy, A. (1996) Size distribution of aerosols released from heated LWR fuel rods. *J. Aerosol Sci.*, **27** (Suppl. 1), 467–468.
94. Czitrovszky, A., Csonka, P.L., Jani, P., Ringelhann, Á., and Bobvos, J. (1996) Experimental investigation of altitude dependence of size distribution and concentration of dust particles within the city of Budapest. *J. Aerosol Sci.*, **27** (Suppl. 1), 117–118.
95. Czitrovszky, A., Csonka, P.L., Jani, P., Ringelhann, Á., and Bobvos, J. (1996) Comparison of different measurement methods of airborne dust pollution

within the city of Budapest. *J. Aerosol Sci.*, **27** (Suppl. 1), 19–20.

96. Jani, P., Nagy, A., and Czitrovszky, A. (1996) Aerosol particle size determination using a photon correlation laser Doppler anemometer. *J. Aerosol Sci.*, **27** (Suppl. 1), 531–532.
97. Czitrovszky, A., Csonka, P.L., Jani, P., Ringelhann, Á., and Bobvos, J. (1996) Measurement of airborne dust pollution within the city of Budapest. Proc. 15th Annual Conference of the American Association for Aerosol Research, Orlando, FL, p. 210.
98. Czitrovszky, A., Csonka, P.L., Jani, P., Ringelhann, Á., and Bobvos, J. (1996) Complex measurement of airborne dust pollution within the city of Budapest. *Proc. SPIE*, **2965**, 103–112.
99. Czitrovszky, A. and Jani, P. (1994) Design and application of the LPC-1-200 liquidborne particle counter. *J. Aerosol Sci.*, **25** (Suppl. 1), 447–448.
100. Czitrovszky, A., Mosoni, T., and Jani, P. (1995) Calibration and application examples of the LPC-1-200 liquid-borne particle counter. *J. Aerosol Sci.*, **26** (Suppl. 1), 791–792.
101. Czitrovszky, A. and Jani, P. (1996) in *Aerosols Dynamics, Behavior and Measurement* (ed. A. Czitrovszky), KFKI, Budapest, pp. 71–75 (in Hungarian).
102. Kashcheev, V.A., Poluektov, P.P., Semikin, A.N., Czitrovszky, A., and Jani, P. (1992) Measurement of aerosol particle charge and size distribution by a laser particle anslyser. *Laser Phys.*, **2** (4), 613–616.
103. Nagy, A., Szymanski, W.W., Golczewski, A., Gál, P., and Czitrovszky, A. (2007) Numerical and experimental study of the performance of the Dual Wavelength Optical Particle Spectrometer (DWOPS). *J. Aerosol Sci.*, **38**, 467–478.
104. Szymanski, W.W., Nagy, A., Czitrovszky, A., and Jani, P. (2002) A new method for the simultaneous measurement of aerosol particle size, complex refractive index and particle density. *Meas. Sci. Technol.*, **13**, 303–307.
105. Czitrovszky, A., Jani, P., Nagy, A., Schindler, C., and Szymanski, W.W. (2000) An approach to a simultaneous assessment of aerosol particle size and refractive index by multiple angle detection. *J. Aerosol Sci.*, **31**, 761–762.
106. Nagy, A., Szymanski, W.W., Czitrovszky, A., and Jani, P. (2001) Modeling of a new optical aerosol particle analyzer for the simultaneous measurement of size, complex refractive index and density. *J. Aerosol Sci.*, **32**, 83–84.
107. Nagy, A., Szymanski, W.W., Golczewski, A., Gál, P., and Czitrovszky, A. (2007) Numerical and experimental study of the performance of the Dual Wavelength Optical Particle Spectrometer (DWOPS). *J. Aerosol Sci.*, **38** (4), 467–478.
108. Szymanski, W.W., Nagy, A., and Czitrovszky, A. (2009) Optical particle spectrometry problems and perspectives. *J. Quant. Spectrosc. Radiat. Transfer*, **110**, 918–929.
109. Czitrovszky, A., Nagy, A., and Jani, P. (1999) Development of a new particle counter for simultaneous measurement of the size distribution, concentration and estimation of the shape-factor of liquid-borne particles. *Proc. SPIE*, **3749**, 574–575.
110. Par-Tec 200/300, Instruction manual, Lasentec LSI Inc., 523 North Belt East, Suite 100, Houston, TX 77066, USA.
111. TSI Mod. 3775 Laser Particle Counter, Instruction manual, TSI Inc., 500 Cardigan Road, P.O. Box 64394, St. Paul, MN 55164, USA.
112. Malvern Mod. 3600 Particle Sizer, Instruction manual, Malvern Instruments Inc., 187 Oaks Road, Framingham, MA 01701, USA.
113. Hiac-Royco Mod. 5100 Aerosol Particle Counter, Instruction manual, Hiac-Royco Inc., 141 Jefferson Drive, Menlo Park, CA 94025, USA.
114. Climet Mod. CI 6400 Airborne Particle Counter, Instruction manual, Climet Instruments Co., P.O. Box 151, Redlands, CA 92373, USA.
115. PMS Mod. LPC-525 Laser Particle Counter, Instruction manual, Particle Measuring System Inc., 1855 South

57th Court, Boulder, CO 80301, USA.

116. APC-032A Airborne Particle Counter, Instruction manual, Technoorg-Linda Ltd, Rozsa u. 24, Budapest 1077, Hungary.

117. LQB-1-200 Liquidborne Particle Counter, Instruction manual, Technoorg-Linda Ltd, Rozsa u. 24, Budapest 1077, Hungary.

118. Topas, Series LAP, Instruction manual, Topas GmbH, Hofmannstrasse 37, Dresden, D-01277, Germany.

119. Dantec, Instruction manual, Measurement Technology Div., Skovlunde, Denmark.

120. Jani, P., Czitrovszky, A., Nagy, A., and Hummel, R. (1999) Investigation of size distribution and concentration of aerosols released at CODEX AIT experiments. *J. Aerosol Sci.*, **30** (Suppl. 1), 101–102.

121. Jani, P., Nagy, A., and Czitrovszky, A. (1999) Nano-particle size distribution measurement in photon correlation experiments. *Proc. SPIE*, **3749**, 458–459.

122. Jani, P., Nagy, A., Lipp, Z., and Czitrovszky, A. (2000) Velosizer–photon correlation LDA system. *J. Aerosol Sci.*, **31**, 390–391.

123. Jani, P., Lipp, Z., Nagy, A., and Czitrovszky, A. (2001) Propagation delay statistics of scattered intensities. *J. Aerosol Sci.*, **32**, 87–88.

124. Jani, P., Koniorczyk, M., Nagy, A., and Czitrovszky, A. (2001) Probability distribution of scattered intensities. *J. Aerosol Sci.*, **32**, 563–564.

125. Jani, P., Nagy, A., Lipp, Z., and Czitrovszky, A. (2001) Simultaneous velocity and size measurement of nanoparticles in photon correlation experiments. *Proc. SPIE*, **4416**, 236–240.

126. Szymanski, W.W., Nagy, A., Czitrovszky, A., and Jani, P. (2002) A new method for the simultaneous measurement of aerosol particle size, complex refractive index and particle density. *Meas. Sci. Technol.*, **13**, 303–307.

127. Jani, P., Koniorczyk, M., Nagy, A., Lipp, Z., Bartal, B., László, A., and Czitrovszky, A. (2002) Probability distribution of scattered intensities. *J. Aerosol Sci.*, **33** (5), 697–704.

128. Czitrovszky, A., Nagy, A., and Jani, P. (2003) Environmental monitoring of the atmospheric pollution by aerosols. *J. Aerosol Sci.*, **34** (Suppl. 1), S953–S954.

129. Jani, P., Nagy, A., and Czitrovszky, A. (2003) Field experiments for the measurement of time of flight statistics of scattered intensities on ensemble of aerosol particles. *J. Aerosol Sci.*, **34** (Suppl. 1), S1209–S1210.

130. Wind, L., Hofer, L., Vrtala, A., Nagy, A., and Szymanski, W.W. (2003) Optical effects caused by non-soluble inclusions in microdroplets and implications for aerosol measurement. *J. Aerosol Sci.*, **34** (Suppl. 1), S1245–S1246.

131. Golczewski, A., Nagy, A., Gal, P., Czitrovszky, A., and Szymanski, W.W. (2004) Performance modelling and response of the Dual-Wavelength Optical Particle Spectrometer (DWOPS). *J. Aerosol Sci.*, **35** (Suppl. 1), S839–S840.

132. Wind, L., Hofer, L., Nagy, A., Winkler, P., Vrtala, A., and Szymanski, W.W. (2004) Light scattering from droplets with inclusions and the impact on optical measurement of aerosols. *J. Aerosol Sci.*, **35**, 1173–1188.

133. Farkas, Á., Balasházy, I., Czitrovszky, A., and Nagy, A. (2005) Simulation of therapeutic and radioaerosol deposition. *J. Aerosol Med.*, **18** (1), 102–104.

134. Hózer, Z., Maróti, L., Windberg, P., Matus, L., Nagy, I., Gyenes, Gy., Horváth, M., Pintér, A., Balaskó, M., Czitrovszky, A., Jani, P., Nagy, A., Prokopiev, O., and Tóth, B. (2006) The behavior of VVER fuel rods tested under severe accident conditions in the CODEX facility. *Nucl. Technol.*, **154** (3), 302–317.

8
The Inverse Problem and Aerosol Measurements

Valery A. Zagaynov

8.1
Introduction

Any measurement is related to the comparison of a particular measured parameter against some standard one. However, in addition to the standard set of requirements applied to any measurement, measurements of aerosol systems have special additional features. Although a precise definition of "aerosol particle" does not exist, everybody understands what it means. On top of this stylistic issue, there is more essential complexity: often aerosol measurements are indirect. One of the main tasks of aerosol science and technology is representative determination of the particle size distribution (PSD). At the same time, this problem is both very acute and ambiguous to solve. There are two obstacles to rectify the problem. First, any monitoring equipment has a defined sensitivity that could leave a substantial quantity of particles not registered. In these cases, the concentration of such particles could be greater than the concentration of counted aerosols. Second, some uncertainty is related to the *inverse problem*, as all *aerosol measurements* are indirect. In particular, any aerosol monitor provides some integrated value of the PSD, which then ought to be mathematically converted into the exact size distribution of the real aerosol.

I should start this chapter by repeating the statement that all aerosol measurements are indirect. For example, measurement of the PSD really means measurement of one of the following aerosol characteristics: aerosol particle light scattering ability, particle Brownian motion, the ability of a particle to acquire electrical charge, and the corresponding mobility of the charged particle in an electric field. The subsequent conversion of the measured parameters into the real PSD requires a number of mathematical steps frequently associated with ambiguity of solution. In addition, any instrument has some detection limits, leading to situations in which particles beyond the sensitivity range are unable to be detected at all, or could be detected with an efficiency of less than 100%. This may significantly distort the real particle size spectrum and concentration. In this chapter, the instrumentation along with theoretical approaches to attack the problem will be discussed.

Aerosols – Science and Technology. Edited by Igor Agranovski

ISBN: 978-3-527-32660-0

In many situations, very sophisticated instrumentation and mathematical procedures are required to acquire relevant data and convert them to representative final results. From the mathematical point of view, there is complexity related to the fact that the procedure for solving an inverse problem (transition from experimental data into PSD) results in a situation with final complex equations with complicated, ambiguous or unstable solutions.

During reconstruction of the PSD from measured data, it is very important to know what type of measurements were undertaken: differential or integral. In the first case, the value of interest is measured directly, while in the second case, the particle spectrum is "hidden" in the measured data.

In particular, if the particle size distribution could be represented as a vector $\boldsymbol{x}$ and the value that was actually measured as a vector $\boldsymbol{P}$, then the relation between these values is

$$\boldsymbol{P} = \boldsymbol{A}\boldsymbol{x} \tag{8.1}$$

where $\boldsymbol{A}$ is the response matrix. If matrix $\boldsymbol{A}$ has elements comparable with 1, and the dimensions of the vectors $\boldsymbol{x}$ and $\boldsymbol{P}$ coincide, then Eq. (8.1) will become linear and the unknown values $\boldsymbol{x}$ can be simply calculated from the measured values $\boldsymbol{P}$. Also, if the determinant of the matrix $\boldsymbol{A}$ is different from 0, then the problem may be resolved unambiguously:

$$\boldsymbol{x} = \boldsymbol{A}^{-1}\boldsymbol{P} \tag{8.2}$$

Unfortunately, these mathematically favorable situations only arise for cases in which measurements may be regarded as differential. For other cases, when the determinant of the matrix $\boldsymbol{A}$ is close to 0, the elements of the matrix $\boldsymbol{A}$ are obtained by integral relations, turning the measurements into integral, which leads to instability of solution. To evaluate in advance what type of measurement is undertaken, the concept of *condition numbers* could be applied [1]:

$$C = \boldsymbol{A}\boldsymbol{A}^{-1} \tag{8.3}$$

As one can understand, uncertainties of raw data acquisition are responsible for errors in the reconstructed PSD. The relation between datasets could be represented as

$$\frac{\Delta x}{x} = C\frac{\Delta P}{P} \tag{8.4}$$

Based on this equation, if the condition numbers C are large, then large experimental errors $\Delta P/P$ could result and, correspondingly, significant uncertainties in the reconstruction of the final PSDs $\Delta x/x$ may be expected. The condition numbers are usually related to particular measurement methods. For example, when the particle spectrum is measured by an impactor, the condition numbers are around 1.2. On the other hand, in cases of utilization of cyclones or elucidators, the condition numbers will be two orders of magnitude greater, reaching values of around $C \propto 2 \times 10^2$. The same magnitude of condition numbers also corresponds to measurements made by differential mobility analysis (DMA). Finally, it should be noted that one of the most complicated cases is related to measurements made by diffusion batteries, where condition numbers are on the order of $C \propto (2.0\text{–}5.4) \times 10^4$.

This chapter is focused on consideration of various strategies allowing one to perform conversion of datasets acquired by various aerosol instruments into representative particle characterization parameters. It will also discuss possible sources of errors, and analytic procedures capable to minimize these errors.

8.2
Forms of Representation of Particle Size Distribution

It is very important to select the optimal form for presentation of data on PSD. A final illustration (graphical or tabular) usually contains measured or mathematically reconstructed particle size spectrum and particle concentration. As a rule, the person performing the measurements defines the method of presentation. However, in many situations the form of presentation is restricted by processes governing particular aerosol system. For example, if sources of new particles are uniformly distributed in space and particles are produced from monomers (supersaturated vapor), then the particle size spectrum is defined by the processes of nucleation, condensation, evaporation, coagulation, and diffusion to a system's boundary. Consequently, to turn the raw data into a representative final result, equations governing particular processes occurring in the system have to be solved.

Commonly, the PSD is presented as

$$f(x) = \frac{1}{N_0} \frac{\mathrm{d}N(x)}{\mathrm{d}x} \tag{8.5}$$

where x is the particle geometry (radius, diameter, surface area, and so on), N_0 is the total number of particles in the system, and $N(x)$ is the number of particles with a size less than x. Accordingly, the number of particles within the size range from x to $x + \mathrm{d}x$ will be $\mathrm{d}N = N_0 f(x)\,\mathrm{d}x$, and the PSD function *in this case* has norm equal to 1. Such a definition of the PSD function is convenient because the corresponding moment ($k = 1$) of this distribution function corresponds to the averaged value of the function and so on:

$$\langle x^k \rangle = \int x^k f(x)\,\mathrm{d}x \tag{8.6}$$

At the same time, to find a suitable function is sometimes a very difficult problem. Moreover, systems with sources uniformly distributed in space are very rare, as most of them consist of multiple randomly operating sources non-uniformly distributed across the system. The only way to handle such systems is to average the PSD data by time and space.

A more favorable situation is related to characterization of the PSD in systems where uniformly distributed sources generate new particles. Such systems could be observed in a controlled laboratory environment or for some natural atmospheric conditions [2]. In this situation, the PSD across the entire size range could be mathematically constructed from data obtained for large particles, which are easy

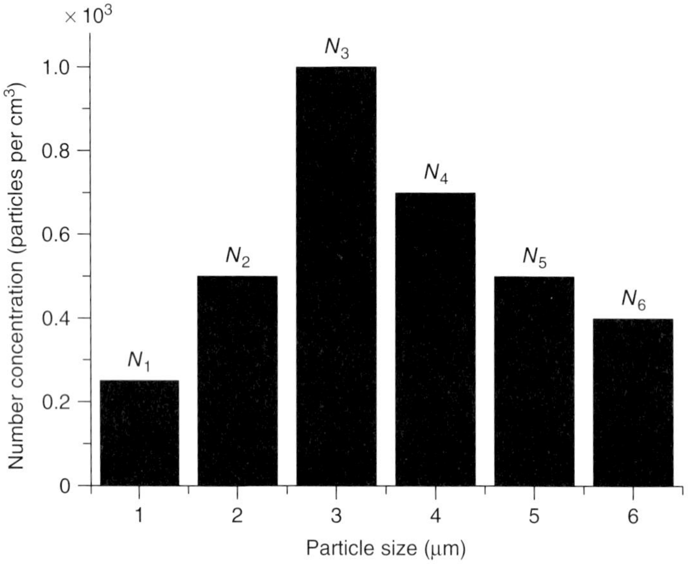

Figure 8.1 Histogram of particle size distribution.

to acquire by relatively trivial instrumentation [2]. However, to perform the task, the mechanism of particle formation ought to be known.

Most commonly, the size distribution is presented by either a frequency distribution or a cumulative distribution. Experimental data as a rule are presented by a histogram, as shown in Figure 8.1. Figure 8.2 shows the particle size distribution or cumulative curve and density of the particle size distribution. Sometimes the word "density" is omitted. The cumulative PSD presentation is convenient for many applications (for example, for design of air quality control equipment), as it shows the total contribution of particles smaller than the particular size of interest.

The above simple and convenient histograms can be directly obtained by some types of instruments, for example, by cascade impactor or optical counter, which are capable of directly quantifying amounts of aerosol within certain size ranges. However, a more complicated approach is required for cases where data acquisition is based on indirect measurement of a certain system parameter with subsequent mathematical construction of the PSD dataset. In this case, some processing of values is required,

$$\langle x\rangle = \int x f(x)\,\mathrm{d}x \tag{8.7}$$

and averaged distribution width,

$$\langle \sigma\rangle = \sqrt{\langle x^2\rangle - \langle x\rangle^2} \tag{8.8}$$

where

$$\langle x^2\rangle = \int x^2 f(x)\,\mathrm{d}x \tag{8.9}$$

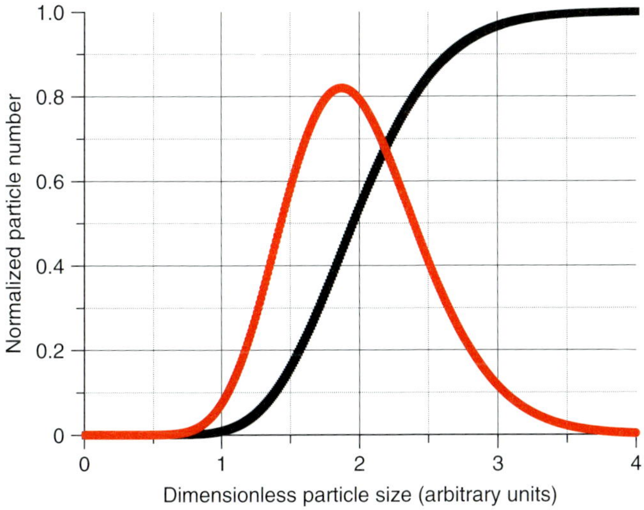

Figure 8.2 Particle size distribution and density of the particle size distribution.

Various approaches to construct theoretical distributions could be used for data presentation for cases when the frequency distribution or cumulative distribution of size have limited use. In the area of aerosol science, most commonly used are the *normal distribution*, log-normal distribution, and gamma distribution (see detailed description in Chapter 1). Now let us try to understand how these particle size distributions can be constructed.

8.3 Differential and Integral Measurements

As discussed above, many types of measurements are indirect. It means that the actual parameter of interest cannot be measured (compared against the corresponding standard) directly. In these cases, some function of this parameter, which could be measured directly, has to be selected, measured, and mathematically converted (for example, integrated) to represent the result of interest:

$$P(n) = \int_0^\infty p(x, n)\psi(x)\,dx \tag{8.10}$$

Here $\psi(x)$ is a function of the PSD (parameter of interest), $P(n)$ is the measured value (for example, particle penetration through diffusion battery), which is dependent on some other parameters (in this particular case, on the number of screens of the diffusion battery's stage), and $p(x, n)$ is the kernel of the integral equation (in this case, the penetration of the particle with size x through a diffusion battery with n screens).

As one can understand, the form of the kernel of the integral equation $p(x, n)$ plays a key role. As discussed, if it is close to a delta or Heaviside step function, then Eq. (8.10) turns into a system of linear equations, which has a simple analytical solution. On the other hand, for example, in the case of diffusion aerosol spectrometry (DAS), to reconstruct the PSD one has to solve a diffusion equation and to use it for integral measurement of the particle penetration through the stage of the DAS. The other representative example of integral measurement is the case when particles deposit due to the influence of an external electric field (for example, in a DMA column). In this case, the reconstruction of the particle size distribution could be done by solving a differential equation describing the particle trajectory as a function of the field parameters [3–6]:

$$\frac{\mathrm{d}r}{\mathrm{d}t} = u_r(r, z) + Z_p E_r(r, z)$$

$$\frac{\mathrm{d}z}{\mathrm{d}t} = u_z(r, z) + Z_p E_z(r, z) \tag{8.11}$$

where r and z are the radial and longitudinal coordinate of the particle in the cylinder, respectively, $u_r(r, z)$ and $u_z(r, z)$ are the particle velocity along the cylinder radius and axis, respectively, Z_p is the electrical mobility, and $E_r(r, z)$ and $E_z(r, z)$ are the electrical intensity along the radial and longitudinal coordinates, respectively.

These examples clearly demonstrate the difference between differential and integral measurements. For differential measurements, the condition number is close to 1. For integral measurements, the deviation of the condition number from 1 becomes appreciable and may reach 10^4. During differential measurements, the problem of conversion of measured data into PSD is stable and directly related to experimental errors only. In the case of integral measurements, the problem becomes unstable, with small experimental errors leading to large resulting uncertainties.

Currently, lack of universal equipment for the characterization of particles in the nanometer size range dictates the need to select an instrument that satisfies the conditions applied to a particular measurement, including consideration of an operational principle, mathematical software, and so on. Let us now consider two main measurement techniques, namely, differential mobility analysis and diffusion aerosol spectrometry, which are the most commonly used for characterization of particles in the nanometer size range.

8.4 Differential Mobility Analysis

Differential mobility analyzers are commonly used for measurements of nanometer sized aerosol particle for a wide range of concentrations [3–6]. Different operation scenarios of the instrument could correspond to either differential or integral type of measurements [3].

The operational principle utilized by the device is based on the motion of charged particles in the gas phase under the action of an electric field. Let us assume that the aerosol particle has an electrical charge q. Then, when it moves in an electric field with strength $\boldsymbol{E}$, it is influenced by the electric force $\boldsymbol{F} = q\boldsymbol{E}$. The action of this force causes some displacement of the particle from its trajectory in the gas carrier, which could be mathematically quantified to find the relation between the electrical and mechanical parameters of the process. In addition, during its motion in the electric field, the particle collides with gas molecules, which adds more uncertainties to the process, requiring consideration of diffusion mechanisms to precisely quantify the process. To describe the motion of a diffusing particle, Eq. (8.11) has to be solved and the geometry of the particle trajectory in the gas phase obtained as the result. At the same time, this approach does not take into account the diffusion type of motion [6]. To account for diffusion, the particle trajectory could be obtained more accurately by introducing a diffusion correction [6]. However, the most accurate outcomes can be achieved by solving the diffusion equation with the electric field acting on the particles taken into account.

Commonly, the DMA is built as a cylindrical capacitor with an electric field created between coaxial surfaces due to the potential difference V_0 across them [7]. Aerosol particles are introduced into the DMA through a narrow slot built at the top of the device between the outer cylinder surface with radius r_2 and the inner cylinder with radius $r_2 - \Delta_r$. The flow rate of the aerosol gas carrier is denoted as Q_a. A buffer gas or filtered air with flow rate Q_{sh} is strategically introduced to the space between the two cylinders simultaneously with the aerosol sample to create a uniform shield covering the sample to ensure equivalent traveling distance for each particle to reach the surface of the inner cylinder. A potential difference is applied, with the positive terminal connected to the outer cylinder and the negative to the inner one, as a rule. During the operation of the device, the aerosol particles drift toward the inner cylinder under the action of the electric field and diffusion, and reach a narrow sampling slot provided at its lower end. As one can understand, for each field strength, only particles with very narrow size distribution would successfully reach the slot, while smaller and larger particles collide with the surface of the inner cylinder above and below the sampling slot. The particles that have successfully reached the slot could then be enumerated by a condensation particle counter (CPC) (see Chapter 7) or used as a monodisperse aerosol stream for laboratory or industrial applications. Also, the device could be used for determination of the PSD of polydisperse aerosol. In this mode, the field strength applied to the DMA is gradually changed within a certain voltage range to enable particles with corresponding sizes to reach the sampling slot. The CPC counts the particles that penetrate through the sampling slot, and the PSD is constructed from the field strength-related counts of particles with corresponding sizes.

The transfer function for the DMA $\Omega(Z_p)$ is related to the probability of the particle with known electric mobility Z_p reaching the sampling slot. The transfer function was presented by Knutson and Whitby [3] as

$$Z_p^* = \frac{Q_{sh}}{2\pi L V_0} \ln\left(\frac{r_2}{r_1}\right) \tag{8.12}$$

Also a correction to the aerosol carrier flow rate Q_a is

$$\Delta Z_p^* = \frac{Q_a}{2\pi L V_0} \ln\left(\frac{r_2}{r_1}\right) \tag{8.13}$$

For the above-mentioned DMA operation in the continuum regime used to acquire information about the PSD of a polydisperse aerosol, some corrections of the transfer function ought to be made in order to account for the time dependence of the potential difference V_0. In this case, the transfer function will also become time dependent, $\Omega(Z_p, t)$, and could be represented as

$$\Omega(Z_p, t) = \frac{\int_{r_1}^{r_2} U_z(r) 2\pi r c_{out}(Z_p, r, t)\, dr}{Q_s c_0(Z_p)} \tag{8.14}$$

where $c_{out}(Z_p, r, t)$ is the concentration of particles with electric mobility Z_p, r is the distance between the cylinder axis and the particle position at time t, $c_0(Z_p)$ is the concentration of particles with electric mobility Z_p at the DMA inlet, and $U_z(r)$ is the velocity distribution of the flow across the cylinder radius. The particle trajectory may be calculated from the equations of particle motion:

$$\begin{aligned} \frac{dz}{dt} &= U_z(r) \\ \frac{dr}{dt} &= Z_p E(r, t) \end{aligned} \tag{8.15}$$

where $E(r, t)$ is the distribution of the electric field in the cylindrical device. This parameter may be time dependent, and is represented in the form

$$E(r, t) = \frac{V(t)}{r \ln(r_1/r_2)} \tag{8.16}$$

The time dependence may be approximated by the following expression:

$$E(r, t) = \frac{r_1}{r} \alpha\, e^{t/\tau} \tag{8.17}$$

where α and τ are constants that depend on the system parameters.

As one can understand, continuous alteration of the potential difference causes instabilities, which do not allow stable flow of charged particles to be established. Such instability creates some additional error source. At the same time, as was shown in [7], if the potential difference change is performed slowly, then it does not influence significantly the correctness of the measurements. The results obtained for both steady-state and slow scanning regimes of measurements are shown in Figure 8.3.

As is seen from the graphs, the particle trajectories obtained for scanning of the potential difference regime have some uncertainties related to their positions. These uncertainties are mainly associated with the width of the aerosol flow. The solid and dashed lines represent the up-scan and down-scan of the potential difference, respectively. When the scanning time is increased, the uncertainties of the measurements are correspondingly decreased. As one can see, the difference in curvature of the trajectory lines obtained for the two regimes of DMA operation

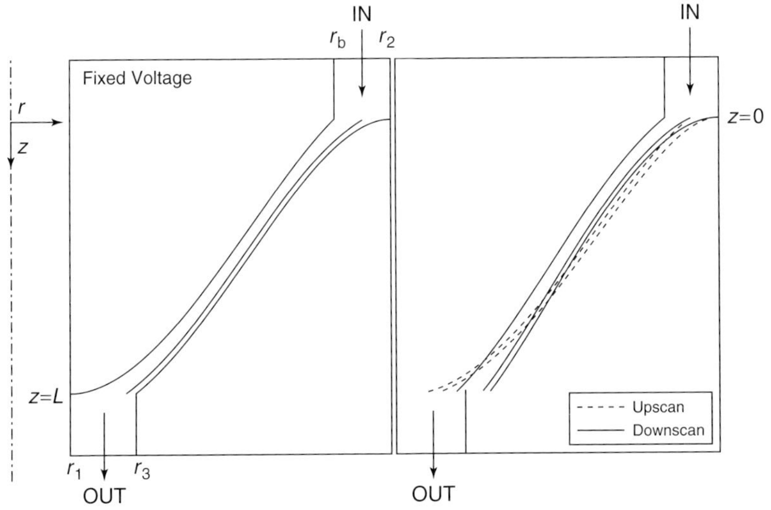

Figure 8.3 Particle trajectories in the cylindrical DMA. The left-hand side represents the results obtained at steady state, and the right-hand side shows the results obtained for the scanning regime with the scan time of 45 s (the ratio of the aerosol flow by diluting air is 1 : 10.)

is the most pronounced at the stage when the particle is getting closer to the inner cylinder. As was shown in [8], the flow may be considered as steady state except for some areas near the surfaces of the inner and outer cylinders. As was demonstrated, at the inlet to the DMA the gas flow is plug flow, and, for a certain length of the device, towards the exit point it becomes viscous flow. Both types of flow were considered for the application in [9]. The difference in the flow behavior was evaluated by the *Monte Carlo method* [6] to evaluate the influence of Brownian motion on the shifting of the particle trajectory. As was found, the diffusion motion leads to widening of the bunch of trajectories of particles with the same charge and size, and respectively the same electric mobility. The analytical approach was used [7] to determine that at relatively long scanning times (90–300 s) all the presented methods give almost the same results, with the largest discrepancy being within several percent. However, when the time of the scan becomes shorter (30–45 s), the discrepancy between the results obtained by different mathematical models reaches 10% and higher. Also, the uncertainties contributed by particle condensation enlargers and by optical counters have to be taken into account to achieve the maximum possible accuracy of the reconstructed PSD [10].

Recently, miniature and multichannel DMA have been proposed [11, 12]. Miniaturization has been achieved by replacing the particle condensation enlarger and optical counter by a system capable of measuring particle charge. Although this approach is very attractive, it contains several drawbacks, mainly related to the need to measure very small charges and currents (10^{-15} A and less). In addition, some particles are able to acquire several elementary charges, which become an additional error source. The suggested use of the charge measuring module as a replacement for the CPC has been investigated before [13], and was not considered as sufficiently accurate, especially for small particles, as the charging efficiency of such particles is very small. Nevertheless, it could provide relatively accurate outcomes for particles with sizes between 50 and 500 nm.

Now, let us evaluate the contribution of diffusion to particle motion in the electric field, which is not commonly taken into account for the PSD reconstruction from raw results of measurements. Let us consider a simplified version of the process, which involves aerosol particles moving between two charged plane plates. It does not have a significant difference compared to the common cylindrical device, but allows one to demonstrate the case much more clearly for the reader.

First, we consider the process of diffusion of a charged particle moving in the gas flow in an electric field. Assuming that every particle has an electrical charge e, radius r, and diffusion coefficient D, the particle motion in the macroscopic gas stream will be composed of two microscopic flows, namely, diffusion and motion under the action of the electric field:

$$\boldsymbol{j} = -D\nabla c + B\boldsymbol{F}c + \boldsymbol{v}c \tag{8.18}$$

where $c(\boldsymbol{r}, t)$ is the particle concentration at the point $\boldsymbol{r}$ at time t, $\boldsymbol{F} = e\boldsymbol{E}$ is the magnitude of the electric field, B is the particle mobility, which is related to the diffusion coefficient as $D = BkT$, and $\boldsymbol{v}$ is the macroscopic flow velocity.

Considering that the distance between the charged plates x_0 is much smaller than the plate size, the electric field between the plates may be assumed to be uniform. If the potential difference between the plates is U, then the electric field intensity is $E = U/x_0$. Considering that the x-axis is directed normally from one plate to another and the y-axis coincides with the gas velocity vector, then Eq. (8.18) may be rewritten as

$$\boldsymbol{j} = \left(-D\frac{\partial c}{\partial x} - D\frac{eU}{x_0 kT}c\right)\boldsymbol{i}_x - D\frac{\partial c}{\partial y}\boldsymbol{i}_y + v_y c\boldsymbol{i}_y \tag{8.19}$$

As one can understand, the negative sign before the second term in brackets means that the directions of the x-axis and the vector $\boldsymbol{E}$ are opposite. Vectors $\boldsymbol{i}_x$ and $\boldsymbol{i}_y$ are unit vectors along the x- and y-axes. To derive a diffusion equation for the particle concentration, let us first write a continuity equation for $c(x, y, t)$, which considering the relation (8.19) becomes

$$\frac{\partial c}{\partial t} + \nabla\boldsymbol{j} = 0 \tag{8.20}$$

For steady-state conditions $\partial c/\partial t = 0$, and substitution of (8.19) into (8.20) gives

$$-D\frac{\partial^2 c}{\partial x^2} - D\frac{eU}{x_0 kT}\frac{\partial c}{\partial x} + v_y\frac{\partial c}{\partial y} = 0 \tag{8.21}$$

In this equation, the particle transport due to diffusion along the y-axis is not taken into account, as it is negligible compared to the transport of the particle by the velocity v_y. To solve Eq. (8.21), some boundary conditions ought to be added. First of all, we assume that the particle is trapped by the plate after collision and stays on the plate surface (some possible deviations from this assumption will be discussed below). In this case the boundary conditions are

$$c(x = 0) = c(x = x_0) = 0 \tag{8.22}$$

To solve Eq. (8.21) we first rewrite it in the form $c(x, y) = c(x)c(y)$, and

$$\frac{D}{v_y c(x)}\left(c''(x) + \frac{eU}{x_0 kT}c'(x)\right) = \frac{c'(y)}{c(y)} = -\lambda^2 \tag{8.23}$$

For the longitudinal direction, the result for $c(y)$ is obtained as

$$c(y) = c_0\,\mathrm{e}^{-\lambda^2 y} \tag{8.24}$$

For the transverse direction, the solution for $c(x)$ is

$$c''(x) + \frac{eU}{x_0 kT}c'(x) + \frac{\lambda^2 v_y}{D}c(x) = 0 \tag{8.25}$$

The characteristic equation for the second-order differential equation (8.25) may be written as

$$z^2 + \frac{eU}{x_0 kT}z + \frac{\lambda^2 v_y}{D} = 0 \tag{8.26}$$

with the solution

$$z_{12} = -\frac{eU}{2x_0 kT} \pm \sqrt{\frac{e^2 U^2}{4x_0^2 k^2 T^2} - \frac{\lambda^2 v_y}{D}} \tag{8.27}$$

The solution for $c(x)$ in Eq. (8.25) may be written as

$$\begin{aligned} c(x) &= c_{01}\,\mathrm{e}^{-(eU/2x_0 kT)x}\sin\left(\sqrt{\frac{\lambda^2 v_y}{D} - \frac{e^2 U^2}{4x_0^2 k^2 T^2}}\,x\right) \\ &\quad + c_{02}\,\mathrm{e}^{-(eE/2kT)x}\cos\left(\sqrt{\frac{\lambda^2 v_y}{D} - \frac{e^2 U^2}{4x_0^2 k^2 T^2}}\,x\right) \end{aligned} \tag{8.28}$$

To satisfy the boundary conditions at $x = 0$, it is necessary to apply $c_{02} = 0$. To satisfy the boundary conditions at $x = x_0$, it is necessary to solve the equation

$$\sin\left(\sqrt{\frac{\lambda^2 v_y}{D} - \frac{e^2 U^2}{4x_0^2 k^2 T^2}}\,x\right) = 0$$

which gives the solution in the form

$$\lambda^2 = \frac{D}{v_y}\left(\frac{\pi^2 n^2}{x_0^2} + \frac{e^2 U^2}{4x_0^2 k^2 T^2}\right) \tag{8.29}$$

Finally, the solution of Eq. (8.21) is

$$\begin{aligned} c(x, y) &= \sum_n c_n \exp\left(-\frac{eU}{2x_0 kT}x\right)\sin\left(\frac{\pi n}{x_0}x\right) \\ &\quad \times \exp\left[-\frac{D}{v_y}\left(\frac{\pi^2 n^2}{x_0^2} + \frac{e^2 U^2}{4x_0^2 k^2 T^2}\right)y\right] \end{aligned} \tag{8.30}$$

To find the series coefficients c_n, the orthogonal properties of the eigenfunction of Eq. (8.21) with weight $e^{(eE/2kT)x}$ can be used. These coefficients are defined by the initial conditions. Let aerosol particles be introduced into the space between the parallel plates through the thin slot with width Δx. The concentration of the particles at this point is taken as c_0. The concentration distribution in the space between the plates can be obtained by solving Eq. (8.30), assuming that the series coefficients c_n are defined for the initial conditions. The concentration distribution at the inlet to the space can be written as

$$c(x, y = 0) = c_0(\Theta(x) - \Theta(x - \Delta x)) = \sum_n c_n \exp\left(-\frac{eU}{2x_0 kT}x\right) \sin\left(\frac{\pi n}{x_0}x\right) \tag{8.31}$$

Taking into consideration orthogonal relations, the series coefficients are calculated as

$$c_n = \frac{4kT\left\{e^{\frac{eU}{2x_0kT}\Delta x}\sin\left(\frac{\pi n}{x_0}\Delta x\right) + \frac{2\pi nkT}{eU}\left[1 - e^{\frac{eU}{2x_0kT}\Delta x}\cos\left(\frac{\pi n}{x_0}\Delta x\right)\right]\right\}}{eU\left(1 + \frac{4\pi^2 n^2 k^2 T^2}{e^2 U^2}\right)} \tag{8.32}$$

Substitution of (8.30) into (8.32) gives

$$c(x, y) = \sum_n \frac{4kT\left\{e^{\frac{eU}{2x_0kT}\Delta x}\sin\left(\frac{\pi n}{x_0}\Delta x\right) + \frac{2\pi nkT}{eU}\left[1 - e^{\frac{eU}{2x_0kT}\Delta x}\cos\left(\frac{\pi n}{x_0}\Delta x\right)\right]\right\}}{eU\left(1 + \frac{4\pi^2 n^2 k^2 T^2}{e^2 U^2}\right)} \times e^{-\frac{eU}{2x_0kT}x}\sin\left(\frac{\pi n}{x_0}x\right)\exp\left[-\frac{D}{v_y}\left(\frac{\pi^2 n^2}{x_0{}^2} + \frac{e^2 U^2}{4x_0^2 k^2 T^2}\right)y\right] \tag{8.33}$$

The solution of Eq. (8.21) with boundary conditions (8.22) and initial conditions (8.31) is shown in Figure 8.4. As is seen, diffusion processes are responsible for widening the bunch of particle trajectories, preventing these trajectories from being presented as precisely defined lines. On this basis, it could be concluded that the diffusion process could introduce significant uncertainties into the consideration of the process of particle motion in an electric field.

8.5 Diffusion Aerosol Spectrometry

Diffusion aerosol spectrometers are commonly used for measurements of aerosol size distribution and concentration. The operational principle of the DAS device is based on the ability of aerosol particles to deviate from gas trajectory lines due to diffusion. This particle-size-related process is responsible for possible particle collision with the surface along the gas carrier pathway. If the geometry of the pathway is known, the probability of a particle of certain size colliding with the surface could be quantified. As one can understand, there is also some likelihood

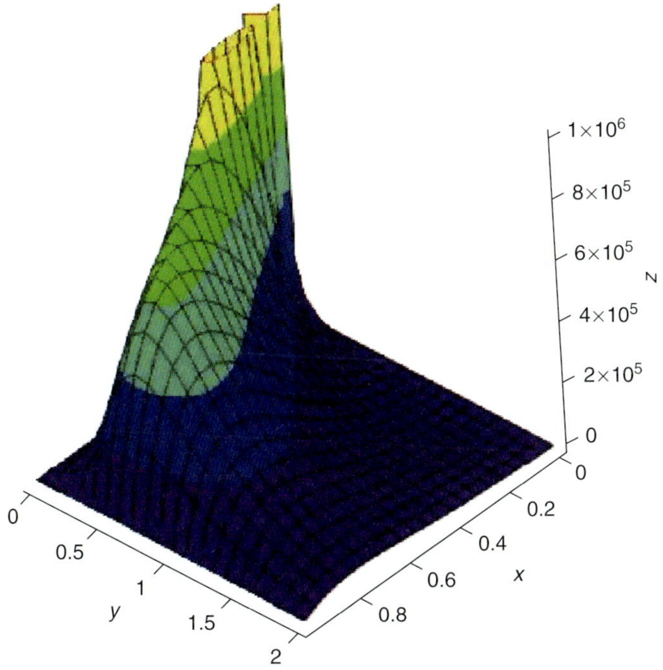

Figure 8.4 Space distribution of particles between charged plates of a capacitor during macroscopic transport. The distance between the parallel plates is 0.5 cm, velocity v_y is 0.1 m s^{-1}, and the diffusion coefficient $D = 10^{-5}$ cm^2 s^{-1}.

for a particle to bounce off the surface after collision [14, 15]. However, at this stage we assume that the particle settles on the surface, and we will come back to the bouncing problem later.

Various studies confirmed the feasibility of DAS for measuring the particle spectrum and concentrations in different environments both in the open atmosphere and inside buildings [16, 17]. The device is capable of providing accurate and reliable data, which was confirmed during the international intercomparison research meeting held in June-July 1993 at Vienna University [18]. Let us now discuss how measurements of PSD and concentration are performed by DAS.

The particle size spectrum is reconstructed from measurements of the penetration of particles with certain sizes through a number of stages with known geometry strategically located in the device. On this basis, to convert the raw data into the particle size distribution, the inverse problem has to be considered.

As was discussed, DAS is able to measure the *diffusion mobility* of particles, which allows the identification of particle size if the diffusion coefficient of the particle in a gas carrier is known. The penetration is defined as a ratio of the particle concentration at the diffusion battery outlet by the concentration at the inlet of the device. There are different designs of diffusion batteries available. However, the two most common ones use either thin cylindrical channels as passages for the aerosol stream, or fine mesh screens installed perpendicular to the direction of motion of the aerosol carrier. Currently, the screen-type diffusion batteries are used more often [19, 20]. To acquire the results for penetration through a diffusion battery with n screens, the particle concentrations before and after the screens are measured and used to obtain the magnitude of penetration (ratio of the concentration after the screens by the concentration before the screens).

Now we can consider a fundamental problem related to a minimal number of diffusion batteries required to acquire sufficient information enabling the reconstruction of the particle size distribution. A single-mode particle size spectrum could be graphed if the average particle size and particle distribution width are known. As a result, to reconstruct a single-mode particle size spectrum, two independent penetration measurements are required, as the two parameters of the particle distribution may be extracted from two independent equations. A more complicated strategy is applied to reconstruct a bimodal distribution, which requires knowledge of five parameters. As one can understand, to calculate them, five independent equations have to be solved, so five independent penetration measurements have to be made.

8.5.1
Raw Measurement Results and their Development – Parameterization of Particle Size Distribution

Practical experience shows that relatively large datasets have to be mathematically processed after measurements by DAS. This means that the exact shape of a PSD curve does not play the decisive role, as very often it has random sources. In this situation, a real PSD can be represented as one of the standard distributions previously discussed in Section 7.2 and in Chapter 1. As we will show later, a diffusion dynamical method commonly used for mathematical analysis of the results acquired by DAS further limits the accuracy of reconstruction of the PSD from the measured data. On this basis, any attempt to graph the PSD containing more than two modes may be considered as an exotic exercise, and our discussion will be confined to bimodal particle size distributions.

It has to be noted that, for the screen-type diffusion batteries, semiempirical formulas [19, 20], based on the theoretical background developed by Kirsh and Stechkina [21, 22], are used:

$$P(n, D) = \mathrm{e}^{-An\,\mathrm{Pe}^{-2/3}} \tag{8.34}$$

Here $A = 4.52$ is a constant dependent on the screen type and other process conditions, n is the number of screens used in the diffusion battery, $\mathrm{Pe} = 2Ru_0/D$ is the Peclet number, which shows the relation between macroscopic and microscopic transport or diffusion, R is channel radius, u_0 is the averaged flow velocity, D is the diffusion coefficient, and $P(n, D)$ is the penetration of particles with diffusion coefficient D through the diffusion battery with n screens.

Expression (8.34) can be used to calculate the diffusion coefficient or corresponding particle radius, if the particles are of the same radius or if the PSD is

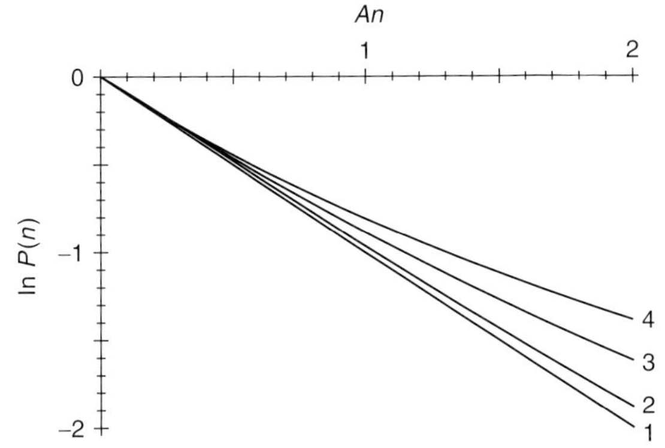

Figure 8.5 Penetration dependence on the dimensionless parameter *An*: very narrow size distribution (curve 1); and distribution parameters $\sigma/\langle D\rangle = 0.2$ (curve 2), 0.5 (curve 3), and 0.8 (curve 4).

monodisperse. Indeed, in practice such particle distributions are rare, and for the more frequently met polydisperse systems a different mathematical approach has to be used. In these situations, the particle size spectrum may be described by a function $F(r)$, where r is the particle radius (8.5), or by $f(D)$, where D is the diffusion coefficient. This function may be defined as $\mathrm{d}N = N_0 f(D)\,\mathrm{d}D$, where $\mathrm{d}N$ is the particle number with diffusion coefficients in the range D to $D + \mathrm{d}D$, and N_0 is the particle concentration in the system. There is a relation between the functions $f(D)$ and $F(r)$, which may be expressed as

$$D = \frac{kT}{6\pi\mu R}\left(1 + a\frac{\lambda}{R} + b\frac{\lambda}{R}\mathrm{e}^{-cR/\lambda}\right) \qquad (8.35)$$

where $a = 1.246$, $b = 0.42$, and $c = 0.87$ are constants, $\lambda = 6.53 \times 10^{-6}\ T/T_0$ is the mean free path length in the gas phase at temperature T, and $T_0 = 296.2$ K.

Particle penetration is related to the PSD as

$$p(n) = \int_0^\infty P(D, n) f(D)\,\mathrm{d}D \qquad (8.36)$$

For example, let us consider the function $p(n)$ for a range of distribution widths and the same averaged particle size. If these curves are plotted as functions of a dimensionless parameter An, the screen number n may be represented as shown in Figure 8.5. As is seen, all the curves start at the point $p(n = 0) = 1$ and then diverge. It has to be noted that the curves do not cross anywhere except at the origin point, meaning that the problem has a unique solution and the inverse problem may be resolved.

The analysis of the experimental data includes the extraction from these curves of information on the distribution parameters. This task may be successfully undertaken, and reliable results obtained, in the case when the curves essentially deviate from each other. On this basis, it can be concluded that the procedure of conversion of the penetration data into size distribution parameters could be done more precisely if the diffusion coefficient values are involved. Then they could be easily converted to corresponding particle sizes.

The problem is solved in parallel for the particle radius,

$$\langle r \rangle = \int_0^\infty r(D) f(D)\, \mathrm{d}D \tag{8.37}$$

and the distribution width,

$$\langle \sigma \rangle = \sqrt{\langle r^2 \rangle - \langle r \rangle^2} \tag{8.38}$$

These parameters provide comprehensive information to graph the particle size distribution.

This approach was proposed in [23], where the final PSD was approximated by a standard Laplace distribution as a function of the diffusion coefficient with subsequent transformation into the corresponding particle size spectrum.

Equation (8.34) [19, 20] may be rewritten as

$$p(n, D) = \mathrm{e}^{-A_1 n D^{2/3}} \tag{8.39}$$

where the constant $A_1 = A/2au_0$. It was supposed that n is large enough to allow its substitution into (8.39) and the following use of the saddle point approximation (8.36) to estimate the value of the integral.

The expression under the integral may be rewritten (8.36) as

$$\mathrm{e}^{F(D)} = \mathrm{e}^{-A_1 n\phi(D) + \ln f(D)} \tag{8.40}$$

where $\phi(D)$ is the exponential part of the function $p(n, D)$, which is dependent on D. Substitution of (8.40) into (8.34) and integration of the saddle point approximation for penetration yields

$$P(n) = \sqrt{\frac{2\pi}{\Delta(D_0)}}\, \mathrm{e}^{F(D_0)} \tag{8.41}$$

and the position of the saddle point is calculated by

$$A_1 n = \frac{f'(D_0)}{f(D_0)\phi'(D_0)} \tag{8.42}$$

Equations (8.41) and (8.42) can now be used to reconstruct the PSD.

8.5.2
Fitting of Penetration Curves

The main issue with the inversion is the unstable solution, which leads to the transformation of small experimental errors into significant uncertainties of the final solution. To overcome this obstacle (Eq. (8.36)), a method of fitting curves could be applied. The main idea of the method is adjustment of the *penetration curves* by experimental results. It could be realized by means of parameterizations of the fitting curves. The penetration of a polydisperse aerosol through a diffusion battery $P(n)$ is represented as a function of two parameters,

$$P(n) = \mathrm{e}^{-a_1 \langle D \rangle^{2/3} n + b\sigma n^{3/2}} \tag{8.43}$$

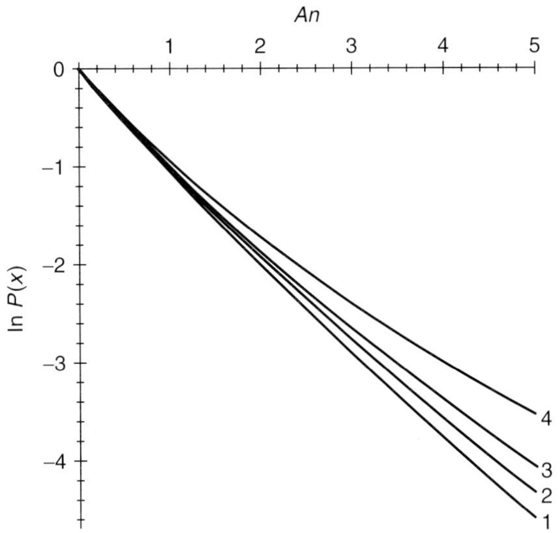

Figure 8.6 Dependence of the penetration on the screen number *An*: narrow distribution width (curve 1); and distribution parameters $\sigma/D = 0.2$ (curve 2), 0.5 (curve 3), and 0.8 (curve 4).

where $\langle D\rangle$ is the average diffusion coefficient, and σ is the average distribution width. The coefficients $a_1 = A(2r_0u_0)^{-2/3}$ and $b = 23.7$ may be identified experimentally, r_0 is the radius of the screen fiber, u_0 is the average flow velocity, and $A = 4.52$ is the fan filter constant defined before.

Figure 8.6 presents the penetration curves as a function of the screen numbers, and $x = a_1n$ was chosen as an independent value. Curves 1, 2, 3, and 4 correspond to relative distribution widths $\sigma/\langle D\rangle = 0.4$, 0.5, 0.6, and 0.8, respectively. These curves resemble the scenario presented in Figure 8.5, meaning that the theoretical curves could be easily fitted to experimental data, which are defined by (8.43).

The uncertainties associated with this strategy are related to experimental errors, because, for integral measurements, diverse types of data acquisition have different errors. To fit theoretical relations with experimental curves, the method of least squares could be used.

This approach is mainly applicable in a case of a single-mode distribution. When a distribution with two or more modes is considered, or when the distribution width is relatively large, then this method cannot be used.

8.5.3
Transformation of the Integral Equation into Nonlinear Algebraic Form

Let us now consider another more reliable method of conversion of the measured data into the PSD. The equation used for transformation of the particle penetration through the diffusion battery with n screens is

$$P(n) = \int_0^\infty e^{-AnD^{2/3}} \phi(D)\, dD \tag{8.44}$$

where $\phi(D)$ is the diffusion coefficient distribution. Simple analysis of this equation shows that, if new variables $x = D^{2/3}$ are introduced, the equation could be simplified, and the particle size increment may be expressed as

$$dN = N_0\phi(D)\, dD = N_0\psi(x)\, dx \tag{8.45}$$

Then Eq. (8.44) may be rewritten as

$$P(n) = \int_0^\infty e^{-Anx} \psi(x)\, dx \tag{8.46}$$

If the dependence of the particle concentration on the particle size is a random variable that is distributed by the defined expression, then any function of this value is also a random variable that is defined by the same expression. This means that the distribution of the variable x could be presented as a gamma distribution with parameters calculated as

$$\psi(x) = \frac{\lambda^\gamma}{\Gamma(\gamma)} e^{-\lambda x} x^{\gamma-1} \tag{8.47}$$

where $\Gamma(\gamma)$ is Euler's gamma function. Then Eq. (8.46) could be presented in the form

$$P(n) = \frac{\lambda^\gamma}{\Gamma(\gamma)} \int_0^\infty e^{-(\lambda+An)x} x^{\gamma-1}\, dx \tag{8.48}$$

After integration, the penetration function could be turned into a very simple form:

$$P(n) = \left(1 + \frac{An}{\lambda}\right)^{-\gamma} \tag{8.49}$$

Considering that there are only two unknowns involved, two experimentally acquired values of the particle penetration would allow one to reconstruct the single-mode PSD by substitution into formula (8.49) and to obtain the parameters λ and γ, which are sufficient to plot the PSD graph. Moreover, the form of the dependence (8.49) allows one to select unknown variables λ and γ, meaning that complex solution of the system of nonlinear equations is not required. In contrast, it turns into simple independent algebraic nonlinear equations. Let us illustrate this approach.

Suppose that two measurements P_1 and P_2 of penetration values corresponding to the screen numbers n_1 and n_2 respectively are made. For convenience, let us introduce new dimensionless variable $\lambda_0 = \lambda/A$. Then Eq. (8.49) may be rewritten in dimensionless form as

$$P(n) = \left(1 + \frac{n}{\lambda_0}\right)^{-\gamma} \tag{8.50}$$

From (8.49), it may be shown that

$$\lambda_0 = \frac{n_2}{(1 + \frac{n_1}{\lambda_0})^{\ln P_2/\ln P_1} - 1} \tag{8.51}$$

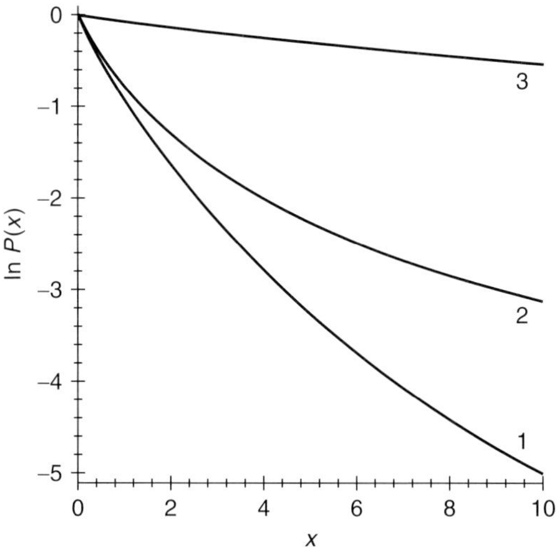

Figure 8.7 Dependence of particle penetration on screen numbers in diffusion batteries. Three penetration curves are shown for the same average particle size and various distribution widths: 0.1 (curve 1), 0.2 (curve 2), and 0.5 (curve 3).

and

$$\gamma = -\frac{\ln P_1}{\ln\left[\left(\frac{n_1}{n_2}\right)(P_2^{-1/\gamma} - 1) + 1\right]} \tag{8.52}$$

Equations (8.51) and (8.52) may be used independently to calculate the parameters λ and γ, respectively. Experimental verification demonstrated that this approach for resolving the inverse problem gives reliable results for cases when the experimental errors of penetration data are not large.

Figure 8.7 shows normalized penetration curves plotted as a function of the dimensionless parameters $x = An$ for three different distributions, which have the same averaged particle size but varying $\sigma/\langle D\rangle$. Figure 8.8 shows the distribution curves for penetration results provided in Figure 8.7. All the distributions are normalized to unity.

Various applications of particle size spectrum measurements and their analysis in different environments, including laboratory experiments and atmospheric studies, can be found in [2, 14–17, 24–26].

8.5.4
Effect of Experimental Errors on Reconstruction of Particle Size Distribution

Undertaking ideal measurements free of significant experimental errors is very difficult in aerosol practice. As experimental errors are unavoidable, we try to evaluate how they could impact on the reconstruction of the PSD. It is also important to evaluate the magnitude of the error, which makes the results meaningless.

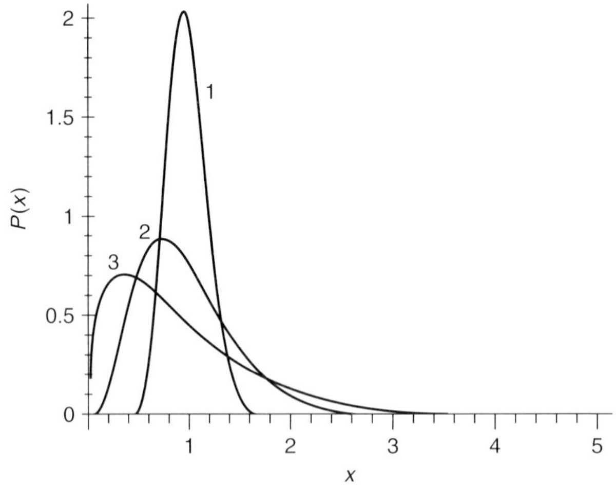

Figure 8.8 Particle radius distribution curves for various distribution widths: 0.1 (curve 1), 0.2 (curve 2), and 0.5 (curve 3). All particle size distributions are normalized to unity.

Let us first analyse Eqs. (8.51) and (8.52). Assuming that the penetration value P_1 is acquired perfectly accurately, the uncertainty would only correspond to that in P_2. As P_1 and P_2 are independent parameters, this assumption would not refine the problem.

The problem may be formulated as follows. For the gamma distribution, definite parameters $\lambda = 4$ and $\gamma = 4$ were selected, meaning that $\langle x \rangle = 1$ and $\sigma = 0.5$. On this basis, using Eq. (8.49) allows one to calculate the penetration values $P(x)$ for $x_1 = 0.5$ and $x_2 = 3$, that is, $P_1 = P(x_1) = 0.624\,295$ and $P_2 = P(x_2) = 0.106\,622$.

It is important to know how the relative error $\Delta P_1/P_1$ impacts on the calculated parameters λ and γ, and especially on the values $\langle x \rangle$ and σ. The results of corresponding investigation are presented in Figures 8.9 and 8.10.

As may be seen from Figure 8.9, the relative errors of the measured penetration values ($\Delta P/P$) lead to deviation of the distribution parameters λ and γ from their former values. In practice, for relative errors $\Delta P/P > 0.05$ the deviation is not very large and measurements have sense. Beyond this limit, the deviations of the reconstructed distribution parameters become large and measurements may be considered as meaningless. At the same time, a departure of parameters λ and γ takes place quite coherently, meaning that, in spite of the fact that there is substantial deviation, their ratio remains relatively stable (see Figure 8.10). As is seen, when parameters λ and γ deviate from their initial values by a factor of 40, the average particle size deviates only by a factor of 0.25. Obviously, the deviation of the distribution width remains very large.

The important conclusion of this consideration is that, if only the average particle size is required from the experimental measurements, then $\Delta P/P = 0.1$ would be considered as a threshold allowing the final outcomes to be treated as meaningful.

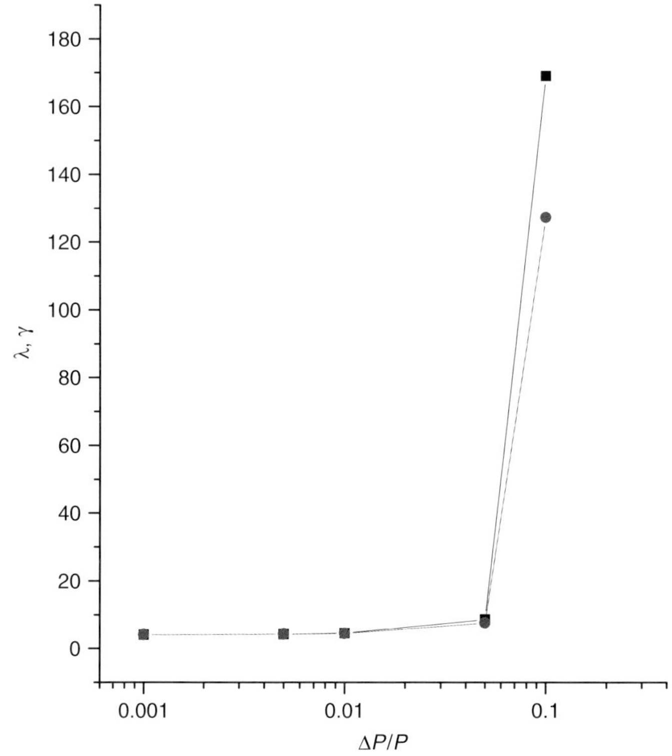

Figure 8.9 Effect of relative errors of the particle penetration measurement on reconstruction of λ (squares) and γ (circles). As one can see, the parameter λ is more sensitive to errors than the parameter γ.

On the other hand, if the average particle size and distribution width are required, then the errors ought to be limited to 0.05.

A similar picture is observed if the relative error $\Delta P/P < 0$. In this case the direction of the deviation of λ and γ remains the same, as is seen in Figure 8.11.

The dependence of the level of uncertainty on reconstructed average particle size and distribution width preserves its shape similarly to the previous case ($\Delta P/P > 0$) only if the graphs are "flipped" as shown in Figure 8.12. As is seen, the tendency of the relation between the calculated λ and γ as a function of $\Delta P/P$ remains the same.

8.5.5
Reconstruction of Bimodal Distributions

In real situations, some aerosol systems could be generated by mixing of different streams produced by alternative sources. As a result, the PSD may have multimodal nature, which reflects several contributing single-mode distributions.

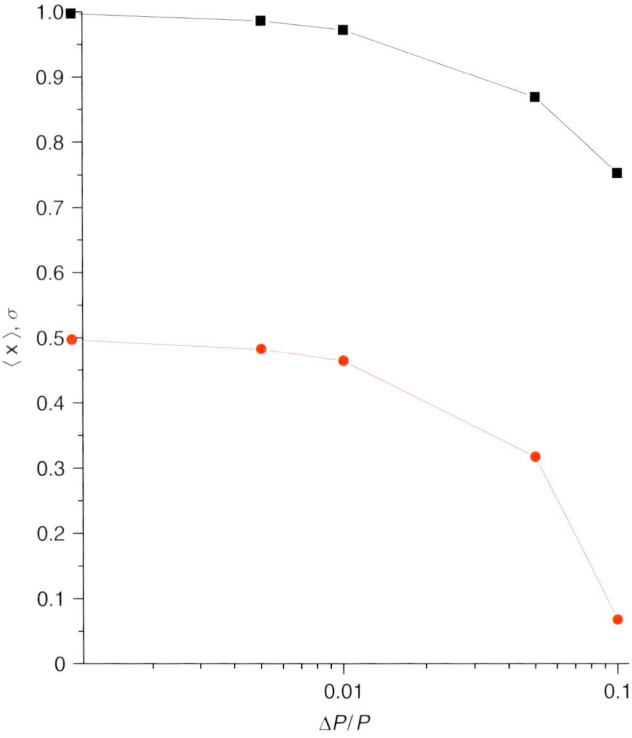

Figure 8.10 Effect of relative errors of the particle penetration measurement on reconstruction of averaged particle size $\langle x \rangle$ (squares) and averaged distribution width σ (circles).

As an example, we can consider a bimodal distribution that is composed of two distributions with average sizes $\langle x_1 \rangle = 2$ and $\langle x_2 \rangle = 1$, and widths $\sigma_1/\langle x_1 \rangle = \frac{1}{4}$ and $\sigma_2/\langle x_2 \rangle = \frac{1}{2}$, as presented in Figure 8.13. Curve 3 is the resulting distribution, which consists of two single-mode distributions represented by curves 1 and 2. The two distinct humps on curve 3 reflect the bimodal nature of this aerosol system. A different scenario is presented in Figure 8.14. The resulting size distribution (curve 3) does not have distinct modes, making it very difficult to state that it has bimodal nature, which could result in misinterpreting curve 3 as a single-mode distribution.

The strategy for the reconstruction of a bimodal distribution from raw experimental results on the particle penetration could now be considered. Let us first consider a bimodal distribution composed of two single-mode distributions:

$$f_0(\lambda_1, \gamma_1, \lambda_2, \gamma_2, x) = \alpha \frac{\gamma_1^{\lambda_1}}{\Gamma(\gamma_1)} e^{-\lambda_1 x} x^{\gamma_1 - 1} + (1 - \alpha) \frac{\gamma_2^{\lambda_2}}{\Gamma(\gamma_2)} e^{-\lambda_2 x} x^{\gamma_2 - 1} \tag{8.53}$$

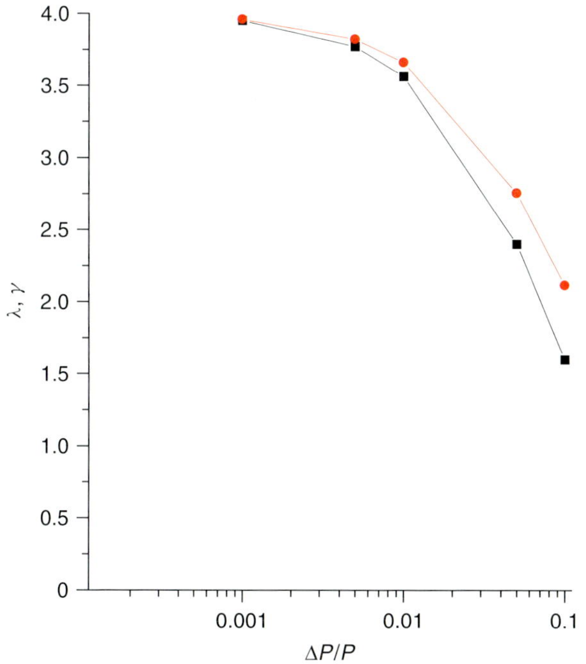

Figure 8.11 Effect of negative relative errors of the particle penetration measurements on reconstructed particle size distribution λ (squares) and γ (circles).

where α represents the first distribution with parameters λ_1 and γ_1, and $(1-\alpha)$ is related to the second distribution with parameters λ_2 and γ_2. Assume that $0 \le \alpha \le 1$, and the resulting distribution $f_0(\lambda_1, \gamma_1, \lambda_2, \gamma_2, x)$ is normalized to unity.

Using Eqs. (8.46) and (8.53), the penetration of the aerosol through the diffusion battery may be written in the form

$$P(n) = \alpha\left(1+\frac{An}{\lambda_1}\right)^{-\gamma_1} + (1-\alpha)\left(1+\frac{An}{\lambda_2}\right)^{-\gamma_2} \tag{8.54}$$

Now the parameters of the combined distribution, that is, λ_1, γ_1, α, λ_2, and γ_2, can be calculated using the measured penetration data: P_1, P_2, P_3, P_4, and P_5.

As discussed, the resulting system of five nonlinear algebraic equations with five unknowns can be routinely solved by well-known contemporary numerical methods. At the same time, it has to be noted that the equations utilize experimental results that were acquired with some uncertainties, making the problem more complicated. In this situation, it is convenient to use the functional relation between the penetration data and unknown variables (8.54).

When x is relatively small and $P(x)$ is comparable to 1, then it is mainly defined by the second term of Eq. (8.54) supposing that $\lambda_1 < \lambda_2$. On the other hand, when $P(x)$ is much smaller than 1, then the dependence is defined by the first term.

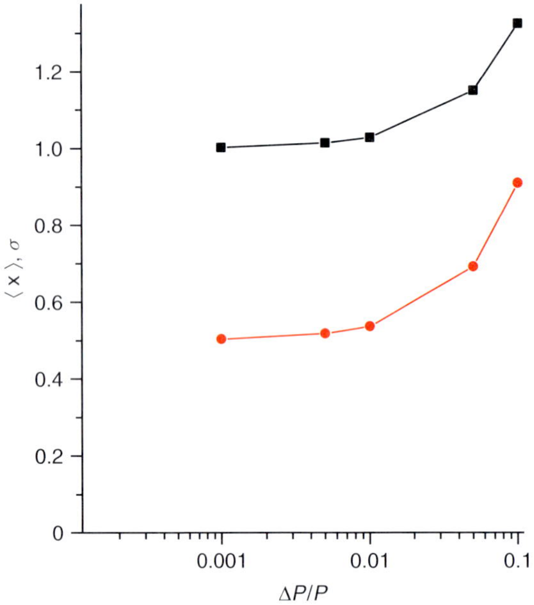

Figure 8.12 Effect of negative relative errors of the particle penetration measurements on reconstructed average particle size $\langle x \rangle$ (squares) and distribution width σ (circles).

At the same time, Eq. (8.54) is linear relatively to α, requiring the penetration value P_3 to be used for calculations.

In summary, first, to estimate values of λ_2 and γ_2, the penetration results P_1 and P_2 are substituted into Eqs. (8.51) and (8.52). Then, a similar substitution of P_4 and P_5 into the same equations allows one to estimate λ_1 and γ_1. Finally, α is obtained by

$$\alpha = \frac{P_3 - (1 + x_3/\lambda_1)^{-\gamma_1}}{(1 + 3x_3/\lambda_1)^{-\gamma_1} - (1 + x_3/\lambda_2)^{\lambda_2}} \tag{8.55}$$

This approach can be used as the first approximation. To refine the results, a standard procedure of serial approximations might be required.

8.5.6
Mathematical Approach to Reconstruct Bimodal Distribution from Particle Penetration Data

Let us now see how the mathematical approach described in the previous section works for reconstruction of bimodal PSD. First, we consider two distributions with parameters $\lambda_1 = 32$, $\gamma_1 = 64$, $\lambda_2 = 4$, $\gamma_2 = 4$, $\alpha = 0.5$, $\langle x_1 \rangle = 2$, $\langle x_2 \rangle = 1$, $\sigma_1 = \frac{1}{4}$, and $\sigma_2 = \frac{1}{2}$.

The graphical presentation of these distributions may be seen in Figure 8.13. As discussed, the humps on curve 3 are very distinctive. Then the penetration values

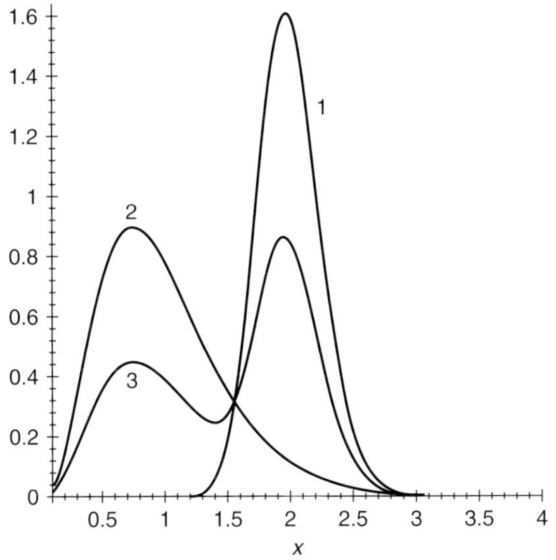

Figure 8.13 Bimodal size particle distribution. Two humps are clearly selected. Here $\langle x_1 \rangle = 2$ and $\langle x_2 \rangle = 1$; $\sigma_1/\langle x_1 \rangle = \frac{1}{4}$ and $\sigma_2/\langle x_2 \rangle = \frac{1}{2}$.

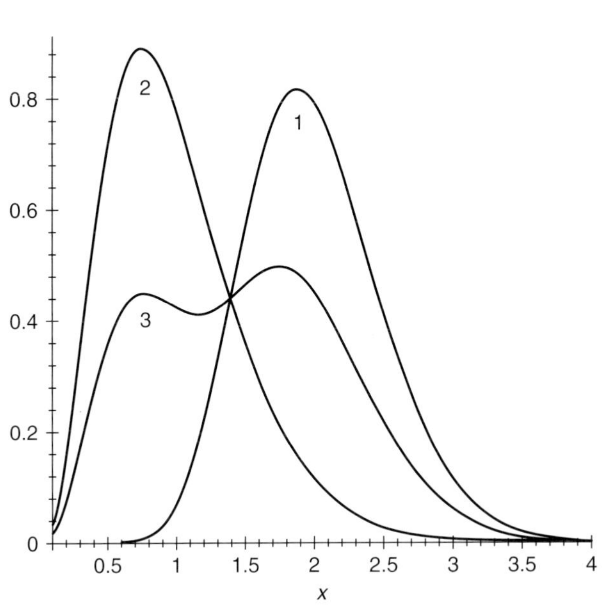

Figure 8.14 Bimodal size particle distribution. Two humps are not selected due to distribution widths.

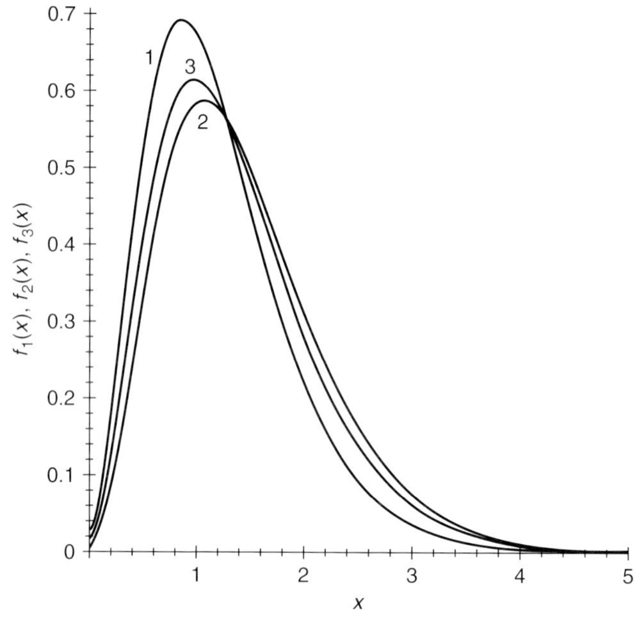

Figure 8.15 Reconstructed PSDs: selected modes (curves 1 and 2) and resulting distribution (curve 3).

through the diffusion batteries may be calculated by Eq. (8.54). For example, at $x = An = 0.5, 1, 3, 20$, and 40, the corresponding magnitudes of P are $P = 0.498$, 0.275, 0.0549, 0.000 386, and 0.000 0342. For this case, the first approximation gives good results for average size. However, for the distribution width, the results are not as good: $\langle x_1 \rangle = 1.52$, $\langle x_2 \rangle = 1.28$, $\sigma_1 = 0.76$, $\sigma_2 = 0.65$, and $\alpha = 0.656$. The distribution widths are too wide, as is seen in Figure 8.15. At the same time, several iterating calculations allow to get desired correctness.

8.5.7
Solution of the Inverse Problem by Regularization Method

There are some other approaches to rectify the inverse problem if the equation is unstable. Some of them are based on regularization of the unstable equation, allowing one to get stable solutions [27–30]. Let us first formulate the equation as an integral *Fredholm equation* [31, 32].

The penetration value through diffusion batteries may be calculated as

$$P_i = \int_0^\infty K(x, n_i) f(x)\, dx + \text{err}_i \tag{8.56}$$

where P_i is the penetration through the battery with n_i screens, $K(x, n_i)$ is the kernel of the integral equation or penetration of the particles with size x through the battery with n_i screens, and $f(x)$ is the PSD to be calculated. Indeed, this is a Fredholm equation of the second type, which was modified from the first type equation by the addition of a term err_i (error of experimental measurements). The problem may then be formulated as

$$P_i = \sum_j K(x_j, n_i) f(x_j) \Delta x_j + \mathrm{err}_i \tag{8.57}$$

where Δx_j is a fraction of the particle size. Introducing the definition $K(x_j, n_i)\Delta x_j = A_{i,j}$ allows Eq. (8.57) to be presented as

$$\mathbf{P} = \hat{\mathbf{A}}\boldsymbol{f} + \mathbf{err} \tag{8.58}$$

where **err** is an error vector. Formally, the solution of the equation may be given as

$$\boldsymbol{f} = \hat{\mathbf{A}}^{-1}\mathbf{P} \tag{8.59}$$

If Eq. (8.58) is linearly independent, then the inverse matrix $\hat{\mathbf{A}}^{-1}$ exists in finite form and the solution $\boldsymbol{f}$ is easy to find. At the same time, the diffusion dynamic method advises that these equations are quasi-independent and $\hat{\mathbf{A}}^{-1}$ turns into a singular matrix. In this case, standard approaches to resolve the problem are inconsistent. In addition, the random nature of the matrix **err** damages the equation. As a rule, this matrix is not defined and alters for each particular measurement.

To overcome this problem, a regularization method proposed by Tikhonov was used. This method can convert an ill-posed problem into a well-posed one. Then, the procedure of smoothing is introduced and regularization by parameter λ could be used. This regularization method may be explained as:

$$\sum_i \left(\frac{P_i - \int_0^\infty K(x, n_i) f(x) \mathrm{d}x}{\mathrm{err}_i} \right) + \lambda J = R(\lambda) + \lambda J(\lambda) \tag{8.60}$$

which has to be minimized. The regularization criteria are as follows:

1) norm of distribution function

$$J = \int_0^\infty (f(x))^2 \, \mathrm{d}x \tag{8.61}$$

2) norm of second derivative

$$J = \int_0^\infty (f''(x))^2 \, \mathrm{d}x \tag{8.62}$$

3) entropy of distribution function

$$J = \int_0^\infty f(x) \ln(f(x)) \, \mathrm{d}x \tag{8.63}$$

Taking the smoothing function in this form [27] allows one to reconstruct the particle size distribution from penetration data containing large errors. This approach may be combined with a method of nonlinear regularization called the L-curve method [28]. For log-normal functions,

$$f(r) = \frac{1}{N} \sum_{i=1}^{n} h_i \exp\left[-\frac{(\log(r) - \log\langle r_i \rangle)^2}{2(\log(\sigma_i))^2} \right] \tag{8.64}$$

where N is the total particle concentration, $\langle r_i \rangle$ is the average particle size of corresponding fraction, σ_i is the average distribution width (deviation), and h_i is the weight of the corresponding function.

In this case, the regularization problem is reduced to the minimization of the functional (8.60), and the value

$$J = \int (f''(x))^2 \, dx \tag{8.65}$$

of the force distribution function is to be smooth.

The above methods may be used to reconstruct the distribution function. Some other examples may also be found in other works [29, 30]. In these works the authors concluded that, although the presented methods of handling inverse problem with regards to the diffusion battery are capable of precisely reconstructing bimodal distributions, they could not be used to reconstruct trimodal distributions. On the other hand, distributions with three or more modes are not of great research and technological interest in the area of aerosol science.

8.6
Conclusions

All the discussed approaches to handle the inverse problem for DAS and DMA in the diffusion dynamic method have inherent advantages and drawbacks. To accurately reconstruct the PSD from penetration data, complicated mathematical tools have to be used, leading to increased impact of errors of measurements on the reconstructed particle size distribution. These procedures could be stabilized [29, 30, 33] if some restrictions are applied [34]. The main one is related to some changes occurring between *trapping efficiency* and Peclet number,

$$\eta = 2.7 \, \mathrm{Pe}^{-2/3} \tag{8.66}$$

where $\mathrm{Pe} = 2aU/D$ is the Peclet number, η is the trapping efficiency of the particle by a cylinder of radius a, and D is the diffusion coefficient of the particle. When $\mathrm{Pe} < 10$, Eq. (8.66) is not valid. This means that, during measurements undertaken by the diffusion dynamical method, the linear velocity of the flow has to exceed $\mathrm{Pe} > 10$.

The other important issue is the assumption that the particle perfectly adheres to the surface upon collision. It has been shown that very small particles with sizes around 1 nm [14, 15] may rebound from the surface, meaning that the sticking probability of these particles to solid surface deviates from unity. It was further discussed [14, 15] that, in the case when the linear size of the surface is much greater than the mean free path length, trapping of collided particles is almost independent of the sticking probability. On the other hand, when particles are trapped by fibers with sizes comparable with the mean free path length, used as screens in the diffusion battery, then the problem becomes more complicated from a mathematical point [14, 15, 35].

In the case when DMA or DAS are involved for measurement of the particle concentration, the low detection limit for particle size is about 1–3 nm. This sensitivity limit depends on the chemical composition of the vapor used for particle enlargement [36] and on the method of enlargement. To increase the sensitivity of the equipment, the method of particle counting may be modified by replacement of the CPC by devices capable of measuring the electrical charge of the particle [12]. However, the applicability of this method is limited, as measurements of ultra-low current are associated with significant errors.

References

1. Farzanah, F.F., Kaplan, C.R., Yu, P.Y., Hong, J., and Gentry, J.W. (1985) Condition numbers as criteria for evaluation of atmospheric aerosol measurement techniques. *Environ. Sci. Technol.*, **19**, (2), 121–126.
2. Zagaynov, V.A., Lushnikov, A.A., Nikitin, O.N., Kravchenko, P.E., Khodhzer, T.V., and Petryanov, I.V. (1989) Background aerosol over Baikal lake. *Dokl. Acad. Sci. USSR*, **308**, (5), 1087–1090.
3. Knutson, E.O. and Whitby, K.T. (1975) Aerosol classification by electric mobility: apparatus, theory and applications. *J. Aerosol Sci.*, **6**, 443–451.
4. Stoltzenburg, M.R. (1988) An ultra fine aerosol size distribution measuring system. Ph.D. Thesis, University of Minnesota.
5. Wang, S.C. and Flagan, R.C. (1990) Scanning electrical mobility spectrometer. *Aerosol Sci. Technol.*, **13**, 230–240.
6. Hagwood, C. (1999) The DMA transfer function with Brownian motion: a trajectory/Monte-Carlo approach. *Aerosol Sci. Technol.*, **30**, 40–61.
7. Mamakos, F., Ntziachristos, L., and Samaras, Z. (2008) Differential mobility analyser transfer function in scanning mode. *J. Aerosol Sci.*, **39**, (3), 227–243.
8. Chen, D. and Pui, D. (1997) Numerical modelling of the performance of differential mobility analysers for nanometer aerosol measurements. *J. Aerosol Sci.*, **28**, 985–1004.
9. Collins, D., Cocker, D., Flagan, R., and Seinfeld, J. (2004) The scanning DMA transfer function. *Aerosol Sci. Technol.*, **39**, 833–850.
10. Shah, S.D. and Cocker, D.R. (2005) A fast scanning mobility particle spectrometer for monitoring transient particle size distribution. *Aerosol Sci. Technol.*, **39**, 519–526.
11. Ranjan, M. and Dhaniyala, S. (2008) A new miniature electrical aerosol spectrometer: theory and design. *J. Aerosol Sci.*, **39**, (8), 710–722.
12. Ranjan, M. and Dhaniyala, S. (2007) A new miniature electrical aerosol spectrometer (MEAS): experimental characterization. *J. Aerosol Sci.*, **38**, 950–963.
13. Tammet, H., Mirme, A., and Tamm, E. (2002) Electrical aerosol spectrometer of Tartu University. *Atmos. Res.*, **62**, 315–324.
14. Zagaynov, V.A., Sutugin, A.G., Petryanov-Sokolov, I.V., and Lushnikov, A.A. (1976) Sticking probability of molecular clusters to solid surfaces. *J. Aerosol Sci.*, **7**, 389.
15. Lushnikov, A.A., Zagaynov, V.A., and Sutugin, A.G. (1977) Boundary conditions to diffusion equation. *Chem. Phys. Lett.*, **47**, 578.
16. Julanov, Yu.V., Zagaynov, V.A., Lushnikov, A.A., Lyubovtseva, Yu.S., Nevsky, I.A., and Stulov, L.D. (1986) High disperse and submicron aerosol of arid zone. *Izv. Acad. Sci. USSR, Ser. Phys. Atmos. Oceans*, **22**, (5), 488–495.
17. Zagaynov, V.A. (1992) Atmospheric aerosol of east Siberia. *J. Aerosol Sci.*, **23**, (Suppl. 1), S1011–S1014.
18. Ankilov, A., Baklanov, A., Colhoun, M., Enderle, K.-H., Filipikova, D., Gras, J., Julanov, Yu., Lindner, A., Lushnikov, A.A., Majerowicz, A.E.,

Mavliev, R., McGovern, F., Mirme, A., O'Connor, T.C., Podzimek, J., Preining, O., Reischl, G.P., Rudolf, R., Sem, G., Szymanski, W.W., Tamm, E., Wagner, P.E., Winklmayr, W., and Zagaynov, V.A. (1994) Workshop on Intercomparison of Condensation Nuclei and Aerosol Particle Counters, Vienna, 1993: an overview. *J. Aerosol Sci.*, **25**, (Suppl. 1), S533–S534.

19. Cheng, Y.S. and Yeh, H.C. (1980) Theory of a screen-type diffusion battery. *J. Aerosol Sci.*, **11**, 313–320.
20. Yeh, H.C., Cheng, Y.S., and Orman, M.M. (1982) Evaluation of various types of wire screens as diffusion battery cells. *J. Colloid Interface Sci.*, **86**, (1), 12–16.
21. Kirsch, A.A. and Stechkina, I.B. (1969) A diffusional method for the determination of the size of condensation nuclei. *Proceedings of the 7th International Conference on Condensation and Ice Nuclei, Prague and Vienna, 18–24 September.*
22. Kirsch, A.A. and Stechkina, I.B. (1978) *Fundamentals of Aerosol Science*, ed. D.T. Shaw, John Wiley & Sons, Inc., New York, chapter 4, p. 165.
23. Lushnikov, A.A. and Zagaynov, V.A. (1990) On diffusion dynamical method of the particle size analysis. *J. Aerosol Sci.*, **21**, (Suppl. 1), S163–S165.
24. Zagaynov, V.A., Julanov, Yu.V., Lushnikov, A.A., Stulov, L.D., Osidze, I.G., and Tsitsikishvily, M.S. (1989) Diurnal variations of parameters of atmospheric aerosol of mountain region. *Izv. Acad. Sci. USSR, Ser. Phys. Atmos. Oceans*, **23**, (12), 1323–1329.
25. Zagaynov, V.A., Julanov, Yu.V., Lushnikov, A.A., Osidze, I.G., and Tsitsikishvily, M.S. (1987) High disperse aerosol induced by sun radiation, in *Proceedings of First All-Russian Conference on Photochemical Processes in the Atmosphere, Moscow*, vol. 1, pp. 34–35.
26. Zagaynov, V.A., Churkin, S.L., and Ogorodnikov, B.I. (1992) Investigation of disperse composition and concentration of aerosols in atmosphere of 30-kilometer zone near ChAES. *Saving Environ., Ecol. Probl. Control Qual. Prod.*, **1**, 25–31.
27. Yee, E. (1989) On the interpretation of diffusion battery data. *J. Aerosol Sci.*, **20**, 797–811.
28. Lloyd, J.J., Taylor, C.J., Lawson, R.S., and Shields, R.A. (1997) The use of the L-curve method in the inversion of diffusion battery data. *J. Aerosol Sci.*, **28**, (7), 1251–1264.
29. Bashurova, V.S., Koutzenogii, K.P., Pusep, A.Y., and Shokhirev, N.V. (1991) Determination of atmospheric aerosol size distribution function from screen diffusion battery data: mathematical aspects. *J. Aerosol Sci.*, **22**, (3), 373–388.
30. Bashurova, V.S., Dreiling, V., Hodger, T.V., Jaenicke, R., Koutsenogii, K.P., Koutsenogii, P.K., Kraemer, M., Makarov, V.I., Obolkin, V.A., Potjomkin, V.L., and Puser, A.Y. (1992) Measurements of atmospheric condensation nuclei size distribution in Siberia. *J. Aerosol Sci.*, **23**, (2), 191–199.
31. Tikhonov, A.N. and Arsenin, V.Y. (1977) *Solutions of Ill-Posed Problems*, John Wiley & Sons, Inc., New York.
32. Kandlikar, M. and Ramachandran, G. (1999) Inverse method for analysing aerosol spectrometer measurements: a critical review. *J. Aerosol Sci.*, **30**, (4), 413–438.
33. Zagaynov, V.A. (2006) Diffusion spectrometer for diagnosing nanoparticles in the gas phase. *Nanotechnics*, **1**, 141–146.
34. Kirsch, A.A., Zagnitko, A.V., and Chechuev, P.V. (1981) On diffusion method of measuring particle size of submicron aerosols. *Zh. Fiz. Khim.*, **55**, (12), 3034–3037.
35. van Gulijk, C., Bal, E., and Schmidt-Ott, A. (2009) Experimental evidence of reduced sticking of nanoparticles on a metal grid. *J. Aerosol Sci.*, **40**, (4), 362–369.
36. O'Dowd, C.D., Aalto, P.P., Yoon, Y.J., and Hameri, K. (2004) The use of the pulse height analyser ultrafine condensation particle counter (PHA-UCPC) technique applied to sizing of nucleation mode particles of differing chemical composition. *J. Aerosol Sci.*, **35**, (2), 205–216.
37. Cooper, D.W. and Spielman, I.A. (1976) *Atmos. Environ.*, **10**, 723–729.

38. Ono-Ogasawara, M., Myojo, T., and Kobayashi, S. (2009) A nanoparticle sampler incorporating differential mobility analyzers and its application at a road-side near heavy traffic in Kawasaki, Japan. *Aerosol Air Qual. Res.*, **9**, (2), 290–304.

39. Seinfeld, J.H. and Pandis, S.N. (1998) *Atmospheric Chemistry and Physics*, John Wiley & Sons, Inc., New York.

Part III
Aerosol Removal

Aerosols – Science and Technology. Edited by Igor Agranovski

ISBN: 978-3-527-32660-0

9
History of Development and Present State of Polymeric Fine-Fiber Unwoven Petryanov Filter Materials for Aerosol Entrapment

Bogdan F. Sadovsky

There are a large number of methods and devices for entrapment of aerosol particles. Their use depends on the parameters of the disperse phase, the composition of the dispersion medium, the required degree of purification, the capabilities of the aspiration systems, and so on. In the case of fine aerosols, *fibrous filters* are the most simple, reliable, and economically sound for the purification of air and process gases [1]. This is especially important when it is necessary to purify *toxic aerosol* toxic, *radioactive aerosol* radioactive, dangerous, and harmful aerosols. The substances used by the present-day industry to manufacture fibrous materials have a wide range of properties. These substances include glass, basalt, quartz, various polymeric materials, cellulose, asbestos, metals, and so on [1, 2]. Petryanov filters (PFs) occupy a special place in this long list of materials [3]. They provide a high degree of removal of fine particles, including the most penetrating ones. At the same time, their *aerodynamic resistance* is comparatively small, which confirms the economic expediency of their use not only for the purification of incoming and exhaust air and gases, but also in respirators for personnel protection in harmful industries.

PFs present plates, cloths, or rolls of unwoven polymeric material consisting of ultrathin homogeneous fibers arranged in two-dimensional disorder, placed freely on the substrate plane (usually gauze) and unbound in places of contact with each other. Owing to the loose layering and the presence of static charge, the bulk porosity of such materials reaches 90–98%. This provides low aerodynamic resistance at the usual industrial filtration rates of 1–10 cm/s.

The thickness of the layer of PF filtering material can vary from fractions of a millimeter to several millimeters. When such filtering layers are pressed, their thickness sharply decreases to the thickness of office paper; however, the resistance to gas flow sharply increases at the same time.

The surface density, that is, the weight of unit area for different PF materials, ranges from 10 to 50 g/m^2. As these are unwoven light polymeric cloths, they have greater flexibility and elasticity, while preserving their structure. This property provides for their use in different articles, in particular, in individual protection devices for the respiratory organs [4]. Another important quality of PF materials

Aerosols – Science and Technology. Edited by Igor Agranovski

ISBN: 978-3-527-32660-0

is the absence of debris or scraps of fibers, which can contaminate the dispersion medium. Individual fibers have the length of hundreds of meters, and their strength corresponds to that of the polymer.

The high efficiency of PFs is explained by the thickness of the fibers. It is known that, the thinner the fiber, the higher the *particle removal efficiency* [5, 6]. Usually, the thickness of the standard fibers used for manufacturing PFs varies between 0.6 and 7 μm. Fibers with the diameter of 1.5 μm are the most widely used. Such fibers do not create large aerodynamic resistance at low packing density but provide high efficiency of particle entrapment.

The large quantity of products that are manufactured under the name of "Petryanov filters" results not only from the great demand for and applications of these materials, but also from the relative simplicity of the process and instruments used in this method.

The method of electrohydrodynamic formation of polymeric fibers, sometimes called *electrospinning*, was developed in the USSR by Petryanov, Rosenblum, and Fuchs in 1937 [7, 8]. Academician Igor Vasil'evich Petryanov [9] made the greatest contribution to the development and industrial implementation of the technique, and to the following studies of the properties of the materials produced. That is why in the 1950s, for their merits in the area of industrial and scientific applications and the great benefit of using these polymeric unwoven fine-fiber materials by industry workers as well as the contribution to the preservation of the country's ecology, they were named Petryanov filters.

As it turned out, in the course of the studies of PF materials, their areas of application include almost all sections of existing industries. Their high efficiency, low aerodynamic resistance, low weight, high elasticity, and the possibility of their use in aggressive media and individual protection devices for respiratory organs, for analytical purposes, and in a number of untraditional cases where filtration is needed, attracted proper attention to PF materials and products based on them. At present, hundreds of thousands (even millions) of cubic meters of both incoming and process air are subjected to fine purification on PF filters in the atomic and chemical industries, medicine, and biotechnology; PF materials are used for the protection of human respiratory organs [3, 4, 7, 10].

The theory of *aerosol filtration* by fibrous filters is now quite fully described, and, knowing the parameters of the filtering layer and the filtration conditions, the efficiency of filtration *filter efficiency* and the resistance of the fibrous layer can be theoretically calculated for particles of any size [5, 6, 10, 11]. The theory and experience demonstrate that the resistance of fine-fiber PF materials to air flow is proportional to the filtration rate. For the most popular material PFP-15, this regularity is preserved at rates up to approximately 10 m/s, for PFP-70 and its analogs up to approximately 3 m/s [12]. When several layers of PF material are used, their total resistance will be equal to the sum of the resistances of the initial layers.

The filter efficiency for aerosol removal is usually taken as the sum of the different mechanisms of particle capture by a single fiber with subsequent use of this parameter for determination of the total filter efficiency [5]. A number of mechanisms, including diffusion, electrostatic removal, inertia, gravitation, and

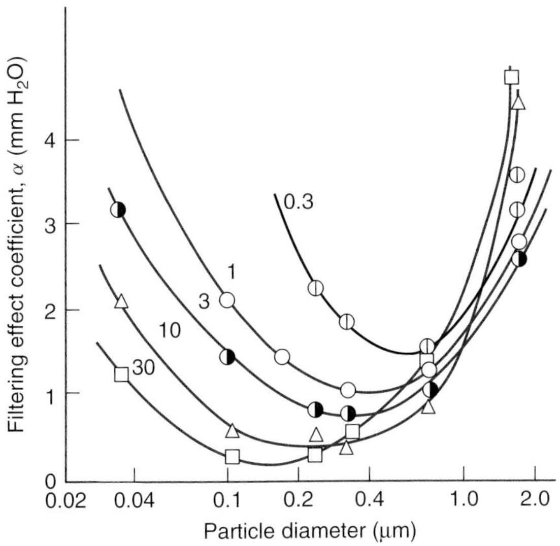

Figure 9.1 The dependence of filtering effect coefficient of PFP-25 material on aerosol particle size at different filtration rates. The numbers on the curves are air velocity (cm/s).

interception, could contribute to particle removal, depending on the properties of the filter medium, particle, and gas carrier.

It is convenient to evaluate the quality of the filtering material by the value of the *α-coefficient* of the filtering effect. The expression for the representation of this coefficient can be written as $K = 10^{-[\Delta p]}$, where $[\Delta p]$ is the standard resistance of the material at a flow rate of 1 cm/s, and K is the particle breakthrough. The value of α is convenient for operational evaluation of a filtering material. For breakthrough of the aerosol through a filter layer with $[\Delta p] = 1$ mm H_2O, if $\alpha = 1$, then $K = 10\%$; if $\alpha = 2$, $K = 1\%$, if $\alpha = 3$, $K = 0.1\%$, and so on.

Figure 9.1 shows the trend of the curves characterizing the degree of particle removal by Petryanov filter PFP-25 for different air flow rates. All the curves have a minimum. This means that at certain flow rates, depending on the sizes of particles, their breakthrough will be maximal. High flow rates of up to 2 m/s and more are often used in stationary systems for air *air purification* and *gas purification* as well as for aerosol sampling onto analytical filters. Under these conditions, the effect of the inertial mechanism of particle entrapment significantly increases, as shown in Figure 9.2. These curves were obtained on monodisperse aerosols. Similar to the curves in Figure 9.1, there are minima on the curves, meaning that there are regions of increased breakthrough of particles, which should be taken into account when determining the requirements for aerosol removal systems [13].

Application of PF materials under high-temperature conditions is determined by the composition of the polymer of which the PF fibers are made. Under negative

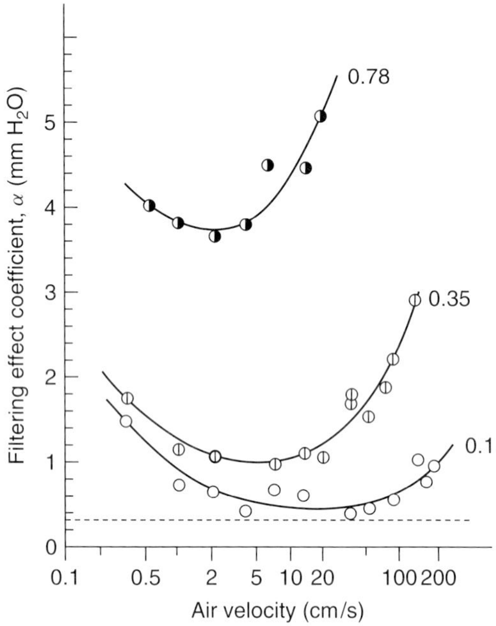

Figure 9.2 The dependence of filtering effect coefficient of PFP-25 material on air velocity for different particle sizes. The numbers on the curves are particle radius (μm).

temperature conditions, PF fibers preserve their structure and can be used for the filtration of both gas and liquid flows, down to the temperature of liquid helium.

Perchlorovinyl is the most widely used for standard atmospheric temperatures. This polymer is used quite universally, as its high chemical stability in both alkaline and acidic media allows for filtration of *aggressive aerosols*. At temperatures of up to 60 °C perchlorovinyl has no competitors. Fibrous PF materials of acetyl cellulose, polycarbonate, polyacrylonitrile, polyfluorostyrene, polyarylate, and other polymers can be used at high temperatures.

The effect of pressure on the efficiency of entrapment of aerosol particles is associated mainly with the change in the viscosity of the dispersion medium. However, the efficiency change is noticeable only at very high pressures and low vacuum.

The range of PF materials developed up to now includes about 50 names [8]. However, the main product manufactured at millions of square meters is the material designated PFP-15-1.5. In this designation, the first two letters denote Petryanov filter, and the third and fourth (if present) letters denote the name of the polymer of which the ultrathin fibers are made – for example, P means perchlorovinyl, A is cellulose diacetate, C is polystyrene, AN is polyacrylonitrile, AR is polyarylate, FS is polytrifluorostyrene, ID is polyimide, SF is polysulfone, C is polycarbonate, and so on. The first set of digits after the letter symbol(s) and the hyphen show the values of the mean or the maximal diameter of the fiber in tenths of a micrometer. A fraction, for example, 3/20, denotes a mixture of fibers with different mean diameters. Finally, the last digit(s) after the second hyphen denote the aerodynamic resistance of the fibrous layer in millimeters water column (mm H_2O) to air flow at the rate of 1 cm/s.

PF materials are used mainly in three areas: (i) fine purification of air and gases for industrial and technical purposes; (ii) manufacture of respirators in individual protection devices for respiratory organs; and (iii) analytical purposes. A large amount of PF material is used in the electrical device manufacturing industry, in particular, as separators of chemical current sources, mainly cadmium–nickel accumulators.

The first mass application of PF materials took place at the end of the 1930s when they were used in gas-masks and were called war filters [8]. In industry, PFs are mainly used at nuclear facilities to provide protection of the atmosphere against radioactive aerosols produced extensively in practically all the stages of raw radioactive material processing up to the end-product.

The first filters presented reservoirs in which PF materials were used in the form of hoses on frameworks. The second stage of application of PF materials was the creation of filters of LAIK type [10, 14] in which PF material was laid between Π-shaped frames. Separators were placed between the layers of filtering material. In filters of this type, it was possible to develop a large filtration area, up to 100 m^2 per 1 m^3 of volume. The embodiments of filters differed depending on the purpose, productivity, and operating conditions [10]. Large amounts of PF material were used for input ventilation and for purification of hazardous industrial wastes.

Frequently, PF materials have been used to provide the pure, practically sterile, atmospheric medium required in facilities involved in the production of bioproducts or ultra-pure chemicals, in microelectronics, and many other areas. Hose filters are used to operate allergen-free rooms in medical institutions to ensure reliable protection against allergens. PF materials were used in live stream-sterilized filters for antibiotic production. We could give dozens of examples of various applications of PF materials for gas purification [8].

Owing to its unique properties, the standard PF material of PFP-15-1.5 grade is mainly used in respirators of "Lepestok" type. This material carries large electrostatic charge on the fibers and preserves it for several years. Efficient, light, and convenient individual protection devices for respiratory organs were created due to the above fact in combination with small aerodynamic resistance [3, 4, 15]. Weighing only 10 g, the "Lepestok" respirator provided protection against radioactive and other toxic aerosols of all sizes with 99.5% efficiency at a respiratory resistance of only 3–4 mm H_2O. The above efficiency is provided for the most penetrating particles – that is, all the rest are entrapped with a higher efficiency. The PF material, which does not irritate the skin, is used in the respirator. The "Lepestok" respirator is intended for a single 8 h shift when used to protect against radioactive and toxic aerosols. For some aerosols and coarse dust, it can be used many times.

If the marking on the respirator of the basic model is "Lepestok-200," it means that it provides protection and reduces the concentration of inhaled aerosols by 200 times. A certain portion of the manufactured "Lepestok" respirators were created based on the PFP-70-0.5 and PFP-70-0.2 materials, of which "Lepestok-40" and "Lepestok-5" respirators are made. At lower respiratory resistance, they reduce the concentration of aerosols with a diameter of 0.3 μm by factors of 40 and 5, respectively. This means that they can be used under conditions of lower air pollution and lower concentrations of radioactive and toxic aerosols.

Various designs of respirators with PF materials have been developed and are used nowadays. There exists a diversity of powder–gas respirators "Lepestok-A," "Lepestok-Apam," "Snezhok-KM," and others for protection in the case when, besides aerosol, the air contains radioactive or toxic gases such as iodine, ruthenium, and so on.

"Lepestok" respirators displayed high qualities when they were used during the accident at the Chernobyl Nuclear Power Plant (see Chapter 6). Independent US examination [13] confirmed the advantage of PF materials over those used for respirators and analytical purposes abroad.

Widely used analytical aerosol filters (AFAs) were developed based on PF materials [6, 16]. They are made in the form of disks with a working surface of 3, 10, or 20 cm^2. They can be used to perform weight, radiometric, chemical, bacterial, radiospectrometric, radiographic, and dispersed analyses. Analytical filters are completed with protective paper rings to exclude contamination of the material edges in the filter holder. AFA filters are placed into a filter holder together with protective rings, and after sampling these rings are disposed of.

The filter marking reflects its purpose, and the polymer of which the fibers and the working surface are made. For instance, AFA-HA-20 filters are used for chemical (radiochemical) analysis; they are made of cellulose acetate and have a working surface of 20 cm^2. AFA-PRMP-20 filters are used in the atomic industry. These filters are intended for radiometric analysis of aerosols. They are made of perchlorovinyl and have a working surface of 20 cm^2. Radiometric AFA-RSP filters are made of perchlorovinyl and are used for alpha-radiating aerosols. The front layer of such a filter consists of fibers with a diameter of 0.3–0.5 μm.

Analytical sorbing filters of AFAS type can be used to analyze polonium, ruthenium, iodine, the vapors of acids and alkalis, hydrogen sulfide, ammonia, and other substances. The amount of fine sorbent (charcoal) in such a filter makes up several milligrams per square centimeter. The filter efficiency is more than 99% for the vapor phase and not less than 96% for the most penetrating particles (Figure 9.3) [6]. Usually, a pack of multilayer filters consists of three analytical AFA filters and three sorbing AFAS filters. Such filters were used to analyze the radioactive aerosols that emerged as a result of the Chernobyl disaster [6, 7, 12]. Filtering analytical tapes are manufactured to select and analyze aerosols in radiometric and spectrometric devices. They were created on the basis of PF filtering material and present tapes of solid even layers of fibrous material made of ultrathin perchlorovinyl fibers glued together. The tape edges are sintered into a 5 mm-wide blister. The total width of the manufactured tapes comes to 50 or 25 mm. The tapes are supplied in rolls 50 or 20 m long. Filtering tapes of three types (NEL-3, NEL-4, and LFS-2) are manufactured for specific purposes [2, 6, 8].

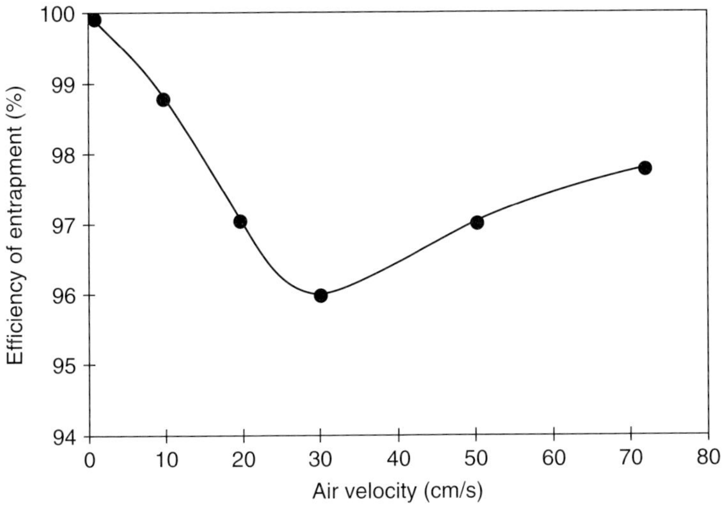

Figure 9.3 The dependence of efficiency of entrapment of standard oil fog (the particle radius is 0.15–0.17 μm) on the flow rate for AFAS-I filters.

The sorption filtering SFL tapes are used for complex entrapment of aerosols and vapors of radioactive and hazardous chemical substances [6, 8].

The application of PF materials is not limited to the area of filtration of aerodispersion systems and the removal of harmful admixtures from gas media. Owing to their universal properties, thin-fiber unwoven polymeric materials can be used in cases when their structure and physico-chemical characteristics allow the solution of problems in electrochemical, electronic, biological, and medical areas, fulfilling the tasks of filtration of liquids, and the creation of heat-insulating materials.

For instance, on the basis of PF materials, systems for the purification of aviation fuel and the control of moisture content of the fuel have been developed and used [7]. In the food industry, PF materials provide filtration of juices, wines, and oils [7, 8]. There is some experience of using PF for filtration of photoemulsions in the area of the production of light conductors, in the production of liquid oxygen and nitrogen, and for the filtration of vapor–liquid systems where PF materials also provide heat-insulating protection. Owing to the fact that the pore diameter in PF materials is close to the mean free path of air molecules, they are excellent insulators. For instance, thin gloves made of PF material allowed manual adjustment of devices at a temperature of −50 °C to be carried out. In medicine, the materials are used for sterile dressings, and, with sprayed drugs, for the treatment of open wounds. There is experience of treating burn surfaces by direct application of ultrathin fibers to damaged spots. The capabilities of PF materials are really unlimited.

References

1. (a) White, P. and Smith, S. (eds) (1964) *High-Efficiency Air Filtration*, Butterworths, London; (b) White, P. and Smith, S. (eds) (1967) *High-Efficiency Air Filtration*, Atomizdat, Moscow.
2. Ogorodnikov, B.I. (1973) *Entrapment of Radioactive Aerosols by Fibrous Filtering Materials*, Central Research Institute Atominform, Moscow.
3. Petryanov, I.V., Kozlov, V.I., Basmanov, P.I., and Ogorodnikov, B.I. (1968) *PF Fibrous Filtering Materials*, Znaniye, Moscow.
4. Petryanov, I.V., Koshcheev, V.S. *et al.* (1984) *"Lepestok" (Light Respirators)*, Nauka, Moscow.
5. Fuchs, N.A. (1955) *Mechanics of Aerosols*, Publishing House of USSR Academy of Sciences, Moscow.
6. Budyka, A.K. and Borisov, N.B. (2008) *Fibrous Filters for Air Pollution Control*, Izdat, Moscow.
7. Druzhinin, E.A. (2007) *Production and Properties of Petryanov Filtering Materials Made of Ultrathin Polymeric Fibers*, Izdat, Moscow.
8. Filatov, Y.N. (1997) *Electric Formation of Fibrous Materials (EFF-Process)*, Oil and Gas Publishers, Moscow.
9. Petryanov, I.V. (1999) *About Myself and My Work, About Him and His Work*, Series "Makers of Nuclear Age", Izdat, Moscow.
10. Basmanov, P.I., Kirichenko, V.N., Filatov, Y.N., and Yurov, Y.L. (2003) *High-Efficiency Removal of Aerosols from Gases Using Petryanov Filters*, Nauka, Moscow.
11. Fuchs, N.A. and Stechkina, I.B. (1962) On the theory of fibrous aerosol filters. *USSR Acad. Sci. Rep.*, **147** (5), 1144.
12. Ogorodnikov, B.I. and Basmanov, P.I. (1965) *The Basis of Application of Fibrous Filtering PF Materials. Radioactive Isotopes in Atmosphere and Their Use in Meteorology*, Atomizdat, Moscow.
13. Ogorodnikov, B.I. *et al.* (2008) *Radioactive Aerosols of the Site Ukrytiye, 1986–2006*, Ukrainian NAS, Chernobyl.
14. Kirichenko, V.N., Yurov, Y.L., Efimov, I.M., and Petryanov-Sokolov, I.V. (1993) Flow and filtration of gases through penetrable walls of channels. *USSR Acad. Sci. Rep.*, **329** (5), 562.
15. Basmanov, P.I., Kaminsky, S.L., Korobeinikova, A.V., and Trubitsyna, M.E. (2002) Individual Protection Devices, Manual, Publishing House Russian Art, St. Petersburg.
16. Basmanov, P.I. and Borisov, N.B. (1970) *AFA Filters (Catalogue–Handbook)*, Atomizdat, Moscow.

10
Deposition of Aerosol Nanoparticles in Model Fibrous Filters

Vasily A. Kirsch and Alexander A. Kirsch

10.1
Introduction

Investigation of the deposition of aerosol nanoparticles in fibrous filters is of great importance for the theory and practice of gas cleaning. The main physical mechanism of *particle deposition* within modern highly efficient fine fibrous filters is the Brownian shift of particles from the streamlines toward the surface of the fibers. For the efficient deposition of nanoparticles, especially upon *ultra-fine fibers*, along with diffusion, other effects are significant: interception of particles of finite size, action of external forces, confined flow gradient, and gas slip on the surface of ultra-fine fibers. Qualitative studies of the process of filtration of gases carrying submicrometer particles have a long history, but no theoretical description for this multi-parameter problem has yet been developed. The difficulties are mainly related to an undefined filter structure and, thus, undefined flow field of the real filter. In order to exclude the influence of any inhomogeneity of the inner structure of the filters, theoretical and experimental investigations are conducted on *model filters* with known *flow field* [1–3]. As model filters, single or multiple rows of parallel circular cylinders, oriented perpendicular to the flow direction, have been used for many years. Intensive theoretical investigations of the flow of aerosols in model filters at low *Reynolds numbers* were performed in [4–9] during the past decade.

The necessity for conducting such investigations stems from the fact that, for computations of particle deposition, some knowledge of the actual flow field is needed, for regions both near to and far from the fibers. The pressure drop and the particle deposition on fibers are strongly dependent on the mutual complex arrangement of the fibers within the filter. Advances in computational fluid dynamics (CFD) make it possible to model any kind of flow in very complex geometries, but the fiber arrangement in real filters cannot be precisely quantified by any independent method. Thus, if the geometry is unknown, CFD cannot be used. Moreover, for experimental investigation of nanoparticle deposition, precise knowledge of the *geometric parameters* of the filters is also required. Only model filters provide such information. Finally, no universal approach is known for

Aerosols – Science and Technology. Edited by Igor Agranovski

ISBN: 978-3-527-32660-0

the description of hydrodynamics in modern filters having fibers with diameters comparable with the mean free path of the gas molecules.

It should be noted that, starting with the work of Langmuir [10], the theory of filtration was constructed based on the model of a single fiber, initially isolated, placed in an infinite *Lamb flow* [10, 11], and later it was based on the *Stokes flow* (*creeping flow*) and on the so-called *cell model* [12]. In the cell model, the fiber is located at the center of the fluid concentric envelope. The influence of neighboring fibers is accounted for by the boundary conditions imposed on the outer cell boundary. It was assumed that the concentration is remixed and thus is uniform before every subsequent layer of fibers. This is not usually the case in practice, and hence another practically insuperable obstacle (even for model filters) arises, namely, the difficulty in providing corresponding precise experimental conditions. The required remix of the concentration behind any row of fibers is not attained because of the small Brownian mobility of particles, even for submicrometer sizes. The diffusional trace behind the fibers is always present. It is handled with the help of numerical modeling, enabling one to quantify a field of concentration and to predict the fiber *collection efficiency* if the inner arrangement of the fibers is known. If the inner structure is not known, then no averaging approaches, generalizations, or analogies are applicable.

As shown by our investigations, and emphasized in [2], the resistance to flow of two filters may differ by more than 100%, even if they are similar in appearance, have the same thickness and porosity, are made of fibers of equal diameter, are optically uniform, do not have hidden or open defects, and are tested under the same conditions, but the fibers in the suspension had different dispersion. Thus, the collection efficiency will be different in the same manner for these filters. Such a sharp dependence of the filtering properties of fibrous materials on the inner structure is caused by the varying flow field within the filters, which is a problem of hydrodynamics at low Reynolds numbers [2].

A filter with a regular hexagonal arrangement of parallel fibers placed normally to the flow direction was found to be a convenient model. As was identified experimentally and numerically, the flow field in this filter [13] coincides very closely with a simple analytical solution suggested by Kuwabara [14] within the framework of the cell model. The dimensionless *stream function* $\boldsymbol{\Psi}$ for the flow in the cell (Figure 10.1) expressed via the dimensionless polar coordinates r, θ is [14]

$$\begin{aligned}
\boldsymbol{\Psi} &= \left(Ar^{-1} + Br + Cr\ln r + Dr^3\right)\sin\theta \\
A &= (2-\alpha)\,C/4, \quad B = -\,(1-\alpha)\,C/2 \\
C &= \left(-0.5\ln\alpha - 0.75 + \alpha - 0.25\alpha^2\right)^{-1} \quad D = -\alpha C/4
\end{aligned} \tag{10.1}$$

The components of the velocity vector $\boldsymbol{u}$ are:

$$u_r = -r^{-1}\partial\boldsymbol{\Psi}/\partial\theta \quad u_\theta = \partial\boldsymbol{\Psi}/\partial r \tag{10.2}$$

The dimensionless cell radius is related to the *filter packing density* α as $b/a = \alpha^{-1/2}$. Here the fiber radius a, constant uniform velocity of incoming flow u, and uniform inlet concentration of aerosol particles are chosen as scaling parameters.

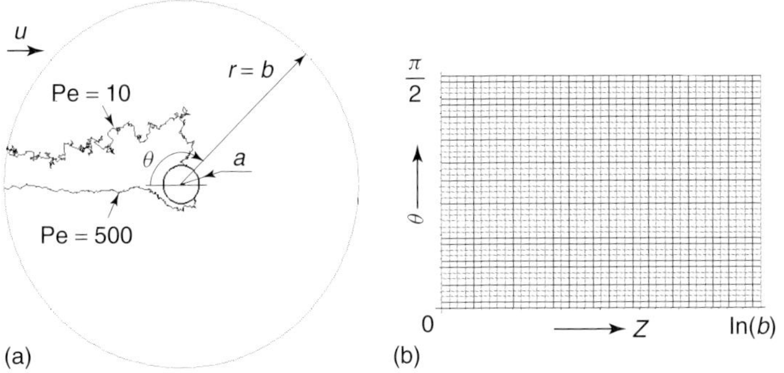

Figure 10.1 (a) Examples of stochastic particle trajectories in the cell with $\alpha = 0.01$ [15]. (b) Computational rectangular region with uniform grid obtained from the coaxial cell (a) with the help of the conformal transform $r = \exp(z)$.

Experimental and theoretical investigations devoted to the Kuwabara cell model are described with details in [1–3].

The remix of concentration behind (before) any layer of fibers is expected for particles with high diffusional mobility only. The only known experimental data on the deposition of nanoparticles with radius lower than 10 nm in model filters with known flow field were obtained in two works [16, 17]. The conditions in these experiments correspond to the regime of purely diffusional deposition, when the contributions from all other mechanisms of deposition are negligibly small.

Formulas for the collection efficiency of point-like particles (nanoparticles) were found by Natanson for an isolated cylinder [11] and by Fuchs and Stechkina for a system of parallel cylinders [12]:

$$\eta = 2.9k^{-1/3}\mathrm{Pe}^{-2/3} \tag{10.3}$$

where η is the fiber collection efficiency, and $\mathrm{Pe} = 2au/D$ is the diffusional Peclet number. The hydrodynamic factor k in Eq. (10.3) is related to the drag force F^* acting on unit length of the fiber:

$$F^* = 4\pi\mu u/k \tag{10.4}$$

where μ is the gas dynamic viscosity, and $*$ denotes dimensional variables. Here $k = 2 - \ln \mathrm{Re}$ for an isolated cylinder, and $k = -0.5\ln\alpha - 0.75 + \alpha - 0.25\alpha^2$ for a hexagonal lattice (and for the Kuwabara cell) at low Reynolds numbers, $\mathrm{Re} \ll \sqrt{\alpha}$.

The fiber collection efficiency is found by integration of the normal component of the density of the overall flux of particles j_r over the surface of the fiber at $r = 1$:

$$\eta = 2\mathrm{Pe}^{-1}\int_0^{\pi} j_r(1,\theta)\mathrm{d}\theta \tag{10.5}$$

The filter efficiency E is given by the formula

$$E = 1 - \exp(-2aL\eta) \tag{10.6}$$

where $L = lH, l = \alpha/\pi a^2$ is the fiber length per unit volume of the filter, and H is the filter thickness. The density of the total flux of particles consists of diffusion and convection components, that is,

$$\boldsymbol{j} = -2\mathrm{Pe}^{-1}\nabla n + \boldsymbol{u}n \qquad (10.7)$$

where $\boldsymbol{u}$ is the flow velocity vector, and n is the dimensionless particle concentration, found from the stationary *convection–diffusion equation*:

$$\nabla \boldsymbol{j} = 2\mathrm{Pe}^{-1}\Delta n - (\boldsymbol{u}\cdot\nabla)n = 0 \qquad (10.8)$$

Rewritten in polar coordinates, Eq. (10.8) is

$$\frac{2}{\mathrm{Pe}}\left(\frac{\partial^2 n}{\partial r^2} + \frac{1}{r}\frac{\partial n}{\partial r} + \frac{1}{r^2}\frac{\partial^2 n}{\partial \theta^2}\right) - u_r\frac{\partial n}{\partial r} - \frac{u_\theta}{r}\frac{\partial n}{\partial \theta} = 0 \qquad (10.9)$$

It is not possible to solve Eq. (10.9) analytically for the case when the flow velocities are functions of the coordinates. At high Peclet numbers, the contribution of diffusion to the tangential flux of particles is small, and it is possible to omit the term $\partial^2 n/\partial\theta^2$ in Eq. (10.9). As a result, the elliptic equation is reduced to a parabolic one, which has been commonly considered in the theory of aerosol filtration.

Formula (10.3) was obtained in the *boundary-layer approximation* (at conditions $\eta \ll 1$ and $\mathrm{Pe} \gg 1$) by an expansion in terms of a small parameter – the thickness of the boundary layer, $\delta \sim Pe^{-1/3}$. This formula was found to agree down to $\mathrm{Pe} \approx 1$ with experimental data obtained with monodisperse nanoparticles with radii varying in the range $r_\mathrm{p} = 1.5$–8 nm in the model filter with hexagonal structure with $\alpha = 0.01$ [16]. Later experiments [18] on the deposition of nanoparticles in model filters with low packing density have shown that the dependence is valid down to smaller values of Pe, namely, to $\mathrm{Pe} = 0.1$. The next term of the expansion in δ was accounted for in the solution performed by Stechkina [19]. Hence the sharper increase of the fiber collection efficiency with decrease of Pe was shown to be

$$\eta = 2.9k^{-1/3}\mathrm{Pe}^{-2/3} + 0.624\mathrm{Pe}^{-1} \qquad (10.10)$$

However, the accuracy of experimental observations was too low to compare with Eq. (10.10) in the range of small Pe. It is important to note that the functional dependence (Eq. (10.4)) remains unchanged down to small Pe and high fiber collection efficiencies but only for highly porous model filters. For dense fibrous filters the dependence $\eta \sim \mathrm{Pe}^{-2/3}$ for $\mathrm{Pe} \sim 1$ is not valid. This was found experimentally for rows of parallel fibers in [17], which reported that, for $\mathrm{Pe} \ll 1$, the fiber collection efficiency tends to its limit, $\eta = h/a$, where h is half the distance between neighboring fibers. Later, the existence of this geometrical limit was confirmed analytically and numerically [15, 20–23]. Analytical formulas were derived for the particle collection efficiency for the case of $\mathrm{Pe} \ll 1$ for the cell model flow field [20], and for flow past a row of parallel fibers [21]. Later, the deposition of nanoparticles within loose and dense model filters for a wide range of Peclet numbers was studied [15, 23], where Eq. (10.9) was solved numerically with and without taking account of the term $\partial^2 n/\partial\theta^2$.

This chapter will discuss the diffusional deposition of point particles in model filters with different structures, composed of different fibers, including ultra-fine, elliptic, strip-like, permeable porous, and composite (covered with porous permeable layers) fibers.

10.2
Results of Numerical Modeling of Nanoparticle Deposition in Two-Dimensional Model Filters

10.2.1
Fiber Collection Efficiency at High Peclet Number: Cell Model Approach

The convection–diffusion equation (Eq. (10.9)) was solved by the method of finite differences in a symmetric half-cell $[1\dots b, 0\dots \pi]$ (Figure 10.1a) at the conditions of full adsorption on the fiber surface and uniform concentration at the outer cell:

$$n(1,\theta) = 0 \quad \text{and} \quad n(b,\theta) = 1 \tag{10.11}$$

For better convergence of the solution, it is convenient to find the concentration at the front stagnation line $\theta = 0$ (the angle is directed clockwise),

$$n(r,0) = c(r) \tag{10.12}$$

where $c(r)$ is the solution of the following *ordinary differential equation:*

$$\frac{2}{\mathrm{Pe}}\left(\frac{\partial^2 c(r)}{\partial r^2} + \frac{1}{r}\frac{\partial c(r)}{\partial r}\right) - u_r(r,0)\frac{\partial c(r)}{\partial r} = 0$$
$$c(1) = 0, \quad c(b) = 1$$

For the region of the diffusional trace at $\theta = \pi$, the approximate condition for the concentration is $n(r,\pi) = 0$. After the change of variable $r = \exp(z)$, the problem is solved on the rectangular region (Figure 10.1b). Equation (10.9) assumes the form

$$\frac{\partial^2 n}{\partial z^2} + \frac{\partial^2 n}{\partial \theta^2} - \frac{\partial}{\partial z}\left(c^{(1)} n\right) - \frac{\partial}{\partial \theta}\left(c^{(2)} n\right) = 0$$
$$c^{(1)} = e^z u_z\left(e^z,\theta\right)\frac{\mathrm{Pe}}{2} \quad c^{(2)} = e^z u_\theta\left(e^z,\theta\right)\frac{\mathrm{Pe}}{2} \tag{10.13}$$

Here, the uniform grid in the rectangular region corresponds to the non-uniform grid that becomes gradually denser toward the fiber surface. The given approach provides a better resolution of the boundary layer adjacent to the fiber, and is able to compute the fiber collection efficiency with higher accuracy, which is important for the case of small collection efficiencies at high Peclet numbers (thin diffusional boundary layer).

At high Pe the convection–diffusion equation is singularly perturbed, and its solution varies sharply within a thin boundary layer near the fiber and through the diffusional trace behind the fiber, being almost unchanged within the region of uniform concentration. For numerical solution of such equations, special methods

Table 10.1 Fiber collection efficiencies for the fiber-in-cell model [15]: η^e is found by solving the elliptic convection–diffusion equation (Eq. (10.9)); η^p is found from the solution of the parabolic equation (Eq. (10.9)), where $\partial^2 n/\partial\theta^2 \approx 0$; and η is found by the formula (10.3) at Pe > 30 and formula (10.10) at Pe ≤ 30.

α	Pe	3000	300	30	3
0.01	η^e	0.0121	0.0574	0.2773	1.3989
	η^p	0.0117	0.0556	0.2760	1.4090
	η	0.0120	0.0558	0.2796	1.4095
0.3	η^e	0.0244	0.1134	0.5238	–
	η^p	0.0240	0.1130	0.5230	–
	η	0.0276	0.1300	0.6145	–

should be used [24, 25]. We have employed a monotonic conservative scheme of second order defined on the five-point stencil on the shifted grid [25]. This scheme was used in [15, 23, 26]. The resulting system of linear algebraic equations was solved by the *matrix sweep method* (*Thomas algorithm*) [27]. Here, the method of lines [28] is also applicable, which reduces the dimension of the problem. The boundary-value problem for the partial differential equation is reduced to the boundary-value problem for a system of ordinary differential equations.

The comparison of the fiber collection efficiencies obtained from the solution of the elliptic and parabolic convection–diffusion equations and calculated by Eqs. (10.3) and (10.10) is presented in Table 10.1.

The comparison of the computed fiber collection efficiencies with experimental data [16] is given in Figures 10.2 and 10.3. The experiments [16] were performed at low Reynolds numbers, when Eq. (10.1) is applicable. For the case of loose filters with small α, the neglect of tangential diffusion (neglect of the term with $\partial^2 n/\partial\theta^2$) in Eq. (10.9) practically does not affect the answer for intermediate and large values of Pe. For dense filters, the solution of the parabolic convection–diffusion equation at Pe < 100 gives a lower estimate for the fiber collection efficiency (curve 4, Figure 10.2) relative to that found from the elliptic equation (curve 3). From the

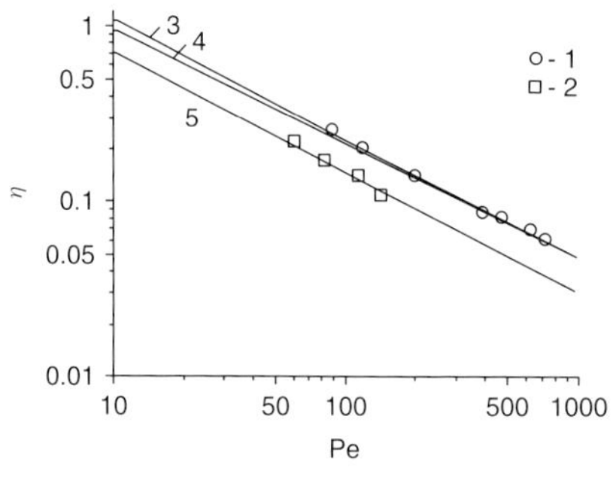

Figure 10.2 Dependence of the fiber collection efficiency on the Peclet number for model filters with packing density $\alpha = 0.27$ (3, 4) and $\alpha = 0.05$ (5) [15]. The experimental data [16] are for $r_p = 5.5$ nm (1) and $r_p = 7$ nm (2).

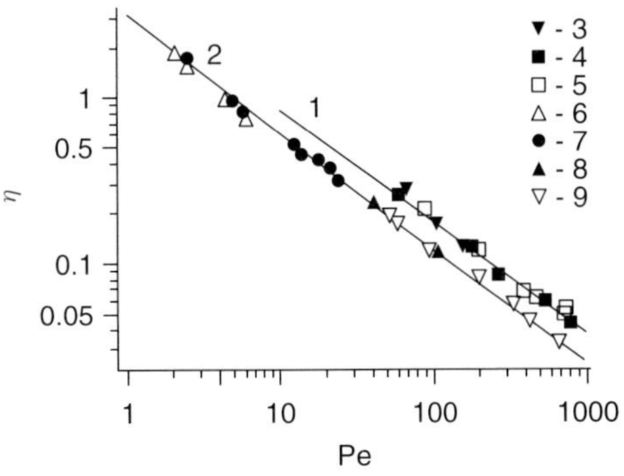

Figure 10.3 Dependence of the fiber collection efficiency on the Peclet number for model filters with packing density $\alpha = 0.13$ (1) and $\alpha = 0.01$ (2) [15]. The experimental data [16] are for $r_p = 4.1$ nm (3), $r_p = 7.0$ nm (4), $r_p = 5.5$ nm (5), $r_p = 1.5$ nm (6), $r_p = 1.8$ nm (7), $r_p = 6.0$ nm (8), and $r_p = 8.3$ nm (9).

figures one can see a good agreement of the computed results with the experiments, including the case of fairly dense fiber lattices ($\alpha = 0.27$). From Table 10.1 it is also seen that the obtained fiber collection efficiency values η agree well with those computed by Eqs. (10.3) and (10.10).

The use of the cell model for estimating the fiber collection efficiency for model filters has difficulties connected with the selection of the proper boundary condition at the outer cell boundary. The commonly used condition of uniform concentration $n(b, \theta) = 1$ is not applicable for the region of the diffusional trace. However, as shown by simulations, the use of this approximate condition far from the cylinder at the entire circular boundary at high and intermediate Pe numbers does not affect the field of the concentration near the fiber and thus the fiber collection efficiency.

10.2.2
Fiber Collection Efficiency at Low Peclet Number: Row of Fibers Approach

For the investigation of the deposition of particles with high diffusional mobility at low Pe, a more suitable model filter was considered, that is, a row of parallel fibers (Figure 10.4). The field of the concentration in the row of fibers was found from numerical solution of the convection–diffusion equation written in rectangular coordinates x, y:

$$\frac{\partial^2 n}{\partial x^2} + \frac{\partial^2 n}{\partial y^2} - \frac{\partial}{\partial x}\left(c^{(1)} n\right) - \frac{\partial}{\partial y}\left(c^{(2)} n\right) = 0$$

$$c^{(1)} = u_x \frac{\text{Pe}}{2}, \quad c^{(2)} = u_y \frac{\text{Pe}}{2} \tag{10.14}$$

Here, the linear scales are a, u, and n_0. The following boundary conditions were used: $n = 1$ at $x = -X$ ($\Gamma 1$), $n = 0$ at $x^2 + y^2 \leq 1$ ($\Gamma 2$), $\partial n/\partial y = 0$ at $y = \pm h/a$ ($\Gamma 3$), and $\partial n/\partial x = 0$ at $x = X(\Gamma 4)$, where $2X$ is the length of the computational cell.

The convective velocities are given by the analytical solution of Miyagi [29], who gave the answer for velocity components as a series expansion with only the first few terms published. The velocity is not zero exactly on the fiber surface, and the corresponding error increases with the packing density. It should be noted that the use of semi-analytical methods of boundary collocation [30] (where the no-slip or

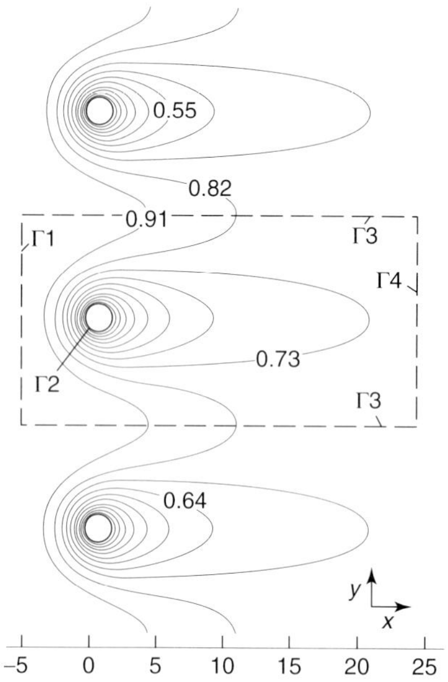

Figure 10.4 Field of concentration in a row of fibers: $h/a = 7.506$, Pe = 1 [15]. The computational region is depicted schematically by dashed lines.

slip boundary conditions on the fiber surface are exactly satisfied) provides better accuracy for the velocities near the fiber surface.

Equation (10.14) was solved on the elongated rectangular region with the help of the iterative domain decomposition method [31] on the rectangular uniform grid with boundary conditions of the high-order interpolation type for the nodes adjacent to the boundary nodes on (or within) the curved fiber surface. The finite-difference scheme from [25] was used. In order to minimize the error on the fiber surface and to resolve the concentration boundary layer properly, the solution obtained was improved on the fine grid within the coaxial circular domain containing the fiber by the algorithm described previously for the cell model.

Given in Figure 10.5 are the experimental results [17], where monodisperse nanoparticles with radii $r_p = 0.7$–2 nm have been deposited in model filters (rows of parallel wires) with $2a = 8.9$ μm and $2h = 66.8$ μm. Here, these data are compared with the results of computations by Eq. (10.3), where for an isolated row of fibers the hydrodynamic factor was expressed as [29]

$$k = \frac{4\pi}{F} \tag{10.15}$$

where

$$F = 8\pi\left(1 - 2\ln 2t + \frac{2}{3}t^2 - \frac{1}{9}t^4 + \frac{8}{135}t^6 - \frac{53}{1350}t^8 + \cdots\right)^{-1}$$

$$t = \frac{\pi a}{2h} \tag{10.16}$$

In this figure, curve 1 was plotted by the simulation results on the basis of Eq. (10.14). As is seen from Figure 10.5, agreement between theory and experiment is observed over the whole range of Peclet numbers.

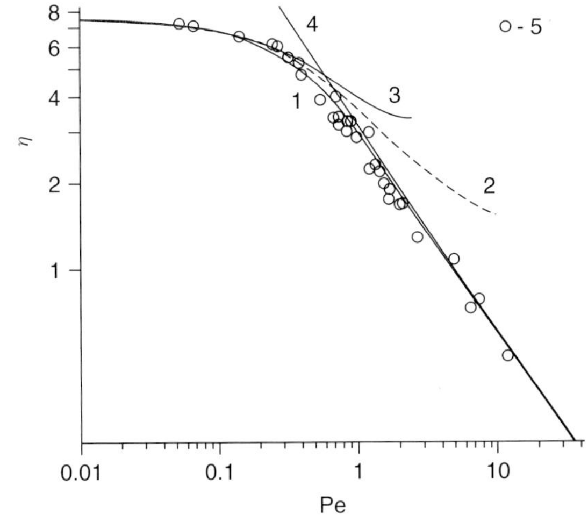

Figure 10.5 Fiber collection efficiency versus Peclet number for the model filter (a row of parallel fibers) [15]. Curves 1 and 2 are plotted from the numerical solution of the convection–diffusion equation (10.14) with Miyagi convective velocities [29] (1) and for the uniform flow approximation, $\boldsymbol{u} = \{1, 0\}$ (2). Curve 3 is plotted by Eq. (10.17), and curve 4 by Eqs. (10.3) and (10.15). Points 5 are experimental data [17].

The case of low Peclet numbers (curves 2 and 3), when diffusion prevails over convection, is interesting not only for aerosol filtration but also for adsorption and catalytic gas cleaning from molecular impurities. Curve 2 was calculated by Eq. (10.14) for a uniform velocity field, $\boldsymbol{u} = (1, 0)$. Curve 3 was calculated by the formula derived in [21] for the fiber collection efficiency for particles diffusing in a uniform stream:

$$\eta = \frac{2\pi}{\mathrm{Pe}}\left[K_0\left(\frac{\mathrm{Pe}}{4}\right) + 2\sum_{m=1}^{\infty} K_0\left(\frac{m\mathrm{Pe}h}{2}\right)\right]^{-1} \tag{10.17}$$

where $K_0(z)$ is the modified Bessel function of imaginary argument. It is seen from Figure 10.5 that, at $\mathrm{Pe} < 0.3$, curves 1–3 coincide with one another and with experiment.

It is clear from the results presented that deposition at low Peclet numbers is governed by a functional dependence different from that for deposition at intermediate and high Peclet numbers. The published formulas [1–3] may be inapplicable here. It is thus necessary to take this into account when one considers polydisperse filters, where the regime of deposition of nanoparticles upon fibers with smaller diameters may correspond to the case of small Peclet numbers, $\mathrm{Pe} < 1$. Then, using the mean fiber radius in computations would not necessarily lead to the right solution.

It should be noted that, at high Peclet numbers $\mathrm{Pe} \gg 1$, the diffusional deposition of nanoparticles within polydisperse model filters can be estimated using the average fiber radius, even if the fibers are very different. This was shown experimentally and theoretically in [2].

10.2.3 Deposition of Nanoparticles upon Ultra-Fine Fibers

The problem of the deposition of nanoparticles upon fine fibers with radii comparable to the mean free path of gas molecules is one of the most difficult and important for high-efficiency gas filtration, for both theory and practice. There are no direct experimental measurements of the collection efficiency for model filters in the literature. Such experiments are very difficult to perform. But we know from experimental data with real filters that the effect of the gas slip on fibers leads to an increase in the collection efficiency and to a decrease in the *filter resistance* to flow. This effect is governed by the dimensionless Knudsen number, $\mathrm{Kn} = \lambda/a$, the ratio of the mean free path of the gas molecules, λ, to the fiber radius, a. For nanofibrous filters, Knudsen numbers belong to the intermediate region of values, $\mathrm{Kn} \sim 1$. Here, the theoretical approaches developed for limits of viscous and free-molecular flows are not valid. Nevertheless, for many years estimates for the collection efficiency were made within the framework of the cell model and for the Stokes flow with slip boundary condition at the fiber surface. The slip correction obtained for the hydrodynamic factor is proportional to $\mathrm{Kn}/(1+\mathrm{Kn})$ [32], while from experiments on the resistance of the model and real filters it follows that the correction is linearly dependent on Kn. The corresponding analytical expression for the drag force in the row, obtained in the linear approximation on Kn, is [32, 33]

$$F^{-1} = F_0^{-1} + \frac{\tau}{4\pi}\left(1 - \frac{2}{3}t^2\right)\mathrm{Kn} \qquad (10.18)$$

where F_0 is given by Eq. (10.16). Here $t = \pi a/2h$, $2h$ is the distance between the axes of the parallel fibers in the row, and $\tau = 1.147$ is the coefficient of isothermal slip, which is related to the interactions of the gas molecules with the fiber surface. Formula (10.18) was derived for $\mathrm{Kn} \ll 1$ and $t \ll 1$. The linear dependence $F^{-1}(\mathrm{Kn})$, as shown experimentally, is unchanged up to $\mathrm{Kn} < 1$ [33–35]. The theoretical confirmation of the experimental fact about the linearity of $F^1(\mathrm{Kn})$ was given in [5], where the solution was found within the framework of Bhatnagar–Gross–Krook (BGK) gas kinetic theory, and where the velocity field was found together with the formula for the fiber drag force, similar to Eq. (10.18):

$$F^{-1} = F_0^{-1} + \frac{\tau}{4\pi} f(\alpha)\mathrm{Kn} = F_0^{-1} + \frac{1.27 - 3\alpha}{4\pi}\mathrm{Kn} \qquad (10.19)$$

In Figure 10.6 the comparison is given for the drag forces, calculated by Eqs. (10.18) and (10.19), with the results obtained from measurements of the rarefied gas permeability through a uniform row of parallel gold wires (diameter $2a = 8.9\ \mu\mathrm{m}$, inter-fiber distance $2h = 62.7\ \mu\mathrm{m}$, $\mathrm{Re} < 0.05$) [33]. As one can see, a good agreement between theory and experiment was achieved. Moreover, the lines plotted by Eqs. (10.18) and (10.19) coincide. Here curve 1 was found in [36] from the solution of the Stokes equation with the slip boundary condition at the fiber

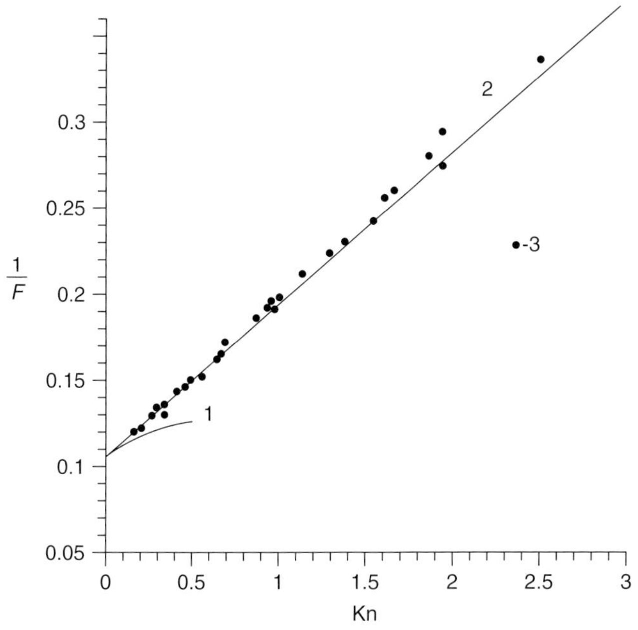

Figure 10.6 Dependence of the reciprocal value 1/*F* of the drag force on the Knudsen number for a fiber in a uniform row [36]: (1) from the direct solution of the Stokes equation with slip boundary conditions; (2) by Eqs. (10.18) and (10.19), and the tangent line to curve 1; and (3) experimental data [33].

surface:

$$u_\xi = 0, \quad u_t = \tau \,\mathrm{Kn}\frac{\partial u}{\partial \xi}$$

where u_ξ and u_t are the normal and tangential components of the velocity. Here, ξ and t are the normal and tangent lines to the fiber surface. In this work, the method of fundamental solutions ("Stokeslets" or point forces) [37] was used. The same results for the velocities and for the drag force were found by us in [36], where the semi-analytical boundary collocation method [30] was employed. It is seen from Figure 10.6 that curve 1 plotted for the solution found by the slip-flow approach agrees with experiment at $\mathrm{Kn} \ll 1$, or more precisely at $\mathrm{Kn} \ll 0.1$ only. It is interesting to note that the plot of the derivative of curve 1, calculated at the initial point $\mathrm{Kn} = 0$, coincides with the line plotted by Eq. (10.18).

From comparison with experiment, it is seen that the flow field found for the creeping viscous flow with slip boundary conditions is inapplicable for computations of nanoparticle deposition upon ultra-fine fibers. Therefore, for modeling the deposition of nanoparticles upon fine fibers, the flow field of [5] was used. This choice is based on the fact that the drag force found in [5] agrees with experiment for intermediate Knudsen numbers, $\mathrm{Kn} > 1$. The computations in [5]

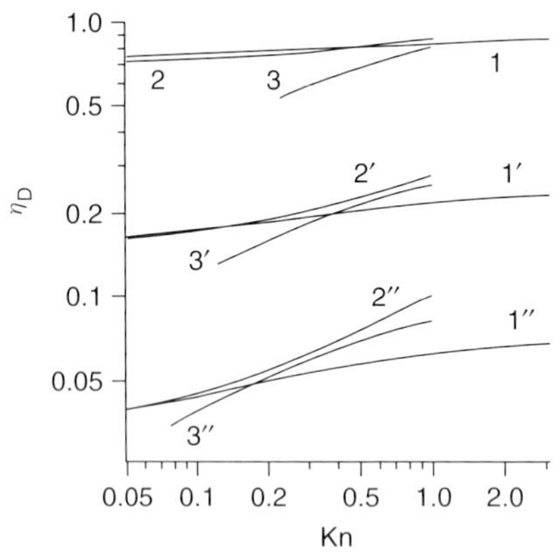

Figure 10.7 Dependence of the fiber collection efficiency for point particles ($R = 0$) on the Knudsen number at various Pe numbers [26]: Pe = 10 (1–3), Pe = 100(1′–3′), and Pe = 1000(1″–3″); $\alpha = 0.0625$. Curves 1–1″ are calculated by Eqs. (10.5) and (10.9); 2–2″ by Eq. (10.20); and 3–3″ by Eq. (10.21).

were performed up to Kn = 10 for several values of the filter packing density. The corrected expressions for the components of the velocity for the flow in the Kuwabara cell are given for $\alpha = 1/36$ and $\alpha = 1/16$ in [26]. In Figure 10.7 the corresponding computed curves for the fiber collection efficiencies are compared with those plotted by approximate analytical formulas:

$$\eta_D = 2.9k_1^{-1/3}\text{Pe}^{-2/3}\left(1 + 0.39k_1^{-1/3}\text{Pe}^{1/3}\text{Kn}\right) \tag{10.20}$$

$$\eta_D = 3.2k_1^{-1/2}\text{Pe}^{-1/2}\left(\tau\,\text{Kn}\right)^{1/2} \tag{10.21}$$

which are valid for $\delta_D \sim \text{Pe}^{-1/3} \gg \text{Kn}$ and $\delta_D \sim \text{Pe}^{-1/2} \ll \text{Kn}$, respectively. Here δ_D is the thickness of the diffusion boundary layer. These formulas were obtained in [38] and in [2] within the framework of the Kuwabara cell model for $\text{Pe} \gg 1$, $\text{Kn} \ll 1$, and $\eta_D \ll 1$. It is seen from Figure 10.7 that curves 2 plotted by Eq. (10.20) coincide with curves 1. Thus, Eq. (10.20) can be used to estimate the fiber collection efficiency. At small Peclet numbers $\text{Pe} \ll 1$, the flow has a small effect on the collection efficiency. For this case, another approach similar to Fuchs' absorbing sphere method [39] should be used for estimation of the efficiency of deposition of nanoparticles upon nanofibers at low velocities.

10.2.4
Deposition of Nanoparticles on Fibers with Non-Circular Cross-Section

The interest in the problem of the deposition of nanoparticles upon non-circular fibers is connected with the widespread use of filters produced by electrospinning. These fibers have a cross-section of dumbbell shape. The first studies on the flow field and diffusional deposition of nanoparticles on model filters with a regular arrangement of non-circular fibers were published in recent papers [36, 40]. It was shown that fibers with a dumbbell shape (Figure 10.8a) can be approximated by fibers with an elliptical cross-section (Figure 10.8b), or by a pair of fibers close together (Figure 10.8c). The computations have shown that the bridge length along the symmetry axis (gap) has no influence on the drag force. In Figure 10.9 the comparison between the drag forces of different fibers having the same

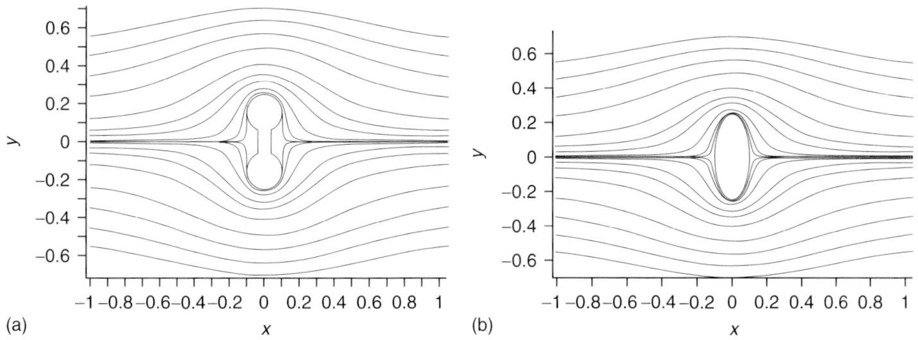

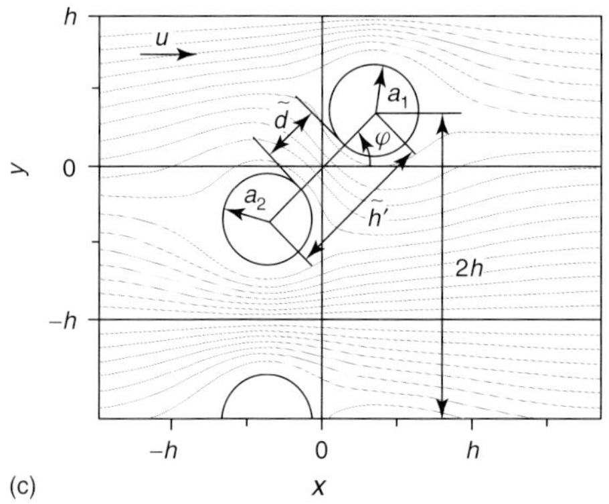

Figure 10.8 (a) Streamlines near a fiber with a dumbbell-shaped cross-section: radii of fibers connected by the bridge are $b = 0.1$, bridge length along the symmetry axis (gap) is $d = 0.1$, and bridge thickness is $\delta = 0.08$. (b) Streamlines near an elliptic fiber with the same mid-section (the minor axis is equal to $2b$). The streamlines correspond to stream function values of $\Psi = \pm[5\times10^{-5}, 5\times10^{-4}, 5\times10^{-3}, 2.5\times10^{-2}, 0.05; 0.1\text{–}0.5$ with step 0.1] [40]. (c) Schematic representation of the Stokes flow around a single periodic row composed of double parallel fibers with radii a_1 and a_2, rotated relative to an advancing flow by an angle $\varphi = 45^\circ$: $\tilde{h}'$ is the inter-fiber distance in a pair, $\tilde{d}$ is the inter-fiber gap, and $2h$ is the distance between even (odd) fibers of neighboring pairs. The streamlines correspond to stream function values lying in the range $\Psi = [-1, 0.1, 1]h$, and $a/h = 0.3$ [40].

mid-section height and the drag force for the pair of fibers is given. Here, the mid-section height is scaled to the fiber radius. For the pair of fibers, the height of the mid-section is $l_m = 2 + \tilde{h}'/a$ (Figure 10.8c). The results of our direct simulations for elliptical fibers coincide with computations using analytical expressions derived in [41].

The hydrodynamic flow field for a row of fibers of non-circular cross-section was found in [36, 40], where the biharmonic equation for the stream function was solved by the boundary method of the fundamental solutions ("Stokeslets" or point forces) [37]. The field of concentration was found from the numerical solution of the convection–diffusion equation by a method similar to that described in Section 10.2.2.

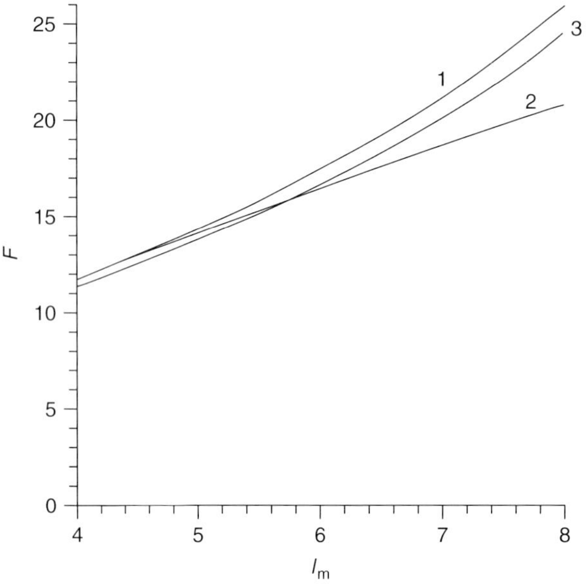

Figure 10.9 Resistance forces of fibers with different shapes of cross-sections as functions of the width of the mid-section: (1) a dumbbell-shaped fiber; (2) a pair of parallel fibers (radius b is fixed, and the gap is varied); and (3) an elliptic fiber (the semi-minor axis is equal to b); $\varphi = 90^\circ$, $b = 0.1$, and the dumbbell bridge thickness is $\delta = 0.04$ [40].

The corresponding results of the computations for the fiber collection efficiency at $b = 0.2$ are given in Figure 10.10. As seen from the figure, the maximum collection efficiency of the elliptic fiber corresponds to the parallel orientation of the major axis relative to the flow direction. If the major axis is turned to an angle φ (Figure 10.8c), then the collection efficiency tends to decrease. Moreover, this in turn causes a change in the functional dependence of the fiber collection efficiency versus Peclet number.

Shown in Figure 10.11 are curves for the dependence $\eta(\mathrm{Pe})$ plotted in logarithmic coordinates, which are straight lines for a wide range of Pe: $\eta = A\mathrm{Pe}^{-m}$. When the flow is parallel to the major axis at any ratio of the axes, the slope of the straight lines corresponds to $m = 2/3$. The same value of m is known for the circular cylinder at high and intermediate Peclet numbers. With the increase of the angle φ, the slope changes. In the limit of the transverse flow of the highly elongated ellipse, $a/b \to 0$, the exponent tends to the value of $m = 3/4$, which is consistent with the theory of diffusion transfer toward a thin plate [42]. In this regard the work of Ushakova *et al.* should be cited (see fig10.3 in [43]), where the measured fiber collection efficiency for particles with $r_p = 41$ nm is precisely described by a dependence of the form $\eta \sim u^{-3/4}$ for a range of velocities $u = 2$–8 cm/s. In this work the diffusional deposition of nanoparticles on an FPP-70 filter made from electrospun fibers (with preliminary neutralized charges) was studied. The FPP-70

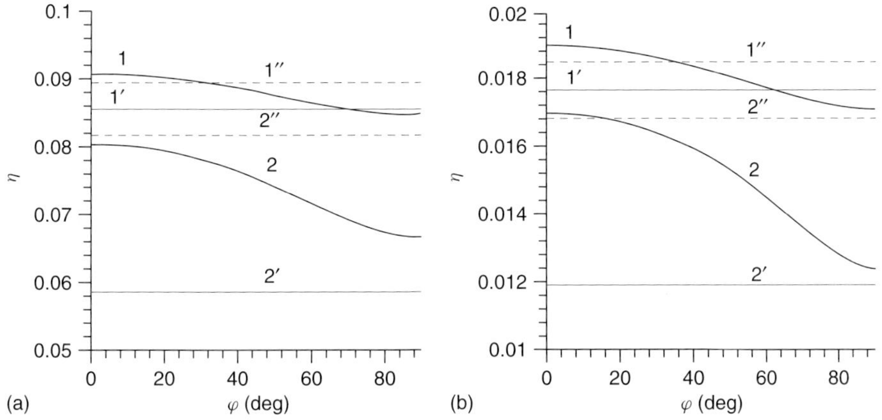

Figure 10.10 Elliptic fiber collection efficiencies due to diffusion of nanoparticles versus the turning angle of the major axis for (*a*) $Pe_h = hu/D = 100$ and (*b*) $Pe_h = 1000$, at $a/b = 0.5$ (1), $a/b = 0.1$ (2), and $b = 0.2$. Solid lines denote fibers of equal cross-sectional area; dotted lines correspond to fibers of the same perimeter.

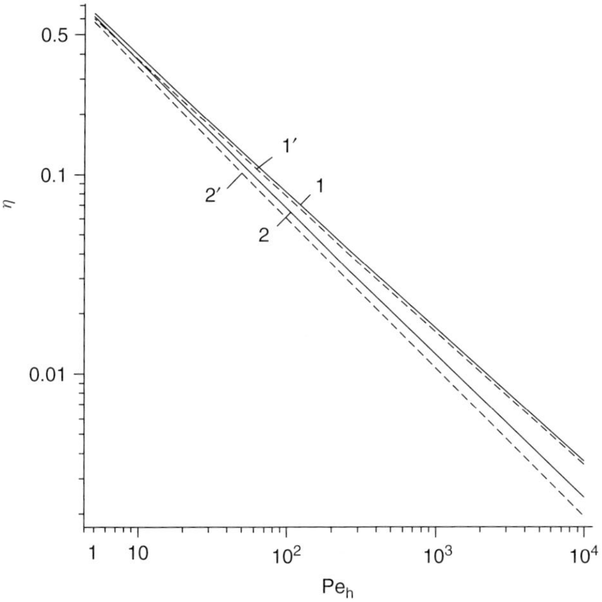

Figure 10.11 Elliptic fiber collection efficiencies due to diffusion of nanoparticles versus the Peclet number at $a = 0.02$ and $b = 0.2$ (1, 2) compared with that for a thin plate with $b = 0.2$ and $a/b \to 0$ ($1', 2'$). Lines 1 and $1'$ are for longitudinal flow past ellipse and plate at $\varphi = 0$; lines 2 and $2'$ are for transverse flow at $\varphi = \pi/2$.

filters consist of equal strip-like electrospun fibers with a width that is several (four to five) times greater than the strip thickness. The major axes of these fibers are oriented normally to the flow direction.

The change in the functional dependence of the fiber collection efficiency on the Peclet number is also seen for a highly porous model filter consisting of pairs of fibers in contact. It was shown in [40] that the collection efficiency for these fibers is practically the same as that for an elliptical fiber with ratio of axes $a/b = 1/2$. The slopes of the lines of their $\eta(\mathrm{Pe})$ dependences coincide also. This slope is somewhat greater for pairs of fibers in a row than for fibers in a uniform row (for the latter case the exponent is $m = 2/3$). For the row with $2a = b = 0.11$ the exponent is $m \cong 0.706$. The calculated fiber collection efficiencies agree with published experimental data on model filters with fiber doublets (Figure 10.12). Since there is an increase of the slope with b/a for elliptical fibers having major axis normal to the flow, the exponent should tend to the value $m = 3/4$ with increase of the gap between fibers in the pair.

10.2.5
Deposition of Nanoparticles on Porous and Composite Fibers

Another topic of investigation is concerned with the diffusional deposition of nanoparticles in model filters consisting of porous permeable fibers and fibers

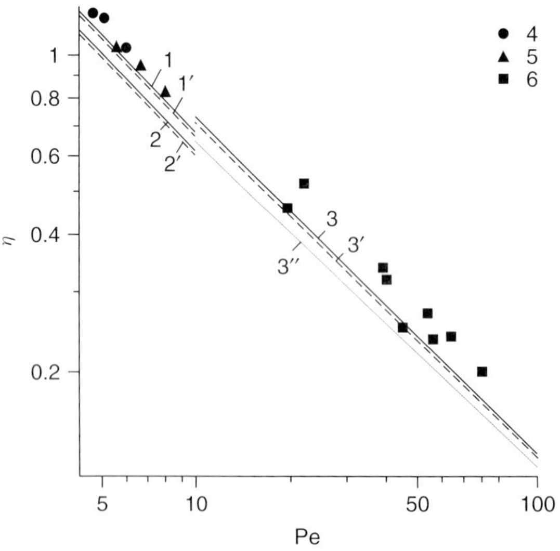

Figure 10.12 Calculated (1–3) and experimental (4–6) [2] dependences of the fiber collection efficiency on the Peclet number for a pair of contacting fibers. Lines 1′–3′ are for elliptic fibers ($2a$ is the minor axis, $4a$ is the major axis); line 3″ is calculation through Eq. (10.5) for a single fiber in a uniform row: 1, 1′ and 4 are for $2h = 2$ mm, $2a = 0.15$ mm, $h' = 0.15$ mm; 2, 2′ and 5 are for $2h = 1$ mm, $2a = 0.043$ mm, $h' = 0.043$ mm; and 3, 3′, 3″ and 6 are for $2h = 1$ mm, $2a = 0.11$ mm, $h' = 0.11$ mm [40].

covered with porous permeable shells [44–46]. The porous shells on fibers, for instance, fur from whiskers, contribute significantly to the collection efficiency with small additional resistance to the air flow. The calculations were done for the cell model and for a row of parallel fibers arranged normally to the flow direction. The flow fields for these models were found in [47, 48] by the combined solution of the Stokes and *Brinkman equations* with the method of boundary collocation [30, 49]. In Figures 10.13 and 10.14 are shown the computed velocity profiles on the gap between the fibers in the row for different permeability parameters of the porous material of the fibers and porous shells.

On the basis of the numerical solution of the convection–diffusion equation, the fiber collection efficiencies were calculated for fibers in the Kuwabara cell and for a single row of fibers over wide ranges of Peclet number, particle radius, ratio of fiber diameter to distance between fiber axes, permeability of the porous material of the shells (layers) (Figure 10.15) [46], and permeability of the porous fibers [45]. It was assumed that the particles entering the porous shells or porous fibers are trapped there completely. The full adsorption boundary condition at the outer porous surface was used.

From Figure 10.15 it follows that at Pe $>$ 1 the fiber collection efficiency is growing with increase of the permeability of the porous shells, tending for Pe $\to \infty$ to its limit, which equals the flux past the porous shell. At Pe $<$ 1 the role of permeability

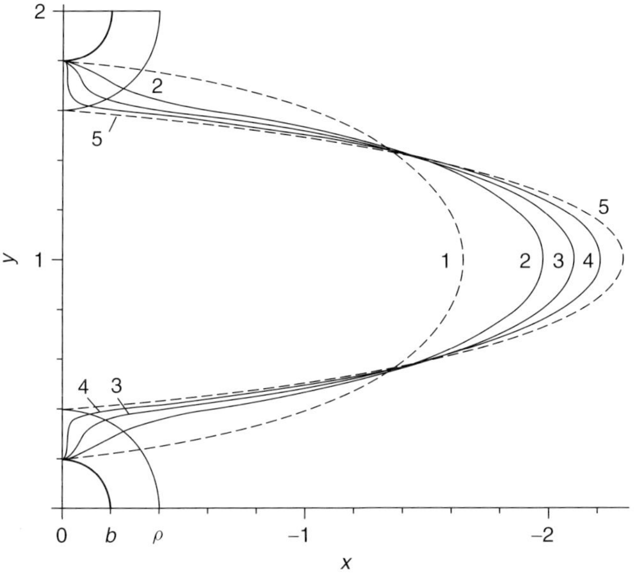

Figure 10.13 Velocity profiles in a row of cylinders with porous shells of radius $\rho = 2b$ at $x = 0$ for different permeability parameters: Brinkman parameter $S = 0$ (1), 15 (2), 25 (3), 50 (4), and ∞ (5), for $a/h = 0.2$. The Brinkman parameter is given by $S = h/\sqrt{\kappa}$, where $2h$ is the distance between the axes of neighboring fibers, and κ is the permeability of the porous layer [48].

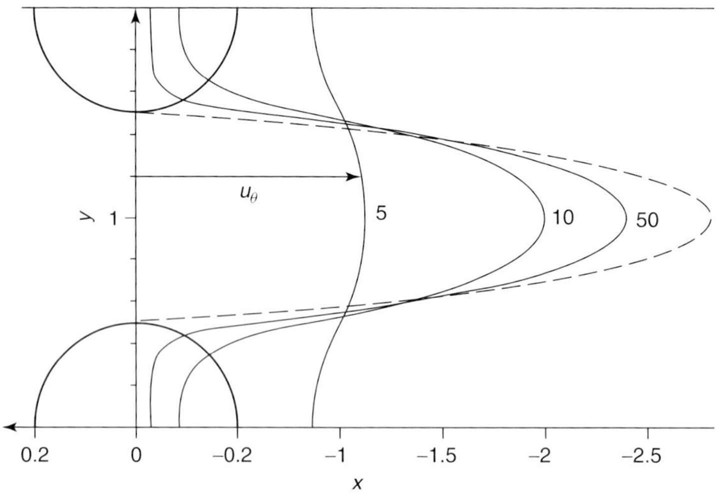

Figure 10.14 Velocity profiles in a row of porous cylinders at x = 0, and $a/h = 0.5$. The numbers on the lines are the S values [47].

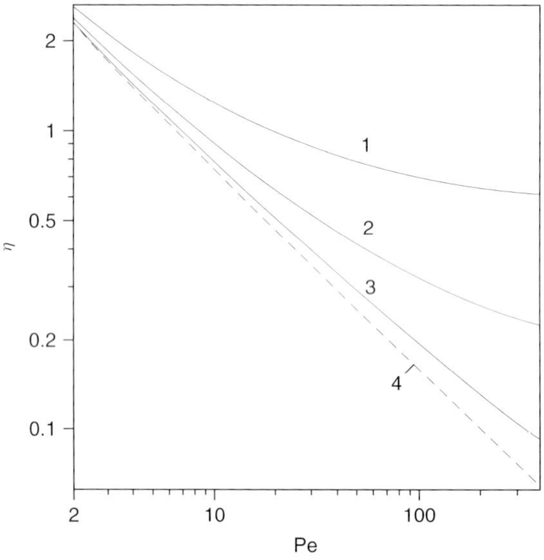

Figure 10.15 Fiber collection efficiencies versus the Peclet number for fibers with a porous shell, for $\alpha = 0.05$ and different permeability parameters: $S = 1.5$ (1), 5 (2), 15 (3), and ∞ (4). The dimensionless radius of the shell is $\rho = a/a_0 = 2$, fiber radius $a_0 = 5$ μm, $r_p = 0.01$ μm, and $u = 5$ cm/s [45].

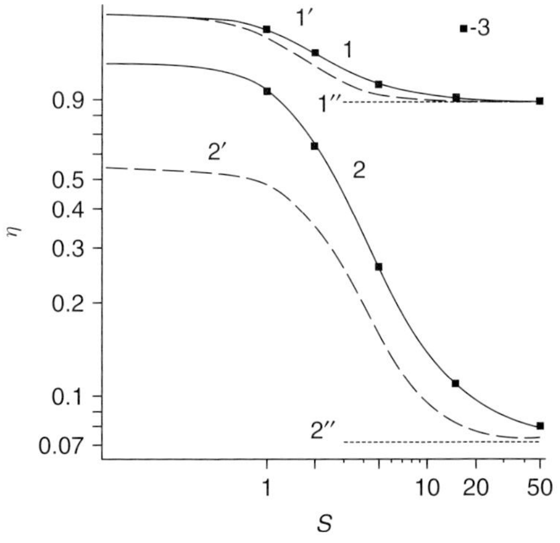

Figure 10.16 Dependence of the fiber collection efficiency of porous fibers in a cell on the permeability parameter S at various particle sizes: 1, 1′ and 1″ are for $r_p = 10$ nm; and 2, 2′ and 2″ are for $r_p = 150$ nm. The field of velocities (1, 2) is found from the Brinkman equation, and (1′, 2′) from the Darcy equation; (1″, 2″) impermeable fiber; and (3) a row of porous fibers; $u = 5$ cm/s, $a = 1$ μm, and $\alpha = 0.05$ [46].

on particle deposition decreases, and at Pe $\to 0$ the collection efficiency tends to the geometric limit for a row of impervious fibers, $\eta = h/a$.

It was also shown in [45] that the use of the *Darcy equation* instead of the Brinkman equation gives underestimated porous fiber collection efficiencies (Figure 10.16). The agreement between the two approaches is seen for η only at $S \ll 1$ and $S \gg 1$. Examples of the streamlines obtained near and within porous fibers calculated for the Brinkman and Darcy equations are shown in Figure 10.17.

In conclusion of this section, it should be noted that here we have considered only nanoparticles whose size is small compared with that of the fiber, that is, point-like particles. Taking account of the finite size of bigger particles (interception effect) is required, especially in problems of deposition upon fine fibers [50]. For that case, one has to take into account three effects: (i) that the particles are moving within the *Knudsen layer* near the fiber surface [51]; (ii) that, on approaching the surface, they are coming into the zone of the action of the retarded *Casimir–van der Waals attractive forces* [50, 52]; and (iii) that small charges on the particles [53], as well as gravity, especially for submicrometer particles of high density of material [54], may have some effect on the deposition. Taking simultaneous account of these mechanisms of deposition is a most difficult problem, since the dependence of the fiber collection efficiency versus particle radius at the given velocity u goes through a minimum, in which all these effects are comparable and non-additive. These questions have been considered in detail in the cited articles.

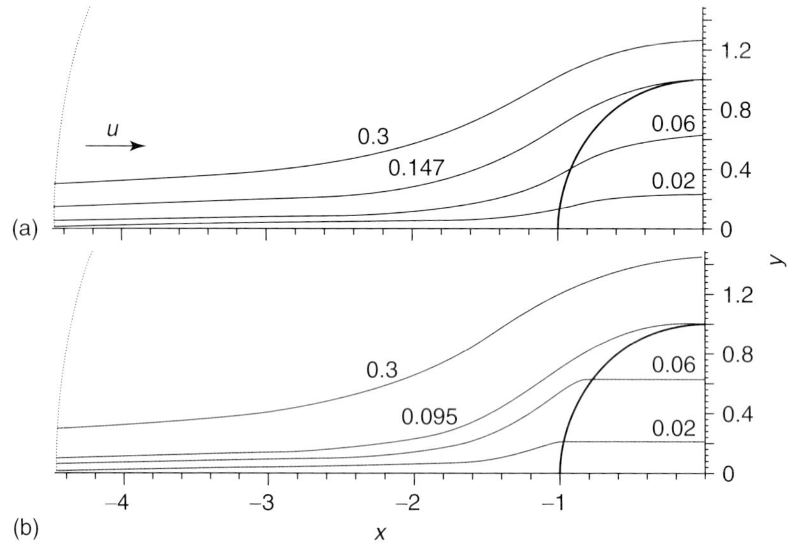

Figure 10.17 Streamlines flowing around a porous fiber in the Kuwabara cell calculated by (a) the Brinkman equation and (b) the Darcy equation at $\alpha = 0.05$ and $S = 5$ [46].

10.3
Penetration of Nanoparticles through Wire Screen Diffusion Batteries

This next section is devoted to the deposition of nanoparticles in three-dimensional (3D) model filters. The applicability of 3D model filters will be considered for the inverse problem of the determination of the particle size by the penetration measured through diffusion batteries.

10.3.1
Deposition of Nanoparticles in Three-Dimensional Model Filters

As noted in the introduction, the first experiments on the deposition of nanoparticles from the Stokes flow within model fibrous filters having different inner structure were published in [16]. In that work the experimental results obtained with a 3D model filter composed of rows of parallel cylinders inclined in their planes at an arbitrary angle were also reported. This model filter was called the *fan model filter*. Penetration of point-like particles through the fan model filter is independent of the angle of inclination and, in the range of filter packing densities $\alpha = 0.03–0.15$, is also independent of the packing density (within the limits of the experimental errors in [16], which were in the several percent range). The corresponding fiber collection efficiency is described by the simple empirical formula

$$\eta = 2.7\text{Pe}^{-2/3} \tag{10.22}$$

This formula was first advanced for the solution of the inverse problem, for estimation of the mean radius of nanoaerosols from the value of the penetration through 3D fibrous structures [55].

The fiber collection efficiency is related to the particle penetration by the following formula:

$$n/n_0 = \exp(-2a\eta L) \tag{10.23}$$

where n and n_0 are the outlet and inlet particle concentrations, and $L = \alpha H/\pi a^2$ is the fiber length per unit surface area of the filter with thickness H.

After the penetration is found, the coefficient of Brownian diffusion is found, from which the radius of nanoparticles is found from the *Einstein–Millikan–Cunningham formula*:

$$D = \frac{kT\left[1 + A(\lambda/r_p) + B(\lambda/r_p)e^{-b(\lambda/r_p)}\right]}{6\pi\mu r_p}$$

Here k is the Boltzmann constant, T is absolute temperature, $A = 1.246$, $B = 0.42$, $b = 0.87$ [39], λ is the mean free path of the gas molecules, and μ is the gas dynamic viscosity.

The method has gained wide acceptance, for a number of reasons: its simplicity; its applicability for different pressures and gas temperatures; the ability to control the processes of measurement and post-interpretation via computer; compatibility with other methods; and, most importantly, the ability to measure a wide range of particle sizes, in particular, nanoparticle sizes within the sub-nanometer range [56]. Moreover, this acceptance was also due to the fact that the theoretical curves of the integral penetration of polydisperse aerosol particles (having common average geometrical radius) plotted versus a complex function that includes the face flow velocity and character scales, intersect in a thin region, practically at one point near the penetration value $n/n_0 \approx 0.4$. This result was first obtained for parallel plate diffusion batteries with the assumption of a log-normal size distribution of particles in [57]. A similar result was obtained for wire screen batteries [18]. The history of *wire screen diffusion batteries* and their application are given in the detailed review [58].

The absence of any dependence of η on α is seen also for real fibrous highly porous filters. This was found directly by measuring the nanoparticle penetration. The outlet concentration remained constant in the process of filter compression [16]. The authors explained this effect by the features of 3D flow within real fibrous filters. Subsequent experiments on the deposition of nanoparticles within fan model filters with small packing density have expanded the range of applicability of the formula (10.22) toward smaller Peclet numbers, down to Pe $\approx$ 0.1 [18]. The independence of η on α for loose fibrous structures, and the validity of Eq. (10.22) for a wide range of Pe numbers, are considered as empirical facts, while a rigorous theoretical explanation has not yet been given. For the case of dense 3D model filters with $\alpha > 0.15$, the fiber collection efficiency increases with α for Pe $\gg$ 1, as shown in [2].

The problem of deposition of nanoparticles in dense 3D fibrous structures has received no attention before, despite the wide use of dense wire screen diffusion batteries for measuring the diffusion coefficient of nanoparticles [58, 59]. The difficulties in predicting the deposition of particles in mesh screens lie in the absence of formulas for the flow field past screens. In recent work [60] the flow velocities and resistance to flow were found numerically for square screens, arranged from touching rows of parallel cylinders shifted at right angles in their planes (F-type pressure welded screens, DIN 4192, ISO 4783/3 [61]). The resistance to flow of these screens was found to be close to that of the woven A-type screens with the same a/h ratio. The parameter a/h for square woven screens was estimated as $2a/(2a+w)$, where w is the gap between the wires. Given below are the results of computations for the deposition of nanoparticles in welded screens for a wide range of Peclet numbers, $0.01 < \mathrm{Pe} < 2000$, and a comparison of the predicted values with experimental data.

10.3.2
Theory of Particle Deposition on Screens with Square Mesh

The field of the particle concentration is found from numerical solution of the stationary convection–diffusion equation

$$2\mathrm{Pe}^{-1}\Delta n - \boldsymbol{u}\cdot\nabla n = 0 \qquad (10.24)$$

in the computational cell shown in Figure 10.18. Here $\boldsymbol{u} = \{u, v, w\}$ is the dimensionless vector of the convective velocity [60], n is the dimensionless concentration, $\mathrm{Pe} = 2au/D$ is the diffusional Peclet number, Δ is the three-dimensional Laplace operator, and

$$\boldsymbol{u}\cdot\nabla n \equiv u\partial n/\partial x + v\partial n/\partial y + w\partial n/\partial z$$

All the variables are brought to dimensionless form by normalization on the fiber radius a, face flow velocity u, and inlet concentration n_0.

Equation (10.20) was solved on a uniform rectangular grid by a second-order regularized finite-difference scheme similar to that suggested in [25]. The domain

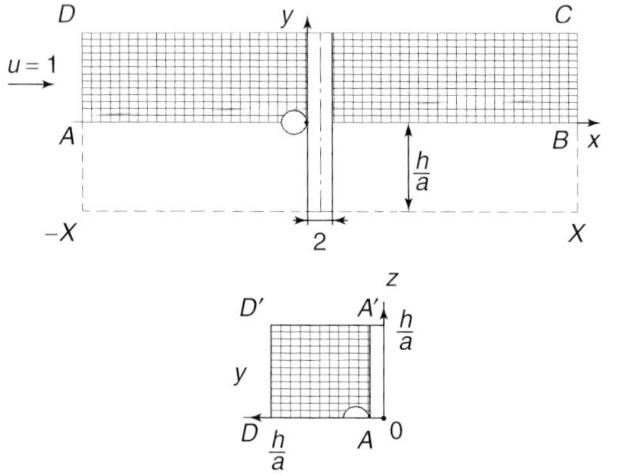

Figure 10.18 Computational cell ABCD; XY and YZ projections. The center of coordinates is placed at the point of contact of cylinders.

decomposition method together with interpolating boundary conditions for the nodes adjacent to the curved boundaries were employed, as in the previously considered two-dimensional case. At the fiber surface the condition of full adsorption was used, $n = 0$. At the inlet at $x = -X$ the uniform concentration condition was imposed, $n = 1$; and at the outlet at $x = X$ the condition of symmetry was used, $\partial n/\partial x = 0$. In the latter case the following conditions were found to be applicable: $n = 0$ for small Peclet numbers, and $n = 1$ for high Peclet numbers. On the upper, lower, left, and right sides of the computational domain, the symmetry conditions were imposed. The half-length of the computational cell, X, was chosen on the basis of the concentration profile on the forward stagnation line OA, found from the ordinary differential equation two-point boundary-value problem

$$2\mathrm{Pe}^{-1}\frac{\partial^2 n}{\partial x^2} - u_x(x)\frac{\partial n}{\partial x} = 0 \tag{10.25}$$

at the corresponding boundary conditions.

The fiber collection efficiency was found by integrating the normal component of the total 3D flux density on the fiber surface, J_r. Written in dimensionless polar coordinates r, z, θ, the fiber collection efficiency per unit length is

$$\eta = \frac{2a}{h}\int_0^{h/a}\int_0^{\pi} J_r(r,\theta,z)r\bigg|_{r=1+r_\mathrm{p}/a} \mathrm{d}z\mathrm{d}\theta \tag{10.26}$$

where $2a$ is the fiber diameter, r_p is the particle radius, and

$$J_r = \frac{2}{\mathrm{Pe}}\frac{\partial n\,(r,\theta,z)}{\partial r} - u_r\,(r,\theta,z)\,n\,(r,\theta,z)$$

For point particles (nanoparticles) the interception parameter is negligibly small, $r_\mathrm{p} \ll a$ and, assuming that $n_r(r{=}1) = 0$, the formula for the fiber collection efficiency reduces to

$$\eta = \frac{4a}{h\mathrm{Pe}}\int_0^{h/a}\int_0^{\pi}\frac{\partial n(r,\theta,z)}{\partial r}\bigg|_{r=1}\mathrm{d}z\mathrm{d}\theta \tag{10.27}$$

10.3.3
Comparison with Experiment

Given in Figure 10.19 are the results of computations for the fiber collection efficiency for screens with a/h [62] corresponding to the experiments of [18]. The details are given in the caption. The dotted line 1 was plotted using empirical formula (10.22). The experimental data of [18] are given by points 5 and 6.

From Figure 10.19 it follows that, for screens with small parameter a/h, the power dependence $\eta(\mathrm{Pe})$ holds down to $\mathrm{Pe} = 0.1$, and the computed curves for screens with a/h that differ by several times are very close together and coincide with the experimental data and with the line plotted by the formula for the fan model filter (10.22).

Next it should be noted that the power dependence of η versus Pe breaks down at low Pe (the higher Pe, the greater is the parameter a/h). Moreover, one can see that

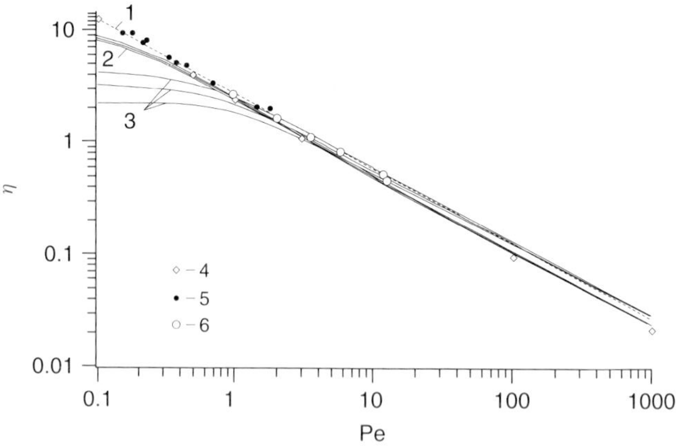

Figure 10.19 Comparison between the computed fiber collection efficiencies for square screens and experimental data of [62]. Curve 1 is plotted by empirical formula (10.22). The families of curves 2 and 3 correspond to η for screens with $a/h = 0.043$ (2) and 0.15 (3), where the upper curves correspond to the front layer of fibers, the lower curves to the rear layer, and the intermediate curves to the average values of η. Points (4) denote computed values of η for screens with $a/h = 0.013$, (5) experiments for $a/h = 0.0132$, and (6) experimental data for $a/h = 0.043$ [18].

the denser the screens, the greater is the difference between the fiber collection efficiency for the front and rear layers of fibers (wires). This is shown in detail in Figure 10.20, where the fiber collection efficiencies are plotted versus Peclet number for screens with $a/h = \pi/12$.

First of all, it is worth noting that the deposition upon fibers in the first layer of fibers relative to the flow direction (Figure 10.20, curve 1) is quite different compared with that for the second (rear) layer (curve 2), and only for small η at $\mathrm{Pe} \gg 1$ do the fiber collection efficiencies become equal. For the dense screens it is typical that the deviation from the power dependence of the form $\eta \sim \mathrm{Pe}^{-2/3}$ is seen at significantly greater Pe, of order unity and greater. From Figure 10.20 it is also seen that for $\mathrm{Pe} \ll 1$ the fiber collection efficiency of the front layer of fibers is independent of Pe, as the fiber collection efficiency for a single row (dotted curve 4) with limit value $\eta = h/a$. For comparison, the theoretical curves are given also for the single row of fibers at $\mathrm{Pe} \ll 1$. Curve 6 is plotted by the formula for the cell model ($h/a = 1/\alpha^{1/2}$) [20]:

$$\eta = \left[\frac{2\pi}{\ln(h/a)}\right]\left[\mathrm{Pe} + \frac{2\pi}{(h/a)\ln(h/a)}\right]^{-1} \tag{10.28}$$

while curve 5 was plotted by Eq. (10.17) for a single row of fibers [21]. It is interesting to note that, for $\eta < 0.1$, that is, for high Peclet numbers, $\mathrm{Pe} \gg 1$, curves 1–4 practically coincide.

Curve 1 shown in Figure 10.21 corresponds to the average fiber collection efficiency with limit value $\eta = h/2a$. Experimental points were obtained on one or two screens with $a/h \approx 0.4$ (details are given in the caption). The data on the particle penetration were taken from the plots published in the cited papers.

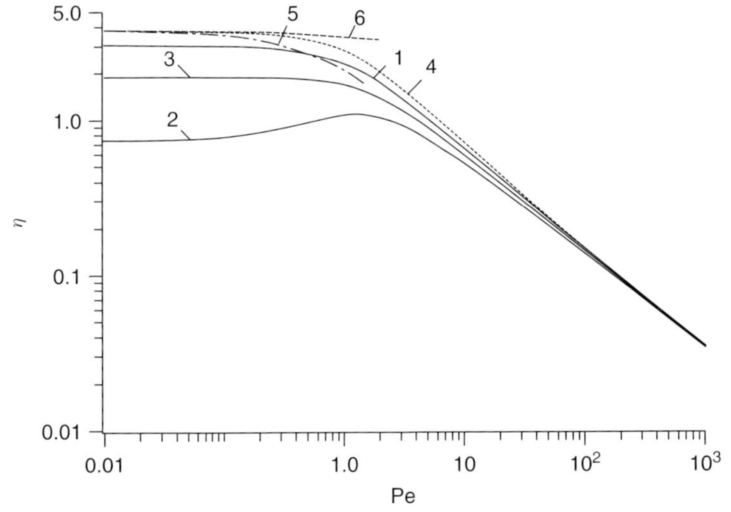

Figure 10.20 Dependence of the fiber collection efficiency for the screen with $a/h = \pi/12$ [62] (a full description is given in the text).

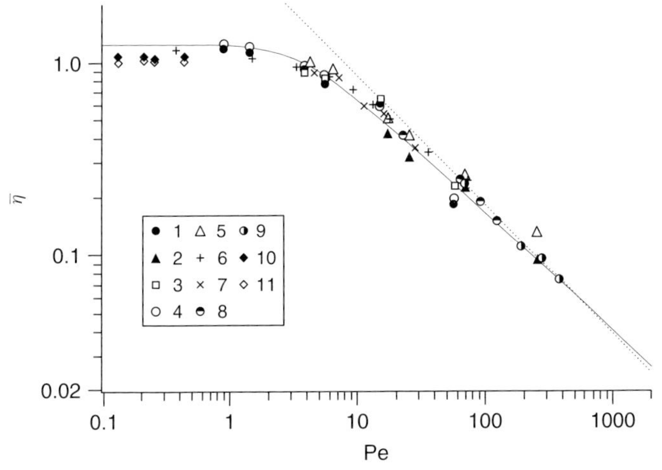

Figure 10.21 Dependence of the average fiber collection efficiency versus the Peclet number computed for screens with $a/h = 0.4$ [62]. The dotted line is plotted by Eq. (10.33). The meaning of the points is as follows. Points 1–5 (see fig. 8 in [63]): $2r_p = 2.4$–20 nm, $2a = 54.6$ μm, $a/h = 0.39$, $\alpha = 0.34$; (1, 2, 3) neutral and (4, 5) charged; (3) three screens and (1, 2, 4, 5) single screen; (1, 3, 4) $u = 1.44$ cm/s and (2, 5) $u = 6.58$ cm/s. Points 6 (see fig. 8 in [64]): $2r_p = 1$–10 nm, $2a = 52$ μm, $a/h = 0.41$, $\alpha = 0.31$; single screen; $u = 3.6$ cm/s. Points 7 (see fig. 7 in [65]): $a/h = 0.41$, $2a = 75$ μm; $u = 12$ cm/s. Points 8, 9 (see fig. 2 in [66]): $a/h = 0.41$, $2a = 75$ μm; (8) $u = 17.5$ cm/s and (9) $u = 52.5$ cm/s. Points 10, 11 (see [67]): $2r_p = 1.6$–2.2 nm, $2a = 25$ μm, $a/h = 0.39$; (10) single screen and (11) two screens; $u = 1$–2 cm/s.

With knowledge of the particle penetration through a *diffusion battery* consisting of m screens, the average fiber collection efficiency is found from the equation for the balance of nanoparticles in the flow with volumetric flow rate Q past the screen with area S:

$$n_0 \frac{Q}{S} - n_0 \frac{Q}{S} l 2a\eta = n_1 \frac{Q}{S} \tag{10.29}$$

where n_0 and n_1 are outlet and inlet concentrations, $l = 1/h$ is the total length of wires per unit surface area of the screen, and $2h$ is the distance between the axes of wires in the row. From Eq. (10.29) the particle penetration through the single screen is found from

$$\frac{n_1}{n_0} = (1 - l2a\eta) \tag{10.30}$$

It was assumed here that the same number of particles is deposited upon any unit length of wire. If the diffusion battery consists of m screens with the same parameters, then the overall length of wires per unit surface area of the diffusion battery is equal to $L = ml$. Let us assume that every subsequent screen is subjected to flux with uniform concentration of particles. Then the particle penetration through m screens is equal to

$$n/n_0 = \left(\frac{n_1}{n_0}\right)^m = (1 - l2a\eta)^m = (1 - Lm^{-1}2a\eta)^m \tag{10.31}$$

where n is the concentration at the outlet of the diffusion battery. From this we derive the expression for the fiber collection efficiency:

$$\eta = \left(\frac{h}{2a}\right)\left[1 - \left(\frac{n}{n_0}\right)^{1/m}\right] \tag{10.32}$$

Since $\lim_{m\to\infty}(1 - b/m)^m = \mathrm{e}^{-b}$, then Eq. (10.31) can be reduced to the form of Eq. (10.23), which is convenient for large m.

The fiber collection efficiencies estimated from the experimental data by the simple formula (10.32) agree well with the theoretical curve for the average η in a wide range of Pe.

It also follows from Figure 10.21 that for Pe $\gg$ 1 the calculated fiber collection efficiencies agree with the empirical relationship (dotted line 2), which was obtained for dense screens at Pe > 20 [59] and repeatedly confirmed [68]:

$$\eta = \frac{2.7\mathrm{Pe}^{-2/3}}{(1 - \alpha)} \tag{10.33}$$

Here it was assumed that the packing density α corresponds to the packing density of the touching screens and equals $\alpha = \pi a/4h$.

In spite of the fact that Eqs. (10.33) and (10.22) have been widely used for the estimation of the nanoparticle diffusion coefficient for several decades, neither the authors of [59] nor their followers gave a definition for the packing density of single screens. It should be noted that, when many screens are used, special thin rings are installed additionally. The whys and wherefores of the estimation of α are usually based on the thickness of single screens, which is greater by 10–20% than its minimal value, being equal to two diameters of the wire (the tortuosity of the wires is not specified).

In connection with the discussion on the accuracy of the diffusional method, it should be noted that the exponential dependence (10.23) is not applicable for processing the measured penetration of nanoparticles obtained with the help of a diffusion battery consisting of one or two screens. The necessity for a single screen is really present when measuring particles of high diffusional mobility, especially for particles within the sub-nanometer range of sizes, for example, radon and thoron progeny. In this case one should use Eq. (10.32).

To achieve better accuracy in the measurement of the diffusion coefficient D, one has to select such conditions of depositions of nanoparticles that correspond to high Peclet numbers, $\mathrm{Pe} \gg 1$, when the power dependence $\eta \sim \mathrm{Pe}^{-2/3}$ holds, bearing in mind that the increase of the velocity and fiber diameter is limited by the condition of low Reynolds numbers ($\mathrm{Re} < 1$), since the relationship between the fiber collection efficiency and the diffusion coefficient was obtained for Stokes flow.

The formula (10.23) is valid for a diffusion battery consisting of a large number of screens, when the effective length of the wires per unit surface area is defined as $L = 2m/(2a + w)$. If the total length of the wires is defined by the method of weighing a number of woven screens, it should be known that the measured actual length of the wires is greater than the calculated value of L.

In conclusion, it should be noted that the curves computed for the average fiber collection efficiency for single dense screens with significantly different step intersect at approximately one point at $\eta \approx 0.8$ and $\mathrm{Pe} \approx 6$ (Figure 10.22). This result can be used for the analysis for polydisperse sub-nanometer particles when measuring the penetration of nanoparticles through a single screen with $a/h = 0.39$, for which, after particle penetration $n/n_0 = 0.38$ [18] is measured (with the adjustment of the corresponding face flow velocity u or fiber radius a), the average coefficient of diffusion is immediately found as $D \approx au/3$.

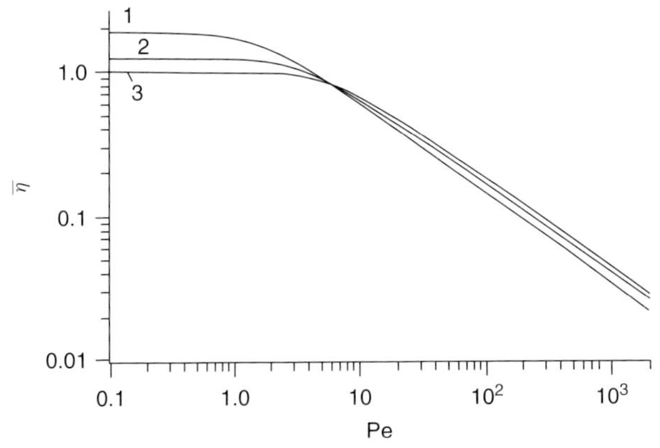

Figure 10.22 The average fiber collection efficiency versus the Peclet number for screens with $a/h = \pi/12(1)$, $a/h = 0.4$ (2), and $a/h = 0.5$ (3).

10.4 Conclusion

The aim of this chapter was to consider the deposition of aerosol nanoparticles in model fibrous filters at low Reynolds numbers. It was shown that recent results for the collection efficiency of particle deposition on fibers found from the numerical solution of the convection–diffusion equation for regular systems of fibers arranged normally to the flow direction agree with known experiments for model filters, with known analytic formulas derived in the boundary-layer approximation at $\mathrm{Pe} \to \infty$, and with the geometric limit for the fiber collection efficiency at $\mathrm{Pe} \to 0$.

Here, the consideration was confined to diffusional deposition of particles from the Stokes flow (creeping flow) at low Reynolds numbers, $\mathrm{Re} \ll 1$, when the streamlines before and behind the mid-section of the fiber are symmetric and the ratio of the filter pressure drop to the face flow velocity is constant and independent of Re. The limiter for the velocity is due to the results of our work conducted over many years, and is related with the practice of highly efficient gas filtration. The latter is performed with high-efficiency fibrous filters at low velocities – first of all, due to their significant resistance to flow; and, second, due to low collection efficiency at high velocities. The significant increase of the velocity to values corresponding to $\mathrm{Re} > 1$ leads to the asymmetry in the streamline pattern and to the increase of the velocity gradient near the front side of the fiber. The deposition of particles should increase. In this case the estimates for the fiber collection efficiency for given values of Pe, found for the Stokes flow regime, are underestimated. It is impossible to find the fiber collection efficiency η analytically for the flow field governed by the full *Navier–Stokes equations*, and thus the numerical methods have been used here for years [69]. But such high velocities are not typical for high-efficiency aerosol filtration.

It should be noted that the collection efficiency due to the diffusion of nanoparticles in real filters is greatly affected by the inner structure of the filters, by the hidden and manifest defects, by variable thickness and packing density, by the presence of double and triple fibers (which interact with the flow as one obstacle), and so on. All these questions, together with questions of the optimization of filter parameters taking account of their clogging, have been considered in numerous papers published in *Colloid Journal* by the authors of this short review.

This review of the experimental and theoretical results of nanoparticle deposition from a gas flow upon the fibers of model filters has shown the possibility of further deeper investigations. The results reviewed could be helpful for the solution of problems of diffusion transport arising in other fields, such as electrochemistry, sorption, and catalysis.

We have considered point-like particles. Taking account of the finite size of bigger particles (interception effect) is connected with the consideration of external forces, for example, the retarded Casimir–van der Waals attractive forces [50, 52], and also requires the presence of the Knudsen layer on the surface of the fiber of any radius, since the thickness of this layer is comparable with the particle size. The simultaneous consideration of mechanisms of deposition of particles of finite size is beyond the scope of our present review.

Acknowledgements

The authors are grateful to Dr I.B. Stechkina for helpful discussions.

References

1. Davies, C.N. (1973) *Air Filtration*, Academic Press, London.
2. Kirsch, A.A. and Stechkina, I.B. (1978) in *Fundamentals of Aerosol Science* (ed. D.T. Shaw), Wiley-Interscience, New York, pp. 165–256.
3. Brown, R.C. (1993) *Aerosol Filtration*, Pergamon Press, Oxford.
4. Chernyakov, A.L. and Kirsch, A.A. (1997) Resistance force in a three-dimensional grid formed by cylindrical fibers. *Colloid J.*, **59** (5), 647–657.
5. Roldughin, V.I., Kirsch, A.A., and Emel'yanenko, A.M. (1999) Simulation of aerosol filters at intermediate Knudsen numbers. *Colloid J.*, **61** (4), 492–504.
6. Chernyakov, A.L. (1998) Fluid flow through three-dimensional fibrous porous media. *J. Exp. Theor. Phys.*, **86** (6), 1156–1166.
7. Chernyakov, A.L., Lebedev, M.N., Stechkina, I.B., and Kirsch, A.A. (1998) Hydrodynamic resistance of the row of parallel polydisperse fibers. *Colloid J.*, **60** (1), 91–103.
8. Chernyakov, A.L. and Kirsch, A.A. (1999) Growth in hydrodynamic resistance of aerosol fibrous filter during deposition of liquid dispersed phase. *Colloid J.*, **60** (6), 791–796.
9. Chernyakov, A.L. and Kirsch, A.A. (2001) The effect of long-range fiber space and orientation correlations on the hydrodynamic resistance and diffusion deposition in fibrous filters. *Colloid J.*, **63** (4), 506–510.
10. Langmuir, I. (1942) Report on smokes and filters. OSRD Report No. 865. Office of Scientific Research and Development, Office of Technical Services, Washington, DC.
11. Natanson, G.L. (1957) Diffusional deposition of aerosols on a streamlined cylinder with a small capture coefficient. *Proc. Acad. Sci. SSSR, Phys. Chem. Sect. (Dokl. Akad. Nauk SSSR)*, **112** (1), 21–25.
12. Fuchs, N.A. and Stechkina, I.B. (1962) On the theory of aerosol fibrous filters. *Dokl. Akad. Nauk. SSSR.*, **147** (5), 1144 (in Russian).
13. Kirsch, A.A. and Fuchs, N.A. (1967) The fluid flow in a system of parallel cylinders perpendicular to the flow direction at small Reynolds numbers. *J. Phys. Soc. Jpn.*, **22** (5), 1251–1255.
14. Kuwabara, S. (1959) The forces experienced by randomly distributed parallel circular cylinders or spheres in viscous flow at small Reynolds numbers. *J. Phys. Soc. Jpn.*, **14** (4), 527–532.
15. Kirsch, V.A. (2003) Deposition of aerosol nanoparticles in fibrous filters. *Colloid J.*, **65** (6), 726–732.
16. Kirsch, A.A. and Fuchs, N.A. (1968) Studies on fibrous aerosol filters III. Diffusional deposition of aerosols in fibrous filters. *Ann. Occup. Hyg.*, **11**, 299–304.
17. Kirsch, A.A. and Chechuev, P.V. (1985) Diffusion deposition of aerosol in fibrous filters at intermediate Peclet numbers. *Aerosol Sci. Technol.*, **4** (1), 11–16.
18. Kirsch, A.A., Zagnit'ko, A.V., and Chechuev, P.V. (1981) On the diffusional method of determination of the sizes of submicron aerosols. *Z. Fyz. Khim. (Russ.J. Phys. Chem.)*, **LV** (12), 3034–3037 (in Russian).

19. Stechkina I.B. (1966) Diffusion precipitation of aerosols in fiber filters. *Dokl. Akad. Nauk SSSR.*, **167** (6), 1327–1330.
20. Roldughin, V.I. and Kirsch, A.A. (1995) Diffusional deposition of aerosol particles in model filter at small Peclet numbers. *J. Aerosol Sci.* **26** (Suppl. 1), S731–S732.
21. Chernyakov, A.L., Kirsch, A.A., Roldugin, V.I., and Stechkina, I.B. (2000) Diffusion deposition of aerosol particles on fibrous filters at small Peclet numbers. *Colloid J.*, **62** (4), 490–494.
22. Chernyakov, A.L. and Stechkina, I.B. (2000) Diffusion capture of aerosol particles in a periodic grid of cylindrical fibers at small Peclet numbers. *Colloid J.*, **62** (5), 644–650.
23. Kirsch, V.A. (2005) Deposition of nanoparticles in a model fibrous filter at low Reynolds numbers. *Russ. J. Phys. Chem.*, **79** (12), 2049–2052.
24. Samarskii, A.A. and Vabishchevich, P.N. (1995) Computational heat transfer, in *The Finite Difference Methodology*, John Wiley & Sons, Inc., New York.
25. Berkovsky, D.M. and Polevikov, V.K. (1973) Prandtl number effect on natural convection structure and heat transfer. *Inzhenerno Fizicheskii Zh. (J. Eng. Phys. Thermophys., Engl. Trans.)*, **24** (5), 842–849.
26. Kirsch, V.A. (2004) The deposition of aerosol submicron particles on ultrafine fiber filters. *Colloid J.*, **66** (3), 311–315.
27. Fletcher, C.A.J. (1988) *Computational Techniques for Fluid Dynamics*, Springer, Berlin.
28. Fedorenko, R.P. (1994) *Vvedenie v Vychislitel'nuyu Fiziku (An Introduction to Computational Physics)*, Moskow Fiziku-Technical Institute, Moscow.
29. Miyagi, T. (1958) Viscous flow at low Reynolds number past an infinite row of equal circular cylinders. *J. Phys. Soc. Jpn.*, **13** (5), 493–496.
30. Wang, C.Y. (2002) Stokes slip flow through a grid of circular cylinders. *Phys. Fluids*, **14** (9), 3358–3360.
31. Rice, J.R., Vavalis, E., and Tsompanopoulou, P. (1997) Interface relaxation methods for elliptic differential equations. Report CSD-TR 97-004, Purdue University, Department of Computer Sciences.
32. Lebedev, M.N., Stechkina, I.B., and Chernyakov, A.L. (2000) Viscous flow through polydisperse row of parallel fibers: account for the slip effect. *Colloid J.*, **62** (6), 731–745.
33. Kirsch, A.A., Stechkina, I.B., and Fuchs, N.A. (1971) Effect of gas slip on the pressure drop in a system of parallel cylinders. *J. Colloid Interface Sci.*, **37** (2), 458–461.
34. Wheat, I.A. (1963) The air flow resistance of glass fiber filter paper. *Can. J. Chem. Eng.*, **41** (4), 67–72.
35. Kirsch, A.A., Stechkina, I.B., and Fuchs, N.A. (1973) Effect of gas slip on the pressure drop in fibrous filters. *J. Aerosol Sci.*, **4** (4), 287–293.
36. Kirsch, V.A., Budyka, A.K., and Kirsch, A.A. (2008) Simulation of nanofibrous filters produced by the electrospinning method: 2. The effect of gas slip on the pressure drop. *Colloid J.*, **70** (5), 584–588.
37. Aleksidze, M.A. (1991) *Fundamental'nye Funktsii v Priblizhennykh Resheniyakh Granichnykh Zadach (Fundamental Functions in Approximate Solutions of Boundary Problems)*, Nauka, Moscow.
38. Pich, J. (1965) Die Filtrationstheorie hochdisperser Aerosole. *Staub*, **25** (5), 186–192.
39. Fuchs, N.A. (1964) *The Mechanics of Aerosols* (translated from Russian by R.E. Daisley and M. Fuchs) (ed. C.N. Davies), Pergamon, London.
40. Kirsch, V.A., Budyka, A.K., and Kirsch, A.A. (2008) Simulation of nanofibrous filters produced by the electrospinning method: 1. Pressure drop and deposition of nanoparticles. *Colloid J.*, **70** (5), 574–583.
41. Kuwabara, S. (1959) The forces experienced by a lattice of elliptic cylinders in a uniform flow at small Reynolds numbers. *J. Phys. Soc. Jpn.*, **14** (4), 522–527.
42. Gupalo, Yu.P., Polyanin, A.D., and Ryazantsev, Yu.S. (1985) *Massoteploobmen reagirujuzhih chastiz s potokom (Heat and mass exchange of reacting particles with flow)*, Nauka, Moscow.

43. Ushakova, E.N., Kozlov, V.I., and Petryanov, I.V. (1973) Regularities of aerosol capture in fibrous filtering materials in the diffusion region. *Kolloidn Zh. (Colloid J.)*, **35** (2), 388–391 (in Russian).
44. Kirsch, V.A. (1996) Aerosol filters made of porous fibers. *Colloid J.*, **58**, 737.
45. Kirsch, V.A. (2007) Deposition of aerosol nanoparticles in filters composed of fibers with porous shells. *Colloid J.*, **69** (5), 615–619.
46. Kirsch, V.A. (2007) Deposition of nanoparticles in filters composed of permeable porous fibers. *Colloid J.*, **69** (5), 609–614.
47. Kirsch, V.A. (2006) Stokes flow past periodic rows of porous cylinders. *Theor. Found. Chem. Eng.*, **40** (5), 465–471.
48. Kirsch, V.A. (2006) Stokes flow in periodic systems of parallel cylinders with porous permeable shells. *Colloid J.*, **68** (2), 173–181.
49. Kolodziej, J.A. (1987) Review of application of boundary collocation methods in mechanics of continuous media. *Solid Mech. Arch.*, **12** (4), 187–231.
50. Kirsch, V.A. (2004) The effect of van der Waals forces on the deposition of highly dispersed aerosol particles on ultrafine fibers. *Colloid J.*, **66** (4), 444–450.
51. Roldughin, V.I. and Kirsch, A.A. (2001) Diffusion deposition of finite size particles on fibrous filters at intermediate Knudsen numbers. *Colloid J.*, **63** (5), 619–625.
52. Kirsch, V.A. (2003) Calculation of the van der Waals force between a spherical particle and an infinite cylinder. *Adv. Colloid Interface Sci.*, **104**, 311–324.
53. Kirsch, V.A. and Budyka, A.K. (2008) Deposition of charged submicron aerosol particles in fibrous filters. *Proceedings of the 10th World Filtration Congress*, 14–18 April 2008, Leipzig, Germany, vol. **3**, Filtech Exhibitions, Meerbusch, pp. 461–465.
54. Kirsch, V.A. (2005) Diffusional deposition of heavy submicron aerosol particles on fibrous filters. *Colloid J.*, **67** (3), 313–317.
55. Kirsch, A.A. and Stechkina, I.B. (1969) A diffusional method for the determination of the size of condensation nuclei. *Proceedings of the 7th International Conference on Condensation and Ice Nuclei*, 18–24 September 1969, Academia-Prague, Prague, pp. 284–287.
56. Thomas, J.W. and Hinchcliffe, L.E. (1972) Filtration of 0.001 μm particles by wire screens. *J. Aerosol Sci.*, **3** (5), 387–396.
57. Fuchs, N.A., Stechkina, I.B., and Starosselskii, V.I. (1962) On the determination of particle size distribution in polydisperse aerosols by the diffusion method. *Br. J. Appl. Phys.*, **13** (6), 280–281.
58. Knutson, E.O. (2007) History of diffusion batteries in aerosol measurements. *Aerosol Sci. Technol.*, **31** (2), 83–128.
59. Cheng, Y.S. and Yeh, H.C. (1980) Theory of a screen-type diffusion battery. *J. Aerosol Sci.*, **11** (3), 313–320.
60. Kirsch, V.A. (2006) The viscous drag of three-dimensional model fibrous filters. *Colloid J.*, **68** (3), 261–266.
61. Purchas, D. (1996) *Handbook of Filter Media*, Elsevier, Oxford, p. 249.
62. Kirsch, V.A. and Kirsch, A.A. (2010) Penetration of nanoparticles through wire screen diffusion batteries. *Colloid J.*, **72** (4), in press.
63. Heim, M., Mullins, B.J., Wild, M., Meyer, J., and Kasper, G. (2005) Filtration efficiency of aerosol particles below 20 nanometers. *Aerosol Sci. Technol.*, **39** (8), 782–789.
64. Itani, Y., Cho, S.J., and Emi, H. (1994) Removal of nanometer size particles and ions from air. Proceedings of the 12th ISCC, 1994, Yokohama, Japan, pp. 21–24.
65. Ichisubo, H., Hashimoto, T., Alonso, M., and Kousaka, Y. (1996) Penetration of ultrafine particles and ion clusters through wire screens. *J. Aerosol Sci.*, **24** (3), 119–127.
66. Alonso, M., Kousaka, Y., Hashimoto, T., and Hashimoto, M. (1997) Penetration of nanometer-sized aerosol particles through wire screen and laminar flow tube. *J. Aerosol Sci.*, **27** (4), 471–480.
67. Attoui, M., Agranovski, I.E., and Zagainov, V.A. (2008) Removal of nanoparticles by fine steel meshes. Proceedings of the European Aerosol

Conference 2008, Thessaloniki, Greece, Abstract T03A063O.

68. Scheibel, H.G. and Porstendorfer, J. (1984) Penetration measurements for tube and screen-type diffusion batteries in the ultrafine particle size range. *J. Aerosol Sci.*, **15** (6), 673–682.

69. Emi, H., Kanaoka, C., and Kuwabara, Y. (1982) The diffusion collection efficiency of fibers for aerosol over a wide range of Reynolds numbers. *J. Aerosol Sci.*, **13** (5), 403–413.

11
Filtration of Liquid and Solid Aerosols on Liquid-Coated Filters

Igor E. Agranovski

11.1
Introduction

There are many technologies currently available for the removal of aerosol particles from a range of gas carriers. They include gravitational settling chambers, cyclones, wet scrubbers, electrostatic precipitators, and others [1, 2]. However, in cases when reliable, highly efficient removal of particles within the size range of 0.01–1 μm is required, there are no alternatives to the employment of filters. The use of *filtration processes* has been around for more than 2000 years, and the interested reader is referred to Davies [3] for a historical review of the various development stages. A broad range of filters is currently available to meet almost any environmental and technological requirement. They are used for the removal of broadly sized aerosols of different nature for a range of technological parameters, including gas carrier temperature, humidity, and chemical composition.

For the theoretical evaluation of filter efficiency, a filtration theory based on a single-fiber approach has been developed over the past 60 years [3–8]. It considers the combined action of a number of mechanisms of particle removal on a single filter fiber, including inertia, interception, diffusion, electrostatic precipitation, thermophoresis, and diffusiophoresis, with subsequent derivation of the total efficiency of a filter made out of multiple single fibers randomly arranged to form a medium with certain packing density and thickness (see Chapter 9).

Usually, the filtration of solid particles is associated with dusty air passing through a filter medium, with the purified stream either released to the atmosphere or further utilized for subsequent technological processes. The particles that are removed from the air remain on the surface of the filter and, in the course of time, form a layer, which increases the filter's *hydraulic resistance*. When the resistance reaches a certain value, the filter is regenerated by a reverse air jet, shaking, or oscillating. For some applications, for example, purification of the air intake for motor vehicles, air filters are not usually regenerated but rather are disposed of and replaced by new ones.

Normally, solid particles carried by sufficiently dry air streams are easily removed from the filter by one of the regeneration mechanisms discussed above [7]. However,

Aerosols – Science and Technology. Edited by Igor Agranovski

ISBN: 978-3-527-32660-0

when the air stream is humid or contains highly viscous particles, the regeneration process becomes inefficient (particles adhere strongly to the filter) and the filter can suffer blockage. Sometimes, the possibility of clogging is so high that, to avoid potential losses associated with frequent shutdowns of the technology for maintenance, alternative air pollution control technologies with lower performance have to be used [9].

Liquid aerosol filtration differs significantly from *solid aerosol* removal. The main difference is related to the fact that the collected liquid particles remain on the filter surface and grow as the result of coalescence with new incoming ones. When the droplet reaches a size sufficient for the gravitational force to exceed the force of adhesion acting between the particle and the fibers, the droplet drains from the filter, creating *self-cleaning* conditions of the process. This is the obvious benefit of liquid aerosol removal as compared to *solid filtration* – the filtration device does not usually require any additional regeneration module, making it more cost-effective. On the other hand, liquid on the filter surface could cause substantial *hydrodynamic resistance*. Compared to the porous and air-penetrable structure of the deposits built up on the filter surface during solid particle purification, liquid could entirely clog some filter areas, decreasing the filter surface available for air to pass through.

The filtering materials utilized in aerosol removal processes can be either wettable [10] or non-wettable [11] for liquids with different physical properties. For wettable filters, liquid aerosols collected by the filter spread along the fibers and generate thin films covering each individual fiber. This is shown in Figure 11.1 by the two scanning electron microscope (SEM) images of a wettable fiber, first dry and then coated by oil. Non-wettable filters can be considered as liquid repellents. Droplets do not spread along non-wettable fibers and remain as spheres. This is illustrated in Figure 11.2 by the optical microscope image of a non-wettable fiber, with water droplets clearly visible. These spheres grow as the result of coalescence with new incoming droplets, and when the gravitational force exceeds the force of adhesion, they are detached from the filter and drained.

This chapter considers filtration processes performed by fibrous filters naturally or artificially irrigated by liquids. It discusses the theoretical approach to model the processes, and illustrates various applications of such filters for the monitoring and control of aerosols of mineral and biological nature.

11.2
Wettable Filtration Materials

Removing particles from exhaust gases has become increasingly important from an industrial standpoint because of the need to reduce health hazards, to limit nuisance emission, and to recover valuable products. Two main parameters are considered as filter quality criteria: particle collection efficiency and hydrodynamic resistance across the filter. Therefore, practical fibrous filters usually have high porosity, with the inter-fiber distance being large compared to the fiber diameter. The classic filtration theory, based on a consideration of aerosol particle removal by

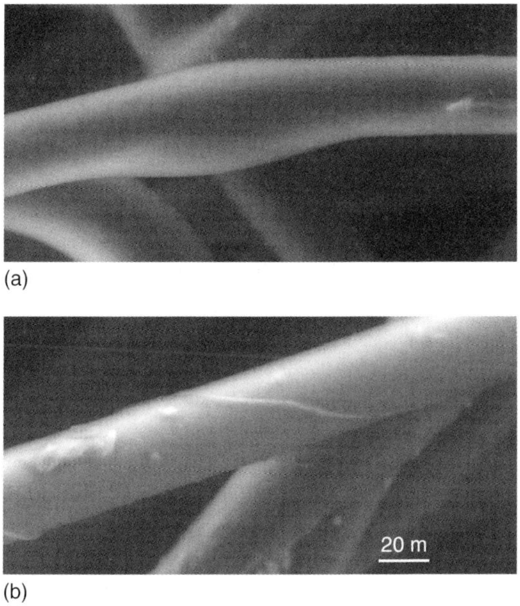

Figure 11.1 Wettable fibers covered by a liquid film: (a) dry fibers and (b) oil-coated fibers.

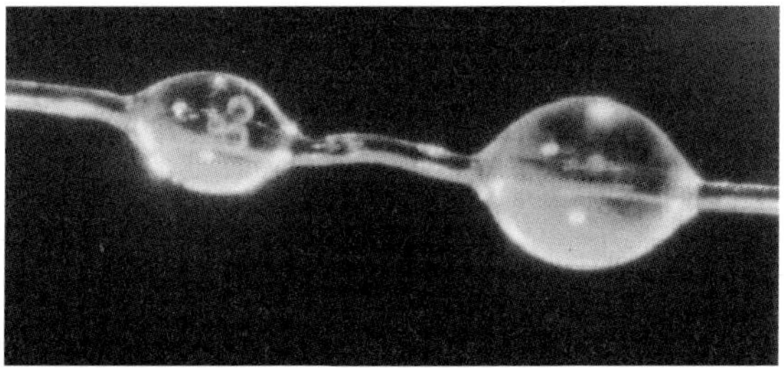

Figure 11.2 A non-wettable fiber.

a single fiber with subsequent reconstruction of the total filter efficiency by taking into account the fibrous filter packing density and the thickness of the medium (see Chapter 9), is capable to closely predict the efficiency and hydrodynamic resistance of the filter.

However, in many situations, the classic theory cannot be used directly for real-world situations, as aerosol particles collected by the filter create porous deposits (in the case of solid particle filtration) or liquid "islands" (in the case of liquid aerosol removal) on the filter surface, significantly changing the parameters of the medium and, correspondingly, causing a substantial discrepancy between actual and predicted process parameters. In the case of solid particle filtration, the deposit of captured particles on the filter surface is random in nature and variable in time, which significantly complicates any effort to develop some theoretical approach realistically predicting the behavior of filtration processes for industrial applications. In contrast, in the case of liquid particle removal, considering the self-cleaning nature of the process, it could be concluded that, in technologically steady-state operation, the liquid on the filter reaches equilibrium conditions, which could be described mathematically if the process parameters along with the physico-chemical characteristics of the collected particles and filter medium are known.

11.2.1 Theoretical Aspects

When liquid aerosols are removed by wettable filters, the collected particles spread along the fibers, creating films coating the entire filter surface. Figure 11.1b shows an image of a wettable fiber coated by an oil film. The image was acquired by environmental SEM, capable of operating at moderate vacuum, ensuring low evaporation of liquid during operation. The oil film smoothly covers the entire fiber surface, which is clearly seen compared to the dry fiber provided in Figure 11.1a.

Numerous studies of the process (see for example [10]) have demonstrated that, although fibers can be coated by relatively thick films during filter operation, there is an *equilibrium film* that cannot be drained from the fiber under the action of gravity. The *equilibrium film thickness* corresponds to the particular process parameters, the fiber characteristics, and the physical properties of the liquid aerosol. A simple procedure could be used to identify the equilibrium thickness of the film. First, a small piece of filter medium with precisely measured parameters (width, length, thickness, fiber diameter, and packing density) is cut and weighed. Then, the filter is submerged into the liquid that is to be used for the following aerosol production (or represents an aerosol material to be removed in an industrial process). After a few minutes, the filter is removed, vertically secured in a frame-type holder, and placed in a closed humidified chamber, allowing liquid to drain (to eliminate evaporation in cases when a volatile liquid is involved). On completion of draining, the filter is carefully weighed again and the equilibrium film thickness is estimated by the following procedure. The total fiber length, f_L, in the filter sample is

$$f_L = \frac{W_{DF}}{\rho_{FM}\pi r_f^2} \tag{11.1}$$

where W_{DF} is the dry filter weight, ρ_{FM} is the filter material density, and r_f is the fiber radius. The equilibrium film thickness is then estimated as

$$\delta_E = r_E - r_f = \sqrt{\left(\frac{W_L}{\rho_L f_L \pi} + r_f^2\right)} - r_f = \sqrt{\left(\frac{W_L \rho_{FM} r_f^2}{\rho_L W_{DF}} + r_f^2\right)} - r_f \tag{11.2}$$

where r_E and r_f are, respectively, the equivalent film radius (taken as the sum of fiber and film radii) and fiber radius, W_L is the weight of the liquid remaining on the filter after drainage, and ρ_L is the density of the liquid. This is a very important parameter for prediction of wettable filter efficiency, which should be used for correction of the fiber radius for more realistic determination of the efficiency of the filter proposed for purification of liquid aerosol streams under particular operational conditions. Obviously, especially for cases when highly viscous liquids are involved, this correction could be quite significant.

As one can understand, the hydrodynamic filter resistance is also strongly influenced by the presence of the coating film. Kuwabara [12] derived the following equation that is commonly used for the evaluation of filter resistance:

$$\Delta P = \frac{4\mu c V H}{d_f^2(c - \frac{3}{4} - \frac{1}{2}\ln c - \frac{1}{4}c^2)} \tag{11.3}$$

where H is the thickness of the filter, μ is the dynamic viscosity of the gas, V is the filter face velocity of the aerosol carrier, d_f is the fiber diameter, and c is the filter packing density (ratio of the total volume of fibers to the volume of the filter). In the case of liquid aerosol filtration, the fiber diameter and correspondingly the filter packing density ought to be corrected by the value of the equivalent film thickness.

The equilibrium film should only be considered in cases of immediate drainage of the liquid, which is relevant for specially designed experiments under controlled laboratory conditions. In real situations, the film thickness always exceeds the equilibrium value, as drainage is not an instantaneous process. In this case, the actual operational film thickness ought to be considered. A procedure allowing evaluation of the actual thickness of the film is based on the filter and process parameters and the aspect ratio of the filter. The equation for estimation of the actual film thickness can be written as [10]

$$\delta(y) = \sqrt[3]{\frac{3G_y\mu}{\rho_L^2 g}} \tag{11.4}$$

where $\delta(y)$ is the film thickness on fibers at filter vertical elevation y, μ is the liquid viscosity, ρ_L is the liquid density, g is gravity, and G_y is the total liquid supply at the filter cross-section at height y. The parameter G_y (kg/m s) is the total amount of liquid passing through 1 m of the fiber perimeter at elevation y per second. It is estimated as

$$G_y = \frac{H - y}{H} C_a G_g E_f \tag{11.5}$$

where H is the total height of the filter, C_a is the aerosol concentration at the filter inlet, G_g is the gas flow rate, and E_f is the filter efficiency. The perimeter of fibers at any cross-section of the filter can be taken as

$$P_f = \frac{2WTc_p}{r_f} \tag{11.6}$$

where W and T are width and thickness of the filter, respectively, and c_p is the filter packing density (filter volume fraction occupied by fibers).

As one can understand, designating $y = H$ (top cross-section of the filter) makes the film thickness equal to zero, as no liquid is supplied at that plane. In contrast, all liquid collected by the filter passes through the plane at height $y = 0$, creating a film with the maximum thickness. On this basis, compared to solid particle filtration, the filter height has become a crucial parameter for liquid aerosol removal, because if the design is inappropriate, some lower part of the filter could become blocked

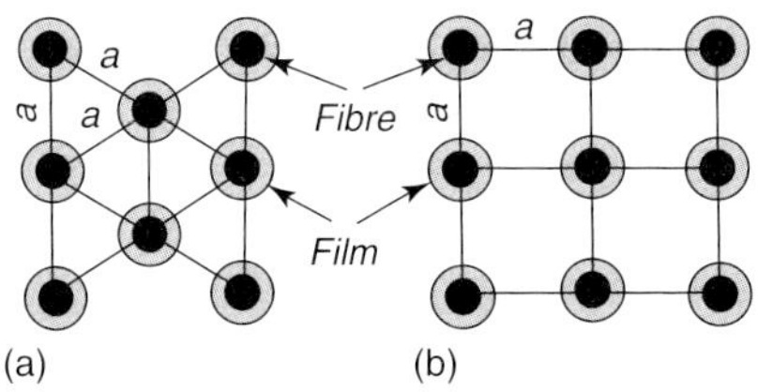

Figure 11.3 Wettable fiber model concept: (a) triangular and (b) rectangular (square) fiber arrangements.

due to the coalescence of films covering neighboring fibers. To avoid filter blockage, some model of the fiber arrangement should be assumed. Most commonly, either triangular or rectangular arrangements are used (see Figure 11.3). Then, based on the selection made, the maximum thickness of the film, to ensure that neighboring films do not coalesce, is, for the triangular model (Figure 11.3a)

$$\delta_{\max} = r_{\mathrm{f}} \left(\sqrt{\frac{2\pi}{c_{\mathrm{p}}\sqrt{3}}} - 2 \right) \tag{11.7}$$

and for the rectangular arrangement (Figure 11.3b)

$$\delta_{\max} = r_{\mathrm{f}} \left(\sqrt{\frac{\pi}{c_{\mathrm{p}}}} - 2 \right) \tag{11.8}$$

It should be noted that the triangular arrangement gives a result that is around 10% higher compared to the rectangular one, so, to be on the safe side, the triangular model of the filter structure is recommended. It should also be noted that the model is based on the assumption that all the fibers are vertical. However, this is certainly not the case for most of the filters used in industry – fibers in real filters are usually touching each other, which enables liquid flow between the fibers, making liquid drainage occur relatively vertically.

Based on the above consideration, the filter aspect ratio (width by height) is one of the main design parameters for the development of wettable filters for liquid aerosol removal [10]. As one can understand, increasing the aspect ratio leads toward a decrease of the film thickness along the filter, minimizing the possibility of its clogging. In addition, variable film thickness along the filter height is responsible for alteration of the hydrodynamic resistance at different vertical points of the medium. Obviously, this phenomenon is responsible for a possible variation of the filter efficiency along the height. This fact has to be taken into account at the stage of filter design. Using Eqs. (11.1)–(11.8) allows filter parameter optimization to avoid undesirable liquid-related non-uniformity effects during operation.

11.2.2 Practical Aspects

Besides liquid aerosol removal, wettable filters can also be used for the removal of solid aerosol particles. These filters do not require any additional regeneration modules if used for liquid aerosol removal. They operate in self-cleaning mode [10]. However, in some industrial processes, liquid particles are accompanied by solid ones. Usually, the solid particles are also washed off efficiently from the filter; however, in some cases, due to insufficient filter irrigation or very high concentration of solids in the air carrier, some blockage of wet filters can occur. In these cases, some additional artificial filter irrigation is required to create a film with appropriate thickness covering the fibers. Such an arrangement has a number of substantial benefits:

1) Artificial irrigation creates self-cleaning regeneration, eliminating the need for any additional regenerating modules [13, 14].
2) The technology allows one to handle viscous sticky particles, for which conventional baghouse filters are usually of limited use due to the high possibility of clogging as a result of their limited regeneration capability. In this case, new incoming particles are captured not by the solid fiber but by the liquid film and can be simply removed by continuous irrigation of the filter [9].
3) The liquid film can significantly reduce particle bouncing, enhancing the filtration efficiency of the medium [15].
4) Using disinfecting liquids could inactivate *biological microorganisms*, minimizing health risks in cases of re-entrainment of *bioaerosols* collected by the filter [16, 17].
5) In cases of limited availability of irrigating liquid due to some technological reasons, the thickness of the film can also be controlled by the filter inclination [18]. In this situation, the efficiency of liquid drainage from the filter is decreased, due to the fact that only a component of the force of gravity acts along the inclined fibers. Practically it is even possible to design filters with adjustable inclination, enabling one to fine-tune the inclination angle directly on site.

Removal of solid particles by liquid-coated fibers can mainly be performed by wettable filters [13]. When the fiber is coated by a liquid, the film surface becomes the primary filtration zone: particles are collected directly onto the liquid and are removed by continuous irrigation. This process minimizes direct contact between the particles and solid fibers, decreasing the possibility of filter blockage. Figure 11.4a shows a wettable filter involved in the process of solid dust particle removal. For comparison, Figure 11.4b shows a non-wettable filter operated under the same process parameters. As can be seen, the wettable filter is able to work in self-cleaning mode, with no evidence of particle accumulation at the surface. On the other hand, the non-wettable filter is entirely clogged by solid dust, and the irrigating liquid flows along the cake, providing minimal regeneration (a liquid jet is clearly seen in the picture).

Unfortunately, the theoretical evaluation of the minimal irrigation rate for a particular filter and process parameters is a very complicated task, which would lead to very limited practical outcomes. As suggested by Agranovski and Shapiro [13], some simple practical experiments can be undertaken to evaluate the minimal irrigation rates for a range of dust loadings. To run the experiment, various quantities of dust are supplied to the filter by the air carrier. Filter irrigation is performed through the nozzle, strategically located to ensure that the liquid efficiently and uniformly covers the filter surface. At the beginning of

(a)
(b)

Figure 11.4 Surface of filters with dust deposits: (a) non-wettable and (b) wettable.

the experiment, a certain dust supply rate is selected and the nozzle operates at the maximum liquid flow rate, which is then gradually decreased down to the point when the filter resistance starts to grow. The liquid flow rate at this point is considered as minimal for that particular dust loading. Obviously, the filter's ability to work at constant resistance indicates that collected dust particles are efficiently "washed off" and do not form any deposit on the filter surface.

This procedure would be even more representative if the filter could be installed in the real industrial process of interest. In this case, a bypass line with filter and nozzle assembly is attached to an exhaust pipe of the industrial process under investigation. A vacuum pump sucks some known quantity of the exhaust gas through the filter irrigated by the nozzle, and the liquid flow is gradually decreased down to the point when the filter resistance starts to grow. The minimal irrigation rate is precisely identified and could be used for design of the scaled version of the filter capable of treating the entire quantity of the exhaust gas.

This procedure was employed to design a filtration technology for the purification of the exhaust gas from galvanizing technology [19]. The sticky nature of the particles produced at the point of contact of chemically pre-treated steel parts with molten zinc eliminated any possibility to use conventional baghouse filters, as filter regeneration by the commonly used shaking and reverse air jet methods was very inefficient, leading to rapid filter clogging. To treat the exhaust gas, low-efficiency wet scrubbers are commonly used in this industry. Of course, a properly designed and operated high-pressure Venturi scrubber can achieve relatively high removal rates for submicrometer particles; however, the resistance of the device in these cases could exceed 30 kPa [1].

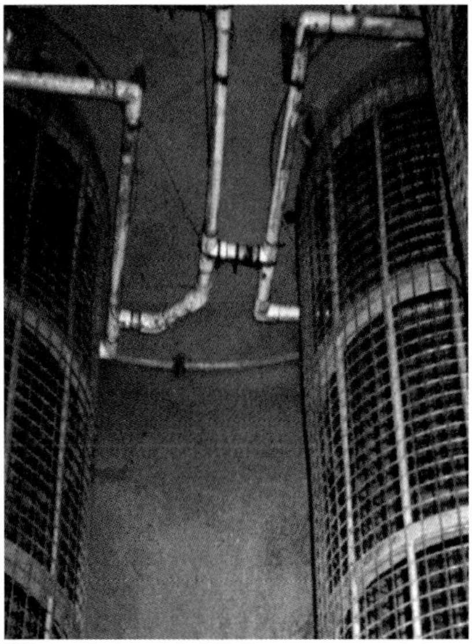

Figure 11.5 Photograph of the filter elements and irrigation system.

A technology involving a wettable filter and spray nozzles strategically installed at the top of the device to provide uniform filter irrigation during the entire operation of the device was proposed (see Figure 11.5). It provided a stable operation that was free of clogging and achieved an efficiency above 98% for the removal of submicrometer particles from the air carrier. In addition, as one can understand, using liquid irrigation provides a very efficient tool for simultaneous removal of gaseous pollutants. Small fiber and, correspondingly, film diameters create a large total area of *gas–liquid interface*, ensuring favorable conditions for mass transfer and gas absorption.

Many industries produce relatively dry and hot exhaust gases, which can evaporate a substantial quantity of the liquid from the surface of the filter coating film, leaving some areas of the filter uncoated. To avoid any direct particle contact with the solid fiber, some additional irrigation might be required. A simple model to evaluate the rate of evaporation for a range of process conditions has been suggested [20]. The model takes into account the physical parameters of the gas and liquid phases, along with the filter structure, to estimate the rate of evaporation of liquid from unit surface area of the filter. As an example, Figure 11.6 shows some results of calculations obtained for three filters with the parameters listed in Table 11.1.

Calculations were done for an air face velocity of 0.5 m/s and for three vapor contents of 10, 20, and 30 g of vapor per kilogram of dry air. As is seen, the amounts of water evaporated from 1 m^2 of the filter are quite substantial, especially in a region where the air becomes hot and dry. Also, owing to the more developed area of the gas–liquid interface, more efficient evaporation is observed from filters with smaller fiber sizes. The moisture content of the air carrier is probably the most influential parameter of the process, with a clear and expected trend toward higher

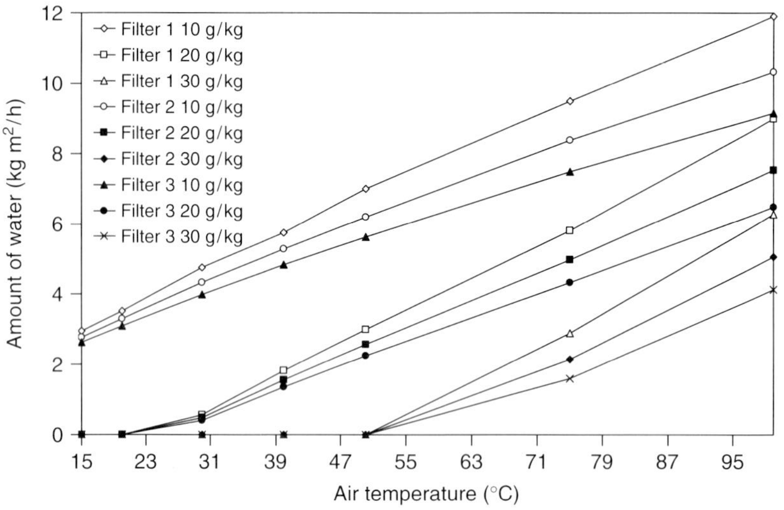

Figure 11.6 Amount of water evaporated from 1 m^2 of the filter by the air carrier depending on the range of temperature and the humidity.

Table 11.1 The parameters of the three filters used in Figure 11.6.

Filter no.	Material	Thickness (mm)	Fiber size (μm)	Specific weight (g/m^2)	Packing density (%)
1	Polyester	3	12	330	12
2	Polyester	3	24	327	12
3	Polyester	3	36	330	12

evaporation rates for drier streams observed. To account for evaporative liquid losses, corresponding amounts of liquid make-up have to be added to the minimal irrigation flow rate evaluated by the procedure discussed above.

A very important characteristic of the liquid film coating the fibers is its ability to minimize particle bouncing after collision of an aerosol particle with the filter surface [20]. There are two classical approaches to describe *particle bounce*. The first defines a critical velocity V_c above which bounce will occur [7],

$$V_c = \frac{\beta}{d_p} = \frac{1}{d_p}\frac{(1 - e_{pl}^2)}{e_{pl}^2}\frac{A}{\pi x^2 (6p_{pl}\rho)^{1/2}} \tag{11.9}$$

where β is a constant for a particular impaction surface, d_p is the particle diameter, e_{pl} is the *coefficient of restitution* (for plastic deformation only), A is the Hamaker constant [21], e_{pl} is the microscopic yield pressure, ρ is the particle density, and x is the distance of separation between the center of mass of the particle and the surface. Hamaker constants are given in the literature for a limited number of elements and compounds [22].

The other method involves the kinetic energy (KE_b) required for bounce to occur when a particle collides with a surface [23–26], the magnitude of which can be calculated as

$$KE_b = \frac{d_p A(1 - e^2)}{2xe^2} \tag{11.10}$$

where e is the coefficient of restitution (plastic and elastic deformation), which is equal to the ratio of the rebound velocity to the approach velocity. The value of e is reported to range from 0.73 to 0.81, although these values were derived using hard impaction surfaces [27]. It is reported that A and e ought to be determined experimentally because it is very hard to determine them theoretically.

For liquid particles, as the particle size diminishes, the surface tension force increases to the point where the particle is so rigid that its behavior is identical to that of a solid particle. Liquid particles are better able to absorb the energy of collision than are solid particles, principally through deformation [7]. It is reported that liquid particles are far more likely to break up on impact than are solid particles. This is principally through shear when the particle impact is not normal to the fiber [28].

Some research into possible particle bounce from liquid-coated filter fibers has been undertaken. Walkenhorst [29] examined the effect of coating model wire filters with vegetable oils, mineral oils, and petroleum jelly (Vaseline), and reported that the former two substances increased the filtration efficiency (particle adhesion) whereas the latter did not. Although oil-coated filters are suitable for laboratory-scale processes, using such liquids in industry would not be advantageous. On the other hand, the technology based on water coating of fibers is feasible for industrial applications, as water for filter irrigation is more readily available and recyclable in industry than is oil [19].

Mullins *et al.* [15] and Boskovic *et al.* [30] experimentally verified that coating the filter could substantially decrease particle bounce after initial collision with the fiber and could minimize any particle motion along the fiber. The experimental procedure consisted of two stages. In the first stage, solid polystyrene latex (PSL) spheres and a liquid aerosol of diethylhexyl sebacate (DEHS) were collected by a dry filter at identical process parameters. In the second stage, the filter fibers were coated by liquid and the experimental procedure used in the first stage was repeated for the coated fiber scenario. As was reported, the removal efficiency of the dry filter was significantly higher for the collection of liquid DEHS particles. Considering that all the process parameters were identical, this difference could only be related to the nature of the aerosol: the kinetic energy that is mainly responsible for particle bounce is absorbed more efficiently by liquid aerosols through particle deformation. This assumption was verified in the second stage of the experimental program when the filter fibers were coated by a liquid. There was negligible difference between the removal of solid and liquid particles, which allowed the conclusion to be drawn that the liquid coating acts to absorb the kinetic energy, minimizing or even eliminating any particle bounce after collision with the fiber. Various liquids, including oils and water, were used, and the effect was always observed, allowing one to assume that a liquid coating can act as an efficient absorber of the energy after impact, minimizing particle bouncing, and correspondingly enhancing filter efficiency, especially in cases of solid particle removal.

11.2.3 Inactivation of Bioaerosols on Fibers Coated by a Disinfectant

Microbial aerosols can cause various human and animal diseases, and their control is becoming a significant scientific and technological topic for consideration. Filtration is considered to be one of the main processes for the removal of biological aerosols from the air, minimizing their concentration in industrial and domestic dwellings. However, the removal of bioaerosol from air carriers does not, on its own, solve the problem of microbial contamination of the ambient air. Considering that in some situations the bioaerosol particles collected on the filter could re-enter back into the air carrier, some disinfection is also required to ensure that no biologically active particles can possibly detach from the filter surface and return to the human-occupied areas. Moreover, in many cases, the dust simultaneously removed by the filter from the air contains microbial nutrients sufficient for colonial growth, which is especially relevant in areas where the air humidity is relatively high.

Liquid-coated wettable filters can be considered as the technology to inactivate the particles that collect or grow on the filter. For this purpose, the fibers could be coated with some disinfecting liquid capable of inactivating the collected microorganisms, ensuring that, even in cases of some re-entrainment, live biological particles would not reach areas occupied by humans and animals.

One suitable candidate for this role – *Melaleuca alternifolia* (tea tree) oil, widely used as a topical antiseptic – was employed as the coating liquid for the inactivation of environmental microbes, including viruses, bacteria [16], and fungi [17], collected on the filter surface. Figure 11.7 shows some results for the inactivation of various microbial strains on tea tree oil-coated fibers. As is seen, the lowest (worst) inactivation capability of tea tree oil was demonstrated on treatment of *Aspergillus versicolor* and *Rhizopus* fungal spores, with almost 50% survival after 60 minutes of treatment. Rapid inactivation was observed for bacterial strains, especially for stress-sensitive *Escherichia coli* and *Pseudomonas fluorescens*. Even the robust bacterium *Bacillus subtilis* var. *niger* lost 90% of viable particles after 15 minutes of treatment and was entirely disinfected within 30 minutes of the experimental run. However, the best performance of the technology was achieved on treatment of *viral aerosols* represented by the *Influenza* virus. All the collected virus particles are inactivated almost immediately, which is especially important considering that the pathogenic nature of these microbes is the highest out of all the bioaerosols.

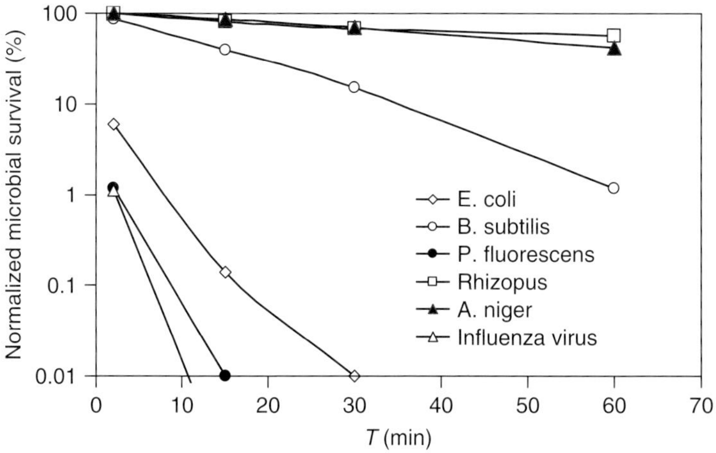

Figure 11.7 Inactivation of biological aerosols on a filter coated with tea tree oil.

11.3 Non-Wettable Filtration Materials

11.3.1 Theoretical Aspects

Compared to wettable filters, liquid aerosol particles collected on non-wettable filters do not spread along the fibers. They remain as spherical objects and grow due to coalescence with new incoming particles up to a point when the force of gravity exceeds the force of adhesion and the droplet drains from the filter surface. A microscopic photograph of droplets on a non-wettable model filter is shown in Figure 11.8. It can be seen that droplets are suspended on multiple neighboring fibers, significantly changing the filter's physical parameters and correspondingly its hydrodynamic and performance characteristics.

As observed during experiments, the droplets grow on the surface and immediately before detachment undergo some oscillations. Such oscillations

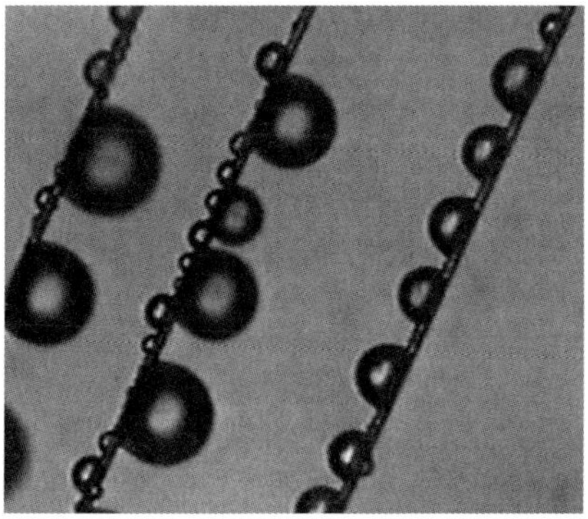

Figure 11.8 Non-wettable model fibers.

enhance the drainage efficiency, adding extra force to overcome adhesion. To model the non-wettable filter, the size of the droplet on the fibers becomes crucial to characterize the filter parameters. In addition, the maximum size of droplet surviving on the filter surface depends on the gas velocity and liquid physico-chemical properties. Some laboratory experiments could be performed to find the maximum droplet size on the filter surface for particular process and filter parameters. The easiest way is to photograph the face surface of the filter during operation, with subsequent analysis of the image with the aid of magnifying equipment. To take such a photograph, one of the walls of the filter casing needs to be transparent.

A different experiment was described by Agranovski and Braddock [11] to evaluate the droplet size. They used a cathetometer to remotely measure the droplet size immediately before detachment. To perform the measurements, the cathetometer was pointed to the face filter surface through a transparent window in the filter holder. Some random growing droplet on the filter surface was chosen and observed up to the moment of commencement of its oscillations and subsequent detachment. At that stage, the size of the droplet was recorded. The results of these measurements are presented in Figure 11.9. As may be seen, the maximum droplet size coincides with the boundary of the laminar flow regime, explaining the droplet oscillations, which are initiated by the chaotic nature of the gas streamlines. The droplet oscillations were comprehensively investigated by Mullins and his colleagues [31, 32]. The experimental results were used to develop some mathematical relations describing oscillation patterns of clamshell- and barrel-shaped droplets.

Considering the random placement and size of the droplets on the filter surface during liquid aerosol removal processes, modeling non-wettable filters is a more challenging task than modeling wettable ones, where the films are usually quite stable during steady-state operation. Here we discuss a semi-empirical procedure that allows the filter performance for liquid aerosol removal to be predicted.

Let us first assume that some part of the filter is blocked by liquid, and further assume that each drop with diameter D_d is located in an imaginary cubic cell with side length D_dX. Here X is the scale factor, which represents the ratio of the side length of the cell by the diameter of the drop D_d. The size of the cell is taken in

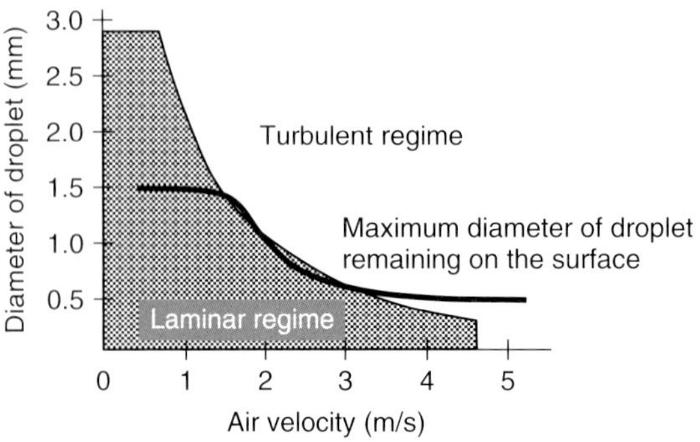

Figure 11.9 Maximum diameter of droplet on the filter surface at different air velocities.

such a way that the ratio $\pi D_d^2/(D_d X)^2$ is equal to the ratio of the area of the filter blocked by liquid by the total area of the filter. The total surface area of fibers in the liquid free cell space can be calculated if the packing density of the filter is known. The packing density of the filter is the ratio of the volume of the fibers by the total volume of the filter, that is,

$$c = \frac{\sum V_{\text{fiber}}}{V_{\text{cell}}} \tag{11.11}$$

where $\sum V_{\text{fiber}}$ is the total volume of the fibers in the cell. The total volume of the fibers may also be represented in the form

$$\sum V_{\text{fiber}} = \frac{\pi d_f^2}{4} L_c \tag{11.12}$$

where L_c is the total length of the fibers in the cell. However, $S_c = d_f L_c$ is the projection of the total front cross-sectional area of the fibers in the cell, so we can rewrite Eq. (11.12) as

$$\sum V_{\text{fiber}} = \frac{\pi S_c d_f}{4} \tag{11.13}$$

But S_c can also be represented as

$$S_c = L_c d_f = \left(\frac{X^3 D_d^3 c}{\pi d_f^2/4}\right) d_f = \frac{4X^3 D_d^3 c}{\pi d_f} \tag{11.14}$$

Some amount of the fiber is located inside the liquid drop, and is inaccessible for the aerosol stream, which can only flow around the drop. To calculate the front cross-sectional area of the fibers in the drop, we can use the following expression:

$$S_{\text{fibre in drop}} = \left(\frac{(\pi D_d^3/6) \times c}{\pi d_f^2/4}\right) d_f = \frac{2D_d^3 c}{3d_f} \tag{11.15}$$

Now we introduce a new parameter S_e, the equivalent surface area of the filter, which represents the sum of the frontal area of the fibers and the droplet in the cell exposed to the gas stream, that is, the area where the filtration process occurs. This parameter is calculated as the total front cross-sectional area of the fibers in the cell plus the cross-sectional area of the drop and minus the front cross-sectional area of the fibers located in the drop:

$$S_e = \frac{4X^3 D_d^3 c}{\pi d_f} + \frac{\pi D_d^2}{4} - \frac{2D_d^3 c}{3d_f} \tag{11.16}$$

Using S_e allows the evaluation of an equivalent filter packing density, c_e, which represents a realistic combined packing density of the fibers and liquid deposits on the filter:

$$c_e = \frac{4X^3 D_d c + 0.785\pi d_f - 0.66\pi D_d c}{4X^3 D_d} \tag{11.17}$$

Analysis of Eq. (11.17) suggests a very interesting conclusion: the equivalent packing density could be larger or smaller than the packing density of the dry filter depending on the filter parameters, liquid loading, and droplet size (relates to the filtration velocity). Considering that filtration efficiency is proportional to the packing density, the filter performance could also vary from lower to higher performance as compared to the dry filter scenario.

11.3.2
Practical Aspects of Non-Wettable Filter Design

Considering that a number of unknowns are involved in the suggested model, in order to apply it to a non-wettable industrial filter design, some prior laboratory investigations are required. Two major parameters ought to be obtained for the particular filter and liquid of interest for a range of gas velocities: area of filter blocked by liquid, and maximum size of droplet capable of remaining on the filter. Considering a linear relation between the gas velocity and the pressure drop, the area of the filter blocked by water could be evaluated by comparison of the dry filter resistance against the resistance of the filter acquired during steady-state liquid aerosol removal. In particular, the liquid of interest is aerosolized and passed through the filter. At the beginning of the process, the pressure grows due to liquid accumulation on the dry filter up to the point when the amounts of liquid supply and drainage become equal, after which no further pressure increase is observed.

Interestingly, the concentration of the liquid aerosol could substantially alter the pressure drop only for low-concentration streams. In contrast, for highly concentrated gas carriers, even significant alterations of the aerosol content cause only minor variations in the pressure drop. Theoretically, this might not be observed for highly viscous liquids – however, even experiments with relatively viscous oils verified its validity.

As discussed before, the easiest way to acquire information about the maximum droplet size sustained on the filter is to take photographs of the filter face under steady-state process conditions and to measure the droplet sizes on the images. The photographs ought to be taken through a transparent window strategically made in the filter holder to ensure maximum possible filter area coverage. The final product of this procedure is a graph similar to that obtained for water droplets (Figure 11.9) by Agranovski and Braddock [11]. Using the results of these experiments allows one to estimate the equivalent packing density by Eq. (11.17) and more realistically to predict the filter efficiency and resistance for a particular application.

11.4
Filtration on a Porous Medium Submerged into a Liquid

11.4.1
Theoretical Approach

A special case of the removal of an aerosol by a wet filter is based on passing a gas stream through a porous medium submerged into a liquid. It was shown [33–35] that aerosol particles can be removed efficiently by such bubbling devices, and the efficiency of filtration does not depend significantly on the height of the liquid layer above the filter (obviously, it ought to be ensured that the porous medium is fully submerged). The low dependence on the height of the liquid layer suggests that the majority of particles are being removed from the gas during passage inside the porous medium. Figure 11.10 illustrates the bubbling process at various gas flow rates. As can be seen, altering the gas flow rate does not significantly increase the bubble size, increasing only the number of bubble release points at the filter surface.

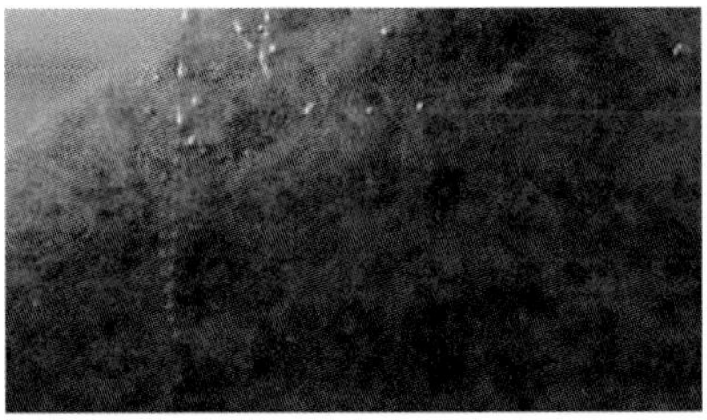

(a)

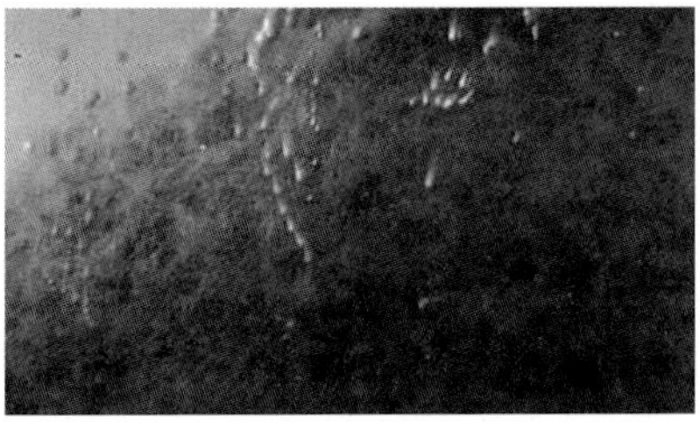

(b)

(c)

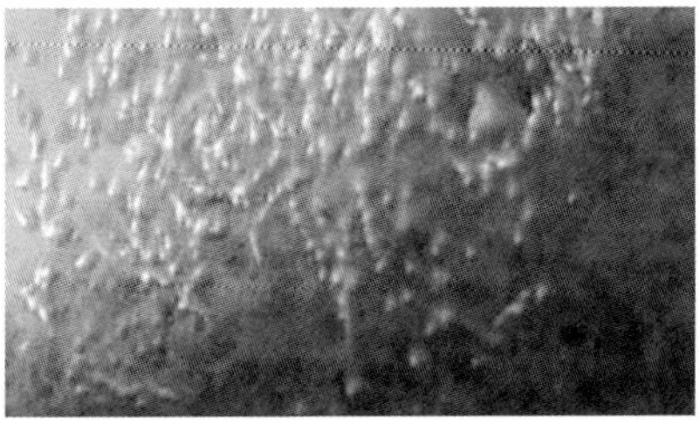

(d)

Figure 11.10 Bubbling of the air through the filter submerged into liquid: (a) to (d) show the air flow rates 0.01, 0.03, 0.1 and 0.5 m^3/m^2 s.

Particle removal occurs when the gas carrier is split into a number of bubbles or jets at the entry to the porous medium and travels along narrow and tortuous paths inside the filter. Using wettable fibers ensures that no direct contact of the gas with the solid fiber occurs, eliminating any possibility of settlement of particles directly on the fiber, causing blockages. In contrast, the particles are removed by the liquid coating of the submerged filter and remain in the liquid, enabling the technology to run in self-cleaning mode. For heavily polluted industrial streams, some fresh liquid make-up could be used to control the particle concentration in the collecting liquid. However, for low-concentration aerosol streams, the technology could run for extended time periods with no make-up liquid required.

The following model is commonly used to evaluate the particle removal efficiency on a bubble wall. During the upward motion of a gas bubble through a liquid, air circulation proceeds inside the bubble [5]. For small bubbles with a Reynolds number, Re, between 1 and 700, the velocity of the circulating gas at the surface of the bubble, U_τ, is given approximately by

$$U_\tau = 1.5 V_b \sin\theta \tag{11.18}$$

where V_b is the velocity of rise of the bubble and θ is a latitude angle measured from zero at the direction of rise.

The aerosol particles are considerably smaller than the bubbles and the interception effect of filtration can be neglected. In this case, the absorption of aerosols in bubbling is mainly due to inertial deposition and diffusion [5]. Commonly, the efficiency of inertial deposition, η_i, of spherical particles in the *Stokes law* region is estimated as the ratio of the number of particles deposited in the filter, to the total concentration of particles before filtration. Thus

$$\eta_i = \frac{3 V_b d_p^2 \rho_p C}{500\ \mu_g d_b^2} \tag{11.19}$$

where d_p is the diameter of the particle, μ_g the dynamic viscosity of the gas, ρ_p the density of the particle, d_b the diameter of the bubble, and C the *Cunningham correction* coefficient. For the efficiency of diffusion deposition, η_D, Fuchs [5] showed that

$$\eta_D = 0.03\sqrt{\frac{D}{V_b d_b^3}} \tag{11.20}$$

where D is the particle diffusion coefficient. The above equations for the estimation of η_i and η_D are based on a distance of 1 cm travelled by the bubble.

One of the main benefits of the suggested arrangement is related to the fact that the presence of liquid along with a substantial area of gas–liquid interface during passage of the gas through narrow channels makes the technology very efficient for the simultaneous removal of particulate and gaseous pollutants for a range of industrial and domestic applications. The easiest way to use the suggested technology is to retrofit it into some existing plate scrubber technology by submerging a porous medium with required characteristics into an irrigation liquid on the top of a heat–mass exchange sieve plate. A rigid restricting metal mesh ought to be installed above the porous medium to ensure that no medium displacement can occur.

To predict the process efficiency, the gas flow in porous media ought to be characterized. Considering that the process is internal, no direct visualization is possible, so more sophisticated instrumental techniques are required. One good candidate able to visualize the process is nuclear magnetic resonance (NMR) [36, 37]. Considering the dynamic nature of the process, the following procedure could be used (see Figure 11.11). A filter is fixed stationary in a hermetic plastic cylindrical capsule (the material of the capsule has to be chosen to be transparent for NMR) with an external diameter suitable to fit in the core of an available NMR imaging machine. The air is supplied by a pump through the bottom of the capsule and bubbles through the wet porous medium securely fixed in the capsule to ensure no displacement throughout the entire run. The water level above the filter is maintained by a syringe and controlled by a high-resolution differential manometer. Any evaporative losses, detected by a decrease in the pressure drop across the medium–liquid system, are made up by fresh water from the syringe to ensure the stationary process conditions required for 3–4 hours to enable NMR scanning of the entire filter in the vertical direction with the minimal possible step. A perforated plate with a small opened area (usually 3–10%) is located below the porous medium to ensure that there are no liquid losses due to fall under gravity. However, for filters with relatively high packing density, liquid remains on the surface due to the pressure gradient and no additional perforated plate is required.

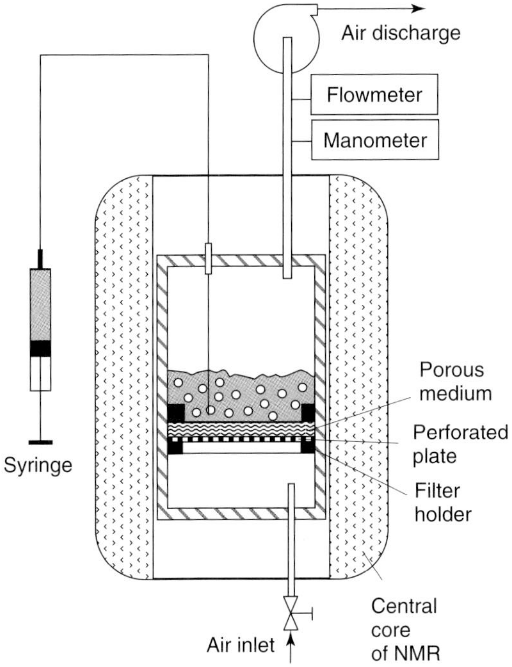

Figure 11.11 Use of NMR for visualization of the process.

Figure 11.12 shows an example of the sequential NMR images made for a 6 mm thick polypropylene porous medium consisting of 12 μm fibers packed with the density of 11%. A series of horizontal images was taken along the filter, with a vertical resolution of 375 μm between them. The black areas in the figure represent air and air pathways (transparent for NMR), while the white areas represent water – the shades of gray represent mixtures of air and water. Figure 11.13 shows a vertical slice through the same filter oriented along the axis of the larger pathway identified in Figure 11.12. The presence of an air pathway throughout the filter is clearly seen in the images. This also indicates some blind pockets of air in other parts of the filter. These are maintained by air pressure and by surface tension at the interface, which is supported by the polypropylene fibers transparent for the NMR sequencing.

Using the results of NMR imaging allows one to trace all the air channels and to calculate their total area and consequently the gas velocity. Within realistic conditions of the process, the air flow is essentially laminar and inviscid, and flow paths through the filter have liquid walls formed by a balance between surface tension and pressure. The measured pressure drops across the filter are small; as a consequence, the air flow can be assumed to be incompressible, or at constant

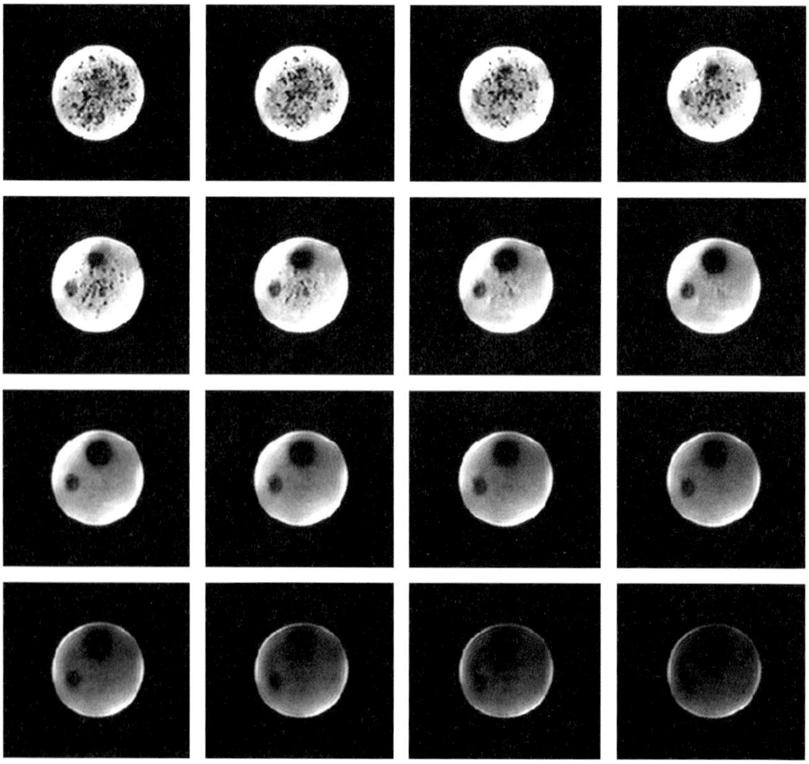

Figure 11.12 Sequential horizontal NMR images across the porous medium.

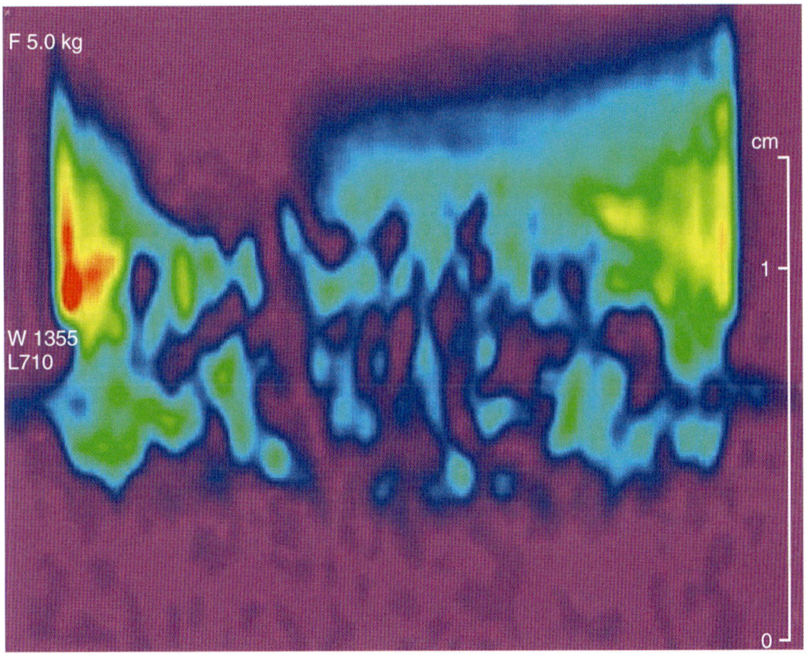

Figure 11.13 Vertical NMR image across the porous medium.

density. Finally, the flow from the air pump into the lower part of the filter cylinder is steady and assessed to be irrotational. With these assumptions, the flow is ideal and

$$\boldsymbol{u} = -\nabla\phi \tag{11.21}$$

where φ is the potential function, $\boldsymbol{u}$ is the velocity vector, and ∇ is the gradient operator. The continuity equation then yields

$$\nabla^2\phi = 0 \tag{11.22}$$

which is the *Laplace equation*. An ideal two-dimensional flow also leads to a stream function ψ that also satisfies the Laplace equation when vorticity is not generated in the flow field. The boundary conditions for inviscid flow are given by ψ constant on the boundaries.

Now, the outline of the air passes obtained by NMR is digitized and assumed to have cylindrical symmetry. The stream function is set to zero on the left-hand boundary and to unity on the right-hand boundary of the air tube. This represents a normalized total air flow of unity up the cylinder, which may be easily scaled. The air flow at the ends of the cylinder is assumed to be uniform. The Laplace equation (Eq. (11.22)) can now be solved numerically. Figure 11.14 shows solutions for the streamlines given by $\psi = c$(constant), with $0 \leq c \leq 1$. There is significant curvature of the streamlines as the air flow adjusts to enter the air tube, with the highest curvatures being in the mouth of the air tube. The regions of curvature of

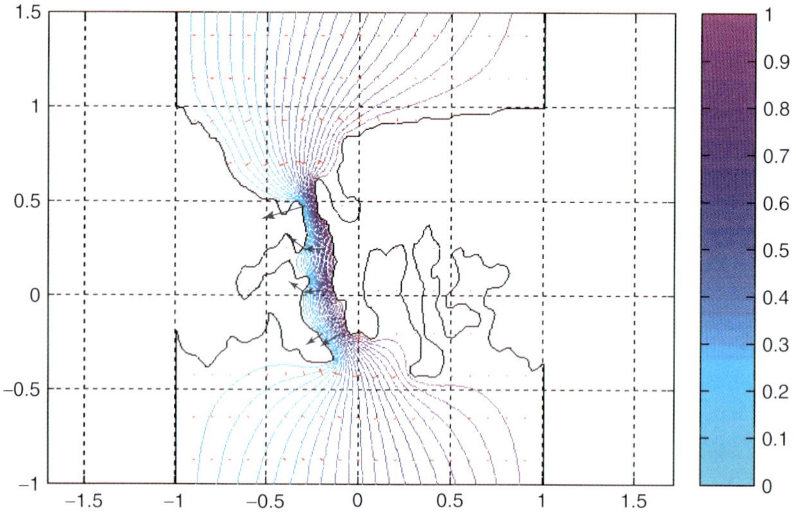

Figure 11.14 Model of gas flow inside the porous medium.

the streamlines on entry to the air tube represent areas where inertial effects will cause deposition of particles on the wet sections of the filter. These regions provide the highest *capture rates* of aerosol particles in the filter.

The deposition of aerosol particles in this flow field can now be predicted. Considering the relatively high gas velocity and short residence time inside the porous medium, the inertial mechanism of particle removal dominates in this process. For the laminar flow regime, the equation of motion for the particles is

$$m\ddot{\boldsymbol{r}} = -m\boldsymbol{g} + F\left(\boldsymbol{u}(\boldsymbol{r}) - \dot{\boldsymbol{r}}\right) \tag{11.23}$$

where $\boldsymbol{r}$ is the position of a particle in the flow field at time t specified by $\boldsymbol{u}(\boldsymbol{r})$, m is the mass of the particle, $\boldsymbol{g}$ is gravity, and F is the drag coefficient. We assume that the particles are spherical, so that $F = 24/R_{\mathrm{p}}$.

The particle trajectory equation (Eq. (11.23)) is an ordinary differential equation, which can be solved numerically using the results obtained from solving the Laplace equation (Eq. (11.22)). The boundary conditions correspond to the release of individual particles at representative points at the edge of the flow field, that is, along the bottom boundary of the filter. At release, the particles can be assumed to be moving with the velocity of the flow. We can also assume that each particle that touches the wall of the air path is removed from the air carrier. Note that the walls of the air pathway are formed by liquid surfaces held in place by the fibers of the filter, so bouncing and re-aerosolization effects are negligibly small and do not need to be taken into account.

For verification of the approach, Eq. (11.23) was solved for particles uniformly distributed across the filter entry point, and the calculations assumed that the particles did not interact. The resulting particle trajectories were followed until the particle hits the liquid boundary shown in Figure 11.14. A few series of calculations

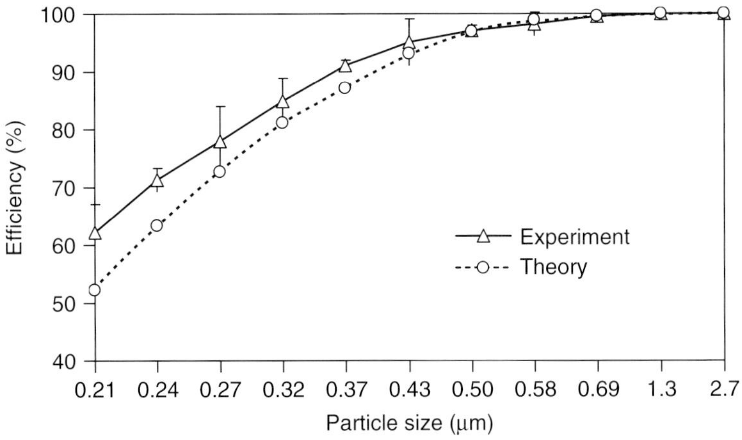

Figure 11.15 Removal of aerosol by the bubbling process.

were made for a number of particle sizes, and the results of these calculations are presented in Figure 11.15. The experimental curve obtained for the air face velocity of 0.3 m/s (the same as that used for calculating the theoretical results) is also presented in Figure 11.15. As is seen, the agreement between theoretical and experimental results is satisfactory. Some discrepancy between the results could be explained by the fact that, for theoretical modeling, fibers located in the air channels are not taken into account, but they could also contribute toward the removal of particles. Also, these calculations are used solely for illustration of the approach and do not include other collection mechanisms, such as interception and diffusion, which could also contribute slightly to the collection efficiency values.

11.4.2
Application of the Technique for Viable Bioaerosol Monitoring

Bioaerosols in occupational and residential environments are known to cause various health effects, which include (with examples in brackets) infections (acute viral infection, legionellosis, tuberculosis), hypersensitivity (allergies, allergic rhinitis, asthma, hypersensitivity pneumonitis), toxic reactions (humidifier fever, stachybotryotoxicosis), irritations (sore throat), and inflammatory responses (sinusitis, conjunctivitis). The growing concern for human exposure to bioaerosols has created a demand for advanced, more reliable, and more efficient monitoring methods.

Most of the currently available bioaerosol sampling devices are based on dry filtration, impaction onto agar, or impingement into liquid. Owing to the strong desiccation effects and corresponding low microbial recovery rates associated with dry filtration, this method is not recommended for elongated viable bioaerosol monitoring procedures and can mainly be employed for total microorganism enumeration by microscopic research methods (see Chapter 12).

The selectivity of methods based on direct collection of airborne microbes onto agar makes them of limited use for comprehensive monitoring of ambient air – some species may not turn into a culture, as different microorganisms require different types of nutrients. Also, for unknown concentrations of bacteria in the air, the agar plate may become overloaded, which reduces the accuracy of the subsequent colony count or makes counting impossible.

Direct collection of viable aerosol particles into liquid has always been a preferred method of monitoring, as it allows the application of various analytical procedures to obtain the most comprehensive information both qualitatively and quantitatively (see Chapter 12). After collection, the liquid can be serially diluted and cultured on various agars to identify the nature and amount of cultivable microorganisms. It can also be used for endotoxin determinations, as well as for immunologic, genetic, and viral analyses. Various impingers are most commonly used for such collection; however, achieving sufficient physical collection efficiency requires a very high sampling velocity (up to 300 m/s), which usually results in violent bubbling of the collection fluid and substantial physical stress on the microorganisms under investigation. Owing to the high-speed impingement and violent bubbling, conventional impingers, such as the AGI-30 (Ace Glass Inc., Vineland, NJ, USA) may lose a considerable amount of the collection fluid within very short sampling periods (up to 2 h) (see Chapter 12).

To address these issues, a new personal sampler has been designed to collect viable airborne microorganisms. The operational principle of this sampler is based on the method of bubbling of the gas carrier through a porous medium submerged into a liquid, as described in the previous part of this chapter.

The proposed personal sampler is shown in Figure 11.16. The device consists of two parts. The inner part contains a porous medium hermetically fixed to the bottom by a restrictor. Before use, 40 ml of collecting liquid (for example, sterilized water) is added to the outer part of the sampler, ensuring that, after assembling, the porous medium is entirely submerged into it. Air inlets are located peripherally, allowing air to enter over a 180° range. A pen-type clamp for attaching the device to the user's lapel is molded at the back wall of the sampler.

During operation, the air enters the device through six inlet pipes, strategically designed to ensure that no liquid spills even during dynamic human activities, and moves downward through the slot created between two parts after assembling. The dimensions of the external and internal cases as well as the size and number of air inlets were chosen to minimize the deposition of particles on the walls after they enter the sampler but before they reach the collection fluid. Upon reaching the bottom of the sampler and coming in contact with the collection fluid, the air turns 180° and passes through the porous medium submerged in the fluid. Particles, including viable airborne microorganisms, are collected by the wet medium. The effluent air then moves through the internal case and leaves the device through an outlet pipe connected to a vacuum pump by flexible tubing. A battery-operated, portable sampling pump is utilized to run the sampler at up to 4 l/min flow rate for up to 8 h.

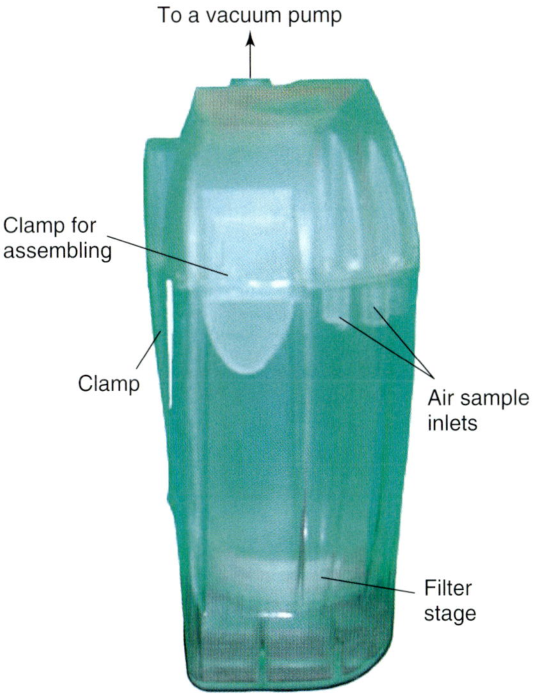

Figure 11.16 Personal bioaerosol sampler.

The proposed device addresses a number of drawbacks related to currently used bioaerosol monitoring devices. First, it utilizes a highly efficient filtration stage to collect bioaerosols within a wide range of sizes. However, compared to the collection of bioaerosol by dry filtration, viable microbial particles are not inactivated due to desiccation, as collection occurs directly into liquid. In addition, use of the relevant microbial maintenance liquid as the collection medium could keep biological particles alive for extended time periods. Second, compared to conventional bioaerosol impingers operated at subsonic velocity to achieve some reasonable collection efficiency, the proposed technology runs efficiently at very low gas velocities, which ensures non-violent bubbling and, as a result, low physical stress conditions of microbial sampling. In addition, non-violent bubbling significantly minimizes liquid losses due to droplet removal, enabling the device to operate continuously for at least 8 h.

The results of various investigations of the device have been published over the past seven years [38–44]. It was found that the device is able to provide significantly higher recovery of bacteria and fungi compared to the other techniques currently used for bioaerosol monitoring. However, the main benefit of this technology is based on its ability to monitor viable airborne viruses, which is not achievable by any other previously discussed conventional technique. Table 11.2 shows the summarized results on the recovery of various microorganisms used for laboratory and field tests in various research programs.

Table 11.2 Recovery rates achieved by the personal sampler for various microorganisms.

Microorganism	Recovery rate (%)
Escherichia coli (bacteria)	57
Bacillus subtilis (bacteria)	95
Pseudomonas fluorescens (bacteria)	61
Rhizopus (fungi)	88
Aspergillus versicolor (fungi)	97
Aspergillus niger (fungi)	93
Influenza strain H5N1 (virus)	24
Influenza strain H3N2 (virus)	20
Influenza strain H11N9 (virus)	23
Mumps (virus)	24
Measles (virus)	22
SARS (virus)	24
Vaccinia (virus)	89

References

1. Cooper, C.D. and Alley, F.C. (eds) (1994) *Air Pollution Control: A Design Approach*, Waveland Press, Long Grove, IL.
2. Cheremisinoff, P.N. (1993) *Air Pollution Control and Design for Industry*, Marcel Dekker, New York, p. 293.
3. Davies, C.N. (ed.) (1973) *Air Filtration*, Academic Press, London.
4. Langmuir, I. (1942) Report on smokes and filters. OSRD Report No. 865, Section I, Part IV Office of Scientific Research and Development, Office of Technical Services, Washington, DC.
5. Fuchs, N.A. (ed.) (1964) *The Mechanics of Aerosols*, Pergamon Press, Oxford.
6. Kirsch, A. and Stechkina I.B. (1978) The theory of aerosol filtration with fibrous filters, in *Fundamentals of Aerosol Science* (ed. D.T. Shaw), John Wiley & Sons, Inc., New York, pp. 165–256.
7. Brown, R.C. (1993) *Air Filtration: An Integrated Approach to the Theory and Applications of Fibrous Filters*, Pergamon Press, Oxford.
8. Friedlander, S.K. (2000) *Smoke, Dust and Haze: Fundamentals of Aerosol Dynamics*, Oxford University Press, New York.
9. Agranovski, I. and Whitcombe, J. (2000) Utilisation of wet fibrous media for filtration of sticky aerosol particles. *J. Aerosol Sci.*, **31** (Suppl.), 204.
10. Agranovski, I. and Braddock, R. (1998) Filtration of mists on wettable fibrous filters. *AIChE J.*, **44**, 2775–2783.
11. Agranovski, I. and Braddock, R. (1998) Filtration of mists on nonwettable fibrous filters. *AIChE J.*, **44**, 2784–2791.
12. Kuwabara, S. (1959) The forces experienced by randomly distributed parallel circular cylinders or spheres in viscous flow at small Reynolds numbers. *J. Phys. Soc. Jpn.*, **14**, 527–532.
13. Agranovski, I. and Shapiro, M. (2001) Clogging of wet filters by dust particles. *J. Aerosol Sci.*, **32**, 1009–1020.
14. Agranovski, I. and Shapiro, M. (2001) Clogging of wet filters as a result of drying. *Chem. Eng. Technol.*, **24** (4), 387–391.
15. Mullins, B.J., Agranovski, I.E., and Braddock, R.D. (2003) Particle bounce during filtration of particles on wet and dry filters. *Aerosol Sci. Technol.*, **37**, 1–14.

16. Pyankov, O., Agranovski, I., Huang, R., and Mullins, B. (2008) Removal of biological aerosols on oil coated fibres. *CLEAN – Soil, Air, Water*, **36** (5), 609–614.
17. Huang, R., Pyankov, O.V., Yu, B., and Agranovski, I.E. (2010) Inactivation of fungal spores collected on fibrous filters by *Melaleuca alternifolia* (Tea Tree Oil). *Aerosol Sci. Tech.*, **44** (4), 262–268.
18. Agranovski, I., Braddock, R., Jarvis, D., and Myojo, T. (2002) Inclined wettable filter for mist purification. *Chem. Eng. J.*, **89**, 229–238.
19. Agranovski, I. and Whitcombe, J. (2001) Case study on the practical use of wettable filters in the removal of submicron particles. *Chem. Eng. Technol.*, **24** (3), 513–517.
20. Ellenbecker, M.J., Leith, D., and Price, J.M. (1980) Impaction and particle bounce at high Stokes numbers. *J. Air Pollut. Control Assoc.*, **30**, 1224–1227.
21. Hamaker, H.C. (1937) The London–van der Waals attraction between spherical particles. *Physica*, **4**, 1058–1072.
22. Tsai, C.-J., Pui, D.Y.H., and Liu, B.Y.H. (1991) Elastic flattening and particle adhesion. *Aerosol Sci. Technol.*, **15**, 293–255.
23. Dahneke, B. (1971) The capture of aerosol particles by surfaces. *J. Colloid Interface Sci.*, **37**, 342–353.
24. Dahneke, B. (1973) Measurements of the bouncing of small latex spheres. *J. Colloid Interface Sci.*, **45** (3), 584–590.
25. Dahneke, B. (1975) Further measurements of the bouncing of small latex spheres. *J. Colloid Interface Sci.*, **51** (1), 58–65.
26. Dahneke, B. (1995) Particle bounce or capture – search for an adequate theory: I. Conservation-of-energy model for a simple collision process. *Aerosol Sci. Technol.*, **23**, 25–39.
27. Wall, S., John, W., Wang, H.-C., and Goren, S. (1990) Measurements of kinetic energy loss for particles impacting surfaces. *Aerosol Sci. Technol.*, **12**, 926–946.
28. Gillespie, T. and Rideal, E. (1955) On the adhesion of drops and particles on impact at solid surfaces. *J. Colloid Sci.*, **10**, 281–298.
29. Walkenhorst, W. (1974) Investigations on the degree of adhesion of dust particles. *Staub Reinhalt. Luft*, **34**, 149–153.
30. Boskovic, L., Agranovski, I., and Braddock, R. (2007) Filtration of nanoparticles with different shape on oil coated fibres. *J. Aerosol Sci.*, **38** (12), 1220–1229.
31. Mullins, B., Braddock, R., Agranovski, I., Cropp, R., and O'Leary, R. (2005) Observation and modelling of clamshell droplets on vertical fibres subjected to gravitational and drag forces. *J. Colloid Interface Sci.*, **284**, 245–254.
32. Mullins, B., Braddock, R., Agranovski, I., and Cropp, R. (2006) Observation and modelling of barrel droplets on vertical fibres subjected to gravitational and drag forces. *J. Colloid Interface Sci.*, **300**, 704–712.
33. Agranovski, I., Myojo, T., and Braddock, R. (1999) Removal of ultra-small particles by bubbling. *Aerosol Sci. Technol.*, **31**, 249–257.
34. Agranovski, I., Myojo, T., and Braddock, R. (2001) Comparative study of the performance of nine filters utilized in filtration of aerosols by bubbling. *Aerosol Sci. Technol.*, **35**, 852–859.
35. Agranovski, I., Braddock, R., and Myojo, T. (2002) Removal of aerosols by bubbling through porous media submerged in organic liquid. *Chem. Eng. Sci.*, **57**, 3141–3147.
36. Agranovski, I., Braddock, R., Kristensen, N., Crozier, S., and Myojo, T. (2001) Model for gas–liquid flow through wet porous medium. *Chem. Eng. Technol.*, **24** (11), 1151–1155.
37. Agranovski, I., Braddock, R., Crozier, S., Whittaker, A., Minty, S., and Myojo, T. (2003) Study of multiphase flow in submersed porous materials. *Sep. Purif. Technol.*, **30**, 129–137.
38. Agranovski, I., Agranovski, V., Grinshpun, S., Reponen, T., and Willeke, K. (2002) Collection of airborne microorganisms into liquid by bubbling through porous medium. *Aerosol Sci. Technol.*, **36**, 502–509.
39. Agranovski, I., Safatov, A., Borodulin, A., Petrishchenko, V.,

Pyankov, O., Sergeev, A., Agafonov, A., Ignatiev, G., Sergeev, A.A., and Agranovski, V. (2004) Natural decay of viruses in bubbling processes utilized for personal bioaerosol monitoring. *Appl. Environ. Microbiol.*, **70**, 6963–6967.

40. Agranovski, I., Safatov, A., Pyankov, O., Sergeev, A., Agafonov, A., Ignatiev, G., Ryabchikova, E., Borodulin, A., Sergeev, A.A., Doerr, H., Rubenau, F., and Agranovski, V. (2004) Monitoring of viable airborne SARS virus. *Atmos. Environ.*, **38**, 3879–3884.
41. Agranovski, I., Safatov, A., Borodulin, A., Pyankov, O., Petrishchenko, V., Sergeev, A., Sergeev, A.A., Grinshpun, S., and Agranovski, V. (2005) New personal sampler for viable airborne viruses: feasibility study. *J. Aerosol Sci.*, **36** (5–6), 609–617.
42. Agranovski, I., Safatov, A., Pyankov, O., Sergeev, A., Sergeev, A., and Grinshpun, S. (2005) Long-term personal sampling of viable airborne viruses. *Aerosol Sci. Technol.*, **39**, 912–918.
43. Agranovski, I., Safatov, A., Sergeev, A.A., Pyankov, O., Petrishchenko, V., Mikheev, M., and Sergeev, A.N. (2006) Rapid detection of airborne viruses by personal bioaerosol sampler combined with the PCR device. *Atmos. Environ.*, **40**, 3924–3929.
44. Agranovski, I. (2007) Personal sampler for viable airborne microorganisms; review of the main development stages. *CLEAN – Soil, Air, Water*, **35**, 111–117.

Part IV
Atmospheric and Biological Aerosols

Aerosols – Science and Technology. Edited by Igor Agranovski

ISBN: 978-3-527-32660-0

12
Atmospheric Aerosols

Lev S. Ivlev

12.1
General Concepts

Atmospheric aerosols are of great importance in most physical and physico-chemical processes in the atmosphere, such as:

1) water *phase transitions* (development of cloud, fog, and mist);
2) *solar radiation balance* (heating and cooling of air layers and the Earth's surface, meteorological visibility, and greenhouse effect) and air movement;
3) transfers of solids and water (volcanic eruptions, dust storms, rainfall, various kinds of industrial and radioactive pollution);
4) heterogeneous catalytic chemical and photochemical processes and reactions [1].

The first three factors are fundamental in creating the movement of air masses on various scales, including tornadoes, convective clouds, typhoons, hurricanes, and so on [2]. Industrial and accidental ejections of harmful substances (aerosols and other additional gases) into the atmosphere can cause a serious risk for the environment and population. Above-ground testing of nuclear weapons in the twentieth century changed the balance of radioactive substances in the atmosphere and caused *nuclear precipitation*. The emission of chlorine- and bromine-containing *freons*, as well as of nitrogen compounds, impact on the aerosols in the troposphere and the ozone layer of the Earth's atmosphere. This has a global aspect, together with long-term processes of changing the chemical and dispersed composition of pollutants, the transfer of substances in the troposphere and stratosphere, the influence of pollution on the *mass balance* of substances, and temperature regimes in the atmosphere [3]. Aerosol particles play an important role in the processes of water vapor condensation and thus precipitation. Therefore, in meteorology, they are called condensation nuclei, regardless of their physical and chemical properties [4].

The classification of atmospheric aerosols is based on their method of generation, nature, and characteristic size [5]. The main types of atmospheric aerosols are dust, smoke, fog, and smog. Dust is composed of solid particles, usually produced in

Aerosols – Science and Technology. Edited by Igor Agranovski

ISBN: 978-3-527-32660-0

the mechanical crushing of solids (explosions, mining, and so on), and in the drying of drops containing dissolved substances or particles (for example, salt). Smoke is commonly produced in the combustion of volatile substances, as well as in chemical and photochemical reactions. The dimensions of smoke particles can vary from a few nanometers to ~5 μm. Fog is composed of liquid drops formed by condensation of water. The droplets of fog are usually up to 10 μm in diameter and can also include dissolved substances or solid particles. Smog is produced as the result of the interaction between natural fog and gaseous pollution containing smoke particles. Droplets, as well as small solid particles suspended in the air, are often referred to as *haze*, which is a combination of the major classes of atmospheric aerosols mentioned above.

Aerosol particles in the atmosphere have a wide range of sizes, from particles consisting of several molecules (clusters) with a diameter of 1 nm to large dust particles and micro-drops of water with a diameter of several tens of micrometers. Exact determination of the largest aerosol particles in the air is difficult because, under different circumstances, particles of the same size can precipitate or remain in the air for extended time periods. For example, particles of volcanic ash, dust, and sand storms with up to 100 μm diameter can remain in the air for a long time.

The classification based on size characteristics defines three classes of aerosol particles: small/fine (*Aitken particles*), with radius $r \leq 0.1$ μm; medium (*Junge particles*), with radius $0.1\ \mu m < r < 1\ \mu m$; and giant, with radius $r \geq 1$ μm. Particles can also be distinguished by their form: isometric, with sizes approximately identical in all three directions; plates, with one size much less than the other two; and fibers or chains, extended in one direction. The form of plates is frequently observed in dust particles, and chain aggregates are formed in combustion processes. Solid aerosol particles in general have a non-spherical form, so they are attributed some average size. It is convenient to use the "*equivalent aerodynamic diameter*" concept, in which an actual particle is represented by a spherical particle with a density of 1 g/cm^3 that behaves similarly to the particle of interest.

Of particular interest are so-called *fractal aggregates* (see details in Chapter 1), which are formed, for example, during nucleation and coagulation of smoke particles or ice crystals in clouds. Usually, they are friable structures with an uncertain relation between their linear dimensions and density (flakes, snowflakes). In this case it is necessary to bring in an additional variable, called the *fractal dimension* of the particles [2].

A typical size distribution of particles in the atmosphere is shown in Figure 12.1. As can be seen, the concentration peak is observed for particles between 0.01 and 0.1 μm. Near the Earth's surface, the concentration, N, of particles in this size range varies from 10 cm^{-3} (above the ocean) to 10^3 cm^{-3} (above the land). The atmospheric aerosol size distribution is subject to strong changes, especially due to processes of evaporation, condensation, and coagulation (see Chapter 1).

Aerosol particles in the atmosphere can be either neutral or charged (ions) [6]. Ions in the upper atmosphere are formed by the interaction of elementary particles and short-wave radiation with gases. At lower altitudes, the main ionizers are radioactive emission from the Earth's crust and cosmic rays; near the Earth's

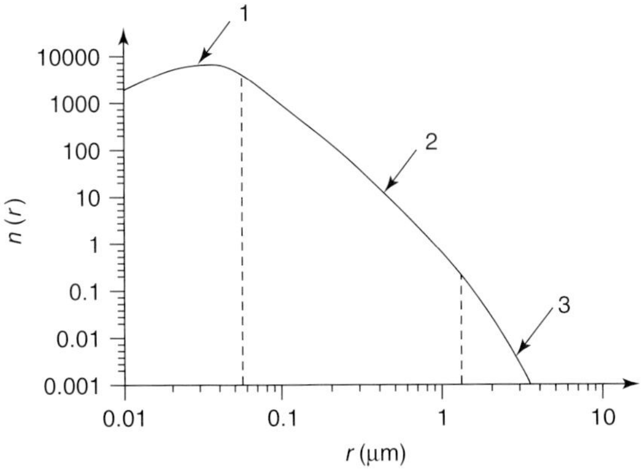

Figure 12.1 Size spectrum of aerosol particles in the atmosphere ($N \approx 3 \times 10^4$ cm^{-3}): (1) Aitken particles; (2) Junge particles; and (3) giant particles.

surface 1.5–1.9 ion pairs are formed per second per cubic centimeter. At a height of 15 km this number reaches its highest value of 225–300 s^{-1} cm^{-3} with subsequent decrease of formation efficiency at higher altitudes [7].

In cloudless systems, charged particles (ions) are contained in the fine dispersed fraction. In the simplest classification, two classes of ions are considered: heavy (or large) and light ions. The heavy ions coincide in size with the condensation nuclei and have a radius from 7×10^{-3} to 1×10^{-1} μm. Light ions are charged clusters consisting of several molecules or atoms ($r \cong 7 \times 10^{-3}$ μm).

Ions are characterized (Table 12.1) by their mobility $\mu_{\pm}$ (cm^2/s V) $= a(e/m)(\lambda_u/V_T)$, where a is a numerical coefficient (0.5–1.0), the parameter e/m is the ion's charge-to-mass ratio, λ_u is the average length of the ion's mean free path, and V_T is an average speed of the ion's thermal movement. The mobility of light ions is 1–2 cm^2/s V, so that the mobility of negative ions is somewhat higher than that of positive ones. The presence of water molecules in the air causes some decrease in the ion mobility.

Ions have been classified according to their physical mechanisms of formation and evolution in the atmosphere, as follows.

Table 12.1 Classification of ions (according to J. Bricard).

Type of ion	$\mu_{\pm}$ (cm^2/s V)	R (10^8 cm)
Light	1–2	7–10
Average	0.01–0.001	80–250
Heavy	0.001–0.00025	250–550
Ultra-heavy	<0.00025	>550

1) Molecular ions are usually single molecules or parts of molecules carrying one or several elementary charges. Their lifetime (τ) depends on the concentration and type of the molecules with the dipole moment. The lifetime of molecular ions decreases with increase in their concentration. Their average minimum lifetime in the boundary layer of the atmosphere under standard conditions is 10^{-10}–10^{-11} s.
2) Light ions are formed by merging of molecular ions carrying a charge of $\pm$ (1–4)e with easily polarizable molecules carrying a large dipole moment, for example, water molecules. In this situation, the ion is surrounded by up to four neutral molecules ($Ze \leq 4$). By analogy with ions in aqueous solution, such ions can be called *hydrated ionic complexes*. The mobility of light ions is in the range 0.5–5 cm^2/s V. The lifetime of a light ion is usually from 1 to more than 100 s.
3) An ion could reach intermediate size if the number of surrounding ionic molecules increases ($Ze \cong 10$). At normal atmospheric conditions the mobility of such ions is within the range of 0.05–0.5 cm^2/s V.
4) The ions become more stable if they recombine with neutral aerosol particles. These objects are called heavy (complex) ions or *Langevin ions*. The range of mobility of such ions is from 0.001 cm^2/s V to infinitesimal quantities, and their sizes are from hundredths (Aitken particles) to tens of micrometers (dusts, hydrometeors).

12.2 Atmospheric Aerosols of Different Nature

The main sources of atmospheric aerosols are the land surface, seas and oceans, meteoritic flows, forest fires, chemical and photochemical reactions in the atmosphere, and various human activities.

12.2.1 Soil Aerosols

The most significant source of aerosol particles is soils, including the surfaces of steppes, deserts, and mountains [3]. Soils contribute about 50% of all aerosol particles in the atmosphere, which represents more than 500 billion kilograms per year. Soil dust rises and moves with winds at a speed of several kilometers per hour [8]. Even a light breeze above a ploughed field raises clouds of dust. The most efficiently aerosolized are loess soils, which are composed of fine dust, settled from the atmosphere over prolonged time periods. Dust storms in the steppe and desert regions of North America exemplify this kind of process.

Analysis of satellite images shows that enormous amounts of dust are conceived in the western Sahara desert. The gigantic dust cloud produced in the regions of the Takla-Makan and Gobi deserts in Asia as the result of the dust storms in April 1976 reached the Arctic and was observed in Alaska. The speed of the dust cloud was approximately 80 km/h and it moved about 4000 tonnes of dust per hour. As a result, for five days of moving across Alaska, it transferred about half a million tonnes of Asian dust.

The presence of quartz, silica, alumina, carbonates, calcites, and oxides of iron is common for aerosols of soil origin. The organic matter in aerosols of soil origin usually does not exceed 10%. On the other hand, soil-originated dust is enriched with the oxides of iron and manganese, which is not the case for their sources. This could possibly be explained by selectivity of physical and chemical aerosol formation processes. A very similar composition of dust is found above the various different open areas of the Earth's surface, namely mountains, deserts, semi-deserts, and steppes.

The size distribution of soil aerosols covers the submicrometer region – however, the maximum can sometimes be observed within the range of $1\ \mu m \leq r \leq 5\ \mu m$. Particles with radii less than 0.1 μm are usually formed by crystallization of salts dissolved in the groundwater. Some estimates show that the number of such particles could reach $10^4 - 10^5\ cm^{-2}$. Very intensive aerosol generation is frequently observed in solonchak soils. Some measurements carried out in the Kara-kum desert allowed the contribution of solonchak-soil-originated aerosols to the total atmospheric aerosol content to be evaluated as approximately 20–30% [9].

Mineral particles, which enter the atmosphere above deserts due to a process of *saltation*, are an essential component of dust aerosols. Initially, the process of saltation was related to the "explosion" of soil particles leading to the production of aerosol particles. The intensity of the process was correlated with the velocity-related shear of the wind on the surface of the soil. In addition, previously the process was considered to be oscillatory. However, it is now agreed that the model of saltation is monotonic in relation to the wind-velocity-related shear [10].

The important characteristic of dust aerosols is related to their ability to undergo distant transfer from the source. For this reason, during dust storms, dust aerosols become an essential component of the tropospheric particle composition above the ocean. Applying factor and cluster analyses of "reverse" trajectories, it is possible to recognize sources and types of aerosols sometimes generated at substantial distances from the ocean.

The deserts of northwestern China cover a territory of $1.3 \times 10^6\ km^2$ and are located at elevated altitudes in the zone of west–east transfer. This region of deserts is characterized by a dry climate, which produces favorable conditions for aerosol formation and subsequent release to the atmosphere, especially during the frequent dust storms. The region is the main source of dust aerosol supply to the Pacific Ocean in the Northern Hemisphere in spring and early summer.

The maximum observed concentration of atmospheric aerosols exceeds $260\ \mu g/m^3$, with approximately 82% being dust particles. They contain Al, Ca, Fe, K, Mn, Si, and Ti, with relative weight concentrations of 7, 6, 4, 2, 0.1, 32, and 1%, respectively. The contribution of local or background aerosols is usually only 11%, whereas the remaining 89% are the products of distant transfer.

An electron microscope study using the method of cluster analysis of samples collected in the Negev desert (Israel) in the summer and winter 1996–1997 showed the presence of 11 classes of particles. The summer tests were dominated by sulfates and mineral dusts, the winter ones by particles of sea salt and industrial aerosols. Interestingly, the submicrometer fraction was substantially composed of particles of the second origin, which in the case of the presence of the sulfate component could be related to distant transfer.

The presence of dust particles in the atmosphere influences processes in the atmosphere and even in the oceans. For example, dust aerosols are responsible for a change in the concentration of sea *phytoplankton*. Some carbon compounds formed in the processes of photosynthesis, including particulate organic carbon (POC), particulate inorganic carbon (PIC), and also dissolved organic carbon (DOC), undergo transfer at depths below 100 m. This process was called the "*biological pump*" and it plays an important role in the global *carbon cycle* [11].

Some measurements have shown an approximate doubling of biomass in the ocean mixed layer within two weeks after passage of a dust cloud from the Gobi desert. This fact could be interpreted as the process of "fertilizing" the ocean, caused by iron in the particles of dust stimulating the formation of organic carbon in the form of particles [12].

During aerosol studies in Tadzhikistan in August–September 1981, along with chemical analysis and monitoring of meteorological conditions, optical and microphysical particle characterization (scattering at $\lambda = 0.55\ \mu m$, ultraviolet–visible absorption), and electron microscope analysis were performed. It was found that the value of the scattering coefficient, $\sigma(\lambda = 0.5\ \mu m)$, was increased by an order of magnitude during dust storm events. Thermo-optical analysis showed that the proportion of volatile components in submicrometer particles was increased during storms. The beginning of a dust storm was accompanied by a sharp increase of concentration of submicrometer particles with $a > 0.4\ \mu m$ (Figure 12.2) and by a decrease of concentration and size of the highly dispersed fraction as a result of their precipitation into the submicrometer fraction. The productivity of the source of the highly dispersed particles was approximately $10^3\ cm^{-3}s^{-1}$ and practically all of them precipitated on submicrometer aerosols. Analysis of the infrared spectra reports that the main components of the submicrometer aerosols are clay minerals, soot, and organic matter. An increase of the mineral particle concentration during dust storms leads to intensive sedimentation of organic matter and soot on these nuclei, which changes the morphology of the particles, sharply increases absorption, and changes the spectral aerosol characteristics.

Some studies of the transfer of Sahara dust aerosols across the Atlantic Ocean show a strong correlation ($r^2 = 0.93$) between the concentrations of non-salt *sulfate aerosols* (SO_4^{2-}) and dust, with the corresponding ranges 0.5–4.2 and 0.9–257 $\mu g/m^3$, respectively. Daily variations in the microstructure of aerosols of both types were found to be insignificant. However, the concentration of large sulfate aerosol particles with an *aerodynamic diameter* exceeding 1 μm varies substantially (from 21% to 73%).

As was visualized from satellites, the maximum values of large particles containing SO_4^{2-} above the ocean are associated with the transfer of Sahara dust, and the minimum values with the arrival of air masses from the central region of the

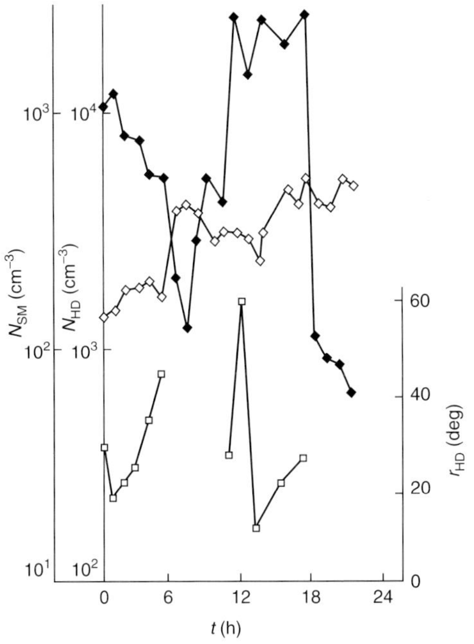

Figure 12.2 Change of concentration of highly dispersed aerosols, N_{HD} (open diamonds), and submicrometer aerosols, N_{SM} (filled diamonds), and of the modal radius of highly dispersed particles, r_{HD} (open squares), during a dust storm in Dushanbe (23 September 1981).

North Atlantic, where the concentration of dust is less than 0.9 $\mu g/m^3$. Elevated amounts of sulfate-containing particles are supposed to be the consequence of the heterogeneous reactions that occur in the atmosphere above North Africa between SO_2, which is contained in polluted air from Western Europe, and Sahara dust.

12.2.2 Marine Aerosols

A substantial part of the Earth's surface is covered by water, which determines the structure and properties of aerosol formation processes in the lower layers of the atmosphere. The air above an ocean carries a substantial amount of water vapor, which can condense or evaporate from the aerosol particle depending on ambient conditions. The air temperature profile above the ocean, and correspondingly the aerosol composition at various altitudes, are determined by interaction of solar radiation with the surface layer of the ocean [13].

At low relative humidity, a *marine aerosol* particle could appear as a solid sea salt nucleus with adsorbed or even absorbed molecules of water. In the initial formation stages, a marine aerosol particle could have a solid core coated by water–salt solution. Finally, the marine aerosols could be micro-droplets of water–salt solution of different concentration depending on the relative humidity and dew point.

The sea surface is the second powerful source of aerosol particles, which contributes approximately 20–30% of the total mass of dispersed phase in the atmosphere. Some estimates show that the global release of marine-originated aerosols exceeds the total release of all natural and anthropogenic aerosols and reached 3300 Tg in the year 2000. Some projections indicate that the emission of marine salt to the atmosphere will reach the huge value of 5880 Tg/year by the year 2100, which could cause an alteration to the solar radiation reaching the ground by approximately 0.8 W/m^2. However, one should consider that the lifetime of salt particles is usually shorter compared to that of particles of alternative origins.

The chemical composition of the mineral components of marine aerosols is quite similar to the chemical composition of the sea water dry residue: NaCl 78%, $MgCl_2$ 11%, and $CaSO_4$, Na_2SO_4, plus K_2SO_4 11% in total. The concentration of the salt particles above the ocean could reach 100 cm^{-3}, but on average is around 1 cm^{-3}. The maximum of the salt particle size distribution is usually observed at around 0.3–0.4 μm.

The basic processes for the formation of marine aerosol include the following:

1) wind generation, in particular, the bubble mechanism of small particle formation, when particles are produced as the result of a bubble burst;
2) evaporation of molecules, ions, and aqueous clusters from the sea surface;
3) formation of particles due to exchange of electric charge between the sea surface and the near-water layer of the atmosphere.

Organic surfactants and oils are concentrated at the surface of the ocean in a thin layer (about 1 μm). Bubble burst at the sea surface leads to the formation and release of organic aerosol matter and sea salts. The organic matter plays an important role in the optical phenomena at the near-water layer of the atmosphere. The concentrations of sea aerosols above the exposed surface of the water depend on the temperature and salinity of the water, and the temperature and humidity of the air. However, the major factor that influences the formation and transfer of aerosols is the wind speed. The intensity of the generation of sea aerosol particles is determined by the wind speed, by the presence of surf zones, and by the temperature conditions of the near-surface layer of water. The generation of particles during sputtering of drops occurs when the wind speed exceeds 7 m/s and reaches its maximum at 15 m/s. The concentration of sea salt in the air shows a clear dependence on the wind speed and is usually within 2–50 $\mu g/m^3$. The average salt content of particles lies within the range 10^{-15}–10^{-14} g, which corresponds to a droplet size of 0.2–0.5 μm. The results of experiments show that a significant contribution to the aerosol spectrum is also made by smaller particles with a salt content of 10^{-16}–10^{-15} g, which corresponds to a diameter of 0.1–0.2 μm.

Two mechanisms determine the wind-generated formation of aerosols: (i) generation of particles from fragments of the liquid film at the point of bubble release from the liquid, termed film drops (FDs); and (ii) formation of particles as the result of evaporation of drops ejected to the atmosphere under the action of the wind, termed jet drops (JDs). FD particles are characterized by an exponential dependence on the wind speed, and their formation is less dependent on the wind speed compared to the JD aerosols. An increased concentration of FD particles at wind speeds of less than 5 m/s testifies to the possibility of existence of some other bubble formation mechanism, which functions most intensively during the daytime, clear weather, and during ice melting events. JD particles consist mainly of organic components; however, the salt component of these particles increases when the wind speed exceeds 12 m/s, making salt the dominating substance. The JD contribution becomes significant at wind speeds above 5 m/s.

The generation of drops and their subsequent drying could cause salt crystal fracture. Some experimental investigations of salt crystal stability confirmed the possibility that sodium chloride crystal fracture could be achieved by ultraviolet irradiation, by an electric field (500 V/cm), and by heating particles 10–15 °C higher than the air temperature (to 30–40 °C). The effect of salt particle fracture was distinctively revealed for a relative humidity below 30–35%.

Although the chemical composition of the particles ought to correspond to the composition of the salts dissolved in the sea water, in the full-scale experiments some enrichment of the content by various elements is observed. This could be explained by the fact that bubble burst occurs at the surface microlayer, where the elemental composition could be different from that in the bulk ocean. Mass spectrometry of this type of particle shows that the prevailing component of organic films is palmitic acid, whereas all other *fatty acids* play an insignificant role. These aerosols have an essential influence on the chemical composition of the near-water layer of the atmosphere. As one can understand, during frequent cases of oil spills, the concentration of organic aerosols sharply grows and significantly dominates over aerosols of other origins.

Morphological particle structure corresponds to the structure of sulfate particles. A considerable proportion of larger particles ($d > 0.2$ μm) also have the morphological structure of sulfate particles. Consequently, in the air above the sea surface, an intensive generation of aerosol matter from gaseous sulfur compounds, including SO_2, H_2S, and $(CH_3)_2S$, occurs. However, gas-phase oxidation to SO_3 does not explain the existing experimental data on the size distribution of *sulfuric acid* and sulfate aerosols, particularly the presence of sulfate particles with $d > 0.2$ μm at concentrations exceeding 5 cm^{-3}. Solar radiation accelerates the processes of conversion of sulfur compounds into sulfurous anhydride, but it is not sufficient to explain the observed particle size distribution, as the concentration of particles with diameter less than 0.1 μm has to be considerably higher. The layer-by-layer determination of the chemical composition of such particles by spectroscopy showed that the nuclei of the particles frequently contain NaCl, which makes it possible to take for granted that the oxidation of sulfur compounds occurs in essence in the drops containing dissolved sea salts. The sulfate particles in the size range of 0.3–2 μm are the products of hetero-phase reactions, while particles with $d < 0.1$ μm are the products of the gas-phase reactions of oxidation.

The formation of sulfate aerosols above the ocean is determined by contributions from two sources: (i) biogenic sources of gaseous compounds of sulfur, mainly dimethyl sulfide (DMS); and (ii) anthropogenic sources, predominantly SO_2. Both sources are characterized by strong time- and latitude-related changes. The biogenic contribution is usually evaluated by consideration of the concentration of methane sulfonate (MSA), which is one of the steady products of hydroxyl, OH, oxidation of DMS.

The most essential source of sulfur products in remote areas of the oceanic atmosphere is DMS, which is formed in the ocean by biological processes and released to the atmosphere in a form of cloud condensation nuclei (CCN). Numerical modeling verifies the fact that approximately 30–50% of DMS is converted into SO_2. Non-salt sulfates are contained mainly in super-micrometer salt particles, contributing 35% (±10%) in summer and 58% (±22%) in winter to their mass. The analysis of samples from Mace Head, on the west coast of Ireland, showed that 30% of non-salt sulfate in summer appears due to gas-phase transformation of DMS emission from the ocean.

A field study of the physical, chemical, and nucleation properties of atmospheric aerosols above the southwestern sector of the Pacific Ocean (to the south of Australia) was commenced in 1995 [14]. The prevailing component of the aerosols in the region is sea salt (90% of particles with a diameter of more than 130 nm, and 70% of particles with a diameter of more than 80 nm). Fifty per cent of particles with a diameter larger than 160 nm contain organic components bonded with sea salt.

Aerial measurements of water-soluble ions above the Pacific Ocean in the Southern Hemisphere reported unpredicted low values of total ionic composition in the troposphere (2–12 km) in spite of the influence of significant emissions from biomass combustion processes coming from the west [15].

The influence of anthropogenic activities on the atmosphere and properties of aerosols are noticeable even at remote oceanic locations. At altitudes higher than 3 km, the products of biomass combustion were observed. About 11–46% of sulfate aerosols with a diameter of more than 100 nm contained soot sourced from biomass combustion in South Africa [16].

12.2.3 Volcanic Aerosols

Volcanoes that eject into the atmosphere colossal quantities of materials are powerful sources of aerosols. Approximately 800–900 volcanoes are considered to be active, and around 20–30 volcanoes erupt annually, emitting into the atmosphere approximately 3×10^9 tonnes of ash. The most powerful eruptions of ash ejected into the atmosphere were: 175 km^3 (Tambora, Indonesia, 1815); 20 km^3 (Krakatau, Sunda Strait, Indonesia, 1883); 10 km^3 (Santa Maria, Guatemala, 1902); 21 km^3 (Katmai, Alaska, USA, 1912); 20 km^3 (Cerro-Azul, Chile, 1932); and 0.5 km^3 (El Chichon, Mexico, 1982).

Volcanic eruptions make a significant contribution to the optical characteristics of the atmosphere, because, according to actinometrical observations, the products of eruption remain in the stratosphere for more than a year [17]. It is remarkable that clouds of gas and ash rise to enormous altitudes (Krakatau to 60 km, El Chichon to 37 km)! The cloud produced by the Krakatau eruption took three years to settle. The products of the El Chichon eruption were observed in Italy and Japan six months after the event and caused a temperature decrease in the Northern Hemisphere by 0.5 °C for three consecutive years. Primary dust particles descend to the troposphere for several months depending on their sizes and heights of ejection. *Sulfur dioxide* and *carbonyl sulfide* undergo chemical and photochemical reactions with atmospheric gases and aerosols, forming sulfuric acid and sulfate aerosol particles.

Another reason for the prolonged existence of aerosols in the upper layers of the atmosphere is sequential eruption of one or more volcanoes following the initial eruption. The strongest volcanic eruptions of the present era were El Chichon (March 1982) and Pinatubo (June 1991), with a time difference of nine years. Some light detecting and ranging (LIDAR) measurements in Western Europe showed a sharp increase of the aerosol component during periods of three years after the eruptions (Figure 12.3). Detailed analysis of the variations in the values of the integral *aerosol back-scattering coefficient* during these eruptions shows high flare and geomagnetic activity (geomagnetic disturbance storm time index $D_{st} > 100$). It should be noted that the El Chichon eruption was preceded by volcanic eruption of

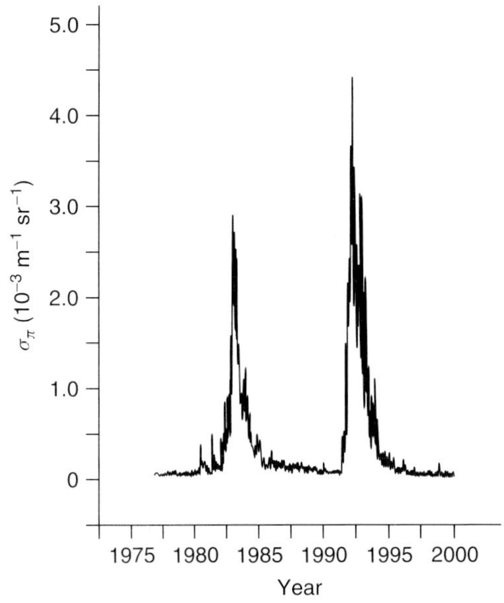

Figure 12.3 Integral aerosol back-scattering coefficients (m^{-1} sr^{-1}) in the years after the volcanic eruptions of El-Chichon (1982) and Pinatubo (1991).

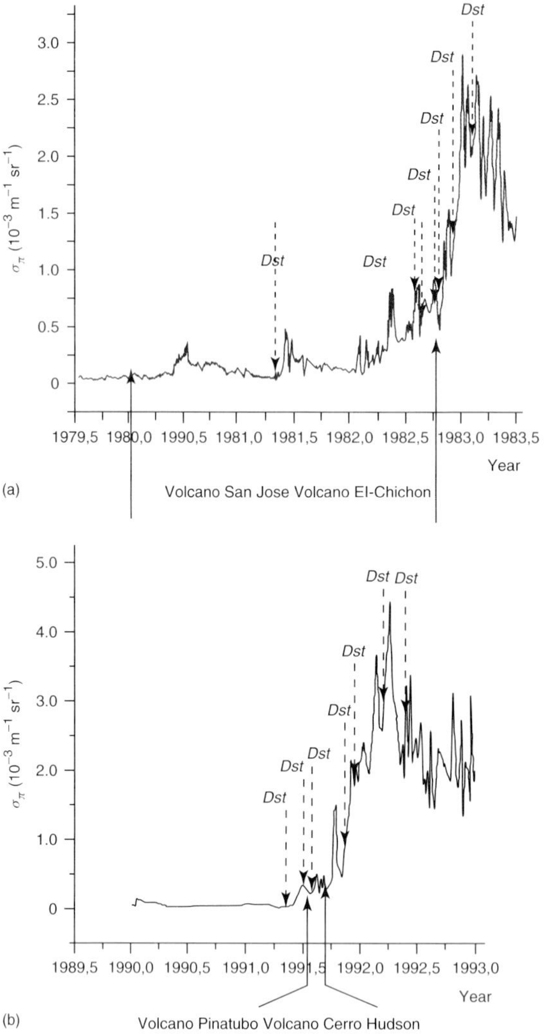

Figure 12.4 Variations of integral aerosol back-scattering coefficient at $\lambda = 694$ nm and geomagnetic index $D_{st} > 100$ as a result of the volcanic eruptions of (a) El-Chichon and (b) Pinatubo.

Mount St Helens (USA, May 1980), and Pinatubo activity was accompanied by the eruption of the Cerro Hudson volcano (Chile, August 1991) (Figure 12.4) [18].

Analysis of the chemical composition of the erupted smoke and dust shows that their major content compounds are silica (60–80%), sulfates (10–30%), calcites (3–10%), and products containing aluminum (0–20%) and iron (1–10%).

The mechanisms of aerosol formation from the gas phase are very diverse. Monitoring of aerosols along the coastline in the vicinity of the volcano Kilauea

(Hawaii, USA) revealed traces of small gas compounds (SGCs), which could be formed by the interaction of water and volcanic lava on entering the ocean [19].

Substantial contamination of the atmosphere by sulfur dioxide and carbonyl sulfide during eruptions could strongly increase the concentrations of sulfuric acid and sulfates in the lower stratosphere. For example, in the explosive volcanic eruption of Pinatubo that occurred on 14–16 June 1991, some 14–26 Mt of gaseous SO_2 was injected into the stratosphere to a height of 30 km, producing 30 Mt of sulfuric acid aerosol over the next 35 days. The volcano-originated cloud moved to the west, and three weeks after the eruption completely girded the Earth. During approximately two weeks, the cloud crossed the equator and reached latitude 10 °S. During the first one to two months the substantial mass of the aerosol cloud was concentrated in the latitude band of 20 °S to 300 °N, and thus formed a tropical reservoir of aerosol substances under the strong influence of quasi-two-year fluctuations (QTFs). This reservoir is either steady or unstable depending on phase (eastern or western) of the QTF. (The eruption occurred during the eastern phase and therefore the tropical maximum of the layer of eruptive aerosols was steady. Only after three to four months did aerosols begin to extend toward the middle latitudes of the Southern Hemisphere.) Comparatively rapid motion of aerosols toward average and high latitudes of the Northern Hemisphere occurred at heights below 20 km. Through gaps in the tropopause and by means of gravitational precipitation, the aerosol from the stratosphere was transferred to the troposphere. The global mass of sulfate volcanic aerosols reached its maximum in October 1991 and then decreased exponentially for the following years [20, 21].

The high speed of nucleation is characteristic of volcanic aerosols at the layer near the tropopause (at the bottom of the main volcanic aerosol layer) during the first year after eruption. During this time period, it is much higher than for sulfate aerosols from ground-based sources – however, during the second year after eruption, the speeds of nucleation became comparable. The volcanic aerosols influence (by means of uniform nucleation) the formation of cirrus clouds and their global evolution. A statistically significant correlation between volcanic activity and climatic characteristics has been observed [12].

Aerosol measurements of the atmospheric near-ground layer were carried out during 1974–1981 by the aerosol laboratory SRIP (Research Institute of Physics) of Saint Petersburg State University near active volcanoes in the Kamchatka region (Tolbachik, Klyuchevskoy, Gorelyi, Karymskiy, and Mutnovskiy) and in 1994–1995 in Mexico. Particle size distributions were determined by photoelectric counter and electron microscope analysis of particles collected by a cascade impactor and filter (Table 12.2) [22, 23]. The smoke and dust compositions for different volcanoes show that their major contents are silica (60–80%), sulfates (30–10%), calcites (3–10%), and products of aluminum (0–20%) and iron (1–10%). However, some differences were observed in more detailed investigations [24].

Some alterations of the chemical composition of aerosol particles at different altitudes are common for volcanic eruptions. An increase in concentration of the moderately volatile elements, including arsenic, selenium, lead, cadmium, and zinc, is detected in small particles. On the other hand, elements characteristic

Table 12.2 Enrichment coefficients K_X for different elements in aerosols during the eruption of the volcano Popocatepetl (reference element Si).

Element	29/12/1994	06/01/1995	14/01/1995	21/01/1995	27/01/1995	27-28/01/1995	28/01/1995
Fe	0.24	0.228	0.72	0.36	0.40	0.40	0.32
Al	1.04	1.20	1.64	1.36	1.12	1.24	1.60
Ca	0.48	0.52	0.84	0.68	0.68	0.72	0.72
S	1.92	340	432	364	180	92	436
P	≤4	≤7	21	≤7	≤4	≤6	6.8
Cl	28.8	44	144	100	26.4	52	68
K	0.16	0.20	0.20	0.12	0.32	0.28	0.28
Ti	0.36	0.24	0.48	0.52	0.48	0.48	0.44
Cr	0.72	0.44	3.00	2.16	0.60	1.24	0.76
Mn	0.12	0.22	0.60	0.15	0.27	0.32	0.24
Ni	0.37	0.33	2.24	0.80	0.56	0.92	0.80
Cu	8.4	2.08	18.0	3.16	0.96	1.36	2.40
Zn	1.12	0.64	7.2	2.32	1.56	2.64	2.44
Ga	0.23	0.19	2.44	1.56	0.52	≤0.44	1.20
Se	13.6	18.0	104	28	6.4	1.8	44
Br	10.4	12.4	48	19.8	22.4	12.8	9.2
Rb	0.18	0.22	2.16	1.32	0.48	0.64	1.08
Sr	0.40	0.40	0.44	0.92	0.56	0.68	0.44
Y	≤0.54	1.08	≤296	3.44	≤0.72	≤1.20	3.52
Zn	≤0.34	0.27	1.52	0.72	0.72	0.72	0.52
Hg	≤180	≤260	≤740	≤360	≤180	≤240	≤330
Pb	5.2	7.6	36	13.6	7.2	12.8	11.6

of the magma, including silicon, calcium, scandium, titanium, iron, zinc, and thorium, are observed in the larger particles. This is explained by the fact that the particles in the upper atmospheric layers are the products of ejected magma and are not related to the destroyed apex cone of the volcano. Also, different stages of eruption generate particles with different chemical and elemental compositions of volcanic substances.

12.2.4
Aerosols *In situ* – Secondary Aerosols

Photochemical and chemical reactions in the atmosphere are responsible for the generation of dispersed nanoparticles, which are so-called *secondary aerosols*. They are formed not only from organic compounds, but also from sulfur dioxide, hydrogen sulfide, carbonyl sulfide, DMS, ammonia, the oxides of nitrogen and some other gases, which react with oxidizers like ozone and different radicals, or with water vapor and aerosol particles, which mainly act as catalysts [25].

The conversion of sulfur dioxide, mainly released by industry and anaerobic bacteria [26], and hydrogen sulfide, released to the atmosphere by vegetation and decomposing organic materials, in approximately equal quantities of 10^8 tonnes per year, produces approximately 4×10^8 tonnes of aerosols containing $SO_4{}^{2-}$. The emission of sulfur products from other sources is much less significant [27, 28].

Some evaluation of possible nitrogen oxide sources suggests that the total annual amount of these materials released to the atmosphere is around 4×10^7 tonnes. However, only ~10–25% of these gases forms aerosol particles. A substantial portion of these gases is oxidized to acid anhydrides, which are dissolved in cloud droplets.

There are three main mechanisms of atmospheric aerosol formation from the gas phase, and these will be considered in the next three subsections.

12.2.4.1 Photochemical Oxidation – Heterogeneous Reactions

This process occurs over dry areas of the Earth in the upper troposphere layer [29]. Every hour, 0.03% of SO_2 is photochemically oxidized in pure air. The methodology of calculation of the concentration of aerosol predecessors such as OH^-, H_2SO_4, and HNO_3 is based on direct measurements of gases, meteorological values, and ultraviolet radiation. The theoretical results show good agreement with the observed concentrations of nanoparticles, which makes it possible to estimate the contribution of the ternary nucleating system H_2SO_4–NH_3–H_2O to the atmospheric aerosols generated from the gas phase. It was found that it does not exceed 50%.

The following set of reactions of transformation of sulfur dioxide to sulfuric acid is considered:

$$SO_2 + OH^- \rightarrow HOSO^{2-}$$
$$HOSO_2^- + O_2 \rightarrow HO_2^- + SO_3$$
$$SO_3 + H_2O + M \rightarrow H_2SO_4 + M$$

The gas-phase conversion of sulfur dioxide occurs in the course of reactions with different radicals. The concentration of radicals, in particular, the concentration of hydroxyl radical, limits the intensity of the process of the production of volatile aerosol-forming compounds (VAFCs). Most of the OH^- is produced in reactions of water molecules with metastable oxygen, $O(^1D)$, mainly sourced in the troposphere from the products of the photolysis of ozone:

$$O_3 \xrightarrow{h\nu} O_2(^1\Delta_g) + O(^1D) \qquad \lambda \leq 310\,\text{nm}$$

Stationary concentrations of metastable oxygen and hydroxyl radical were calculated by a model proposed by Isidorov [30]. As was found, the concentrations of $O(^1D)$, OH^-, H_2SO_4, and HNO_3 were very close to the concentrations observed in the atmosphere in full-scale experiments.

As far as nitric acid is concerned, it participates in the processes of heterogeneous condensation along with some volatile organic compounds (VOCs), rather than in the homogeneous nucleation. To achieve a reasonable speed of nucleation in binary and ternary systems with nitric acid as one of the components, its concentration has to be on the order $10^{16}-10^{18}$ cm^{-3}, which is beyond any realistic values in the Earth's atmosphere.

12.2.4.2 Catalytic Oxidation in the Presence of Heavy Metals

The speed of this type of reaction strongly depends on the presence of suitable catalysts (ions of *heavy metals*) and favorable values of pH, and can be sufficiently high in contaminated air. The reaction takes place in both dry air and cloud droplets.

12.2.4.3 Reaction of Ammonia with Sulfur Dioxide in the Presence of Water Droplets (Reaction of Cloud Droplets)

The speed of formation of sulfate particles in the SO_2-NH_3 reaction depends on the NH_3 supply, which ensures elevated pH values, and creates favorable conditions for the reaction to occur. The mechanism of formation of ammonium sulfate is effective only in the presence of liquid water, that is, in the regions of cloud and fog. The particles of ammonium sulfate can also remain in the air after water evaporation. The initial nuclei of ammonium sulfate have radii of approximately 30 nm and turn into drops with a size of 1 μm.

12.2.5 Biogenic Small Gas Compounds and Aerosols

Organic materials in the atmosphere either emitted from the Earth's surface (processes of combustion and dispersion) or produced *in situ* in the atmosphere as a result of gas-phase oxidation reactions of VOCs are essential components of atmospheric aerosols.

The main source of organic materials that are partially transformed into aerosol particles is vegetation. According to Went [31], 10^8 tonnes of *terpene*-based products and oxidized hydrocarbons are released annually into the atmosphere, which creates a natural aerosol background of approximately 3–6 μg/m^3. In areas with high NO_x concentration, rapid reactions of VOCs with ozone and HO and NO_3 radicals noticeably contribute to the formation of organic aerosols, which substantially influences radiation transfer in the atmosphere.

In the first stages of VOC oxidation in the troposphere, the condensation of products with low volatility leads to secondary organic aerosol (SOA) formation. The main predecessors of SOA are likely to be mono- and sesqui-terpenes and aromatic compounds. Most frequently in the atmosphere, mono-terpene hydrocarbons are observed as α- and β-terpenes, limonene, and sabinene. The contribution of biogenic VOCs to troposphere aerosol formation strongly depends on climate, type of vegetation, and other factors.

Consideration of the processes involving organic gases is complicated because of the lack of reliable information about release rates and the absence of adequate information about reaction kinetics. For a reliable theoretical description of tropospheric hydrocarbon products (and the kinetics of the processes) from the regional to the global scale, three approaches are commonly used: (i) surrogate SGCs; (ii) united molecule; and (iii) united structure of chemical processes (USCPs). The last two approaches are most popular due to their higher reliability. USCP is considered to be the most attractive approach, since it ensures the involvement of a minimal number of SGCs and reactions. Also, a new united model, called carbon bond mechanism–Zaveri (CBM-Z), based on the earlier developed method of calculation of intermolecular carbon products (called CBM-IV), allowed the implementation of numerical simulation over a wide range of time scales [32]. To obtain information about the formation of aerosols out of 14 VOCs, Griffin and colleagues [32] conducted experiments in open "smog" chambers. They showed that, for a concentration of organic matter within 5–40 $\mu g/m^3$, the produced aerosol component varied within: 17–67% for sesqui-terpenes, 2–23% for cyclic dienes, 2–15% for *bicyclic alkenes*, and 2–6% in the case of the acyclic triene of ocimene. The experiments with bicyclic alkenes (α-pinene, β-pinene, Δ^3-karene, and sabinene) were conducted in the dark. Oxidation was performed by either ozone or NO_3 radicals. An exceptionally intensive formation of aerosols from β-pinene, sabinene, and Δ^3-karene was demonstrated for the case when NO_3 radicals were involved.

Organic compounds in the atmosphere beyond the polar circles play an important role as tracers of distant transfer and determine the chemical composition of the atmosphere and snow cover. Table 12.3 shows the results of observations of the chemical composition of *Arctic aerosols* in Alert, Canada, in February–June 1991. As can be seen, the carbon content varies from 2.4% to 11%.

Organic compounds in the water-soluble aerosols do not exceed 20%. The concentration of dibasic acid aerosol correlates with the concentration of ozone-destroying Br and I aerosol. Some observations demonstrated the presence of lipids, including *n-alkanes* (C_{18}–C_{32}), polynuclear aromatic hydrocarbon (PAH) products, *n-alcohols* (C_{13}–C_{30}), fatty acids (C_7–C_3), and α,ω-*dicarboxylic acids* with long chains (C_6–C_{26}), in Arctic aerosols. These data confirm the correlation between natural and anthropogenic sources of aerosol organic matter.

Chemical analysis of snow samples showed that the main natural sources of light carboxylic acid are biogenic emission from vegetation and oxidation of hydrocarbon products in the atmosphere. The chemical composition of the snow is strongly influenced by volcanic eruptions and forest fires. The observational data testify to the wide prevalence of organic aerosols in the atmosphere above the ocean and above the land. About 10–20% of organic aerosol contains cloud condensation nuclei [33]. Organic matter also makes up a substantial part of sea salt aerosols (about 10%).

A model of the chemical composition of organic aerosols in the atmosphere assumes that the particles contain liquid nuclei covered with a hydrophobic organic monolayer of substances of biogenic origin. Any alterations in organic surface layer are caused by interaction with atmospheric radicals, which convert the inert hydrophobic film into a chemically and optically active hydrophilic layer that influences the radiation balance and causes climate alterations [34].

Table 12.3 Chemical content of Arctic aerosols.

Components	Concentration	
	Range	Average value
Aerosols, total concentration (ng/m^3)	2500–9100	5200
Total carbon, Tc (ng/m^3)	88–639	359
Total nitrogen, TN (ng/m^3)	16–154	86
Weight ratio, TC/TN	2.4–7.1	4.5
TC/atmosphere (%)	2.4–11.1	6.8
TN/atmosphere (%)	0.48–2.4	1.6
Water-soluble organic carbon, WSOC (ng/m^3)	47–300	186
WSOC/TC (%)	30–72	53
WSOC/atmosphere (%)	1.2–5.5	3.4
Dicarboxylic acids, C_2–C_{11} (ng/m^3)	7.4–84.5	36.6
Ketonic acids, C_2–C_6 (ng/m^3)	0.76–8.9	3.7
α-Dicarboxyl, C_2–C_3 (ng/m^3)	0.05–2.8	0.88
Dibasic acids, C/TC (%)	1.5–9.1	3.8
Ketonic acids, C/TC (%)	0.18–0.78	0.34
α-Dicarboxyl, C/TC (%)	0.019–0.17	0.073
n-Alkanes, C_{18}–C_{35} (ng/m^3)	0.15–2.7	0.85
Peroxiacetilnitrate, PAN (ng/m^3)	0.0002–0.85	0.11
n-Alcohols (ng/m^3)	0.24–0.95	0.50
Fatty acids, C_7–C_{32} (ng/m^3)	1.3–6.5	3.2
Long-chain dibasic acids, C_{22}–C_{26} (ng/m^3)	0.074–0.56	0.27

The "processing" of organic aerosols in the atmosphere leads to the transfer of small organic fragments into the troposphere, where they play an essential role in the homogeneous chemistry of the atmosphere. The coating layer on the particle surface consists of substances supplied naturally from the Earth's surface or from biomass combustion processes.

A significant contribution to atmospheric aerosols originates from vegetation fires, which in middle and high latitudes reaches about 4%, and in some years 12%, of the total aerosol mass, and contains 9% and 20% of black carbon (BC) and particulate organic matter (POM), respectively. For example, boreal forest fires produce 20–598 Gg/year (BC) and 0.37–11.8 Tg/year (POM), and grass fires in Mongolia produce 62 Gg/year (BC) and 0.4 Tg/year (POM).

Incomplete combustion is common in the real environment due to stoichiometrically insufficient quantities of oxygen in the combustion zone. It could also occur due to transfer of reactants into colder regions, where an abrupt deceleration of the combustion reaction or even its total stoppage may be observed ("quenching" phenomenon) [35].

Some studies of organic aerosols in the atmospheric near-ground layer in Eurasia along the Trans-Siberian railroad at distances of a few thousand kilometers have been conducted since 1995 (the "TROICA" project). Daily variations of aerosol and

SGC concentrations were obtained for different types of soils and vegetations, and rates of CO and CH_4 production for different ecosystems were evaluated [36]. As was shown, the maximum flow of CH_4 from soils in the dry areas of Eastern Siberia reached 70 ± 35 μmol/m^{2}h. Although CH_4 emissions from tundra in the latitude band of 67–77 °N are similar to those observed at considerably lower latitudes, the boreal wetlands of Siberia in the latitude band of 50–60 °N contribute significantly to the global amount of methane.

12.3 Temporal and Dimensional Structure of Atmospheric Aerosols

12.3.1 Aerosols in the Troposphere

The composition of atmospheric aerosols for a number of regions of the former Soviet Union is shown in Table 12.4, which shows data obtained by averaging the experimental results that exist for each of the regions [37]. The lowest concentrations of the elements were observed in the Ural region, Western Siberia and Eastern Siberia, which demonstrate a close trend at total levels of 13.5, 13.3, and 14.2 μg/m^3 respectively. The largest values were observed in the Central Asian regions: Tadzhikistan 31.8 μg/m^3 and Kyrgyzstan 28.2 μg/m^3. An unexpectedly high concentration of aerosols was detected in the Kamchatka region, 22 μg/m^3. The difference between the regions is more significant for the group of terrigenous elements, reaching a value of 5 for Kyrgyzstan/Far East.

12.3.1.1 Terrigenous Elements

The share of terrigenous elements increases from the European area of the Russian Federation to Kazakhstan, and then tends to decrease (Table 12.5). In the Central Asian regions, the concentration of terrigenous elements alters significantly, reaching a maximum in Turkmenistan (40%). From Western Siberia to the Far East region, the share of terrigenous elements decreases monotonically, reaching a minimum of 9.9% in the Far East. In Kamchatka, almost a twofold increase in the relative content of terrigenous elements is observed in comparison with the Far East.

12.3.1.2 The Group of Ions

This group in Table 12.5 all regions is the largest contributor to the total aerosol mass. Its share in the aerosol is minimal in Kyrgyzstan (56.2%) and maximal in the Far East region (89%) and Kamchatka (80.5%). The composition of aerosols from different sources changes significantly. Continental sources contribute elements from the lithosphere with a relative content approximately corresponding to their concentrations in the Earth's crust. However, this assumption is not valid for all the elements. Thus, for instance, the content of silicon in the aerosol varies from 3% in Kamchatka to almost 27% in Kazakhstan. Moreover, the content of 26.7% in aerosols of Kazakhstan is practically equal to the silicon concentration in the

Table 12.4 Average concentrations (μg/Mm^3) of elements and ions in the composition of the atmospheric aerosols above the regions of the former Soviet Union.

Element	European Russia	Ural	Kazakhstan	Turkmenistan	Tadzhikistan	Uzbekistan	Kyrgyzstan	Western Siberia	Eastern Siberia	Far East	Kamchatka
Si	3.8950	2.0618	4.8914	1.3813	2.2378	1.4290	5.9067	2.6217	2.3619	0.6014	0.6749
Al	0.8654	0.6377	0.9748	2.4970	0.4218	0.8346	1.9614	0.7596	0.2703	0.5756	0.1319
Fe	0.4956	0.7334	0.4206	0.2588	1.5975	0.1283	0.8476	0.3379	0.2142	0.1562	1.9931
Mg	0.1606	0.1781	0.2472	0.4528	1.5818	0.2287	0.1274	0.1304	0.0896	0.3149	0.0906
Ca	0.3372	0.7110	0.3011	2.9871	–	3.0150	0.5022	0.1538	0.2093	0.2129	0.9086
Cu	0.0862	0.0640	0.0781	0.1029	–	0.2200	0.2105	0.0762	0.0387	0.0775	0.0363
Ba			0.1150	–	–	–	–	0.0339	–	–	0.0870
Ti	0.0318	0.0934	0.0110	0.0302	0.1260	0.0169	0.0225	0.033	0.0283	0.0227	0.0855
Mn	0.0168	0.1100	0.0255	0.0654	0.1511	0.0215	0.0212	0.0123	0.0161	0.0162	0.0151
Cr	0.0470	0.0598	0.0593	0.1610	0.0453	0.0095	0.1200	0.0347	0.0388	0.0160	0.0580
Mo	0.0237	0.0021	0.0129	0.0063	0.0041	0.0036	–	0.004	0.0240	–	0.0058
Ag	0.0027	0.0141	0.0043	0.0002	–	–	0.0049	0.0043	0.0014	–	0.0154
Pb	0.0724	0.0483	0.0145	0.0035	0.0129	0.0190	0.0110	0.0176	0.0068	0.0185	0.0060
Ni	0.0307	0.0506	0.0457	0.0519	0.0534	0.0412	0.1410	0.0446	0.0551	0.0083	0.0364
Zn	0.0256	0.0492	0.0244	0.0549	–	2.7815	2.4600	0.0333	0.1418	0.0380	0.0810
B	0.0108	0.0110	0.0042	–	–	–	–	0.0679	0.4327	–	0.0672
V	0.0145	–	0.0302	0.0320	0.0156	0.0038	–	0.0056	0.0022	0.0016	0.0045
NO_3	4.9571	1.6867	3.1613	1.6335	9.8715	3.0365	9.6500	2.2814	1.3260	1.7433	–
Na	0.8520	0.8100	0.4183	1.5807	1.2689	0.7823	0.0630	0.6803	0.4072	0.4767	0.9150
Cl	4.3102	2.8150	4.0318	2.4424	2.1174	3.5380	1.7250	1.7229	4.0518	5.9811	0.2484
SO_4	0.4398	1.0350	0.5532	1.1609	1.1187	0.5143	–	0.6628	0.4264	0.2650	0.2400
K	0.1255	0.1175	0.0383	0.4243	0.4819	0.2179	0.0900	0.1128	0.1451	0.1010	0.8565
Br	1.3800	1.6908	1.8440	3.7679	7.2802	4.4711	2.4600	2.1817	2.8887	7.2267	13.770
NH_4	1.5749	0.4950	1.0096	0.8454	3.3764	2.4614	1.8800	1.2372	1.0262	0.9422	1.7313
Total	19.821	13.530	18.317	19.940	31.762	23.774	28.204	13.302	14.203	18.796	22.058

Table 12.5 Relative element composition of aerosols in the regions of the former Soviet Union.

Element	European Russia	Ural	Kazakhstan	Turkmenistan	Tadzhikistan	Uzbekistan	Kyrgyzstan	Western Siberia	Eastern Siberia	Far East	Kamchatka
Si	19.65	15.24	26.7	6.01	7.29	7.05	20.96	19.71	16.63	3.2	3.06
Al	4.37	4.71	5.32	3.51	13.18	1.33	6.96	5.71	1.9	3.06	0.6
Fe	2.5	5.42	2.3	0.54	1.37	5.03	3.01	2.54	1.51	0.83	9.04
Mg	0.81	5.26	1.35	12.68	15.77	–	1.78	0.98	1.47	1.13	4.12
Ca	1.7	1.32	1.64	0.96	2.39	4.98	0.45	1.16	0.63	1.68	0.41
Cu	0.43	0.47	0.43	0.93	0.54	–	0.75	0.57	0.27	0.41	0.16
Ba	–	–	0.63	–	–	–	–	0.25	–	–	0.39
W	0.33	0.41	–	–	–	–	–	0.39	–	–	–
Ti	0.16	0.69	0.06	0.07	0.16	0.4	0.08	0.25	0.2	0.12	0.39
Mn	0.08	0.81	0.14	0.09	0.35	0.48	0.08	0.09	0.11	0.09	0.07
Cr	0.24	0.44	0.32	0.04	0.85	0.14	0.43	0.26	0.27	0.09	0.26
Mo	0.12	0.02	0.07	0.02	0.03	0.01	–	0.03	0.17	–	0.03
Ag	0.01	0.1	0.02	–	–	–	0.02	0.03	0.01	–	0.07
Pb	0.37	0.36	0.08	0.08	0.02	0.04	0.04	0.13	0.05	0.1	0.03
Ni	0.15	0.37	0.25	0.17	0.27	0.17	0.5	0.34	0.39	0.04	0.17
Zn	0.13	0.36	0.13	11.7	0.29	–	0.73	0.25	1	0.2	0.37
B	0.05	0.08	0.02	–	–	–	–	0.51	3.05	–	0.3
V	0.07	–	0.16	0.02	0.17	0.05	–	0.04	0.02	0.01	0.02
NO_3	25.1	12.47	17.26	12.77	8.62	31.08	34.17	17.15	9.34	9.27	–
Na	4.3	5.99	2.28	3.29	8.35	3.99	0.22	5.11	2.87	2.54	4.15
Cl	21.75	20.8	22.01	14.88	12.9	6.67	6.12	12.95	28.53	31.82	1.13
SO_4	2.22	7.65	3.02	2.16	0.85	3.52	–	4.98	3	1.41	1.09
K	0.63	0.87	0.21	0.92	2.24	1.52	0.32	0.85	1.02	0.54	3.88
Br	6.96	12.5	10.07	18.81	19.89	22.92	8.73	16.4	20.34	38.45	62.42
NH_4	7.95	3.66	5.51	10.35	4.46	10.63	6.67	9.3	7.23	5.01	7.85

Earth's crust. For other regions the content of silicon in aerosols is lower compared to its percentage in the Earth's crust. This picture is observed in the Far East, Kamchatka, Uzbekistan, Turkmenistan, and Tadzhikistan.

Quantitatively, the deviation of the content of an element in the aerosol from its content in the Earth's crust is evaluated by using the *enrichment coefficient* concept:

$$K_X = \frac{(X/Fe)_{aer}}{(X/Fe)_{Ec}}$$

Here K_X is the enrichment coefficient of the element X; and $(X/Fe)_{aer.}$ and $(X/Fe)_{Ec}$ are the ratios of the concentrations of element X and Fe in aerosols and in the Earth's crust, respectively. Aluminum is often used as the base element in calculations. Terrigenous elements have values of enrichment coefficients close to 1. Values $K_X \geq 10$ could be caused by an essential influence of some other sources, including anthropogenic and oceanic. Also, elevated values of the enrichment coefficient could be significantly influenced by aerosol transformation in the atmosphere, fractionation, and particle removal from the atmospheric environment. For the elements Si, Al, Fe, Mg, Ca, Ti, Mn, Cr, and Zn, enrichment coefficients close to 1, and usually not exceeding 10, point to their natural origin. However, iron, calcium, and chromium are exceptions in their contributions to the aerosols in the Kamchatka region, where their enrichment coefficients reach 39.2, 18.4, and 17.6, respectively.

Also, it is possible to distinguish those elements whose concentrations change synchronously either in space or in time. Table 12.6 shows the main elements present in atmospheric aerosols along with a number of other elements that correlate in concentration statistically significantly. For example, iron has a significant correlation with three elements, Ca, Ag, and Ti, and with four ions, K^+, Cl^-, Br^-, and Na^+. The largest pair *correlation coefficient* for iron is observed with the potassium ion, 0.9452, which corresponds to the 0.995 significance level. The latitudinal alteration of potassium concentration generally corresponds to the alteration of iron concentration, with the only discrepancy observed in the area from Kazakhstan to Eastern Siberia, where the potassium concentration increases faster.

Calcium has an almost identical correlation in latitudinal variation with iron ($r = 0.9048$) and titanium ($r = 0.8992$). The smallest concentration of calcium is

Table 12.6 Number of concentration correlations for elements.

Fe	Al	Ag	K	Ca	Ti
7	6	6	6	5	5
Cl	Si	NO_3	Br	Cu	Mn
5	4	4	4	3	3
Zn	V	Na	Mo	NH_4	Mg
3	3	3	2	2	1
Cr	Pb	B	SO_4	Ni	
1	1	1	1	0	

observed in Western Siberia, and the highest in Kamchatka. The highest concentration of titanium is registered in the Urals, and the lowest in Kazakhstan. The correlation coefficient of iron with the sodium ion is equal to 0.6974. The greatest values of sodium ion concentration are observed in aerosols of the European area of the Russian Federation and Kamchatka. This is explained by the fact that sodium traditionally originates from the sea, and the quantity of sea air masses passing these regions is the largest compared to other regions of the former Soviet Union. The minimal concentrations of sodium are observed in Kazakhstan and Eastern Siberia.

The latitudinal changes in the chlorine ion concentration from the European area of the Russian Federation to Kamchatka are opposite compared to iron. The correlation coefficient between them is equal to 0.7832. It is remarkable that a similar pattern is observed also in the sodium–chlorine pair. The smallest values for chloride concentration are registered in Kamchatka, where their absolute content is considerably lower than in other regions.

Aluminum concentration in the latitudinal direction has a positive correlation with copper, silicon, vanadium, nitrate ion, and bromide ion, and a negative correlation with zinc and the potassium cation. The concentrations of aluminum in all these cases are close to each other and are in the interval of 0.6–1 $\mu g/m^3$. In Eastern Siberia the aluminum concentration is around 0.27 $\mu g/m^3$, and in Kamchatka it is approximately 0.13 $\mu g/m^3$ (the absolute minimum of aluminum concentration is observed in Kamchatka). The correlation of aluminum with silicon in the latitudinal direction can be explained by the fact that they are the most common terrigenous elements in nature. It is difficult to explain the correlation of aluminum and copper in terms of the natural origin of these elements only. Copper, in contrast to aluminum, appears in the atmosphere in large quantities in aerosol particles of anthropogenic origin. A high correlation of aluminum with the NO_3^- ion is also observed. Moreover, the latter also correlates well with silicon in the latitudinal direction.

Increased concentrations of silver are registered in the Ural region and Kamchatka. It was observed that the concentrations of silver in atmospheric aerosols are similar in magnitude in Kazakhstan and Western Siberia. However, the value is approximately 10 times lower compared to the Urals and Kamchatka. The smallest content of silver is observed in the aerosols of Eastern Siberia. The absolute concentration of silver in the entire latitudinal direction is 100 times lower than the concentration of the terrigenous elements – iron and calcium – that correlate with silver. Silver contributes only 10^{-5}% of the total mass of atmospheric aerosols.

Manganese in the latitudinal direction correlates with vanadium, sulfate anions, and ammonium cation. The correlation of manganese with ions is poor. The manganese concentration increases in the Ural region by almost 10 times. In other regions the manganese concentrations are close to each other.

The formation of sulfate ions in the atmosphere occurs according to the following mechanism. Sulfur dioxide, SO_2, dissolves in atmospheric water, forming sulfurous acid:

$$SO_2 + H_2O = H_2SO_3$$

The solution of sulfurous acid is slowly oxidized into sulfuric acid:

$$2H_2SO_3 + O_2 = 2H_2SO_4$$

Sulfuric acid is a strong electrolyte, so in dilute aqueous solution it dissociates:

$$H_2SO_4 = 2H^+ + SO_4^-$$

The sulfate ion reacts with metal cations, for example, manganese, and with the ammonium cation. In the first case metal sulfates are formed, and in the second case ammonium sulfate $(NH_4)_2SO_4$. This formation mechanism explains the rapid increase of manganese concentration in aerosols in the Ural region where all the required conditions for manganese sulfate formation in the atmosphere are present: manganese in substantial quantity, anthropogenic ejections of sulfur dioxide, and sufficient amount of atmospheric moisture.

The above mechanism of sulfate formation is not the only one that explains the presence of sulfates in aerosol particles. Sulfates are widespread in nature. The most common are sodium sulfate (Na_2SO_4), potassium sulfate (K_2SO_4), magnesium sulfate ($MgSO_4$), which is contained in sea water, and calcium sulfate ($CaSO_4$), which is found in nature in gypsum. Thus, atmospheric sulfates could be formed directly in the atmosphere, or could originate from the continents and sea. In the latitudinal direction, sulfate concentration has two maxima, which are observed in the Ural region and Western Siberia. Most likely, this is the result of anthropogenic sulfur dioxide from metallurgical foundries, which are frequently met in these regions. The sulfate concentration decreases monotonically in the eastern direction.

Ammonium (NH_4^+) does not correlate well with sulfates. However, its correlation in the latitudinal direction is observed with manganese and the nitrate ion, NO_3^-. Most likely, these two ions represent ammonium nitrate, NH_4NO_3, which is formed in the reaction of ammonia NH_3 (commonly present in the atmosphere) with water, leading to the formation of ammonium ions and nitrogen oxides. The lowest concentrations of ammonium ions, as well as nitrate ions, are observed in the Ural region, and the highest in the European area of the Russian Federation and Kamchatka.

The elements and ions that are found in atmospheric aerosols form various groups correlating with each other in the latitudinal direction. The elements that have the largest numbers of such correlations are of terrigenous origin, and also the potassium, chlorine, and nitrate ions.

The total concentration of terrigenous elements in essence reflects the special physical geographical features of the regions. On the other hand, the total concentrations of microelements and ions do not have this characteristic, which is possibly explained by their considerable dispersion in the atmosphere.

The group of ions makes the greatest contribution to the relative chemical composition of the aerosol. The elemental composition of aerosols and the absolute content of elements and ions have maximum diversity in the Central Asian regions.

Data analysis on aerosol pollution from 1980 to 1990 reveals a definite tendency toward a decrease of the aerosol content during this period. Figure 12.5 summarizes

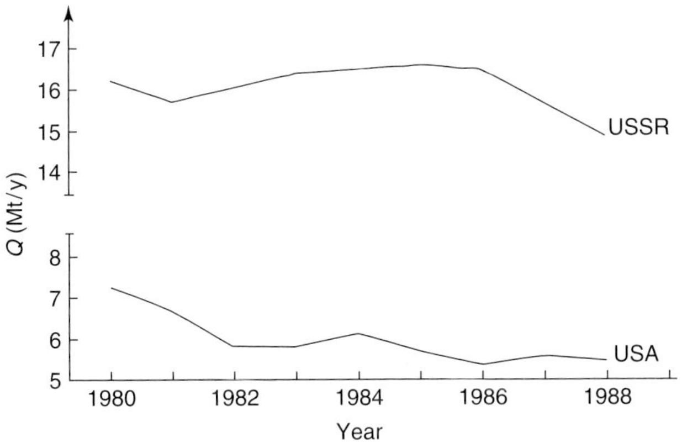

Figure 12.5 Emission of aerosols in the USSR and USA, 1980–1988.

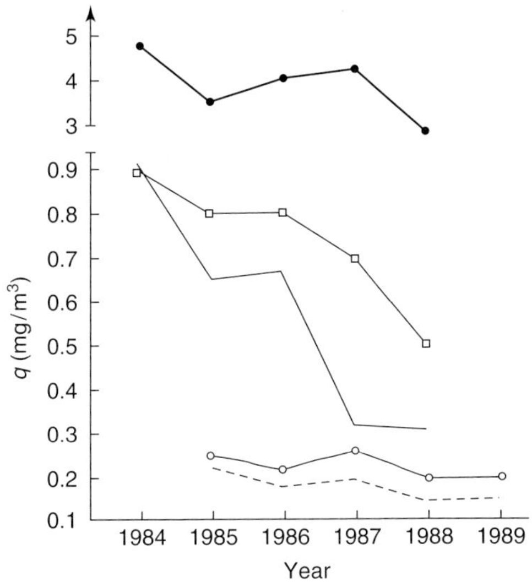

Figure 12.6 Concentration of suspended matter in various cities of the former USSR, 1984–1989: Omsk (full circles); Kaliningrad (open squares); Novosibirsk (full line); Tolyatti (open circles); Karaganda (dashed line).

data published in different sources on the total concentration of suspended matter in the former USSR and USA. Figure 12.6 shows the total levels of atmospheric aerosols in various cities across the former USSR territory. Some decrease in the aerosol concentration was also apparent in Calcutta, Athens, Madrid, and Milan. Some analysis of the results allows one to draw the conclusion that the decrease of the aerosol concentration does not occur because of a decrease of particles of anthropogenic origin, since their emission to the atmosphere did not decrease in the 1980s. Instead, it could be assumed that some background aerosol resulting from natural processes substantially decreased, and the observed results justify the long-standing assertion of the existence of atmospheric aerosols. Most likely, the decrease in aerosol concentration observed in the lower atmosphere is caused by natural processes, and the fluctuation period is close to the known 11-year cycle.

The aerosol concentration trend in the mid-1980s could be caused by circulation processes, in particular by an increase in the frequency and intensity of the western zonal circulation accompanied by an increase in frequency of the meridional circulation, without any substantial change in its intensity. This phenomenon is also confirmed by an increase of the *Blinova index,* which represents the ratio of the linear velocity of the air along the latitude direction by the distance to the axis of the Earth. In 1984–1989 this index grew from 34 to 42, that is, the intensity of zonal circulation in this period was increased. An increase in intensity of the western zonal flow above the Ural mountains could form a high-altitude pressure crest playing a blocking role in atmospheric motion. As a result, air masses from the Arctic Ocean start to move along ultra-polar trajectories toward Western Siberia. The intensity increase of the western zonal circulation was accompanied by more frequent occurrence of blocking processes in the Ural region, which was increased from 16% in 1983 to 30% in 1988. On this basis, it could be concluded that the aerosol concentration behavior is characterized by two simultaneously acting processes: (i) increase in the periodicity of the moderate-scale air masses entering from the Atlantic Ocean along zonal trajectories; and (ii) pressure-blockage-related trajectory alteration of Arctic air masses, making them move to Western Siberia from the Kara Sea, not across the European part of Russia, but along ultra-polar (meridional) paths ensuring higher purity of air.

A similar conclusion could also be reached based on an analysis of the chemical composition of aerosols. The concentrations of sea-related NH_4^+ and Na^+ in the aerosol mass were substantially increased during the period of time considered (Figure 12.7). In addition, the increase in these components excludes from consideration any possibility of their anthropogenic origin.

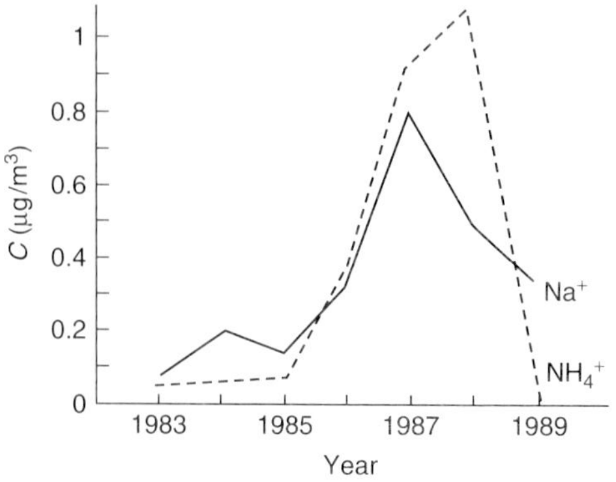

Figure 12.7 Average annual concentrations of Na^+ and NH_4^+ ions in Western Siberia aerosol, 1983–1989.

12.4
Aerosols in the Stratosphere

Direct measurements of aerosols in the stratosphere by *aerostatic impactors* were started in the late 1950s by Junge's aerosol group [38]. For the following years, the instrumentation was dynamically developed by several research groups. In the 1960s, an *aerostatic photoelectric counter* with two size ranges, $r > 0.15$ and $0.25\ \mu m$, was produced [39]. With the aid of this counter, a substantial amount of data on the vertical structure of aerosols at heights of up to 30–35 km for different climatic regions was acquired. The impactor's operational principle is based on particle settlement onto a continuously moving substrate, located near a specially designed nozzle capable of creating a subsonic velocity for the air sample passing through the device. The position of the particle on the substrate determines the height of sampling, as it is associated with the duration of the ascent of the equipment carrying the aerostat. The accuracy of the sampling height determination is ± 200 m, which mainly relates to the finite width of the nozzle, deviation of the particle from the center of the jet, and the accuracy of the radar tracing the flight time of the aerostat. Errors in determination of the particle concentration mainly depend on the particle size and air sample intake altitude. Parallel measurements using other instruments (photoelectric counter, nephelometer, and so on) significantly improve the quality and reliability of the data.

Since 1987, a two-stage impactor has mainly been used for aerostatic measurements. The first stage is used for precipitation of large particles, and also for protection of the second-stage substrate from contamination during the violent landing of the equipment carrier (aerostat). As a result, in many cases, data acquired by the first stage of the instrument are not taken into consideration due to the low reliability. After careful selection and testing in a large number of experiments, polyvinyl-formaldehyde (formvar) was selected as the most suitable material for the substrate. It is resistant to the chemically active substances found in atmospheric aerosols, possesses strong adhesive properties, and has sufficient mechanical strength.

In 1975 at the base of the Central Aerological Observatory near Ryl'sk City (USSR) and in summer 1976 in the area of Laramie (USA), large-scale aerostatic measurements of atmospheric aerosols were undertaken. Vertical profiles of atmospheric aerosols were obtained by a filter trap, impactor (USSR), and optical particle counter (USA). The possible influence of aerosols on the penetration of long-wave radiation was examined by actinometrical sounding of the atmosphere, taking into consideration the vertical distribution of ozone and other active gas components of the atmosphere. These studies were continued in 1978 and the following years.

Measurements of particles with a radius $r > 0.2$ and $0.25\ \mu m$ by impactor, and particles with $r > 0.15$ and $0.25\ \mu m$ by dust photoelectric probe, demonstrated the good agreement of the results acquired in the troposphere. The discrepancy was slightly larger at higher elevations – namely, in the lower stratosphere. Both methods were capable of registering *Junge's layer* at heights from 17 to 22 km. The background content of stratospheric aerosol particles in Junge's layer was measured to be approximately 1–3 cm^{-3} (for particles with $r > 0.15$ μm), reaching a maximum concentration in the lowest layers of the tropopause.

The concentration of aerosols in the lower stratosphere is characterized by maximum magnitudes and a layered fine structure of the vertical distribution. The presence of a permanent and long-lasting aerosol layer in the stratosphere was confirmed for both the eastern and western hemispheres of the planet. Simultaneous measurements of condensation nuclei and aerosols showed that the concentration of nuclei in the stratosphere is minimal and comparable with the concentration of particles with radius larger than 0.15 μm. Relative humidity change and the presence of condensation nuclei lead to the formation of both condensation and coagulation stratospheric aerosols. Stratospheric aerosol content has seasonal and latitudinal variations, which, in many cases, are related to atmospheric circulation and volcanic activity. This assumption is supported by comparable concentrations of aerosols at the North and South Poles.

The size distribution of stratospheric aerosols in the range from 0.2 to 2.5 μm is described by the widely used Junge formula [22] with an exponent value of 3.5–4.0 that is relatively constant. Some tests showed the presence of approximately 0.3 $\mu g/m^3$ of Ca and 0.001 $\mu g/m^3$ of Mn. Careful analysis of impactor tests demonstrated that the majority of particles are either sulfuric acid or sulfate drops, depending on the timeframe between the nearest volcanic eruptions. Most of the results characterizing the structure and microphysical characteristics of atmospheric aerosols are obtained by electron microscope analysis of impactor tests. The formvar film used as the collection substrate allows the observation of particles with a diameter down to 5 nm. In the range of sizes below 0.2 μm there is a great uncertainty in the effectiveness of particle capture on the substrate, especially for heights below 10 km: the results could be substantially understated (by up to 2–3 times). The particle size distribution was determined by counting of particles visualized by the microscope. The *particle size* was defined as the average projection diameter without taking into account some possible particle deliquescence, which could overestimate the actual particle size by up to ~20%. Morphological analysis enables the following types of particles to be determined: particles from organic matter, sulfuric acid and sulfate particles, small particle agglomerates, and soot fractals (Table 12.7). It could also determine particles of mixed origin: uniform spherical particles of dense substances (high-temperature condensates, for example micrometeorites), and crystals formed in atmospheric reactions (*in situ*). Chemical and elemental analysis is usually carried out by mass spectrometry, infrared spectroscopy, neutron activation, and X-ray fluorescence. Samples are usually obtained at three altitude ranges: 4–10, 10–15, and higher than 15 km. The height of the tropopause has an essential effect on the vertical profile of the aerosol concentration. Aerosol layers in the tropopause are formed by exchange processes at the boundary of the tropopause and are significantly driven by volcanic activity.

The available data allow one to assume that optically active aerosol layers, similarly to clouds, have radiation-active boundary layers, where a radiant heat inflow

Table 12.7 Average relative contribution (mass %) of various components in finely dispersed and giant aerosol particles.

Type of aerosol	Weight part	
	<2 μm	2–10 μm
Organic aerosols	26 ± 7	18 ± 9
Element carbon	3.2 ± 1.2	<1.0
Sulfates	17 ± 4	11 ± 11
Ammonium	10 ± 2	36 ± 12
Nitrates	25 ± 6	0.3 ± 0.1
Sea salt	1.5 ± 1.3	0.1 ± 0.3
Soil	3.2 ± 1.8	32 ± 11
Small elements	0.3 ± 0.1	10 ± 7
Smoke	0.4 ± 0.2	5.7 ± 3.8
Total	86 ± 4	113 ± 50

changes sign. The optical properties and microstructure of aerosols substantially influence atmospheric radiation conditions, and their fluctuation at the global scale could lead to noticeable climate changes.

Temperature changes resulting from the action of short-wave and long-wave radiation are maximum at the atmospheric boundary layer and decrease with increase of altitude. The presence of large-scale aerosol layers in the atmosphere (some type of Sahara aerosol layer or industrial aerosol) leads to radiation heating at the level of 0.4 °C/h and decrease of the Earth's cooling by long-wave radiation. The presence of stratospheric aerosols and thermal stratification leads to radiation heating by 0.1 °C/h in the 50–100 mbar layer.

The vertical profiles of aerosol concentration obtained in Apatites City by a two-stage impactor (SRIPh LSU) are represented in Figure 12.8. For particles larger than 0.3 μm, no influence of stratospheric aerosol layer was observed (above 10 km). Such an influence is weakly present in the 15.8–18.1 km elevation range and is much more strongly seen at heights above 23.8 km. In contrast, for smaller particles, distinctly expressed layers were revealed. Moreover, for $d > 0.01$ and 0.1 μm, the scenarios are noticeably different. The presence of two layers with elevated aerosol concentrations were reliably observed in the polar stratosphere at heights of 14–16 km and 20–21 km. Electron microscope analysis data showed definite trends in alterations of both size distribution and morphological structure of aerosol particles according to height. A survey of these data testifies that in the Arctic stratosphere the quantity of particles with diameters less than 0.1 μm is much smaller compared to the middle latitudes. The numbers of particles decrease with increase of particle size (Table 12.8). Considerably larger particle numbers representing all size ranges were observed in the lower stratosphere. At heights above 24 km, the contribution of the smallest measured particles (with sizes below 0.03 μm) to the total aerosol content decreased with altitude increase. Obviously,

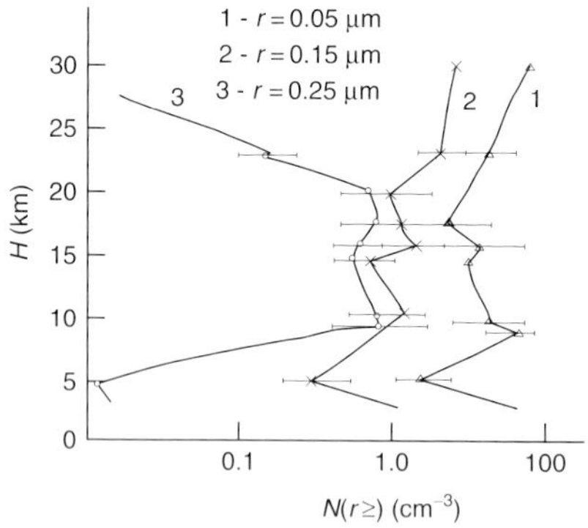

Figure 12.8 Vertical profiles of the aerosol concentration $N(r \geq)$ (cm^{-3}).

Table 12.8 Sizes of aerosol particles at different heights, $\Delta N(d_2 - d_1)$ (cm^{-3}) in Apatites City, Russia.

z (km)	Δd_i (μm)					
	<0.015	0.015–0.03	0.03–0.05	0.05–0.07	0.07–0.10	0.10–0.15
5–10		17 ± 2		3 ± 1		2 ± 1
15–20	46 ± 5	22 ± 3	9 ± 2	4 ± 1.5	3 ± 1	3 ± 1
24–30	0.2 ± 0.1	1 ± 0.5	1.5 ± 0.8	2.5 ± 1.5	3 ± 1.5	4 ± 2
	0.15–0.20	0.20–0.30	0.3–0.5	0.5–0.7	0.7–1.0	>1.0
5–10	2 ± 1	2 ± 1	1 ± 0.7	1 ± 0.7	0.1 ± 0.07	0.1 ± 0.07
15–20	1 ± 0.5	0.5 ± 0.3		0.5 ± 0.3		
24–30	2 ± 1	1.0 ± 0.7		1.5 ± 0.8		

this could be explained by a lack of production sources of particles within these sizes. The particle size range 0.05–0.15 μm is represented by the maximum concentration within the entire range of suspended matter. This phenomenon could be explained by their long lifetime in the stratosphere, which leads to particle condensation and coagulation growth allowing them to reach this range of sizes.

Trends in the change in morphological structure of aerosol particles with height are distinctively observed. In general, relatively large particles of mineral origin, smaller particle conglomerates, and organic particles are frequently observed in the troposphere. Sulfuric acid particles are much less common at these elevations. Approximately 10% of these aerosols could be interpreted as soot particles. The contribution of acidic particles grows with height, reaching 50–60% of the total concentration at elevations of 9–10 km. In essence, as visualized by microscopy, these are spherical liquid particles with submerged small solid inclusions. Interestingly, at this height, even solid particle agglomerates contain sulfuric acid, which is present between the solid components.

In the lower stratosphere, the concentration of spherical sulfuric acid particles increases further. There are almost no particles that contain organic matter and a very low concentration of particles of mineral origin (mainly, volcanogenic particles).

In the upper levels of the atmosphere, particles with irregular shape practically disappear, and mainly particles of sulfuric acid or sulfate origin are found. In addition to acidic and sulfate particles, some interesting structures (that look like spider's webs), made out of ultra-small particles, could be observed.

Aerostatic measurements of aerosols by two-stage impactor were carried out in the spring of 1994 on Heiss Island, Franz Josef Land, Russia (Table 12.9). Unlike previous measurements, sampling was undertaken from a captive aerostat at several height ranges: 25–50, 75–125, and 150–250 m. Unfortunately, no correction for the wind was possible, which could contribute some inaccuracies in height reporting. The main scope of the task was to determine any specific microstructural features of polar aerosols at lower layers of the atmosphere. Some parallel measurements were performed on Ziegler Island, Franz Josef Land, where data were acquired by electrostatic analyzer and photoelectric counters. The results of tests obtained by different instruments demonstrated close agreement on the size distribution of the aerosol particles. A multimodal size distribution structure was

Table 12.9 Distribution of polar aerosol particles, ΔN (liter^{-1}), in the lower atmosphere at various heights above sea level in March 1994 according to size, $d_i – d_{i-1}$ (μm).

$d_i – d_{i-1}$ (μm)	ΔN_i (liter^{-1})					
	Franz Josef Land				Southern Ocean	
	0 m	25–50 m	75–125 m	150–250 m	Atlantic	Pacific
0.015–0.03	2200	1400	608	710	–	–
0.03–0.06	2200	1560	815	816	–	–
0.06–0.10	280	160	101	74	–	–
0.10–0.15	140	75	51	37	–	–
0.15–0.25	80	36	31	24.0	–	–
0.25–0.40	17.0	17.0	24.0	50.0	–	–
0.40–0.50	6.0	5.3	8.0	16.0	1.30	1.0
0.50–0.60	8.5	3.4	1.80	2.7	0.90	0.70
0.60–0.70	7.4	2.9	1.50	2.1	0.70	0.60
0.70–0.80	5.8	2.3	1.20	1.40	0.60	0.40
0.80–0.90	4.2	1.70	0.88	1.40	0.40	0.30
0.90–1.0	3.2	1.30	0.67	1.00	0.30	0.20
1.0–1.5	2.6	1.00	0.60	0.85	0.30	0.20
1.5–2.0	1.90	1.40	0.75	8.6	0.10	0.10
2.0–4.0	0.27	0.20	0.18	0.35	0.07	0.06
4.0–7.0	0.13	0.060	0.055	0.08	0.00	0.00
7.0–10.0	0.006	0.002	0.005	0.004	0.00	0.00

reliably determined. It was found that the size distribution is substantially influenced by the wind direction: southern winds bring in particles of anthropogenic origin in the range 0.4–3.0 μm, whereas eastern winds noticeably decrease the concentration of small particles with diameter less than 0.01 μm. The averaged results of the size distribution of the polar aerosol particles are shown in Table 12.9. Data on the size distribution of Antarctic aerosols near the coast are provided for comparison.

Comparison of the particle size distribution spectra for different high-latitude regions shows that the main differences in particle size are observed for aerosol with diameter less than 0.1 μm. This finding testifies to the domination of photochemical processes of aerosol formation from gases in the atmosphere of the region. Therefore, alterations in particle concentrations for different sampling locations were significant, which was especially related to the giant particles. Some elevational variations in the particle concentration for the entire measured size spectrum were observed in the lower layers of the atmosphere. Even unexpected concentration bursts were detected at high altitudes. Most likely, this phenomenon is caused by special features of the particle transfer processes occurring near the terrestrial surface.

The morphological structure of aerosol particles at different heights testifies to their different origins. Sulfate particles of rosette shape dominate at heights beyond 150 m. On the other hand, at the heights below 150 m, a noticeable part of the aerosol is made up of relatively dense spherical particles of metal oxides with sizes of 0.01–0.03 μm. In the lowest layers of the atmosphere, a wide variety of particle shapes and densities, along with the presence of sulfuric acid, is observed. These results testify to the very weak vertical mixing of the lower layers of the polar troposphere, and its distinct stratification. These findings allow one to consider polar aerosols to be unique and to separate them into a special type.

Research on atmospheric aerosols at heights beyond 35 km is extremely complicated from both technological and methodological points of view, as aerostats cannot be used at those elevations. Most known outcomes were reported by a group of Swedish researchers [40], who undertook measurements in the zone of noctilucent clouds by sampling from a rocket. Also, some impactor tests of aerosols at elevations from 20 to 100 km were undertaken by staff of the Fedynski Laboratory at the Central Aerological Observatory (Russian Federation) [41]. The electron microscope analysis of these tests showed an abrupt change in the morphological particle structure with height. At heights below 25–30 km the picture was similar to the previously discussed aerostatic experiments performed at similar altitudes. With increase of elevation, the appearance of distinctively different spherical particles with diameter smaller than 0.2 μm was observed. Also some particle concentration bursts were detected at heights where the temperature gradient changes. The size distribution at those heights has a distinctive multimodal character.

References

1. Kondratyev, K.Ya. and Ivlev, L.S. (2008) *Climatology of Aerosols and Cloudness,* Prirodnaye i tekhnogennyye aerozoli, vol. 1, S-Petersburg, VVM, p. 555.
2. Ivlev, L.S. and Dovgalyuk, Y.A. (1999) *Physics of Atmospheric Aerosol Systems,* St. Petersburg University, St. Petersburg.
3. Kondratyev, K.Y. (ed.) (1991) *Aerosol and Climate,* Gidrometeoizdat, Leningrad.
4. Matveyev, L.T. (2000) *Physics of the Atmosphere,* 3rd edn, Gidrometeoizdat, St. Petersburg.
5. Green, H. and Lane, V. (1969) *Aerosols – Dusts, Smokes and Fogs,* Russian translation, Khimiya, Leningrad.
6. Ivlev, L.S. (2002) *Basis of Physics. The Formation of Weaether and Climate,* Prirodnaye i tekhnogennyye aerozoli, (part 1, vol. 2), S-Petersburg, VVM, p. 284.
7. Smirnov, V.V. (1992) *Ionization in the Troposphere,* Gidrometeoizdat, St. Petersburg.
8. Grini, A., Zender, C.S., and Colarco, P.R. (2002) Saltation sandblasting behavior during mineral dust aerosol production. *Geophys. Res. Lett.,* **29** (18), 1868.
9. Twomey, S. (1977) *Atmospheric Aerosols,* Elsevier, Amsterdam.
10. Reist, P. (1987) *Aerosols. Introduction into the Theory,* Mir, Moscow.
11. Kondratyev, K.Y. and Isidorov, V.A. (2001) The influence of the combustion of biomass on the chemical composition of the atmosphere. *Opt. Atmos. Ocean,* **14** (12), 107–114.
12. Kondratyev, K.Y., Ivlev, L.S., Krapivin, V.F., and Varotsos, C.A. (2005) *Atmospheric Aerosol Properties, Formation, Processes and Impacts,* Springer Praxis, Chichester.
13. Friedlander, S.K. (1977) *Smoke, Dust and Haze: Fundamental of Aerosol Behavior,* John Wiley & Sons, Inc., New York.
14. Bates, T.S. (1999) First aerosol characterization experiment (ACE-1). Preface. *J. Geophys. Res.,* **104** (D17), 21645–21647.
15. Sinha, P., Hobbs, P.V., Yokelson, R.J., Blake, D.R., Gao, S., and Kirchstetter, T.W. (2003) Distributions of trace gases and aerosols during the dry biomass burning season in southern Africa. *J. Geophys. Res.,* **108** (D17), 4536.
16. Cooke, W.F., Ramasvamy, V., and Kasibhatla, P. (2002) A general circulation model study of the global carbonaceous aerosol distribution. *J. Geophys. Res.,* **107** (D16), 4279.
17. Arias Villanueva, E.I., Ivlev, L.S., and Kudryashov, V.I. (2003) Variation of the aerosol concentration and chemical composition in lower layer of West States Mexico. Proceedings of the Third International Conference on Natural and Anthropogeneous Aerosols, CRI Physics, SPSU, Saint Petersburg (ed. L.S. Ivlev), pp. 54–71.
18. Mironova, I.A. (2005) Influence of the solar activity on the transparency of the atmosphere and the optical properties of aerosol. Thesis for scientific degree of Candidate of Physics and Mathematical Sciences, Saint Petersburg.
19. Cadle, R. (1969) *Hard Particles in the Atmosphere,* Russian translation, Nauka, Moscow.
20. Hansen, J., Lacis, A., Ruedy, K., and Sato, M. (1992) Potential climate impact of Mt. Pinatubo eruption. *Geophys. Res. Lett.,* **19**, 215–218.
21. Hansen, J., Sato, M., Lacis, A., and Ruedy, R. (1997) The missing climate forcing. *Philos. Trans. R. Soc. Lond., Ser. B,* **352** (231–240)
22. Ivlev, L.S. (1982) *Chemical Composition and the Structure of Atmospheric Aerosols,* Leningrad State University, Leningrad.
23. Ivlev, L.S., Galindo, J., and Kudriashov, V.I. (1996) in *Report Centro Universitario de Investigation en Ciencias de la Tierra* (ed. J.Galindo), Universidad de Colima, Colima, Mexico, pp. 257–284.
24. Donchenko, V.K. and Ivlev, L.S. (2003) Identification of different origin aerosols. Proceedings of the Third International Conference on Natural and Anthropogeneous Aerosols, CRI Physics, SPSU, Saint Petersburg (ed. L.S. Ivlev), pp. 41–51.
25. Seinfeld, J.H. and Pandis, S.N. (1998) *Atmospheric Chemistry and Physics. From Air Pollution to Climate Change,* Wiley Interscience, New York.

26. Leyva EŒntreras, A., Ivlev, L.S., Vasilyev, A.V., and Vasilyev, S.L. (2003) Complex aerosol investigation of urban Mexico. Proceedings of the Third International Conference on Natural and Anthropogeneous Aerosols, CRI Physics, SPSU, Saint Petersburg (ed. L.S. Ivlev), pp. 72–74.
27. Metzger, S., Dentener, F., Pandis, S., and Lelieveld J. (2002a) Gas/aerosol partitioning. 1. A computationally efficient model. *J. Geophys. Res.*, **107** (D16), 4312.
28. Metzger, S., Dentener, F., Krol, M., Jeuken, A., and Lelieveld, J. (2002b) Gas/aerosol partitioning. 2. Global modeling results. *J. Geophys. Res.* **107** (D16), 4313.
29. Crutzen, P.J. and Zimmerman, P.H. (1991) The changing photochemistry of the troposphere. *Tellus*, **43A–B**, 136–151.
30. Isidorov, V.A. (2001) *Ecological Chemistry*, Khimiya, St. Petersburg.
31. Went, F.W. (1966) On the nature of Aitken condensation nuclei. *Tellus*, **18** (2–3), 549–556.
32. Griffin, R.J., Nguyen, K., Dabdub, D., and Seinfeld, J.H. (2003) A coupled hydrophobic–hydrophilic model for predicting secondary organic aerosol formation. *J. Atmos. Chem.*, **44** (2), 171–190.
33. Narukawa, M., Kawamura, K., Hatsushika, H., Yamazaki, K., Li, S.-M., Bottenheim, J.W., and Anlauf, K.G. (2003) Measurement of halogenated dicarboxylic acids in the Arctic aerosols at polar sunrise. *J. Atmos. Chem.*, **44**, 323–335.
34. Keller, M., Jacob D., Wofsy S.C., and Harris, R. (1991) Effects of tropical deforestation on global and regional atmospheric chemistry. *Climate Change*, **19**, 139–158.
35. Quinn, P. and Bates, T. (2003) Comparison of regional aerosol chemical and optical properties from the European, Asian, and North American plumes. *IGACtiv Newsl.*, **28**, 24–30.
36. Oberlander, E.A., Brenninkmeijer, C.A.M., Crutzen, P.J., Elansky, N.F., Golitsyn, G.S., Granberg, I.G., Scharffe, D.H., Hofmann, R., Belikov, I.B., Paretzke, H.G., and van Velthoven, P.F.J. (2002) Trace gas measurements along the Trans-Siberian railroad: the TROICA 5 expedition. *J. Geophys. Res.*, **107** (D14), 4206.
37. Tolmachev, G.N. (2000) Horizontal distribution of ion-element content of atmospheric aerosol above the area of USSR. Proceedings of the Second International Conference on Natural and Anthropogeneous Aerosols, CRI Chemistry, SPSU, Saint Petersburg (ed. L.S. Ivlev).
38. Junge, Ch., Chagnon, C.W., and Manson, J. (1961) Stratospheric aerosols. *J. Meteorol.*, **18**, 81–108.
39. Rosen, J.M., Hofmann, D.J., and Laby, J. (1975) Stratospheric aerosol measurements II: the world wide distribution. *J. Atmos. Sci.*, **32**, 1457–1462.
40. Witt, G. (1969) *The Nature of Noctilucent Clouds*, Space Research, vol. IX, North-Holland, Amsterdam, pp. 154–169.
41. Ivlev, L.S. and Fedynsky, A.V. (1975) Studies of aerosol content in the upper atmosphere, in *Physics of Mesosphere and Mesospheric Clouds*, Meteorology Researches, vol. 22, Results of Researches on the International Geophysical Projects, Nauka, Moscow, pp. 17–26.

13
Biological Aerosols

Sergey A. Grinshpun

13.1
Introduction

This chapter addresses *biological aerosols* or bioaerosols – an increasingly visible area of knowledge that combines several disciplines such as *aerosol physics, microbiology, environmental science,* occupational and *public health,* and others. Owing to the space limitations of this book and its objective, no attempt was made to write a comprehensive review on bioaerosol research. Several recently published monographs and textbooks have covered a much greater territory as compared to the scope of this chapter. Nonetheless, in my opinion, this book (as any book on aerosols published these days) would be lacking completeness if the bioaerosol-related material was not included. This chapter is meant to serve as a guide through the fundamentals and some of the recent research accomplishments in the areas of measurement, characterization, and control of *bioaerosol particles*.

In addition to recognizing the importance of the bioaerosol research area, I was excited to have the opportunity to contribute to a project that brought together scholars who were originally trained in Russia and the Eastern European countries during the Cold War and who are presently working at different institutions around the globe.

13.2
History of Bioaerosol Research

The earliest effects of bioaerosols were recorded in ancient history [1]. Microscopic recognition that became available in the seventeenth century helped to establish the first studies on airborne transmission. By the end of the nineteenth century, research involving airborne biological particles was conducted mostly by microbiologists and medical researchers. The exploration was primarily motivated by exploring the transmission pathways of infectious diseases. At the time, the lack of aerosol sampling methods was a major challenge. The first measurement data, reported by Singerson [2], resulted from collecting bioaerosols in room air. In the

Aerosols – Science and Technology. Edited by Igor Agranovski

ISBN: 978-3-527-32660-0

1880s, a glass impinger was used for sampling of *bacterial aerosols* in Robert Koch's laboratory at the Institute of Hygiene in Berlin, Germany [3]. By the mid-1930s, it was recognized that airborne, micrometer-sized particles may carry pathogenic microorganisms that cause infection. For the following several decades, bioaerosol research was motivated primarily by practical needs in three areas: public health, defense, and agriculture. From the 1940s to the 1980s, extensive studies on *biological warfare* (bacterial spores, viruses, and toxins) were conducted in major institutions of the former Soviet Union. During about the same period, similar research programs were established and carried out in the USA and in several other countries. Many of these programs were closed in the 1970s based on international treaties, though some continued through the end of the Cold War.

The extraordinary progress in aerosol research that occurred in the 1960s through 1980s led to the emergence of many subdisciplines, some of which have become "self-sufficient." Bioaerosol research appears to serve as a perfect example of such a discipline. It was initially driven by the need to identify and quantify potentially pathogenic airborne particles generated by people infected with viral or *bacterial illnesses* in indoor and outdoor environments. In the initial phase, the investigators as well as their "customers" were mostly microbiologists and infectious disease experts, and the main challenge was to develop adequate methods and techniques for sampling and analyzing *airborne microorganisms*. By the early 1990s, it was widely realized that progress in measuring and characterizing biological particles and their health effects could be most effectively made by bringing together scientists representing aerosol physics, engineering, and microbiology, as well as environmental and occupational health. For the past two decades, the aerosol research community has enjoyed a rapid increase in peer-reviewed publications and conference presentations related to bioaerosols. Recognizing the increasing importance of bioaerosol research, the *Journal of Aerosol Science* allotted three special issues to the measurement and characterization of biological aerosols – edited by Ho and Griffiths [4], Lacey [5], and Grinshpun and Clark [6]. Further advances in bioaerosol studies have enhanced areas such as indoor and outdoor *air quality assessment*, filtration and *respiratory protection*, investigation and prevention of health effects (including emerging diseases), and biodefense and counter-terrorism research. In 2008, another special issue devoted to a variety of topics in bioaerosol research was published by the then newly established journal *CLEAN – Soil, Air, Water* (formerly *Acta Hydrochimica et Hydrobiologica*) [7], an interdisciplinary journal covering all aspects of sustainability and environmental safety.

The anthrax attack in the USA as well as several recent outbreaks of emerging diseases worldwide – severe acute respiratory syndrome (SARS), avian flu, and swine flu – have produced an additional spike of attention to bioaerosol research and a significant increase in governmental and industrial funding. Several reports have been published in this decade on the assessment of *health hazards* posed by aerosolized biological agents [8–12]. While numerous techniques are presently available for sampling and analyzing airborne biological particles, the measurement and characterization of bioaerosols remains a challenge. Another related area that has recently moved up on the priority list is *air quality control* and exposure reduction in environments contaminated with bioaerosol particles. This includes (but is not limited to) filtration and respiratory protection against airborne microorganisms.

13.3 Main Definitions and Types of Bioaerosol Particles

A *bioaerosol* is defined as an aerosol comprising particles of variable biological origin, for example, *viruses*, bacterial cells and spores, *fungi*, *pollen*, as well as their aggregates and fragments. The definition is often extended to protozoa, excreta, or fragments of insects, skin scales or hair of mammals, or other components, residues or products of organisms, such as bacterial lipopolysaccharides, that is, endotoxins, or fungal *mycotoxins* [1, 13, 14]. Bioaerosols are present in both indoor and outdoor air. According to Jaenicke [15], 5–50% of ambient particles larger than 0.2 μm are of biological origin. Bioaerosol particles are characterized by a broad size spectrum (from nanometers to hundreds of micrometers) and a wide range of number concentrations. They may be aerosolized from different media, for example, water, soil, vegetation, and, of course, living organisms. As knowledge about the physical, chemical, and biological properties of bioaerosols has grown, and more associations have been identified between these properties and various health effects in humans and animals, it has become clear that bioaerosols should be considered as an integral part in the development of a sustainable and safe environment [16].

Generally, the behavior of bioaerosol particles utilizes the same principles as for non-biological ones. Biological particles present a great diversity in dimension, shape, electric charge, and other physical characteristic. In addition, it is often imperative to distinguish between viable and *non-viable bioaerosol* particles. The definition of viability is not explicit [17], but generally regards viable microbial cells as those which are able to reproduce or have metabolic activity [14]. Less than 10% of the airborne bacteria are typically viable [18]; higher viabilities have been reported for fungal spores [19]. Following the same concept, non-viable organisms are unable to reproduce. Survival or biological activity of airborne microorganisms in the environment is an important aspect, which makes bioaerosol characterization different from that of inert aerosols.

Among potentially viable particles, viruses are the smallest, ranging approximately from 0.02 to 0.3 μm. Bacteria and fungal spores cover a size range from about 0.3–0.5 μm up to almost $\sim 10^2$ μm. Pollen, algae, protozoa, and dander are several tens to hundreds of micrometers in diameter [14]. The density of bioaerosol particles is not well characterized. For microbial cells, which consist mainly of water, the density is relatively close to (and often exceeds) 1000 kg/m^3 and depends on the degree of cell hydration, the reserve materials, and the lipid content of the cell [20]. Intact fungal spores and pollen grains are monodisperse aerosol particles that are usually slightly heavier than water; however, they are often aerosolized as aggregates with *particle effective density* below 1000 kg/m^3.

From the aerosol science and microbiology perspectives, the most common and explored types of bioaerosol particles are viruses, bacteria, fungi, and pollen.

Viruses represent a unique type of microorganism because they can reproduce only inside a host cell. Those replicating exclusively within bacterial cells are called *bacteriophages*. The smallest of all microorganisms, they consist of only one type of nucleic acid, either RNA or DNA. There is a lack of information about the size, shape, and density of airborne viral particles. Until very recently, there has been a common believe that only viruses that are attached to larger particles are able to survive while airborne and that high humidity increases the survival rate. Recent studies have challenged this dogma. For instance, laboratory-generated single MS2 virions survived in the air for considerable time periods set for the evaluation of air purification techniques [21] and face-piece respirators [22, 23]. Another study showed that the airborne transmission of influenza virus was (surprisingly) improved under low (<20%) relative humidity [24]. Overall, the potential for airborne transmission of viral infections is still being debated in the scientific literature.

Bacteria are single-celled microorganisms of various shapes (including spherical, rod-shaped, spiral, and others). They are present in air environments as either vegetative cells or endospores. Bacteria may be carried by other aerosol particles, such as water droplet residues, plant materials, or the skin fragments of animals. In relatively clean indoor environments, airborne bacteria range from 1 to 3 μm in *aerodynamic diameter* [25, 26]. Bacteria tend to grow in colonies in their natural habitats. When aerosolized, they are often aggregated as clusters or chains or attached to other materials [27]. There are two groups of bacteria differentiated based on the ability of the cell wall to retain crystal violet dye: Gram-positive (for example, *Bacillus*) and Gram-negative (for example, *Legionella*). Airborne pathogenic bacteria cause various diseases in humans, animals, and plants. Some types, for example, the actinomycetes, may generate spores that have been associated with specific health responses, such as respiratory allergy and asthma, particularly resulting from occupational exposures [28].

Fungi are disseminated by the release of spores that are well adapted to various air environments. They exhibit high resistance to stresses such as high and low temperature, low humidity, and ultraviolet radiation. Ranging generally from 1.5 to 30 μm, fungal spores may be aerosolized as single spores but are more often agglomerated. Most indoor fungal spores have been reported to be 2–4 μm in aerodynamic diameter [26, 29]. A considerable amount of fungal material, such as allergens, *glucan*, and mycotoxins, can also exist in smaller fragments, which are below 1 μm in size [30, 31]. *Fungal aerosols* can cause allergic reactions, asthma, allergic rhinitis, and other health effects.

Airborne pollen grains are produced by plants in large amounts. They are usually resistant to environmental stresses. Pollen grains from different plants vary in size, surface structure, and, to a lesser extent, shape. Limited information is available about their aerodynamic diameters, but the physical size range is approximately 10–100 μm, with many types of pollen grains being between 25 and 50 μm [14], which is considerably larger than the respirable size fraction. Respiratory health effects are associated with pollen because much smaller pollen fragments may contain allergens [32, 33].

13.4 Sources of Biological Particles and their Aerosolization

Bioaerosol particles originate outdoors and indoors. Most bacteria and fungal spores are aerosolized from the surfaces of plants. Various microorganisms, such as Gram-negative bacteria, actinomycetes, and algae, are aerosolized from natural and anthropogenic water reservoirs. Droplets resulting from rain, splashes, or bubbling processes may contain biological particles that remain airborne after the liquid evaporates. Some industrial environments are major sources of bioaerosols, for example, microorganism-contaminated metal-working fluids (MWFs). Agricultural environments produce a high level of fungal and actinomycete spores. Avian and rodent droppings can be a source of viral and fungal agents. A variety of bacterial and viral pathogens are aerosolized in health-care settings.

In most indoor environments, the primary source of airborne bacteria is humans or animals. Specific bioaerosol sources may develop due to microbial growth in buildings, including the heating, ventilation, and air-conditioning (HVAC) systems as well as the building materials. This development occurs primarily due to excessive moisture.

As a result of biological warfare, terrorist attack, or an accident, highly pathogenic microorganisms, for example, *Bacillus anthracis* (bacterial spores causing anthrax) and *Variola major* (smallpox virus), as well as microbial toxins, can be released and subjected to atmospheric transport.

There are two approaches to determining the maximum plausible exposure to bioaerosol particles released by a source. One method is to conduct air sampling in the vicinity of the source for a prolonged period of time (aiming to catch the peak concentration). The other approach is to assess the source. The particle concentration measured with air samplers during specific time intervals does not always adequately represent the maximum bioaerosol concentration levels, especially for spores, due to the irregular and sporadic nature of spore release from sources. On the other hand, the conventional source assessment (for example, bulk sampling or surface sampling of moldy materials) does not quantify the aerosolization potential of the source. In order to estimate the "worst-case scenario" of air contamination with spores aerosolized from indoor surfaces, one should be able to assess the source strength under the conditions most favorable for spore aerosolization.

To address the exposure assessment needs, a novel concept was developed and evaluated [34–36]. The device designed based on this concept – the fungal spore source strength tester (FSSST) – allows the assessment of the potential of aerosolization of spores from contaminated surfaces. The tester is a cup-like device with a square cross-section, A, which is held against the mold-contaminated surface. The spores are released from the surface by small air jets that originate in multiple capillary-like orifices, which are distributed over an internal cross-section and directed toward the source. The aerosolized spores are then collected into an air sampler, attached to the tester. The number of spores, N, obtained with this sampler operated during time t determines the aerosolization rate, N/At. The aerosolization rate of fungal spores from a mold-contaminated material depends on the material properties, air velocity, air humidity, and other factors [35–37].

The source strength assessment with respect to aerosolization of biocontaminant has been utilized beyond fungal spore enumeration. For instance, Adhikari *et al.* [38] used the FSSST to determine the aerosolization rate of β-(1→3)-D-glucan and *endotoxin* in homes in New Orleans affected by hurricanes Katrina and Rita. A significant positive correlation was observed between the aerosolized β-glucan and endotoxin levels. In several studies, the bioaerosol release from sources has been assessed through the source strength evaluation combined with air sampling. Niemeier *et al.* [39] concluded that reliance on one sampling or enumeration method (be it air sampling, source strength assessment, or bulk sampling) might not provide an accurate estimate of fungal contamination of a microenvironment.

The other series of studies addressed the rate of microbial aerosolization from MWFs. Since water-based MWFs are often used in industrial environments, their microbial contamination is common, resulting in the air biocontamination when the fluids are used in the workplace. Health effects associated with human exposure to MWFs include dermatitis, respiratory symptoms, and diseases [40–42]. Evidence of these health effects in workplaces motivated a number of laboratory and field studies of aerosolization from MWFs. Reponen *et al.* [43] reported data on aerosolized particles measured with a photometric mass monitor, optical particle counter, and condensation nucleus counter. Microbial contamination of semi-synthetic and soluble MWFs increased the mass concentration (as determined by the photometric aerosol mass monitor) and the fine particle number concentration (as determined by the condensation nucleus counter). These effects were seen most clearly for the fine size range of particles aerosolized from contaminated semi-synthetic MWF. An increase in the fine particle concentration was attributed, at least partially, to the increase in the microbial cell wall components (fragments) [43–45].

Laboratory-based studies on microbial aerosolization from various sources as well as other studies involving biological aerosols require appropriate techniques for dispersion of microorganisms into the air. Some techniques initially developed for aerosolizing biologically inert particles were adopted for the aerosolization of microorganisms, whereas others were specifically developed for biological particles [46–48].

13.5
Sampling and Collection

A wide variety of *bioaerosol sampling* and analysis methods have been used and new methods are being developed [6, 14, 49–55]. However, no single sampling method is suitable for the collection and analysis of all types of bioaerosols, and no standardized protocols are currently available [55]. Data between studies are often difficult to compare due to differences in sampler design, collection time, air flow rate, and analysis method. In addition, *human exposure limits* have not been established for biological aerosols to the same extent as for chemical hazards, because of the lack of exposure, dose, and response data. This represents the main challenge in utilizing the bioaerosol sampling results for risk assessment.

Depending on the objective of sampling, an appropriate sampling and analysis method can be selected and incorporated into the bioaerosol monitoring design. Measurement of airborne microorganisms with a bioaerosol sampler often aims at documenting the presence of specific sources. Bioaerosol particles are usually removed from the air through active air sampling. While a microorganism is airborne, its motion is governed by the same laws of physics as applied to biologically inert particles. The sampling phases include aspiration from the ambient environment into the inlet of a measurement device, the transport through the inlet to the collection area, and the collection of the bioaerosol on a specific medium. In addition to physical considerations, the microbiological mechanisms are addressed when there is a need to ensure the survival or biological potency of bioaerosol particles during and after sampling. Furthermore, sample handling, storage, and bioaerosol sample analysis are usually different from the procedures applied when sampling biologically inert particles [56]. The overall physical *sampling efficiency* is the product of the aspiration, transmission, and collection efficiencies. Each of these components depends on the particle aerodynamic diameter, wind velocity, and direction, as well as the inlet characteristics, such as the air velocity at the inlet, inlet dimensions, and orientation. The inlet characteristics of several bioaerosol samplers were determined for different types of bioaerosol particles sampled under various conditions [57]. Aerosol sampling of outdoor or indoor microorganisms and aeroallergens is particularly challenging due to the wide size range of the particles (from submicrometer bacteria to pollen grains of about 100 μm). An additional challenge is associated with the variation in the aspiration efficiency of the sampler caused by changing wind conditions, especially in the case of outdoor bioaerosol sampling.

In order to detect and enumerate viable microorganisms, the biological particles must be efficiently removed from the air and collected in a manner that does not affect their viability. Sample collection time is an important parameter in bioaerosol sampling design. Guidelines for the selection of optimal sampling time for various bioaerosol samplers are available [14, 53]. The expected bioaerosol concentration, the quantitation range of the sampler, and the effect of sampling stress on the overall collection efficiency should be considered in determining the sampling time period. It must be sufficiently long to obtain a representative sample of the airborne microorganisms present, without exceeding the upper quantitation limit of the sampler or causing losses in culturability of airborne organisms [14, 55]. An additional complexity in selecting the optimal collection time is associated with the temporal variability of the bioaerosol concentrations, which may reach several orders of magnitude in the same environment.

The three principal collection methods used in quantitative bioaerosol sampling are *impaction, impingement,* and *filtration* [55]. Some alternative techniques, such as *gravitational settling* and *electrostatic precipitation,* have also been employed.

13.5.1
Impaction

Conventional single-stage and multi-stage (cascade) impactors collect bioaerosol particles onto a solid or semi-solid collection surface. An agar medium is used for culture-based analysis, which allows viable microorganisms to be enumerated, whereas an adhesive-coated surface is more appropriate for microscopic analysis, which determines the total bioaerosol concentration. For biological and non-biological particles alike, lower-inertia particles remain airborne and move with the air flow while particles with higher inertia are collected on the substrate. From this perspective, larger pollen grains and fungal spores are likely to be efficiently collected by impaction, but smaller airborne bacteria and especially viruses may not. Another inertia-based mechanism – centrifugal impaction – is also utilized for collecting bioaerosol particles.

A variety of impactor samplers are commercially available [14, 55]. They differ in their nozzle dimensions, jet-to-plate distance, nozzle shape, number of nozzles, and number of stages. If air is drawn through a single nozzle, the shape of the nozzle is usually rectangular and the impactor is referred to as a *slit sampler*. If there are several nozzles, usually circular in shape, the plate with the impaction nozzles resembles a sieve. A *cascade impactor* features several stages with successively smaller nozzles and allows the separation of bioaerosol particles by their aerodynamic diameter. The Andersen six-stage impactor (Graseby Andersen, Smyrna, GA, USA) is widely used for measuring viable (culturable) bioaerosol concentrations in specific particle size ranges [58], primarily bacteria and fungi.

The physical characteristics of the impaction nozzle(s) and the air flow rate are used to calculate the cut size, or d_{50}, of the impaction stage (d_{50} is the particle diameter at which 50% of the particles are collected). The cut size derives from the non-dimensional Stokes number Stk_{50} (see Chapter 1). Given the sharp cut-off characteristics of impactor samplers, d_{50} is generally designated as the particle diameter above which all particles are collected while all those below d_{50} pass through [59, 60]. For efficient collection, it is important to choose an impactor whose d_{50} is below the aerodynamic diameter of the microorganism being collected. The cut size, d_{50}, depends on several parameters, including the ratio of the jet-to-plate distance, S, to the impactor's nozzle size, W. Most of the commonly used bioaerosol impactors meet the conventional Marple's design criterion, so that S/W is greater than the established threshold (which is 1.5 for rectangular nozzles, and 1 for circular ones) [59]. Recent studies have shown that these samplers underestimate the concentration of some measured bioaerosol particles, for example, fungal spores [61]. The d_{50} of several of the most commonly used single-stage impaction-based spore collectors – such as Air-O-Cell (Zefon Analytical Instruments, Inc., Saint Petersburg, FL, USA) and Burkard collectors (Burkard Manufacturing Co. Ltd, Hertfordshire, UK) – is about 2.5 μm or greater. These impactors do not offer efficient collection for all fungal species because some species produce spores of about 1.8–2.5 μm in aerodynamic diameter. An increase in the sampling flow rate – and consequently the *impact velocity* – would help to reduce the d_{50} of these impactors [14, 59, 61–63]. At the same time, the high impact velocity may cause particle bounce, decreasing the actual collection efficiency, particularly for spores; a high-velocity impact also affects the viability of stress-sensitive microorganisms, as discussed below.

An alternative approach was recently developed, in which, contrary to the Marple criterion, the non-dimensional jet-to-plate distance was proposed to be below one [63, 64]. The feasibility of these "imperfectly designed" impactors (relative to the above-quoted Marple criterion) for total spore collection and enumeration has been demonstrated through laboratory and field evaluation. Grinshpun *et al.* [64] evaluated the collection efficiency and spore deposit characteristics of impactors with $S/W < 1$ using real-time *aerosol spectrometry* and different microscopic enumeration methods. In this study, the test impactors were challenged with non-biological polydisperse NaCl aerosol and aerosolized fungal spores of *Cladosporium cladosporioides, Aspergillus versicolor*, and *Penicillium melinii*. It was shown that a relatively small reduction in the jet-to-plate distance of a single-stage, single-nozzle impactor with a tapered inlet nozzle, combined with adding a straight section of sufficient length, can significantly decrease the cut-off size to a level that is sufficient for the efficient collection of spores of all fungal species. Furthermore, it appears that the slit jet design may improve the application of partial spore counting methodologies with respect to those applied to circular deposits. Several impactors with very small jet-to-plate distances became commercially available in 2004–2005 from Environmental Monitoring Systems, Inc. (Charleston, SC, USA), including the Cyclex Bioaerosol Impact Sampler, Cyclex-d cassette, Micro-5 Microcell, Allergenco-d, and others. The smallest d_{50} featured by these impactors is slightly below 1 μm, which makes them suitable for collecting not only all fungal species but also most bacterial species. Much larger pollen grains obviously have inertia sufficient for efficient impaction; however, they may not be retained on the collection surface at 100% because of the particle bounce effect.

The stress imposed by impaction may result in loss of culturability of the collected microorganisms. High impact velocity can also result in metabolic and structural injuries. Survival of bacteria has been found to decrease when the impact velocity increases; bacterial survival also depends on the degree of embedding of bacteria into the collection medium (particularly for impactors) and the sampling time [14, 65, 66]. Increasing the jet-to-plate distance in an impactor can decrease the viability of the collected bacteria, which can be attributed to increased jet dissipation, which desiccates a larger fraction of the agar surface [67]. The effect of mechanical impact stress is more pronounced at higher impact velocities; the *desiccation stress* increases with increase in the sampling time and when embedding is insufficient.

Unlike bacteria, fungi, and pollen, virus particles have very small sizes, which makes them essentially inertia-less in conventional aerosol sampling systems. Thus, inertial impactors, successfully used for collecting bacteria, fungi, and pollen, have a limited application for airborne virions.

13.5.2
Collection into Liquid

Liquid impingement is similar to impaction because the inertial force is the principal force removing the particle from the air. In some impingers, the particle–medium interaction velocity may be even higher than in impactors. However, collection into a liquid (usually a dilute buffer solution) is generally gentler than impaction on an agar surface. As a result of impingement, aggregates of cells may be broken apart, and particles remaining in the airstream may diffuse to the surface of a bubble and be transferred to the collection buffer in this manner. The collection of bioaerosol particles in a liquid medium is a preferred method in various applications because it allows the sample to be divided for subsequent analysis by different methods.

One of the most common bioaerosol impingers – the AGI-30 all-glass impinger sampler (Ace Glass, Inc., Vineland, NJ, USA) – has an impaction distance of 30 mm from the jet to the bottom of the sampler. Other models such as the AGI-4 all-glass impinger feature a shorter distance of 4 mm, which was originally done to improve particle collection efficiency over the AGI-30. However, added sampling stress may result from impaction against the glass bottom of the sampler, resulting in a loss of cell viability [55]. The very high sampling velocity used in impingers designed to collect small bioaerosol particles usually produces a violent motion of the collection fluid. The latter enhances the evaporation of the fluid, causes re-aerosolization of the initially collected bioaerosol particles, and imposes stress on the microorganisms remaining in the collection fluid (consequently reducing their viability over time). These effects have been discussed and quantified in the scientific literature [68–71].

Conventional impingers can only be used with water or liquids that have about the same viscosity as water. To address the above limitations of conventional impingers associated with evaporation and re-aerosolization of the collection fluid, an alternative approach combining impingement into a liquid with centrifugal motion was developed [72]. The swirling aerosol collector – later commercialized as the BioSampler (SKC, Inc., Eighty Four, PA, USA) – can utilize viscous collection fluids, for example, heavy white mineral oil. Having the same inlet geometry and the same air flow rate of 12.5 l/min as AGI impingers, the BioSampler achieves particle collection by drawing aerosol through three nozzles that are directed at an angle toward the inner sampler wall. During normal operation, the liquid swirls upward on the sampler's inner wall and removes collected particles. When used with heavy white mineral oil, the BioSampler can maintain microbial viability and high physical collection efficiency for several hours [70, 71].

Another approach to bioaerosol sampling into a liquid utilizes a porous medium submerged into a liquid layer so that the aspirated air is blown through it [73, 74]. As a result, the aerosol flow is split into many very small bubbles, and particulates are effectively removed on the walls of the impinger's vessel. Stationary and personal prototypes of the new sampler were found to be capable of achieving high physical collection efficiency (>95%) for airborne bacteria and fungi, with a pressure drop lower than that for most conventional bioaerosol samplers. The collection liquid losses due to evaporation and aerosolization did not exceed 18% in 8 h of operation. In addition, the culturability of sampled microorganisms remained high: the recovery rate of stress-sensitive *Pseudomonas fluorescens* bacteria was 61% ± 20%, and for stress-resistant *Bacillus subtilis* bacteria and *A. versicolor* fungal spores it was 95% ± 9% and 97% ± 6%, respectively [74]. The "frit-bubbler" method was also found to be feasible for collecting and enumerating airborne robust viruses [75, 76], although more significant infectivity losses were observed.

Liquid bioaerosol samplers, such as the all-glass impingers, swirling bioaerosol collectors [70–72], as well as the recently developed "frit-bubbler" [73–76], can operate with different fluids and are characterized by a broad range of sampling efficiencies, especially for virions [77]. Furthermore, some studies conducted with airborne viruses report vastly different efficiencies for the same samplers, which may be due to different analytical protocols affecting the viral recovery level. Because they are a particularly diverse group of microorganisms in terms of their response to environmental stress, viruses often require gentle collection, making liquid-based collectors preferable. But it is difficult to reconcile the need for gentle collection with the concurrent need for high shear forces to remove small (often inertia-less) viruses from the sampling air flow. This makes the development of adequate liquid-based samplers for efficient collection of viable viruses especially challenging.

13.5.3 Filter Collection

The collection of bioaerosol particles on filters is commonly used in bioaerosol monitoring, especially if the objective is to determine the total count of microorganisms, regardless of their survival. Some filter-based samplers such as the Button Inhalable Aerosol Sampler (SKC, Inc., Eighty Four, PA, USA) provide better spatial uniformity of the particle deposits as compared to single-stage impactors [62]. The latter is often advantageous, for instance, if the sample is analyzed using microscopy. Filter-based sampling is adaptable to a variety of analytical methods. The filter material and sample extraction method generally affect the performance of a filter-based bioaerosol sampler [78]. Further studies demonstrated that the effect of filter material is more significant for the size range of single virions than for bacteria [79]. Burton *et al.* [79] reported that the effect of filter loading (up to 4 h) did not cause significant change in the physical collection efficiency of the filter tested in their study.

When collecting bioaerosol particles on filters, significant loss of viability may occur, mostly due to desiccation stress during sampling [80–83]. This effect is more pronounced for bacterial vegetative cells and stress-sensitive viruses than for robust bacterial or fungal spores. The viability loss depends on the microbial species, sampling time, and relative humidity [83]. Some samplers collect airborne microorganisms on a gelatin filter to reduce desiccation stress. Recent experiments demonstrated that gelatin filters are particularly applicable for collecting viable viruses [21–23]. Although gelatin filters are adequate for sampling functional viruses, their use may be limited by environmental conditions, for example, in environments featuring very low or very high humidity [77].

13.5.4 Gravitational Settling

Gravity, or depositional sampling, is a non-quantitative collection method, in which an agar medium is exposed to the environment and airborne organisms are collected primarily by gravity [55]. As a result, large particles are more likely to be deposited onto the collection surface than smaller ones [49]. This can lead to misrepresentation of the prevalence of airborne microorganisms and the exclusion of smaller particles from collection [84]. In addition, the gravitational settling method provides the total number (or mass) of the collected bioaerosol particles but does not allow quantification of their airborne concentration, since the volume of air from which the particles originated is unknown. The data derived from this method are not qualitatively or quantitatively accurate and do not compare favorably with those obtained by other bioaerosol sampling methods [84–87].

13.5.5 Electrostatic Precipitation

Conventional sampling methods, such as inertia-based impaction and impingement as well as filter collection, have been shown to adversely affect the viability of the microorganisms being sampled [65, 71]. For instance, in inertial samplers, the particle velocity toward the collection medium is usually tens or hundreds of meters per second. Such a high impact velocity assures high collection efficiency, but is often damaging for the microorganisms that are being collected. Alternative methods include electrostatic precipitation, when bioaerosol particles are charged in the sampler's inlet and exposed to an electric field inside the sampler. This results in their cross-sectional migration and subsequent deposition on the charged plates, from which the microorganisms can be extracted and analyzed. In electrostatic precipitators, the particle velocity component perpendicular to the collection medium is 2–4 orders of magnitude lower than that in bioaerosol impactors and impingers at comparable sampling flow rates [88]. This provides much more "gentle" collection, which is advantageous for stress-sensitive microorganisms.

In the past decade, several studies have been published on the development and validation of electrostatic bioaerosol samplers [88–94]. The method was found generally feasible for collecting bacterial cells and spores. The physical and biological efficiencies were found to depend on the applied voltage, dimensions of the precipitator, flow rate, initial particle charging level, and other factors. The viability losses can be controlled by limiting the initial charge on airborne microorganisms at the inlet.

Electrostatic precipitation may be implemented without additional charging in the sampler's inlet. The new electrostatic precipitator developed by Lee *et al.* [94] had no charging unit. It is able to differentiate positively from negatively charged microorganisms, which adds a signature to the sampled bioaerosol particles. This feature may assist in their identification or differentiation.

As an important "by-product" of the electrostatic sampler's developmental and evaluation effort, significant information has been gathered on the *electrobiological properties* of airborne microorganisms. It was reported [91] that bacteria dispersed from a liquid by pneumatic nebulization have a wide and bipolar *electric charge distribution*. Electric charge imposed on bacteria during aerosolization was found to be a factor affecting their viability. This, on the one hand, limits the use of charging for bioaerosol sampling, but, on the other, facilitates the application of electric charging for environmental control involving inactivation of bacterial cells by imposing high electric charges on them.

Since the electrostatic precipitator is essentially an open channel, low power is required for the sampling flow through it. Also, very little power is needed to create the precipitation voltage across the electrodes. Thus, this method seems feasible for low-power monitoring of airborne microorganisms and their electric charge distributions, for example, in a counter-bioterrorism network.

13.6 Analysis

Different methods are presently available for the detection and enumeration of the collected bioaerosol particles, including traditional methods such as microscopic and culture-based counting as well as other (more advanced) techniques such as biochemical, immunological, and molecular biological assays.

Optical microscopic enumeration is used for particles collected on a glass slide, tape, or appropriate filter. Light microscopy allows counting and, in many cases, identification of larger biological particles such as fungal spores and pollen. The enumeration and identification of bacterial cells is challenging because they are smaller and can be masked by other particles. In addition, bacterial cells are not visible with a light microscope unless stained. Application of a fluorescent stain, for example, acridine orange, allows bacteria to be counted under an epifluorescence microscope [14]. Various microscopic analysis methods have been discussed in detail by Morris [95]. The main limitation of optical microscopy is that it does not distinguish between culturable and non-culturable bioaerosol particles.

Culture-based analysis is used to assess the viability of bacteria and fungi collected directly onto a nutrient agar or transferred onto agar from a liquid or a filter sample. Different media can be used for culturing different viable microorganisms, reflecting the specifics of growth for different species. The results for bacteria and fungi are expressed in colony-forming units (CFUs) per cubic meter of air. An aggregate of two or more cells or spores forms one colony and therefore is regarded as one particle [27]. For viruses, the viability is measured in terms of infectivity of tissue cultures, the embrionated egg, or animals [96]. In the most common assay, monolayers of host cells are inoculated with appropriate dilutions of virus. After a certain incubation period, plaques are produced, indicating infected cells. Since it is assumed that a single viable (infective) virus induces one plaque-forming unit (PFU), the airborne concentration of viable virus is expressed in PFUs per cubic meter of air. Generally, the accuracy of culture-based assays is affected by many factors, including growth medium, incubation temperature, and incubation time.

Immunochemical and biochemical methods are capable of measuring specific biological molecules and/or agents in bioaerosol particles, such as allergens, endotoxins, mycotoxins, β-(1→3)-D-glucan, or DNA. Depending on the subject of analysis, these methods may require gas chromatography, mass spectrometry, high-performance liquid chromatography, or spectrophotometry. Immunochemical assays are used to quantify antigens and allergens utilizing the binding of antibodies on target antigen. The biological particles are most commonly detected with enzyme-linked immunoassay (ELISA). Traditional immunoassays have relatively high detection threshold, which limits their application to samples with very high bioaerosol concentration. A newly developed modification utilizing a fluorescence multiplex assay may have lower detection limits [97] and thus has a good potential for analyzing less concentrated samples.

Health studies concerned with exposure to bioaerosols suggested the important role of endotoxin and β-(1→3)-D-glucan [98]. Bioaerosol samples can be analyzed for endotoxin and β-glucan by using the *Limulus* amoebocyte lysate (LAL) assay method based on enzymatic cascade in amoebocytes isolated from horseshoe crabs (*Limulus*). Recently, β-(1→3)-D-glucan has been identified in fungal fragments aerosolized from moldy building materials [99, 100]. These submicrometer fragments have been demonstrated to produce high respiratory deposition [101] and are likely associated with adverse health outcomes.

Mycotoxins – toxic compounds produced by certain fungi that have been associated with respiratory and other health effects – can be analyzed from air samples by biological, chemical, and immunological assays [102, 103]. Several recent studies featured different approaches to the airborne mycotoxin analysis (see, for example, [104, 105]).

The polymerase chain reaction (PCR) method, which is based on the amplification of a specific DNA or RNA sequence present in the target microorganisms, has been adopted for bioaerosol samples and opened a new avenue for the identification and quantification of airborne viruses, bacteria, and fungi [106–112]. PCR offers a rapid and sensitive way for identifying and quantifying microorganisms. It seems particularly advantageous for slow-growing microorganisms, such as *Mycobacterium tuberculosis*. Competitive PCR or quantitative polymerase chain reaction (QPCR) is capable of determining the microorganism concentration [14].

13.7 Real-Time Measurement of Bioaerosols

The above-described traditional measurement methods involve two steps: (i) particle collection on a substrate; and (ii) analysis of the collected sample. Direct-reading methods that are capable of measuring aerosol particles without separating them from the air can be used for studying biological aerosol particles, especially under controlled laboratory conditions where other particles are eliminated [113]. There is increased interest – particularly in the defense and counter-terrorist communities – in developing direct-reading instruments that can distinguish particles of biological origin from other particles. One commercially available device, the Ultraviolet Aerodynamic Particle Size Spectrometer (UV-APS; TSI, Inc., Saint Paul, MN, USA), is capable of simultaneously measuring the aerodynamic particle size in the range of 0.5–15 μm, the *light scattering intensity*, and the *fluorescence intensity*. The fluorescence measured between 400 and 580 nm upon excitation of the particles by an ultraviolet (UV) laser beam is related to the presence of life-indicating biomolecules [114, 115]. As fluorophores decay rather quickly after the death of a microorganism, UV-APS is best suited for the measurement of viable bioaerosol particles [14].

The area of real-time biodetection has grown remarkably over the past several years, supported primarily by the military and homeland security agencies of the major industrialized countries. Recent advances in aerosol optics, microbiology, and molecular biology have helped to develop novel concepts for rapid detection, identification, and enumeration of aerosolized biological threat agents [10]. Various principles, for example, single-microorganism analysis, have been integrated into automated biodetection systems, which has created the foundation for the development of field-compatible alarm-type biodetector systems capable of continuous long-time operation.

13.8 Purification of Indoor Air Contaminated with Bioaerosol Particles and Respiratory Protection

13.8.1 Air Purification

Numerous techniques have been developed for controlling *aerosol pollutants*, including biological particles. Conventional techniques such as mechanical filtration and electrostatic precipitation have been incorporated into commercial devices of various capacities and efficacies [116]. While being widely and successfully used in indoor air environments, the mechanical devices and electrostatic precipitators are often criticized for their considerable size and power consumption, excessive noise level, and need for routine maintenance (including routine filter replacement and plate cleaning).

In a perfect filtration process, it is expected that the particles are permanently retained after their first contact with a filter fiber or an already captured particle. In reality, the captured particles may re-entrain from the filter into the air stream. Biological particles collected on HVAC filters may grow on the filter material over time, which increases the likelihood of re-aerosolization. Data on the collection of airborne bacteria and fungi on stationary air filters, their survival, and re-entrainment have been reported in the scientific literature [117–119].

As an alternative to conventional filtration and electrostatic precipitation, emission of unipolar ions into the air environment has been considered as a method for indoor air control. At certain ion and particle concentration levels, ions in the air may efficiently charge airborne particles. As a result, unipolarly charged aerosol particles migrate toward indoor surfaces and ultimately deposit on these surfaces. Ion emission has been experimentally evaluated and shown to be capable of reducing the concentration of airborne dust and microorganisms in indoor environments [120–125]. Among several particle charging methods, corona ionization is known to be particularly effective in charging small aerosol particles, including the fine and ultrafine fractions [126–129].

Based on experimental data, Lee *et al.* [123] concluded that the corona discharge ion emitters (either positive or negative), which are capable of creating an ion density of 10^5–10^6 $e^{\pm}$ cm^{-3}, can be efficient in controlling bacteria- and virus-sized aerosol pollutants in indoor air environments, such as a typical office or residential room. Theoretical modeling performed by Mayya *et al.* [130] supports the above finding. At a high ion emission rate, the particle mobility becomes sufficient to develop considerable migration velocities for removing most aerosol pollutants from a room-sized air volume within tens of minutes. The particle removal efficiency was not significantly affected by the particle size, while it increased with increasing ion emission rate and time of emission. Commercially available unipolar ionic air purifiers (e.g., VI-2500, Wein Products, Inc., Los Angeles, CA, USA) operated for 30 minutes were found to remove approximately 97% of 0.1 μm particles and approximately 95% of 1 μm particles from the air in addition to the natural decay effect. In addition, a significant bactericidal effect of unipolar ion emission was reported [131].

While some indoor air purification techniques aim solely at aerosol concentration reduction, others are designed to inactivate viable bioaerosols. Several techniques have recently been developed for bioaerosol viability control, including ion emission, *ozone generation*, *germicidal ultraviolet* (UV) irradiation, and *photooxidation* involving UV radiation and TiO_2 as a photocatalyst. Grinshpun *et al.* [21] studied an indoor air purification technique that combines unipolar ion emission and *photocatalytic oxidation* – promoted by a specially designed radiant catalytic ionization (RCI) cell – in two test chambers (2.75 and 24.3 m^3) using non-biological and biological challenge aerosols. It was observed that the concentration of aerosol particles representing viruses and bacteria decreased about 10 to 100 times more rapidly when the purifier operated as compared to the natural decay. Particle removal occurred due to unipolar ion emission, while the inactivation of viable airborne microorganisms was associated with photocatalytic oxidation. Approximately 90% of initially viable MS2 viruses were inactivated as a result of 10–60 minutes exposure to photocatalytic oxidation. Approximately 75% of viable *B. subtilis* spores were inactivated within 10 minutes, and about 90% or more after 30 minutes.

The response of a specific species to a specific stress determines the efficiency of a specific inactivation method against this species. Different microbial groups provide different responses. For instance, unlike vegetative cells, spores are resistant to heat stress [132]. Spore inactivation by high temperatures has been extensively investigated, but most of these studies involved microorganisms in aqueous or on solid media [132–135]. Very few experiments were conducted with aerosolized spores – but see the studies by Jung *et al.* [136] with fungal spores and by Grinshpun *et al.* [137] with bacterial spores. Thus far, the effect of thermal inactivation of aerosolized viable spores has not been sufficiently characterized and understood, particularly for exposure periods as short as 10^{-2}–1 s. Based on their experimental data, Grinshpun *et al.* [137] suggested that the thermal inactivation of *B. subtilis* spores occurred as a result of heat-induced damage to DNA and denaturation of essential proteins. This damage can be repaired up to a certain level of thermal exposure. However, the self-repair capability diminishes at higher thermal exposures of aerosolized microorganisms, so that the damage becomes totally irreversible. As the thermal processes are believed to be effective, safe, and environmentally friendly for controlling viable bioaerosol particles in continuous-flow settings [136], the microbiological and biochemical aspects of heat-induced damage to aerosolized microorganisms needs to be further investigated, and the relationship between the exposure temperature and exposure time should be established. A special focus should be given to a short-term exposures ($\sim$1 s), which is a unique feature of aerosol systems (in contrast to aqueous or solid media, in or on which microorganisms are typically exposed to stress factors over periods of minutes or hours).

Some commercial air purifiers generate excessive ozone, either as a primary biocidal agent or as a by-product. Their manufacturers claim that ozone helps to reduce exposure to airborne pollutants, including biological particles, but this has not been substantiated by credible scientific investigations. Furthermore, these devices have raised public health concerns [138]. Among various guidelines for ozone exposure, the following thresholds have been specified for occupational environments: 0.2 ppm for 2 h [139], 0.05–0.10 ppm for 8 h [139], 0.1 ppm for 8 h [140], and 0.05 ppm for instantaneous (no time limit specified) exposure [141]. For comparison, EPA established the outdoor air standards at 0.075 ppm (1997) and 0.08 ppm (2008) for 8 h [142] (it has recently been proposed to decrease). To our knowledge, there is no scientific evidence that ozone by itself plays any role in reducing the number of aerosol particles in indoor environments. Ozone generators can inactivate viable microorganisms; however, this inactivation occurs at concentrations far exceeding the health standards [143, 144].

Attempts have been made to combine conventional filtration and ion emission for reducing exposure to indoor aerosol contaminants, including bioaerosols. Grinshpun *et al.* [145] demonstrated that the collection of aerosol particles on a filter material can be significantly enhanced due to continuous unipolar air ionization in the vicinity of the filter. Agranovski *et al.* [146] have shown that the enhancement effect can be achieved for commercial low-efficiency HVAC filters. The air ions with high mobility are deposited on the fibers, forming a macroscopic electric field, which shields out some incoming unipolarly charged particles due to repulsive forces. The effect was confirmed by testing with aerosolized bacterial cells, bacterial and fungal spores, and virus-carrying particles [147].

13.8.2
Respiratory Protection

Particulate face-piece respirators and surgical masks are extensively used to reduce respiratory exposure to various hazardous aerosols, including biological aerosols. Earlier studies, motivated mostly by the need to protect health-care workers against tuberculosis, addressed the performance of filters of respiratory protection devices challenged with airborne bacteria [148–151]. They demonstrated significant differences between the penetration of airborne microorganisms and inert particles, which were largely attributed to particle shape. It was found that for, an aspect ratio of 4, the penetration of rod-shaped bacteria is about half of that of spherical particles of the same aerodynamic diameter [151]. Further reports showed that spore-forming bacteria of *B. subtilis* and vegetative cells of *P. fluorescens* collected on common polypropylene respirator filters were unlikely to grow on this filter even under optimal nutrition and incubation conditions [118]. Similar results were reported for *Mycobacterum smegmatis* [152].

The face-piece filtering respirators used to prevent exposure to bioaerosols are subjected to certification with respect to the collection efficiency of their filters. In the USA, these are certified according to the National Institute for Occupational Safety and Health (NIOSH) regulations [153], and rated as N95 (95% filter collection efficiency), N99 (99%), N100 (99.97%), and so on. The certification test is conducted by challenging the filter with charge-neutralized NaCl aerosol particles of approximately 0.3 μm (mass median aerodynamic diameter) at a flow rate of 85 l/min. The value of 0.3 μm is presently accepted as the most-penetrating particle size (MPPS) for particulate filters. However, numerous investigations have demonstrated that the MPPS can vary considerably from one filter model to another and is dependent on the operational conditions. The use of electrically charged fibers in the respirator filters shifts the MPPS to the nano-scale range of about 30–70 nm [154]. This finding is particularly important given that many virus particles fall in the above size range. The collection efficiency of N95 and N99 respirator filters against aerosolized viruses has recently been investigated by Balazy *et al.* [155] and Eninger *et al.* [23]. These studies have shown that the penetration of virions through N95 and N99 respirators may exceed 5% and 1%, respectively – the levels expected based on their certifications. Surgical masks were found to have much higher filter penetration and thus are generally not efficient in protecting a wearer from a bioaerosol hazard. (It is acknowledged that, in contrast to face-piece filtering respirators, surgical masks were not originally designed to protect the wearer but, instead, were aimed at reducing the number of particles generated by the wearer into the ambient air.)

Existing respirator filter performance testing does not address the viability issue. At the same time, several manufacturers of respiratory protection devices have recently introduced filter materials that are claimed to have antimicrobial properties, that is, can inactivate bacteria and viruses penetrating through the respirator filter. To test these claims, experimental protocols were developed that allow differentiation between the physical and viable filtration when challenging a respirator with bioaerosols [22]. No evidence of an instant inactivation effect has yet been reported; a long-term effect of some biocidal compounds (such as embedded iodine) is being explored.

A novel concept involving continuous emission of unipolar ions in the vicinity of the filter (described in the previous section for stationary filters) was explored for face-piece filtering respirators and surgical masks tested on breathing manikins [156, 157]. It was found to provide tremendous reduction of the particle penetration (up to about 3000-fold), making a respirator filter almost a perfect shield [157]. The most significant enhancement effect is achieved in the viral and bacterial particle size range. The approach is being further explored for application in respiratory protection.

The existing protocols for evaluating the performance of face-piece respirators address the efficiency of filter media but do not consider the role of face-seal leaks. At the same time, respirators and face-masks do not fit perfectly; as a result, face-seal leakage represents a substantial particle penetration pathway that may compete with and even exceed the filter media penetration. By comparing results of the human subject and manikin-based tests with the same breathing pattern, it was found that the number of particles penetrating through face-seal leakage exceeded the filter media penetration by an order of magnitude: it is (on average) ~7-fold greater for 40 nm particles, ~10-fold for 100 nm, and ~20-fold for 1 μm [158]. These findings suggest that the primary effort in improving respirator design should be shifted toward introducing a better fit between the respirator periphery and the face surface.

Recent outbreaks of SARS (caused by a coronavirus), avian flu (H5N1), and swine flu (H1N1), as well as increasing concern about future emerging diseases with respiratory transmission (including the threat of bioterrorism), have considerably enhanced public interest in research related to the control of and protection against bioaerosol particles. To address these challenges, national and international interdisciplinary research efforts are being developed, which involve collaboration among aerosol scientists, microbiologists, and public health experts.

References

1. Hirst, J.M. (1995) Bioaerosols: introduction, retrospect and prospect, in *Bioaerosols Handbook* (eds C.S. Cox and C.M. Wathes), Lewis, Boca Raton, FL, pp. 5–14.
2. Singerson, G. (1870–74) Micro atmospheric researches. *Proc. R. Irish Acad., Second Series, Sci.*, 13–31.
3. Spurny, K. (2001) in *Aerosol Measurement: Principles, Techniques and Applications*, 2nd edn (eds P.A. Baron and K. Willeke), John Wiley & Sons, Inc., New York, Chapter 1, pp. 3–30.
4. Ho, J. and Griffiths, W.D. (eds) (1994) Special Issue. *J. Aerosol Sci.*, **25** (8), 1369–1613.
5. Lacey, J. (ed.) (1997) Special Issue: Sampling and Rapid Assay of Bioaerosols. *J. Aerosol Sci.*, **28** (3), 345–538.
6. Grinshpun, S.A. and Clark, J.M. (eds) (2005) Special Issue: Measurement and Characterization of Bioaerosols. *J. Aerosol Sci.*, **36** (5–6), 553–812.
7. Grinshpun, S.A. and Agranovski, I.E. (eds) (2008) Special Issue: Bioaerosol Research. *CLEAN – Soil, Air, Water*, **36** (7), 535–621.
8. Knobler, S.L., Mahmoud, A.A.F., and Pray, L.A. (eds) (2002) *Biological Threats and Terrorism: Assessing the Science and Response Capabilities*, National Academy Press, Washington, DC.
9. Fong, I.W. and Alibek, K. (eds) (2009) *Bioterrorism and Infectious Agents*, Springer, Berlin.
10. Morrison, D., Milanovich, F., Ivnitski, D., and Austin, T.R. (2005) *Defense Against Bioterror: Detection Technologies, Implementation Strategies and Commercial Opportunities: Proceedings of the NATO Advanced Research Workshop*, Springer, Berlin.
11. National Research Council (2007) *Protecting Building Occupants and Operations from Biological and Chemical Airborne Threats*, National Academies Press, Washington, DC.
12. National Research Council (2008) *A Framework for Assessing the Health Hazard Posed by Bioaerosols*, National Academies Press, Washington, DC.
13. Cox, C.X. and Wathes, C.M. (1995) Editors' introduction, in *Bioaerosols Handbook* (eds C.S. Cox and C.M. Wathes), Lewis, Boca Raton, FL, pp. 3–4.
14. Reponen, T., Nevalainen, A., Willeke, K., and Grinshpun, S.A. (2009) Biological particle sampling, in *Aerosol Measurement: Principles, Techniques, and Applications*, 3nd edn (eds P.A. Baron, K. Willeke, and P. Kulkarni), John Wiley & Sons, Inc., New York, Chapter 24.
15. Jaenicke, R. (2005) Abundance of cellular material and proteins in the atmosphere. *Science*, **308**, 73.
16. Grinshpun, S.A. and Agranovski, I.E. (2008) Editorial – Special Issue: Bioaerosol Research. *CLEAN – Soil, Air, Water*, **36** (7), 541–543.
17. Roszak, D.B. and Colwell, R.R. (1987) Survival strategies of bacteria in the natural environment. *Microbiol. Rev.*, **51**, 365–379.
18. Rinsoz, T., Duquenne, P., Greff-Mirguet, G., and Oppliger, A. (2008) Application of real-time PCR for total airborne bacterial assessment: comparison of epifluorescence microscopy and culture-dependent methods. *Atmos. Environ.*, **42**, 6767–6774.
19. Lee, T., Grinshpun, S.A., Martuzevicius, D., Adhikari, A., Crawford, C., and Reponen, T. (2006) Culturability and concentration of indoor and outdoor airborne fungi in six single-family homes. *Atmos. Environ.*, **40**, 2902–2910.
20. Doetsch, R.N. and Cook, T.M. (1973) *Introduction to Bacteria and their Ecobiology*, University Park Press, Baltimore, MD.
21. Grinshpun, S.A., Adhikari, A., Honda, T., Kim, K.-Y., Toivola, M., Rao, K.S.R., and Reponen, T. (2007) Control of aerosol contaminants in indoor air: combining the particle concentration reduction with microbial inactivation. *Environ. Sci. Technol.*, **41** (2), 606–612.
22. Eninger, R., Adhikari, A., Reponen, T., and Grinshpun, S.A. (2008) Differentiating between physical and viable

penetrations when challenging respirator filters with bioaerosols. *CLEAN – Soil, Air, Water*, **36** (7), 615–621.

23. Eninger, R., Honda, T., Adhikari, A., Heinonen-Tanski, H., Reponen, T., and Grinshpun, S.A. (2008) Filter performance of N99 and N95 facepiece respirators against viruses and ultrafine particles. *Ann. Occup. Hyg.*, **52** (5), 385–396.
24. Lowen, A.C., Mubareka, S., Steel, J., and Palese, P. (2007) Influenza virus transmission is dependent on relative humidity and temperature. *PLoS Pathog.*, **3**, 1470–1476.
25. Nevalainen, A. (1989) Bacterial Aerosols in Indoor Air. Publications of the National Public Health Institute of Finland A3/1989. Academic Dissertation.
26. Górny, R., Dutkiewicz, J., and Krysinska-Traczyk, E. (1999) The size distribution of bacterial and fungal bioaerosols in the indoor air. *Ann. Agric. Environ. Med.*, **6**, 105–113.
27. Eduard, W., Lacey, J., Karlsson, K., Palmgren, U., Ström, G., and Blomquist, G. (1990) Evaluation of methods for enumerating microorganisms in filter samples from highly contaminated occupational environments. *Am. Ind. Hyg. Assoc. J.*, **51**, 427–436.
28. Lacey, J. and Dutkiewicz, J. (1994) Bioaerosols and occupational lung disease. *J. Aerosol Sci.*, **25**, 1371–1404.
29. Reponen, T., Hyvärinen, A., Ruuskanen, J., Raunemaa, T., and Nevalainen, A. (1994) Comparison of concentrations and size distributions of fungal spores in buildings with and without mold problems. *J. Aerosol Sci.*, **25**, 1595–1603.
30. Brasel, T.L., Douglas, D.R., Wilson, S.C., and Straus, D.C. (2005) Detection of airborne *Stachybotrys chartarum* macrocyclic trichothecene mycotoxins on particulates smaller than conidia. *Appl. Environ. Microbiol.*, **71**, 114–122.
31. Reponen, T., Seo, S.-C., Grimsley, F., Lee, T., Crawford, C., and Grinshpun, S.A. (2007) Fungal fragments in moldy houses: a field study in homes in New Orleans and southern Ohio. *Atmos. Environ.*, **41**, 8140–8149.
32. Rantio-Lehtimäki, A. (1995) Aerobiology of pollen and pollen antigens, in *Bioaerosols Handbook* (eds C.S. Cox and C.M. Wathes), Lewis, Boca Raton, FL, pp. 387–406.
33. Miguel, A.G., Taylor, P.E., House, J., Glovsky, M.M., and Flagan, R.C. (2006) Meteorological influence on respirable fragment release from Chinese elm pollen. *Aerosol Sci. Technol.*, **40**, 690–696.
34. Grinshpun, S.A., Górny, R.L., Reponen, T., Willeke, K., Trakumas, S., Hall, P., and Dietrich, D.F. (2002) New method for assessment of potential spore aerosolization from contaminated surfaces. Proceedings of the Sixth International Aerosol Conference, 8–13 September 2002, Taipei, Taiwan (ed. C.S. Wang), vol. 2, pp. 767–768.
35. Sivasubramani, S.K., Niemeier, R.T., Reponen, T., and Grinshpun, S.A. (2004) Assessment of the aerosolization potential for fungal spores in moldy homes. *Indoor Air*, **14**, 405–412.
36. Sivasubramani, S.K., Niemeier, R.T., Reponen, T., and Grinshpun, S.A. (2004) Fungal spore source strength tester: laboratory evaluation of a new concept. *Sci. Total Environ.*, **329**, 75–86.
37. Górny, R.L., Reponen, T., Grinshpun, S.A., and Willeke, K. (2001) Source strength of fungal spore aerosolization from moldy building material. *Atmos. Environ.*, **35**, 4853–4862.
38. Adhikari, A., Jung, J., Reponen, T., Lewis, J.S., DeGrasse, E.C., Grimsley, L.F., Chew, G.L., and Grinshpun, S.A. (2009) Aerosolization of fungi, $(1\rightarrow3)$-β-D-glucan, and endotoxin from flood-affected materials collected in New Orleans homes. *Environ. Res.*, **109**, 215–224.
39. Niemeier, R.T., Sivasubramani, S.K., Reponen, T., and Grinshpun, S.A. (2006) Assessment of fungal contamination in moldy homes: comparison of different methods. *J. Occup. Environ. Hyg.*, **3** (5), 262–273.
40. Popendorf, W., Miller, E.R., Sprince, N.L., Selim, M.S., Thorne, P.S.,

Davis, C.S. *et al.* (1996) The utility of preliminary surveys to detect the cause of acute metalworking fluid hazards. *Am. J. Ind. Med.*, **30**, 744–749.

41. O'Brien, D.M. (2003) Aerosol mapping of a facility with multiple cases of hypersensitivity pneumonitis: demonstration of mist reduction and a possible dose/response relationship. *Appl. Occup. Environ. Hyg.*, **18**, 947–952.
42. Taibjee, S.M. and Foulds, I.S. (2003) Microorganism-induced skin disease in workers exposed to metalworking fluids. *Occup. Med.*, **53**, 483–484.
43. Reponen, T., Wang, H.X., and Grinshpun, S.A. (2005) Effect of microbial contamination of water-based metalworking fluids on the aerosolization of particles and microbial fragments. *J. ASTM Int.*, **2** (8), JAI12838. doi: 10.1520/JAI12838
44. Wang, H., Reponen, T., Adhikari, A., Willeke, K., and Grinshpun, S.A. (2004) Effect of fluid type and microbial properties on the aerosolization of microorganisms from metalworking fluids. *Aerosol Sci. Technol.*, **38** (11), 1139–1148.
45. Wang, H., Reponen, T., Martuzevicius, D., Grinshpun, S.A., and Willeke, K. (2005) Aerosolization of fine particles increases due to microbial contamination of metalworking fluids. *J. Aerosol Sci.*, **36** (5), 721–734.
46. Ulevicius, V., Willeke, K., Grinshpun, S.A., Donnelly, J., Lin, X., and Mainelis, G. (1997) Aerosolization of particles by bubbling liquid: characteristics and generator development. *Aerosol Sci. Technol.*, **26** (2), 175–190.
47. Reponen, T., Willeke, K., Ulevicius, V., Grinshpun, S.A., and Donnelly, J. (1997) Techniques for dispersion of microorganisms into air. *Aerosol Sci. Technol.*, **27** (3), 405–421.
48. Eninger, R., Hogan, C.J., Biswas, P., Adhikari, A., Reponen, T., and Grinshpun, S.A. (2009) Electrospray versus nebulization for aerosolization and filter testing with virus particles. *Aerosol Sci. Technol.*, **43** (4), 298–304.
49. Burge, H.A. and Solomon, W.R. (1987) Sampling and analysis of biological aerosols. *Atmos. Environ.*, **21**, 451–456.
50. Chatigny, M.A., Macher, J.M., Burge, H.A. and Solomon, W.R. (1989) Sampling airborne microorganisms and aeroallergens in (ed S.V. Hering), *Air Sampling Instruments for Evaluation of Atmospheric Contaminants, 7th edition, publ.*, American Conference of Governmental Industrial Hygienists, Inc., pp. 199–220.
51. Flannigan, B. (1997) Air sampling for fungi in indoor environments. *J. Aerosol Sci.*, **28**, 381–392.
52. Leser, T.D., Boye, M., and Hendriksen, N.B. (1995) Survival and activity of *Pseudomonas* sp. strain B13(FR1) in a marine microcosm determined by quantitative PCR and a rRNA-targeting probe and its effect on the indigenous bacterioplankton. *Appl. Environ. Microbiol.*, **61**, 1201–1207.
53. Macher, J.M. (ed.) (1999) *Bioaerosols: Assessment and Control.* American Conference of Governmental Industrial Hygienists, Cincinnati, OH.
54. Muilenberg, M.L. (2003) Sampling devices. *Immunol. Allergy Clin. N. Am.*, **23**, 337–355.
55. Grinshpun, S.A., Buttner, M.P., and Willeke, K. (2007) in *Manual of Environmental Microbiology*, 3nd edn (eds C.J. Hurst, R.L. Crawford, G.R.J.L. Garland, D.A. Lipson, A.L. Mills, and L.D. Stetzenbach), ASM Press, Washington, DC, pp. 939–951.
56. Grinshpun, S.A. and Reponen, T. (2004) Sampling of biological particles from ambient environment: physical principles, efficiency, and exposure assessment. *Int. Aerobiol. Newsl.*, **59**, 1–2.
57. Grinshpun, S.A., Chang, C.W., Nevalainen, A., and Willeke, K. (1994) Inlet characteristics of bioaerosol samplers. *J. Aerosol Sci.*, **25**, 1503–1522.
58. Andersen, A.A. (1958) New sampler for the collection, sizing, and enumeration of viable airborne particles. *J. Bacteriol.*, **76**, 471–484.
59. Marple, V.A., Rubow, K.L., and Olson, B.A. (2001) in *Aerosol Measurement:*

Principles, Techniques and Applications, 3nd edn (eds P.A. Baron and K. Willeke), John Wiley & Sons, Inc., New York, pp. 229–260.

60. Nevalainen, A., Pastuszka, J., Liebhaber, F., and Willeke, K. (1992) Performance of bioaerosol samplers: collection characteristics and sampler design considerations. *Atmos. Environ.*, **26A**, 531–540.
61. (a) Trunov, M., Trakumas, S., Willeke, K., Grinshpun, S.A., and Reponen, T. (2001) Collection of bioaerosol particles by impaction: effect of fungal spore agglomeration and bounce. *Aerosol Sci. Technol.*, **34** (6), 490–498; (b) Trunov, M., Trakumas, S., Willeke, K., Grinshpun, S.A., and Reponen, T. (2001) Collection of bioaerosol particles by impaction: effect of fungal spore agglomeration and bounce. *Aerosol Sci. Technol.*, **35** (1), 617–624.
62. Aizenberg, V., Reponen, T., Grinshpun, S.A., and Willeke, K. (2000) Performance of AIR-O-Cell, Burkard, and button samplers for total enumeration of airborne spores. *Am. Ind. Hyg. Assoc. J.*, **61** (6), 855–864.
63. Grinshpun, S.A., Mainelis, G., Trunov, M., Górny, R.L., Sivasubramani, S.K., Adhikari, A., and Reponen, T. (2005) Collection of airborne spores by circular single-stage impactors with small jet-to-plate distance. *J. Aerosol Sci.*, **36** (5), 575–591.
64. Grinshpun, S.A., Adhikari, A., Cho, S.-H., Kim, K.-Y., Lee, T., and Reponen, T. (2007) A small change in the design of a slit bioaerosol impact or significantly improves its collection characteristics. *J. Environ. Monit.*, **9**, 855–861.
65. Stewart, S.L., Grinshpun, S.A., Willeke, K., Terzieva, S., Ulevicius, V., and Donnelly, J. (1995) Effect of impact stress on microbial recovery on an agar surface. *Appl. Environ. Microbiol.*, **61**, 1232–1239.
66. Terzieva, S., Donnelly, J., Ulevicius, V., Grinshpun, S.A., Willeke, K., Stelma, G.N., and Brenner, K. (1996) Comparison of methods for detection and enumeration of airborne microorganisms collected by liquid impingement. *Appl. Environ. Microbiol.*, **62** (7), 2264–2272.
67. Yao, M. and Mainelis, G. (2007) Analysis of portable impactor performance for enumeration of viable bioaerosols. *J. Occup. Environ. Hyg.*, **4**, 514–524.
68. Grinshpun, S.A., Willeke, K., Ulevicius, V., Juozaitis, A., Terzieva, S., Donnelly, J., Stelma, G.N., and Brenner, K. (1997) Effect of impaction, bounce and reaerosolization on collection efficiency of impingers. *Aerosol Sci. Technol.*, **26**, 326–342.
69. Lin, X., Willeke, K., Ulevicius, V., and Grinshpun, S.A. (1997) Effect of sampling time on the collection efficiency of all-glass impingers. *Am. Ind. Hyg. Assoc. J.*, **58**, 480–488.
70. Lin, X., Reponen, T., Willeke, K., Grinshpun, S.A., Foarde, K.K., and Ensor, D.S. (1999) Long-term sampling of airborne bacteria and fungi into a non-evaporating liquid. *Atmos. Environ.*, **33**, 4291–4298.
71. Lin, X., Reponen, T., Willeke, K., Wang, Z., Grinshpun, S.A., and Trunov, M. (2000) Survival of airborne microorganisms during swirling aerosol collection. *Aerosol Sci. Technol.*, **32** (3), 184–196.
72. Willeke, K., Lin, X., and Grinshpun, S.A. (1998) Improved aerosol collection by combined impaction and centrifugal motion. *Aerosol Sci. Technol.*, **28**, 439–456.
73. Agranovski, I.E., Agranovski, V., Grinshpun, S.A., Reponen, T., and Willeke, K. (2002) Collection of airborne microorganisms into liquid by bubbling through porous medium. *Aerosol Sci. Technol.*, **36** (4), 502–509.
74. Agranovski, I.E., Agranovski, V., Grinshpun, S.A., Reponen, T., and Willeke, K. (2002) Development and evaluation of a new personal sampler for culturable airborne microorganisms. *Atmos. Environ.*, **36** (5), 889–898.
75. Agranovski, I.E., Safatov, A.S., Borodulin, A.I., Pyankov, O.V., Petrishchenko, V.A., Sergeev, A.N., Sergeev, A.A., Agranovski, V., and Grinshpun, S.A. (2005) New personal sampler for viable airborne viruses:

feasibility study. *J. Aerosol Sci.*, **36** (5), 609–617.
76. Agranovski, I.E., Safatov, A.S., Pyankov, O.V., Sergeev, A.A., Sergeev, A.N., and Grinshpun, S.A. (2005) Long-term sampling of viable airborne viruses. *Aerosol Sci. Technol.*, **39** (9), 912–918.
77. Verreault, D., Moineau, S., and Duchaine, C. (2008) Methods for sampling of airborne viruses. *Microbiol. Mol. Biol. Rev.*, **72**, 413–444.
78. Burton, N.C., Adhikari, A., Grinshpun, S.A., Horning, R., and Reponen, T. (2005) The effect of filter material on bioaerosol collection of *Bacillus anthracis* stimulant. *J. Environ. Monit.*, **7** (5), 475–480.
79. Burton, N.C., Grinshpun, S.A., and Reponen, T. (2007) Physical collection efficiency of filter materials for bacteria and viruses. *Ann. Occup. Hyg.*, **51** (2), 143–151.
80. Jensen, P.A., Todd, W.F., Davis, G.N., and Scarpino, P.V. (1992) Evaluation of eight bioaerosol samplers challenged with aerosols of free bacteria. *Am. Ind. Hyg. Assoc. J.*, **53**, 660–667.
81. Lundholm, I.M. (1982) Comparison of methods for quantitative determinations of airborne bacteria and evaluation of total viable counts. *Appl. Environ. Microbiol.*, **44**, 179–183.
82. Thorne, P.S., Kiekhaefer, M.S., Whitten, P., and Donham, K.J. (1992) Comparison of bioaerosol sampling methods in barns housing swine. *Appl. Environ. Microbiol.*, **58**, 2543–2551.
83. Wang, Z., Reponen, T., Grinshpun, S.A., Górny, R.L., and Willeke, K. (2001) Effect of sampling time and air humidity on the bioefficiency of filter samplers for bioaerosol collection. *J. Aerosol Sci.*, **32** (5), 661–674.
84. Solomon, W.R. (1975) Assessing fungus prevalence in domestic interiors. *J. Allergy Clin. Immunol.*, **56**, 235–242.
85. Buttner, M.P. and Stetzenbach, L.D. (1993) Monitoring airborne fungal spores in an experimental indoor environment to evaluate sampling methods and the effects of human activity on air sampling. *Appl. Environ. Microbiol.*, **59**, 219–226.
86. Sayer, W.J., Shean, D.B., and Ghosseiri, J. (1969) Estimation of airborne fungal flora by the Andersen sampler versus the gravity settling plate. *J. Allergy*, **44**, 214–227.
87. Sayer, W.J., MacKnight, N.M., and Wilson, H.W. (1972) Hospital airborne bacteria as estimated by the Andersen sampler versus gravity settling culture plate. *Am. J. Clin. Pathol.*, **58**, 558–562.
88. Mainelis, G., Grinshpun, S.A., Willeke, K., Reponen, T., Ulevicius, V., and Hintz, P. (1999) Collection of airborne microorganisms by electrostatic precipitation. *Aerosol Sci. Technol.*, **30** (2), 127–144.
89. Mainelis, G., Willeke, K., Baron, P., Reponen, T., Grinshpun, S.A., Górny, R.L., and Trakumas, S. (2001) Electrical charges on airborne microorganisms. *J. Aerosol Sci.*, **32** (9), 1087–1110.
90. Mainelis, G., Adhikari, A., Willeke, K., Lee, S.-A., Reponen, T., and Grinshpun, S.A. (2002) Collection of airborne microorganisms by a new electrostatic precipitator. *J. Aerosol Sci.*, **33** (10), 1417–1432.
91. Mainelis, G., Górny, R.L., Trunov, M., Grinshpun, S.A., Baron, P., Yadav, J., and Willeke, K. (2002) Effect of electrical charges and fields on injury and viability of airborne bacteria. *Biotechnol. Bioeng.*, **79** (2), 229–241.
92. Mainelis, G., Willeke, K., Adhikari, A., Reponen, T., and Grinshpun, S.A. (2002) Design and collection efficiency of a new electrostatic precipitator for bioaerosol collection. *Aerosol Sci. Technol.*, **36**, 1073–1085.
93. Mainelis, G., Willeke, K., Baron, P., Grinshpun, S.A., and Reponen, T. (2002) Induction charging and electrostatic classification of micrometer-size particles for investigating the electrobiological properties of airborne microorganisms. *Aerosol Sci. Technol.*, **36** (4), 479–491.
94. Lee, S.-A., Willeke, K., Mainelis, G., Adhikari, A., Wang, H., Reponen, T., and Grinshpun, S.A. (2004) Assessment of electrical charge on airborne microorganisms by a new bioaerosol

sampling method. *J. Occup. Environ. Hyg.*, **1**, 127–138.

95. Morris, K.J. (1995) Modern microscopic methods of bioaerosol analysis, in *Bioaerosols Handbook* (eds C.S. Cox and C.M. Wathes), Lewis, Boca Raton, FL, pp. 285–316.
96. Hensel, A. and Petzoldt, K. (1995) Biological and biochemical analysis of bacteria and viruses, in *Bioaerosols Handbook* (eds C.S. Cox and C.M. Wathes), Lewis, Boca Raton, FL, pp. 335–360.
97. Earle, C.D., King, E.M., Tsay, A., Pittman, K., Saric, B., Vailes, L., Godbout, R., Oliver, K.G., and Chapman, M. (2007) High throughput fluorescent multiplex array for indoor allergen exposure assessment. *J. Allergy Clin. Immunol.*, **119**, 428–433.
98. Rylander, R. (2002) Endotoxin in the air environment – exposure and health effects. *J. Endotoxin Res.*, **8** (4), 241–252.
99. Seo, S.-C., Reponen, T., Levin, L., Borchelt, T., and Grinshpun, S.A. (2008) Aerosolization of particulate $(1\rightarrow3)$-β-D-glucan from moldy materials. *Appl. Environ. Microbiol.*, **74** (3), 585–593.
100. Seo, S.H., Reponen, T., Levin, L., and Grinshpun, S.A. (2009) Size-fractionated $(1\rightarrow3)$-β-D-glucan concentrations aerosolized from different moldy building materials. *Sci. Total Environ.*, **407**, 806–814.
101. Cho, S.-H., Seo, S.-H., Schmechel, D., Grinshpun, S.A., and Reponen, T. (2005) Aerodynamic characteristics and respiratory deposition of fungal fragments. *Atmos. Environ.*, **39**, 5454–5465.
102. Hendry, K.M. and Cole, E.C. (1993) A review of mycotoxins in indoor air. *J. Toxicol. Environ. Health*, **38**, 183–198.
103. Johanning, E. (ed.) (2001) *Bioaerosols, Fungi and Mycotoxins: Health Effects, Assessment, Prevention and Control*, Fungal Research Group, Albany, NY.
104. Charpin-Kadouch, C., Maurel, G., Felipo, R., Queralt, J., Ramadour, M., Dumon, H., Garans, M., Botta, A., and Charpin, D. (2006) Mycotoxin identification in moldy dwellings. *J. Appl. Toxicol.*, **26** (6), 475–479.
105. Wang, Y., Chai, T., Lu, G., Quan, C., Duan, H., Yao, M., Zucker, B.-A., and Schlenker, G. (2008) Simultaneous detection of airborne aflatoxin, ochratoxin and zearalenone in a poultry house by immunoaffinity clean-up and high-performance liquid chromatography. *Environ. Res.*, **107** (2), 139–144.
106. Sawyer, M.H., Chamberlain, C.J., Wu, Y.N., Aintablian, N., and Wallace, M.R. (1994) Detection of *Varicella zoster* virus DNA in air samples from hospital rooms. *J. Infect. Dis.*, **169**, 91–94.
107. Alvarez, A.J., Buttner, M.P., and Stetzenbach, L.D. (1995) PCR for bioaerosol monitoring: sensitivity and environmental interference. *Appl. Environ. Microbiol.*, **61** (10), 3639–3644.
108. Curtis, L., Ross, M., Persky, V., Scheff, P., Wadden, R., Ramakrisnan, V., and Hryhorczuk, D. (2000) Bioaerosol concentrations in the quad cities 1 year after the 1993 Mississippi river floods. *Indoor Built Environ.*, **9**, 35–43.
109. Williams, R.H., Ward, E., and McCartney, H.A. (2001) Methods for integrated air sampling and DNA analysis for detection of airborne fungal spores. *Appl. Environ. Microbiol.*, **67**, 2453–2459.
110. An, H.R., Mainelis, G., and White, L. (2006) Development and calibration of real-time PCR for quantification of airborne microorganisms in air samples. *Atmos. Environ.*, **40**, 7924–7939.
111. Blachere, F.M., Lindsley, W.G., Slaven, J.E., Green, B.J., Anderson, S.E., Chen, B.T., and Beezhold, D.H. (2007) Bioaerosol sampling for the detection of aerosolized influenza virus. *Influenza Other Respir. Viruses*, **1**, 113–120.
112. Oppliger, A., Charriere, N., Droz, P.-O., and Rinsoz, T. (2008) Exposure to bioaerosols in poultry houses at different stages of fattening; use of real-time PCR for airborne bacterial quantification. *Ann. Hyg.*, **52** (5), 405–412.

113. Qian, Y., Willeke, K., Grinshpun, S.A., Donnelly, J., and Coffey, C.C. (1998) Performance of N95 respirators: filtration efficiency for airborne microbial and inert particles. *Am. Ind. Hyg. Assoc. J.*, **59** (2), 128–132.
114. Hairston, P.P., Ho, J., and Quant, F.R. (1997) Design of an instrument for real-time detection of bioaerosols using simultaneous measurement of particle aerodynamic size and intrinsic fluorescence. *J. Aerosol Sci.*, **28**, 471–482.
115. Agranovski, V. and Ristovski, Z.D. (2005) Real-time monitoring of viable bioaerosols: capability of the UVAPS to predict the amount of individual microorganisms in aerosol particles. *J. Aerosol Sci.*, **36**, 665–676.
116. Ludwig, J.F. and Turner, W. (1991) in *Control Strategies in Indoor Air Pollution: A Health Perspective* (eds J.M. Samet and J.D. Spengler), Johns Hopkins University Press, Baltimore, MD, pp. 351–377.
117. Maus, R., Goppelsroder, A., and Umhauer, U. (1997) Viability of bacteria in unused air filter media. *Atmos. Environ.*, **31**, 2305–2310.
118. Wang, Z., Reponen, T., Willeke, K., and Grinshpun, S.A. (1999) Survival of bacteria on respirator filters. *Aerosol Sci. Technol.*, **30** (3), 300–308.
119. Jankowska, E., Reponen, T., Willeke, K., Grinshpun, S.A., and Choi, K.-J. (2000) Collection of fungal spores on air filters and spore reentrainment from filters into air. *J. Aerosol Sci.*, **31** (8), 969–978.
120. Grabarczyk, Z. (2001) Effectiveness of indoor air cleaning with corona ionizers. *J. Electrostat.*, **51–52**, 278–283.
121. Grinshpun, S.A., Mainelis, G., Reponen, T., Willeke, K., Trunov, M.A., and Adhikari, A. (2001) Effect of wearable ionizers on the concentration of respirable airborne particles and microorganisms. *J. Aerosol Sci.*, **32** (Suppl. 1), S335–S336.
122. Grinshpun, S.A., Mainelis, G., Trunov, M., Adhikari, A., Reponen, T., and Willeke, K. (2005) Evaluation of ionic air purifiers for reducing aerosol exposure in confined indoor spaces. *Indoor Air*, **15**, 235–245.
123. Lee, B.U., Yermakov, M., and Grinshpun, S.A. (2004) Removal of fine and ultrafine particles from indoor air environments by the unipolar ion emission. *Atmos. Environ.*, **38**, 4815–4823.
124. Krueger, A.P. and Reed, E.J. (1976) Biological impact of small air ions. *Science*, **193**, 1209–1213.
125. Niu, J.L., Tung, T.C.W., and Burnett, J. (2001) Quantification of dust removal and ozone emission of ionizer air-cleaners by chamber testing. *J. Electrostat.*, **51–52**, 20–24.
126. Adachi, M., Kousaka, Y., and Okuyama, K. (1985) Unipolar and bipolar diffusion charging of ultrafine aerosol particles. *J. Aerosol Sci.*, **16**, 109–124.
127. Büscher, P., Schmidt-Ott, A., and Wiedensohler, A. (1994) Performance of a unipolar "square wave" diffusion charger with variable *nt*-product. *J. Aerosol Sci.*, **25**, 651–663.
128. Wiedensohler, A., Büscher, P., Hansson, H.C., Martinsson, B.G., Stratmann, F., Ferron, G., and Busch, B. (1994) A novel unipolar charger for ultrafine aerosol particles with minimal particle losses. *J. Aerosol Sci.*, **25**, 639–649.
129. Hernandez-Sierra, A., Alguacil, F.J., and Alonso, M. (2003) Unipolar charging of nanometer aerosol particles in a corona ionizer. *J. Aerosol Sci.*, **34**, 733–745.
130. Mayya, Y.S., Sapra, B.K., Khan, A., and Sunny, F. (2004) Aerosol removal by unipolar ionization in indoor environments. *J. Aerosol Sci.*, **35**, 923–941.
131. Grinshpun, S.A., Adhikari, A., Lee, B.U., Trunov, M., Mainelis, G., Yermakov, M., and Reponen, T. (2004) in *Air Pollution: Modeling, Monitoring and Management of Air Pollution* (ed. C.A. Brebbia), WIT Press, Southampton, pp. 689–704.
132. Setlow, P. (2006) Spores of *Bacillus subtilis*: their resistance to and killing by radiation, heat and chemicals. *J. Appl. Microbiol.*, **101**, 514–525.

133. Molin, G. and Östlund, K. (1975) Dry-heat inactivation of *Bacillus subtilis* spores by means of infra-red heating. *Antonie Van Leeuwenhoek (Int. J. Gen. Mol. Microbiol.)*, **41**, 329–335.
134. Pfeifer, J. and Kessler, H.G. (1994) Effect of relative humidity of hot air on the heat resistance of *Bacillus cereus* spores. *J. Appl. Bacteriol.*, 77, 121–128.
135. Nicholson, W.L., Munakata, N., Horneck, G., Melosh, H.J., and Setlow, P. (2000) Resistance of *Bacillus* endospores to extreme terrestrial and extraterrestrial environments. *Microbiol. Mol. Biol. Rev.*, **64**, 548–572.
136. Jung, J.H., Lee, J.E., Lee, C.H., Kim, S.S., and Lee, B.U. (2009) Treatment of fungal bioaerosols by a high-temperature, short-time process in a continuous flow system. *Appl. Environ. Microbiol.*, **75**, 2742–2749.
137. Grinshpun, S.A., Adhikari, A., Li, C., Reponen, T., Yermakov, M., Schoenitz, M., Dreizin, E., Trunov, M., and Mohan, S. (2009) Thermal inactivation of airborne viable *Bacillus subtilis* spores by short-term exposure in axially heated air flow. *J. Aerosol Sci.*, **41**, 352–363.
138. U.S. Environmental Protection Agency (EPA) (2006) Ozone Generators that are Sold as Air Cleaners: an Assessment of Effectiveness and Health Consequences. Available at: *http://www.epa.gov/iaq/pubs/ozonegen.html* (accessed 22 May 2006).
139. ACGIH (2004) *Threshold Limit Values for Chemical Substances and Physical Agents and Biological Exposure Indices*, American Conference of Governmental Industrial Hygienists, Cincinnati, OH, p. 44.
140. Occupational Safety and Health Administration (OSHA) (2006) Regulations (Standards 29 CFR): Air Contaminants – 1915.1000. Available at: *http://www.osha.gov/pls/oshaweb* (accessed 31 May 2006).
141. U.S. Food and Drug Administration (FDA) (1989) Labeling Regulatory Requirements for Medical Devices. HHS Publication FDA 89-4203.
142. U.S. Environmental Protection Agency (EPA). Ozone Air Quality Standards. Available at: http://www.epa.gov/groundlevelozone/standards.html (accessed 2 April 2010)
143. Foarde, K.K., Van Osdell, D.W., and Steiber, R.S. (1997) Investigation of gas-phase ozone as a potential biocide. *Appl. Occup. Environ. Hyg.*, **12**, 535–542.
144. Li, C.-S. and Wang, Y.-C. (2003) Surface germicidal effects of ozone for microorganisms. *Am. Ind. Hyg. Assoc. J.*, **64**, 533–537.
145. Grinshpun, S.A., Lee, B.U., Yermakov, M., and McKay, R. (2004) How to increase the protection factor provided by existing facepiece respirators against airborne viruses: a novel approach. *J. Aerosol Sci.*, **35** (Suppl. 2), 1263–1272.
146. Agranovski, I.E., Huang, R., Pyankov, O.V., Altman, I.S., and Grinshpun, S.A. (2006) Enhancement of the performance of low-efficiency HVAC filters due to continuous unipolar ion emission. *Aerosol Sci. Technol.*, **40** (11), 963–968.
147. Huang, R., Agranovski, I.E., Pyankov, O.V., and Grinshpun, S.A. (2008) Removal of viable bioaerosol particles with a low-efficiency HVAC filter enhanced by continuous emission of unipolar air ions. *Indoor Air*, **18**, 106–112.
148. Dinnen, P. (1971) Microbial filtration by surgical masks. *Surg. Gynecol. Obstet.*, **133**, 812–814.
149. Chen, S.-K., Vesley, D., Brosseau, L.M., and Vincent, J. (1994) Evaluation of single-use masks and respirators for protection of health care workers against mycobacterial aerosols. *Am. J. Infect. Control*, **22**, 65–74.
150. Johnson, B., Martin, D.D., and Resnik, I.G. (1994) Efficacy of selected respiratory protective equipment challenged with *Bacillus subtilis* subsp. *niger*. *Appl. Environ. Microbiol.*, **60**, 2184–2186.
151. Willeke, K., Qian, Y., Donnelly, J., Grinshpun, S.A., and Ulevicius, V. (1996) Penetration of airborne microorganisms through a surgical mask and a dust/mist respirator. *Am. Ind. Hyg. Assoc. J.*, **57** (4), 348–355.

152. Reponen, T., Wang, Z., Willeke, K. and Grinshpun, S.A. (1999) Survival of mycobacteria on N95 personal respirators. *Infect. Control Hosp. Epidemiol.*, **20** (4), 237–241.

153. National Institute for Occupational Safety and Health (1995) Standard 42 CFR Part 84. *Respiratory Protective Devices*. NIOSH, Morgantown, WV.

154. Balazy, A., Toivola, M., Reponen, T., Podgorski, A., Zimmer, A., and Grinshpun, S.A. (2006) Manikin-based performance evaluation of N95 filtering-facepiece respirators challenged with nanoparticles. *Ann. Occup. Hyg.*, **50** (3), 259–269.

155. Balazy, A., Toivola, M., Adhikari, A., Sivasubramani, S.K., Reponen, T., and Grinshpun, S.A. (2006) Do N95 respirators provide 95% protection level against airborne viruses and how adequate are surgical masks? *Am. J. Infect. Control*, **34** (2), 51–57.

156. Lee, B.U., Yermakov, M., and Grinshpun, S.A. (2004) Unipolar ion emission enhances respiratory protection against fine and ultrafine particles. *J. Aerosol Sci.*, **35** (11), 1359–1368.

157. Lee, B.U., Yermakov, M., and Grinshpun, S.A. (2005) Filtering efficiency of N95- and R95-type facepiece respirators, dust-mist facepiece respirators, and surgical masks operating in unipolarly ionized indoor air environments. *Int. J. Aerosol Air Qual. Res.*, **5** (3), 27–40.

158. Grinshpun, S.A., Haruta, H., Eninger, R.M., Reponen, T., McKay, R.T., and Lee, S.-A. (2009) Performance of an N95 filtering facepiece particulate respirator and a surgical mask during human breathing: two pathways for particle penetration. *J. Occup. Environ. Hyg.* **6** (10), 593–603.

14
Atmospheric Bioaerosols

Aleksandr S. Safatov, Galina A. Buryak, Irina S. Andreeva, Sergei E. Olkin, Irina K. Reznikova, Aleksandr N. Sergeev, Boris D. Belan, and Mikhail V. Panchenko

14.1
Introduction

According to the definition of bioaerosol given in the *Bioaerosols Handbook* [1]: "[a] bioaerosol is an aerosol comprising particles of biological origin or activity, which may affect living things through infectivity, allergenicity, toxicity, pharmacological, or other processes. Particle size may range from aerodynamic diameters ca. 0.5 to 100 μm." This definition excludes smaller aerosol particles of biological origin (for example, *nucleation aerosols* consisting of components of biological origin such as isoprenols [2–4]) as well as smaller aerosols carrying biological activity (for example, drug or *toxic nanoaerosols* [5, 6]). It should be noted that viral particles normally having a size of less than 0.5 μm are within the size range specified in the definition because, being obligate parasites, viruses get into the atmosphere together with the system in which they have replicated: the cell or its fragments [7].

For *atmospheric bioaerosols*, which are within the size range from 0.5 to 100 μm, it is expedient to differentiate between particles of non-biogenic origin carrying biological activity and biogenic particles. The first category of aerosols is studied by *aerosol toxicology* (including allergic manifestations) and *aerosol pharmacology*, which study the effects of both natural and artificial bioaerosols on living beings. The second category of atmospheric bioaerosols is studied by *atmospheric aerobiology*. The present chapter will focus upon on the latter category of atmospheric bioaerosols.

Soil, vegetation, water surfaces, humans, and animals can be the sources of biogenic aerosols. According to [8], a total of approximately 56 Tg of bioaerosols with particle size of more than 1 μm get into the atmosphere every year; in Europe alone, the estimated amount of primary biogenic particles getting into the atmosphere is 0.233 Tg/year [9], and that in the world is 18.5 Tg/year [10]. Their composition is determined both by the source [4, 11–23] and by the physical and chemical processes influencing the particles and their components in the process of atmospheric transfer [4, 14–20]. Usually bioaerosols contain various macromolecules or microorganisms as well as different inorganic and low-molecular-weight organic

Aerosols – Science and Technology. Edited by Igor Agranovski

ISBN: 978-3-527-32660-0

components [4, 13, 15, 16, 21–25]. From our point of view, two components of atmospheric bioaerosols are the most important: (i) *viable microorganisms*, which (as distinct from other components) can cause infectious processes; and (ii) *total protein*, the basis of all living organisms on the Earth and a universal marker of an aerosol's biogenic origin. The use of other *chemical markers* to detect different microorganisms or plant pollen in aerosols [22, 24, 26–35] provides more detailed information on particular species (kingdoms, genera) of microorganisms, but it does not cover the whole biogenic material of an aerosol. The organic carbon (OC) present in particles can also be used as a marker of biogenic aerosol [22, 23, 27, 32]; however, with this marker, some low-molecular-weight compounds of photochemical rather than biogenic origin can be considered as the corresponding component of bioaerosol.

The determination of the concentration and type of different bioaerosol components in an atmospheric aerosol under different conditions allows us to answer questions relating to the following:

1) *temporal variations* (daily, annual, long-term) of concentration and composition of atmospheric bioaerosol components;
2) *spatial variations* of concentration and composition of atmospheric bioaerosol components;
3) possible sources of atmospheric bioaerosols and their transfer in the atmosphere; and
4) potential danger of atmospheric bioaerosols for humans and animals.

This chapter presents both our own results and other literature data on the study of atmospheric bioaerosols of biological origin. In 1995, the *Bioaerosols Handbook* [1] and a special issue of the *Journal of Aerosol Science* (volume 25, issue 8) were published. They described in detail the available methods and the results of numerous bioaerosol studies, including those of atmospheric bioaerosols. Later special issues of various journals devoted to the same subject did not cover the whole range of bioaerosol studies in such detail. That is why the main attention in this chapter is paid to work published after 1995; the earlier ones are cited only when the data or concepts that they present are unique.

14.2 Methods of Atmospheric Bioaerosol Research

The study of the characteristics of atmospheric bioaerosols can be performed using remote contactless methods, *in situ* methods that do not require sampling, and laboratory methods that analyze bioaerosol samples in any way.

Methods in the first category include *laser sounding* of the atmosphere. In principle, this can not only provide information on the concentration and dispersity of the aerosol but also detect bioaerosols in it by any marker that determines some optical characteristic(s) of the bioaerosol. However, as the optical characteristics of many bioaerosol components are unknown, and the particles themselves are often multi-component, such methods are not widely used. Here we only wish to mention two works [36, 37] aimed at the detection of specific microorganisms in aerosols.

Methods based on biogenic substance fluorescence were developed to detect and characterize bioaerosols *in situ*. For example, a commercial device for the determination of the concentration and dispersity of bioaerosols – the UV-APS 3314 Ultraviolet Aerodynamic Particle Size Spectrometer (TSI, Inc., Saint Paul, MN, USA) – was created using these methods. It has been used to carry out a number of studies (see, for example, [38, 39]). However, this device (and similar devices from other manufacturers) has some essential disadvantages: (i) the operating temperature range (+10 to +35 °C) makes its all-weather application impossible; (ii) the presence of inorganic material, for example, a dust particle, in the biogenic particles can make the biogenic particles invisible due to absorption of the incident or fluorescent radiation; and (iii) the emission bands of some polycyclic aromatic hydrocarbons (PAHs) lie within the same fluorescence range, which can cause the device to misfunction.

Recently developed *aerosol mass spectrometers* [40–42] not only allow the concentration and dispersity of particles with a maximal diameter of 1–3 μm to be determined *in situ* but also allow information to be obtained on the *chemical composition* of individual particles. Although attempts have been made to use such devices for the detection of biological compounds and microorganisms in aerosol [40–48], at the present time it is not possible to characterize any bioaerosol particles: the extreme range of the detected ion masses make up less than 9000 Da, but the mass of most biological macromolecules is much larger than this.

Other automated methods that allow the detection of specific molecular compounds and/or microorganisms in aerosol are presented in [49–52]. However, such methods cannot be used to detect and identify previously unknown microorganisms and a broad range of macromolecules in the samples. In addition, the equipment used in these methods is often unique and very expensive.

Thus, collecting atmospheric air samples followed by their laboratory analysis is today the most informative method to study the characteristics of atmospheric bioaerosols. Different variants of this method are considered below.

14.2.1
Methods and Equipment for Atmospheric Bioaerosol Sampling

A wide variety of methods used for atmospheric *bioaerosol sampling* and the equipment used in them is presented in the *Bioaerosols Handbook* [1] and in other mainly recent work [53–80]. According to [1, 53–57], these methods can be classified as follows:

- bioaerosol deposition on dishes;
- impaction, which, based on the construction of the impactor, can be divided into

  - single-stage and multi-stage,
  - single-slit and multi-slit,
  - virtual (for the concentration of aerosols or the isolation of target size fractions from them followed by deposition of particles practically by any method);
- impingement (including single- and multi-stage constructions);
- centrifugal (different cyclones and centrifuges);
- thermoprecipitation;
- electroprecipitation (and electroseparation); and
- aerosol deposition on filters.

The reviews [1, 58–62] consider in detail the advantages and disadvantages of each method of bioaerosol sampling and the devices used in them. Therefore, here there is no need to include the information given in these reviews. Let us only emphasize that, for different concrete purposes, it is often expedient to use samplers that are most suitable for testing collected samples with the methods employed.

The following sampling methods were selected from the above list for the studies carried out by the present authors:

- the impingers described in [81] to collect samples of viable bacteria and fungi;
- the personal sampler described in [82] to detect viral aerosols; and
- aerosol deposition on fibrous filters to carry out chemical analyses and determine the genetic material of microorganisms with the polymerase chain reaction (PCR) method.

One of the most important characteristics of samplers is the construction of their intakes (inlets). The fact is that sampling should be isokinetic to provide reliable information on the concentration and disperse composition of aerosols. For this purpose, intakes are designed to maximally maintain isokinetic conditions of sampling at different rates of incident flow [83–85]. The characteristics of the intakes of some bioaerosol samplers widely used in laboratory and natural experiments were described in [84, 85].

Isokinetic sampling from a jet, for example, during aircraft sampling, should be considered separately. Here we should mention the constructions described in [85, 86]. An isokinetic sampler (photographs of the sampler and the airplane are presented in [81]) designed especially for the used type of airplanes [87] to maintain isokinetic conditions [83] was employed by the authors during bioaerosol sampling from an aircraft laboratory [81, 88].

In this research, altitude samples of atmospheric bioaerosols were collected on fiber filters of the AFA-HA type [89] at a flow rate of about 250 l/min and on impingers [81] at a flow rate of about 50 l/min. Fifty milliliters of non-colored Hanks' solution (ICN Biomedicals, Irvine, CA, USA) was poured into the impingers as the sorbing fluid. The *retention efficiency* of this device for aerosols of more than 0.3 μm (minimal size of known bacteria) exceeds 80%, making up a constant value of 90% ± 15% for particles with a diameter of 2 μm. An A-D1-04 pump (JSC "Kot," St. Petersburg, Russia) was used to pump the air samples through the impinger. The pump outlet was connected to overboard air, which provided the required pressure drop for the critical nozzle. Terrestrial ("on-land") samples were collected four times a day on the same impingers and filters at the site of the "Vector" research center (Koltsovo, site A) during a day in the middle of the month to reveal daily variations of the measured values. Also samples were collected once during seven successive days of each season to detect culturable microorganisms in Klyuchi settlement (site B); samples were collected during 30 days to determine average daily concentrations of total protein. In addition, irregular summer measurements of average daily concentrations of total protein and aerosol were performed during the summer period in Zav'yalovo settlement (site C) located not far from the aircraft laboratory runway. The volumetric flow rate through the filters at site A was 50 l/min, and at site B was approximately 250 l/min. Samples for detection of viral aerosols were also collected at site A by deposition on fibrous filters with a volumetric flow rate of 50 l/min and on a personal sampler for viral aerosols [82] with a volumetric flow rate of 4 l/min. Samples were collected successively in the daytime at altitudes of 7000, 5500, 4000, 3000, 2000, 1500, 1000, and 500 m over forestland to the south of Novosibirsk during the last 10 days of each month. The distance from the on-land sampling points to the aircraft laboratory runway is approximately 80 km.

Samples of snow cover, which is a good accumulator of all the pollutants that get into it during the cold season, were collected on sites located both near to and far from powerful anthropogenic sources. A sample of 1 dm^2 for the whole depth of the snow cover was taken from snow with a special sterile sampler. Fresh snow samples were collected on sterile 1 m^2 polyethylene film. The collected samples were thawed under sterile conditions, and each sample was divided into several parts to perform chemical and biological analyses of the samples.

14.2.2
Methods to Analyze the Chemical Composition of Atmospheric Bioaerosols and their Morphology

Let us start by considering the methods used to analyze the chemical composition of atmospheric bioaerosols and their morphology with different variants of microscopic analysis. In the method described in [90, 91] aerosol particles deposited on a specially developed support are stained with a dye, which interacts with the carboxylic groups of the protein. Thus, *light microscopy* allows biogenic particles or biogenic components in particles of complex composition to be detected, which provides data on the concentration and disperse composition of the bioaerosol and the morphology of its particles.

Other microscopic methods, which do not use dyes specific for biogenic material, reviewed in [1, 92–97], are less informative with respect to linking aerosol particles to bioaerosols. However, reliable conclusions concerning the morphology of particles cannot be drawn without using such microscopic methods (both optical and electronor *atomic force microscopy*), with the exception of particles within which typical microorganism structures are observed [13, 92, 96, 98–100]. But if such structures are lacking, it is very difficult to link individual particles to bioaerosols. Even electronography of individual particles (or the application of another method providing information on the elemental composition of particles and/or their surface [101, 102]) does not allow one to link individual particles to bioaerosols with certainty. Also, although the potential of microscopic methods is far from being exhausted, their application for the analysis of aerosols, and bioaerosols in particular, takes too much time even when automatic image processing systems are used [103, 104]. Thus, the application of microscopic methods provides reliable information only on the sizes and morphology of particles, while characterization of their composition requires other methods of analysis.

Before we proceed to the consideration of methods for analysis of the chemical composition of atmospheric bioaerosols, let us dwell upon an important aspect of performing such analyses. Generally speaking, information on the chemical composition of an atmospheric aerosol is necessary for each of its particles. The fact is that, first, aerosols originate from different natural and anthropogenic sources, and, consequently, they have different components and chemical compositions. Aerosols from entirely different sources are present in the atmosphere simultaneously. Second, atmospheric aerosols are unstable. Different physical and chemical processes occur in aerosols under the influence of varying temperature, humidity, and radiation [1, 4, 14–20]. These processes result in the appearance of new components and chemical compounds, and change the dispersity and concentration of particles. Atmospheric aerosols from a powerful source carry a pronounced "signature" of the source's chemical composition [11, 12]. They contain elements, that are present in the chemical composition of a powerful source in the largest amounts [12, 13, 105, 106]. Therefore, bioaerosols can be detected in the analysis of the chemical composition of individual particles. However, as noted above, neither the use of aerosol mass spectrometers nor the application of microscopic methods provide a comprehensive characteristic of the chemical composition of an atmospheric bioaerosol. That is why researchers have to characterize the whole aerosol (or its individual size fractions) by the presence of different markers of biogenic components of bioaerosol and the chemical composition of the whole collected sample of atmospheric aerosol without referring the detected chemical compounds or elements separately to atmospheric bioaerosols and other aerosols.

The various methods summarized in Table 14.1 can be used to analyze the chemical composition of bioaerosol samples. All of them provide information (sometimes in an indirect way) on the presence of bioaerosols in the atmosphere, as they detect different chemical or biological markers of atmospheric bioaerosol. Different methods can provide information on the presence of the same chemical compounds or elements, with a detection threshold required for practical purposes. Therefore, the choice of a concrete method to analyze the chemical composition of bioaerosol samples rests with the researcher. The following methods were used by the present authors in the studies performed.

Table 14.1 Methods that can be used to analyze the chemical composition of bioaerosols after deposition.

Method	Chemical compounds (elements) detected by this method	Peculiarities of the method application	Sensitivity of the method	References
Chromatographic methods				
Gas chromatography using pyrolysis decomposition of particles, including that conjugated with mass spectroscopy	Different biological molecules; allows organic carbon to be determined	–	Not worse than 1 ng in a sample	[107–110]
High-performance liquid chromatography, including that conjugated with mass spectroscopy	Different biological molecules	–	Not worse than 1 ng in a sample	[111–113]
Fluorescent and luminescent spectroscopy	Different biological molecules or microorganisms and plant pollen	–	Up to one particle in 1 liter	[26, 52, 94, 95, 103, 114–127]
Cytometry	Only cell objects	Requires transition of aerosol into fluid	One cell in a sample	[128–132]
Oscillatory spectroscopy including Fourier and X-ray spectroscopy	Different biological molecules	Allows studying the surface layers of particles and their whole volume	Not worse than 1 ng in a sample	[133–139]

(continued overleaf)

Table 14.1 (*Continued*)

Method	Chemical compounds (elements) detected by this method	Peculiarities of the method application	Sensitivity of the method	References
Atomic absorption and atomic emission methods	Element composition of samples; depending on concrete implementation of the method there are limitations in the list of detected elements	Requires bioaerosol deposition on filters	Largely depends on atom excitation method, but not worse than 10^{-9} g in a sample	[23, 73, 140–142]
Mass spectroscopic methods	Different ions of chemical elements and molecules	Allows studying the surface layers of particles and their whole volume	10^{-15} g	[32, 40–48, 107, 143, 144]
Immunochemical methods	Certain epitopes (sites) of macromolecules. Allows determining only the sought macromolecules (microorganisms)	Allows studying the surface layers of particles and specific epitopes within cells	Several microorganisms (their macromolecules)	[50, 145–149]
Methods based on the determination of microorganism genetic material	Genetic material of microorganisms or fragments of different organisms and plants	–	Several copies of the fragment of genetic material	[51, 150–156]

Reviews of methods listed in this table can also be found in [1, 157, 158].

The *fluorimetric method* based on the acquisition of intense fluorescence by a protein after its modification with a fluorogenic reagent was employed to determine the concentration of the total protein in the samples. The reagent 3,4-carboxybenzoyl quinoline-2-carboxyaldehyde (CBQCA), which forms fluorescent derivatives with higher quantum yield than other dyes when reacting with proteins, was used as the modifying reagent [159]. Proteins are determined in the presence of lipids, detergents, and surface-active substances. The detection threshold of the Shimadzu RF-520 spectrofluorimeter for total protein using CBQCA was 0.0005 μg/ml of concentrated sample, and the error of the concentration determination did not exceed 20%. As fluorescence from some PAHs lies in the same wavelength range as that of the total protein, the value referred to PAHs, which was determined independently, was subtracted from the total fluorescence values [160].

Different methods were employed for elemental analysis of the samples. The determination of the elemental composition of altitude samples of atmospheric aerosol was performed using *secondary ion mass spectrometry*, with the element distribution with depth described in detail in [161]. The analysis of terrestrial samples was performed with the *atomic absorption spectroscopy* method using a Shimadzu model AA-6300 device with flame and electrothermal atomizers. Snow samples were subjected to quick thawing followed by concentration on a graphite collector. Then the unified technique for the analysis of the graphite concentrate of microelements developed by the Analytical Laboratory of the Institute of Inorganic Chemistry, Siberian Branch of the Russian Academy of Sciences, was used [162]. The method of varying weights (aliquots of 0.2, 1.0, and 5.0 ml) was employed to take into account the matrix effect caused by the complex composition of the analyzed sample [4]. The elemental analysis of snow samples was performed with the method of *atomic emission spectroscopy* with arc excitation of spectra on a PGS-2 diffraction spectrometer (Carl Zeiss Jena, Germany) [162]. Emission spectra were recorded using a photodiode ruler (NPO "Optoelectronika"); spectral information was processed by a computer with the "ATOM" program for processing spectral analysis data (developed by NPO "Optoelectronika" in cooperation with the Institute of Organic Chemistry, Siberian Branch of the Russian Academy of Sciences).

The concentrations of PAHs in samples of atmospheric air and snow were determined with the method of *high-performance liquid chromatography* using a Spectra Physics SP8800, with a Shimadzu RF-530 fluorescence detector and a Spectra 100 spectrophotometric detector according to the Russian standard [160]. PAHs were extracted from the sample with hexane and chromatographically separated; the signal was recorded with an ultraviolet detector. Identification of the peaks of individual compounds on the chromatogram was carried out by the retention time; the mass of the compound was calculated for each individual PAH. The relative error limits (for 95% confidence level) make up 25% in the whole range of measurements for each PAH.

14.2.3
Methods Used to Detect and Characterize Microorganisms in Atmospheric Bioaerosols

The presence of microorganisms in the air has been investigated for more than 100 years [163]. This is connected with the fact that, besides their "usual" effect on the climate and atmospheric processes [164–166], microorganism aerosols cause a large number of diseases in humans, animals, and plants [167–173]. That is why work devoted to the study of microorganism aerosols in the atmosphere (see, for example, [75, 100, 164, 174–183]) and their transfer over long distances [170, 172, 182, 183] are so numerous.

On the whole, all the methods used to detect and characterize microorganisms in atmospheric bioaerosols can be divided into two main groups: cultural and non-cultural. The first group of methods allows the detection of microorganisms propagating on different media (test systems), that is, preserving their viability. Such analyses usually take two to three days, and in the case of viable viruses and lower fungi up to two to three weeks. The second group of methods does not provide information on the viability of microorganisms – though research has been carried out in this direction (see, e.g., review [184]). However, they can provide information on non-culturable microorganisms and the presence of their fragments, and can allow the identification of microorganisms by typical signs. All these procedures take less than 24 hours, and the fastest methods just 2–3 hours.

The following methods to detect and characterize non-culturable microorganisms in atmospheric bioaerosols can be distinguished.

- **Microscopic methods.** This group of methods includes optical, electron, and atomic force microscopy with or without the use of different dyes, including fluorescent and immunofluorescent ones [98, 185, 186]. Such methods allow direct observation of microorganisms or their fragments in the samples. It should be noted that there are methods to reveal the vital activity of bacteria [114, 128, 187–189] that allow viable bacteria to be identified under a microscope without cultivation.
- **Chromatographic methods.** The methods presented in Table 14.1 allow the identification of some known microorganisms by the composition of certain compounds. However, the sensitivity of currently used chromatographic methods, even those involving mass spectroscopic detection of high-molecular-weight compounds, does not allow the identification of individual microorganisms.
- **Cytometry including flow cytometry.** On the contrary (see Table 14.1), *cytometry* detects individual bacterial cells (spores) in the sample. However, the identification of microorganisms by fluorescence spectra is less detailed and is possible only in the case where the detected microorganism has been encountered and identified previously.
- **Methods used to detect specific antigens or macromolecules of microorganisms.** This group of methods includes different modifications of enzyme-linked immunoassay (ELISA), immunofluorescent, and radioisotope

methods [126, 127, 190]. All of them are based on the reaction of a labeled antibody with the sought receptor on the microorganism surface. The most sensitive modifications of these methods allow several antigens to be detected in a sample. However, unlike cytometry or microscopic methods that detect unknown microorganisms, this group of methods detects only known antigens. Moreover, this group of methods, like the next one, provides information neither on the viability of the microorganism detected with the receptor nor on its integrity.

- **Methods using the analysis of microorganism genetic material.** Different PCR-based methods are now widely used in aerobiology to detect various microorganisms in bioaerosols [151, 153, 154, 187–196]. These methods detect only a few copies of a certain domain of microorganism genetic material in the sample. An approach has been developed that allows the determination of the place of even an unknown microorganism in bacterial systematics using the nucleotide sequence read from a genetic material amplified fragment [155, 197]. However, as noted above, this group of methods provides information neither on the viability of the microorganism genetic material detected in the fragment nor on its integrity.

Only the use of the cultural methods [185, 186, 198, 199] allows us not only to determine the viability of a microorganism, but also to produce its "pure line" for further comprehensive study of its characteristics with different methods, from physical to molecular biological ones. It should be noted that non-specific media can be used to detect viable bacteria or lower fungi in samples of atmospheric bioaerosols, and the detection of viable viruses requires the use of either cell cultures or laboratory animals susceptible to this virus [7]. In the authors' researches, both cultural and non-cultural methods were used to identify and characterize the microorganisms in atmospheric bioaerosols.

Standard methods were employed to determine the concentration of culturable microorganisms. Samples were seeded onto Petri dishes containing agar media as follows: Luria–Bertani (LB) medium or Fish agar (FA). "Fish agar" consists of (g/l): fish pancreas hydrolysed product – 17.9, agar – 12.0, NaCl – 8.0, pH 7.0–7.5. [200] was used to detect saprophyte bacteria; depleted LB medium (diluted 1 : 10) was used to isolate microorganisms inhibited by the excess of organic substances; starch–ammonia agar (SAA) medium [189] was used to detect actinomycetes; soil agar was used for soil microorganisms; and Sabouraud medium [201] was used for fungi and yeast. Successive sample dilutions were prepared as required. The seedings were incubated in a thermostat at a temperature of 28–30 °C for 3–14 days.

An evaluation and description of the cultural–morphological indices of grown colonies of fungi were performed on days 5 and 14 for more complete evaluation of the number of fungi, including those at quiescent stages of development and requiring longer time periods for growth on a medium. Qualitative evaluation included the description of the colonies' morphology for the whole variety of grown fungi and optical microscopy of isolates. Photographs of *conidiophores*, *conidia*, and *mycelium* of some isolates were taken using a Zeiss Axioscop microscope. Standard manuals and classifiers were used for identification of fungi [202, 203]. After determining the isolates up to the genus, some fungal cultures were reseeded on a medium to create a collection of pure cultures and to perform further investigation of their properties.

The morphological features of the detected bacteria were studied visually and with light microscopy. Fixed preparations of Gram-stained cells and live preparations of cell suspensions observed with the phase contrast method were prepared for this purpose. The taxonomic groups of the detected microorganisms were determined according to [189, 204, 205], and the analysis of nucleotide sequences of PCR products corresponding to the fragments of the 16S rDNA gene was performed for some bacteria [206].

The calculation of the number of viable microorganisms in samples expressed in terms of colony-forming units (CFUs) was performed according to standard methods [207]. The number of microorganisms was averaged over three or four parallel samples seeded on four or five different media. Taking into account the volume of atmospheric air samples collected for analysis, the minimal detection threshold for the concentration of viable fungi in the atmosphere was 40 CFU/m^3 for altitude samples and 11 CFU/m^3 for on-land samples. For bacteria, the detection threshold was 100 CFU/m^3 for altitude samples and 28 CFU/m^3 for on-land samples.

The determination of the pathogenic properties of bacteria included testing the studied microorganisms for the presence of hemolytic, plasmocoagulation, fibrinolytic, and gelatinase activities.

- The *hemolytic activity* was determined on blood agar by clarification zones around colonies according to a method described in the literature [205].
- In order to determine the, dry rabbit blood plasma was diluted 1 : 5 with sterile physiological solution and used as a test reaction. Diluted plasma (0.5 ml) was placed into a sterile test tube, and one loop of 18–20 h culture was suspended in it. The test tubes were placed into a thermostat at 37 °C, and the presence of plasma coagulation was observed within 1, 2, 3, 18, and 24 h of incubation.
- In order to reveal the fibrinolytic properties of microorganisms, sterile test tubes were filled with 0.1 ml of citrate plasma, 0.4 ml of physiological solution, 0.25 ml of 18–20 h broth culture of the tested strain, and 0.25 ml of 0.25% calcium chloride solution. The test tubes were shaken and placed into a thermostat at 37 °C for 15–20 minutes. If a clot was formed in the tube (as in the control case, where nutrient medium was added instead of broth culture), the tested culture was considered to have no fibrinolytic properties. If fluidization of the clot was observed within 2 h, the culture was characterized by the presence of fibrinolysin.
- The *gelatinase activity* was determined by seeding microorganisms in beef-peptone broth containing 12% gelatin. A positive reaction was the fluidization of the gelatin column [189].

The growth characteristics of bacteria at increased salt concentration were determined by growing microorganisms on FA medium with increased NaCl concentration (5 or 10%). The growth of colonies provided indirect evidence of bacterial resistance to drying.

The determination of *enzymatic activity* of isolated bacteria was performed by the following tests. The determination of proteolytic (caseinolytic) activity of the tested isolates was performed using milk agar (MA). The composition of MA was as follows: 3% starved agar (tap water + 3% agar), which was sterilized at 1 atm for 30 minutes, and 12% non-fat milk sterilized at 1 atm for 20 minutes. The starved agar was melted and cooled to 50–55 °C before use, thoroughly mixed with sterile milk warmed to the same temperature, and poured into dishes. The tested microorganisms were seeded onto the medium surface by stroking, and incubated under optimal conditions. The presence and size of substrate hydrolysis zones were determined by measuring them in millimeters [189].

Amylolytic activity was determined by the ability of the strains to secrete amylolytic enzymes during growth on SAA medium containing starch. The studied cultures were seeded onto the medium surface by stroking or injection with a bacteriological loop. After incubation, iodine solution was poured onto the dish, which was used to detect colonies producing hydrolysis zones. A positive test result was indicated by the appearance of a colorless area around the growth zone.

The determination of lecithinase and lipase activities of isolates was performed on yolk agar (FA + yolk). Yolk was placed into 400 ml of molten agar and mixed to obtain a uniform suspension under aseptic conditions. Then, the yolk medium was poured into dishes and left until it became hard. The cultures were seeded by stroking onto the medium surface, and incubated for three to seven days. Positive lecithinase reaction was expressed as the appearance of turbid zones at clarification of the yolk agar around colonies of tested cultures [189]. The presence of lipase was revealed by viewing the grown colonies in inclined light. A positive result was indicated by the formation of an oily, glittering and shimmering, or pearl, layer above the colony or around it on the agar surface [189].

The screening of isolates for *lipolytic activity* was performed at room temperature by seeding the strains by injection onto L-agar containing 1% Tween-20 or Tween-40 supplemented with 0.01% $CaCl_2$. The results were determined within three to four days of incubation of the seedings by the presence of turbid zones in the agar around the colonies. The relative activity was determined by measuring the diameters of the colony and the zone [189].

In order to detect *alkaline phosphatase*, 0.3 ml of a suspension in physiological solution (0.85% NaCl) was added to 0.3 ml of substrate solution containing 0.04 M glycine buffer at pH 10.5 and with 0.01 M disodium *n*-nitrophenyl phosphate (Sigma). The mixture was incubated for 3 h at 37 °C. Positive reaction manifested itself as yellow staining of the reaction mixture [204]. The enzyme activity was determined within 3 h of incubation by absorption on a Uniplan apparatus (Russia) with a color filter at a wavelength of 450 nm.

The *endonuclease* activity was determined on a solid medium with thymus DNA and toluidine blue. The reaction was evaluated by the appearance of a bright pink zone around the bacterial colony [205].

When screening the isolates for the presence of *restriction endonucleases*, individual colonies collected from a solid culture were suspended in 100–200 μl of TEN buffer: 0.1 M Tris (tris(hydroxymethyl)aminomethane), pH 7.5, 0.01 M EDTA (ethylenediaminetetraacetate), and 0.05 M NaCl. Lysozyme and Triton X-100 were used to destroy the cell walls of the bacteria. The cell extract obtained was used for analysis of the presence of restriction endonucleases. DNAs of phages λcI857 and T7 were used as substrates for hydrolysis. The electrophoresis of DNA after restriction was performed in 1% agarose (Sigma) [208]. The presence of restriction endonucleases in the microorganism strains was revealed by the appearance of discrete fragments of substrate DNA in the electrophoregram under ultraviolet light.

The concentration of *plasmid DNA* in isolates was determined with the screening method using a standard procedure. Cells from a solid medium were suspended with a loop in 100 μl of buffer (50 mM Tris, pH 8.0, 50 mM Na_2 EDTA, and 15% sucrose), 200 μl of alkaline solution (0.2 N NaOH and 1% sodium dodecyl sulfate (SDS)) and 150 μl of 3 M sodium acetate, pH 5.0, were added, and centrifugation was performed for 5 minutes on a desktop centrifuge. Then 1 ml of 96% ethanol was added to the sediment. The DNA obtained was analyzed in 0.8% agarose in Tris–borate buffer, pH 8.0 [205].

Antibiotic resistance was studied with the disk diffusion method [189], by seeding microorganisms onto solid media followed by application of paper disks (Research Center of Pharmacology, Saint Petersburg, Russia) with the antibiotics widely used in practice: ampicillin (10 μg/disk), neomycin (30 μg/disk), benzylpenicillin (100 U/disk), levomycetin (30 μg/disk), carbenicillin (100 μg/disk), canamycin (30 μg/disk), oleandomycin (15 μg/disk), rifampicin (5 μg/disk), streptomycin (30 μg/disk), polymyxin (300 U), erythromycin (15 μg/disk), lincomycin (15 μg/disk), oxacillin (10 μg/disk), gentamycin (10 μg/disk), and tetracycline (30 μg/disk).

Atmospheric air samples were not tested for the presence of viable viruses. Genetic material of viruses was detected in atmospheric air samples with PCR method. The procedure of DNA isolation and amplification used in the study was described in detail in [209, 210]. The real-time polymerase chain reaction (RT-PCR) method described in [211] was used for RNA isolation and amplification.

For electron microscopy of samples, the polyethyleneglycol PEG-6000 was added to the samples up to a final concentration of 8%, thoroughly mixed, and put on ice for 4 h. After that, centrifugation was performed at 10 000 *g* for 30 minutes at 4 °C. The supernatant was carefully collected, and the sediment was dissolved in NTE buffer: 0.1 M NaCl, 0.01 M Tris HCl, and 0.001 M EDTA, pH = 7.6. Then 0.05 ml of the sample was poured into two test tubes. The precipitate of microorganisms obtained by PEG desalination was fixed with 4% paraformaldehyde solution, fixed once again with 1% osmic acid solution, and dehydrated in ethanol solutions with incremental concentration and acetone. Epon-Araldite was poured into the mixture. Ultrathin sections were prepared on a Reichert–Young microtome and contrasted with uranyl acetate and lead citrate solutions. An H-600 electron microscope (Hitachi, Japan) was used to examine 700–800 Å thick sections.

14.3 Atmospheric Bioaerosol Studies

14.3.1 Time Variation of Concentrations and Composition of Atmospheric Bioaerosol Components

Let us analyze the results obtained in the course of long-term observation of atmospheric bioaerosols in southwestern Siberia. As noted above, we consider the total protein and viable microorganisms to be the most important and informative components of atmospheric bioaerosols.

First of all, it should be noted that, during 10 years of observation, the average annual concentrations of these components measured at altitudes from 500 to 7000 m in southwestern Siberia tended to decrease (Figure 14.1). On average, this decrease is not large, and comes to 7.0% per year for the total protein concentration and 7.9% per year for the concentration of viable microorganisms (note that, as accepted in microbiology, this concentration value here and subsequently in the text is expressed in decimal logarithms of the number of viable microorganisms in 1 m^3 of air). Similar trends in the variation of these values revealed for terrestrial ("on-land") measurements at site A give a decrease by 6.5% per year for the total protein concentration and 8.2% per year for the concentration of viable microorganisms. There are no literature data on such long-term year-round observations of the full concentration and type of viable microorganisms in atmospheric aerosol, though long periods of observations of individual microorganism species and pollen have been described in the literature for different regions of the world [212–217].

Literature data show that various bioaerosols including viable microorganisms can be found in all atmospheric layers: in the surface layer and at altitudes higher than 70 km [1, 179, 218–225]. Their concentration in 1 m^3 of atmospheric air varies from several species to several thousands and even tens of thousands [1, 179, 218–229]. The concentrations of viable microorganisms found in the atmosphere of southwestern Siberia at altitudes up to 7000 m correlate well with the values determined in these works.

The dynamics of the annual variation of the total protein and viable microorganism concentrations for altitude and on-land measurements in southwestern Siberia normalized by the corresponding average annual values are presented in Figures 14.2 and 14.3 and in Table 14.2. Two important aspects should be noted here. First, there exists an expressed dependence of these values for all observation sites over the course of a year. The annual amplitude of variation of the value of the total protein concentration at altitudes of 500–7000 m reaches ±40%, and the concentrations of viable microorganisms reaches ±0.45 on the logarithm scale used.

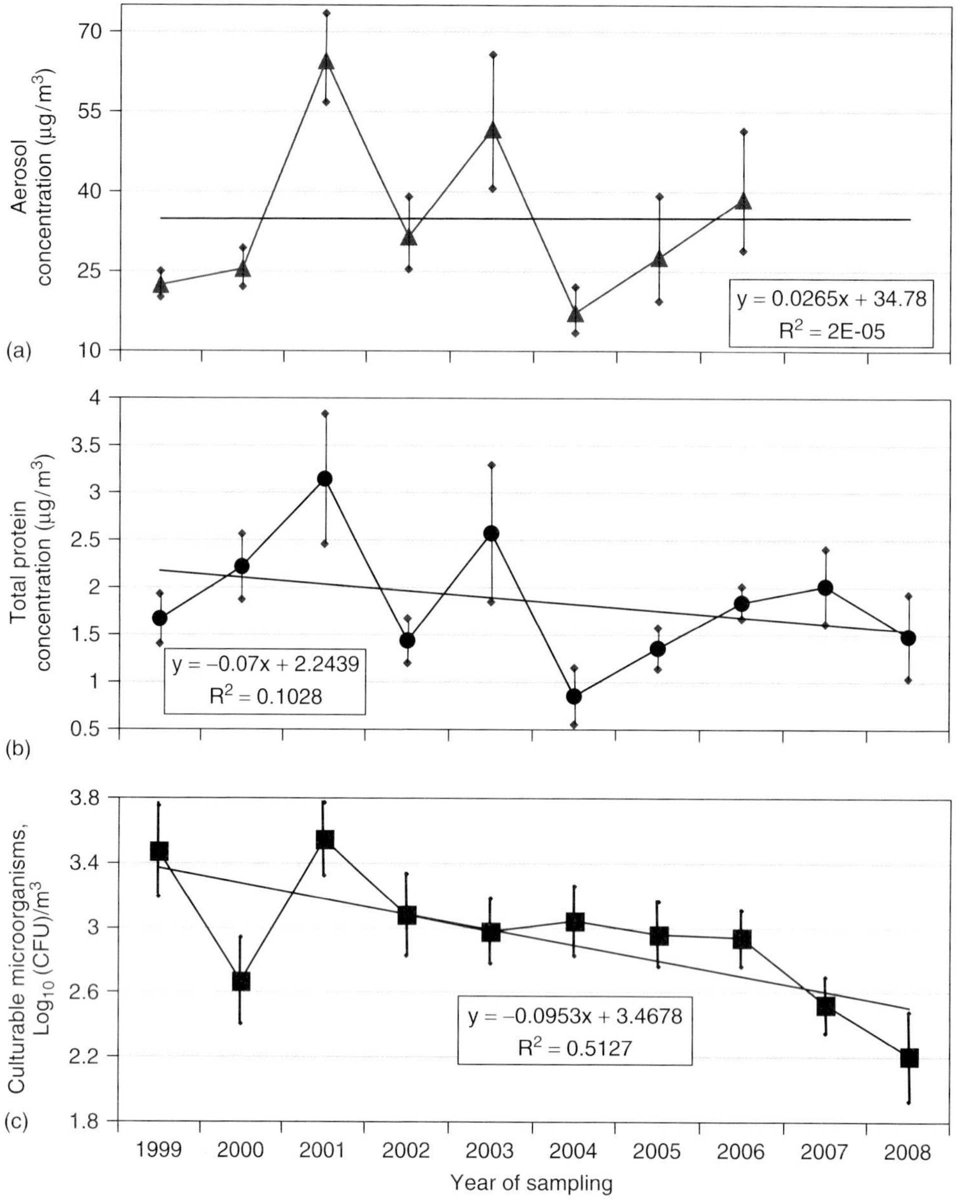

Figure 14.1 Long-term variations in average annual concentrations in the atmosphere of southwestern Siberia at altitudes of 500–7000 m: (a) aerosol (data for 2007–2008 have not been processed); (b) total protein; and (c) microorganisms. Average annual values and their confidence interval at 95% reliability level are presented.

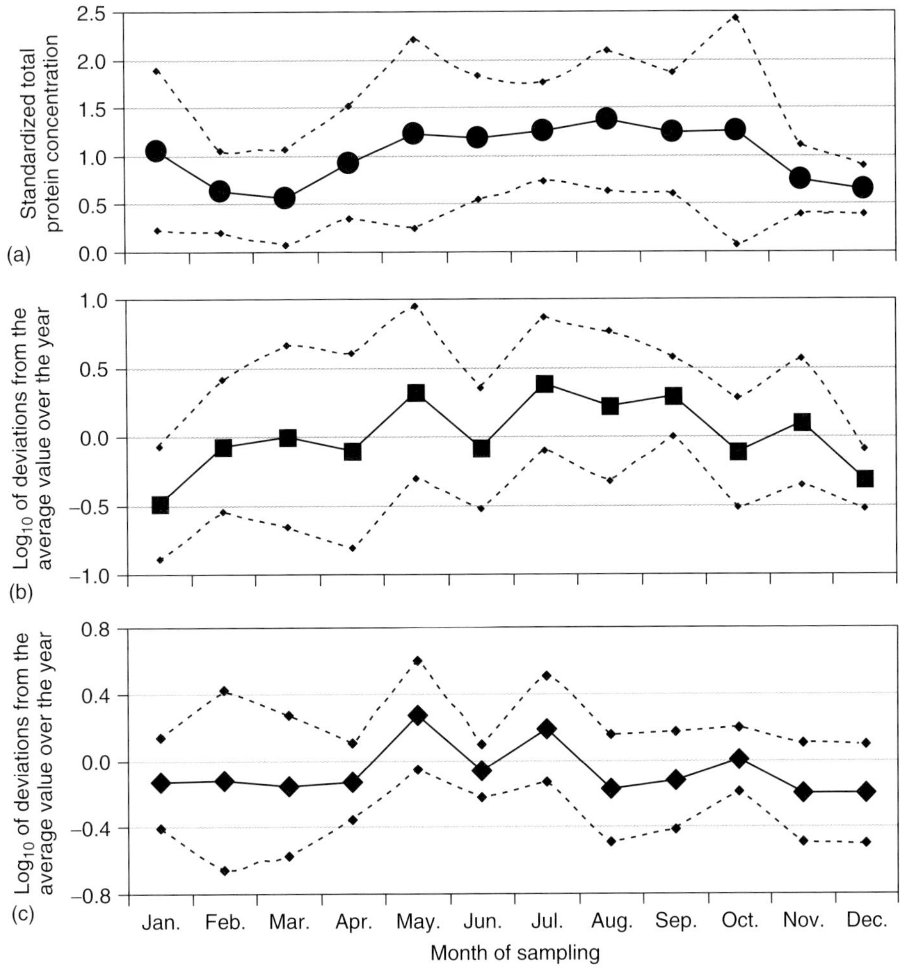

Figure 14.2 Annual variations of concentration normalized over the average annual value in the atmosphere of southwestern Siberia at altitudes of 500–7000 m, averaged over 10 years of observations: (a) total protein; (b) culturable microorganisms; and (c) culturable fungi. Average values and their confidence interval at 95% reliability level are presented.

For on-land measurements at site A, the variation of the total protein concentration reaches ±30%, and that of the viable microorganism concentration reaches ±0.55 on the logarithm scale. The dependences of the concentration of viable bacteria for the whole year follow those for the full number of viable microorganisms, as their concentration is usually larger than that for viable lower fungi. (According to the data for the whole observation period, average annual concentrations of viable bacteria varied within the range of 200–3200 CFU/m^3, while the concentration of

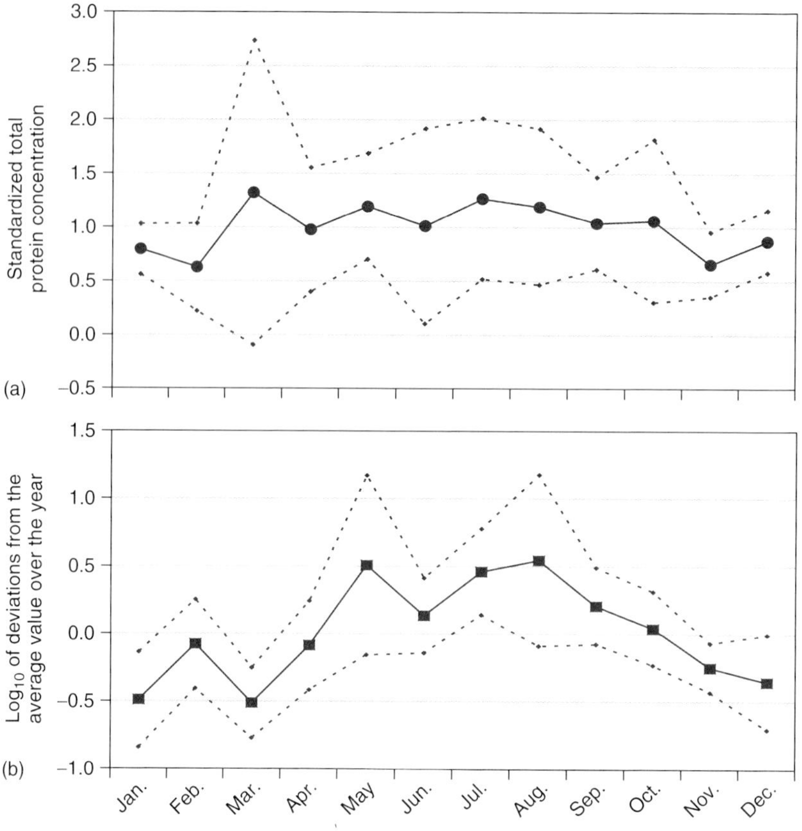

Figure 14.3 Annual variations of concentration normalized over the average annual value for on-land measurements at point A, averaged over eight years of observations: (a) total protein; and (b) culturable microorganisms. Average annual values and their confidence interval at 95% reliability level are presented.

viable fungi varied within the range of 90–200 CFU/m^3.) This is similar to the situation in Marseilles according to [230]. The dependences for the concentration of viable bacteria summarized in [231] for Paris, Moscow, Montreal, and other regions [177] demonstrate similar dynamics of their variation, but with somewhat larger amplitude. A similar variation of concentrations in the atmosphere is also observed for fungi (Figure 14.2C). The annual dynamics of both individual representatives of different genera and the total concentration of fungi are also presented in the literature [137, 177, 216, 217, 230, 232–235]. Thus, the expected dependences, which are in qualitative agreement with data for other regions, were revealed. In the warm season, the region's atmosphere contains more bioaerosols than in the cold season, though, as shown in [177, 216, 217, 232, 235], for some genera of fungi the maximal concentrations are reached in other seasons.

Table 14.2 Average seasonal characteristics of aerosols determined for samples collected at site B.

Year	Characteristic	Winter	Spring	Summer	Autumn
2001	Aerosol mass concentration (μg/m³)	27 ± 11	61 ± 38	34 ± 12	31 ± 17
	Total protein concentration (μg/m³)	0.007 ± 0.003	0.01 ± 0.01	0.24 ± 0.14	0.09 ± 0.09
	Percentage of total protein in aerosol mass	0.03 ± 0.02	0.01 ± 0.01	0.76 ± 0.40	0.24 ± 0.26
	Culturable microorganisms ($\log_{10}$CFU/m³)	2.8 ± 0.4	4.3 ± 0.9	3.2 ± 0.3	4.4 ± 0.8
2002	Aerosol mass concentration (μg/m³)	22 ± 5	52 ± 38	26 ± 10	38 ± 19
	Total protein concentration (μg/m³)	0.08 ± 0.04	0.06 ± 0.06	0.3 ± 0.1	0.3 ± 0.1
	Percentage of total protein in aerosol mass	0.4 ± 0.2	1.1 ± 0.8	1.3 ± 0.6	0.8 ± 0.3
	Culturable microorganisms ($\log_{10}$CFU/m³)	3.0 ± 1.1	2.6 ± 0.8	2.8 ± 0.2	1.2 ± 0.9
2003	Aerosol mass concentration (μg/m³)	26 ± 9	36 ± 22	28 ± 12	25 ± 19
	Total protein concentration (μg/m³)	0.01 ± 0.01	0.7 ± 0.4	0.6 ± 0.3	0.9 ± 0.6
	Percentage of total protein in aerosol mass	0.4 ± 0.5	2.3 ± 1.1	3.0 ± 3.3	4.2 ± 2.9
	Culturable microorganisms ($\log_{10}$CFU/m³)	2.6 ± 0.5	3.4 ± 0.8	3.2 ± 0.3	3.6 ± 0.6
2004	Aerosol mass concentration (μg/m³)	34 ± 16	57 ± 43	18 ± 9	26 ± 13
	Total protein concentration (μg/m³)	0.21 ± 0.12	1.27 ± 1.59	1.48 ± 0.65	0.442 ± 0.22
	Percentage of total protein in aerosol mass	0.79 ± 0.57	1.97 ± 1.32	10.0 ± 5.7	1.77 ± 0.71
	Culturable microorganisms ($\log_{10}$CFU/m³)	2.5 ± 0.3	3.2 ± 1.0	2.7 ± 0.3	2.6 ± 0.4
2005	Aerosol mass concentration (μg/m³)	34 ± 10	39 ± 23	25 ± 12	31 ± 16
	Total protein concentration (μg/m³)	0.07 ± 0.03	0.51 ± 0.34	0.62 ± 0.23	0.18 ± 0.13
	Percentage of total protein in aerosol mass	0.25 ± 0.15	1.41 ± 0.87	3.13 ± 2.32	0.71 ± 0.61
	Culturable microorganisms ($\log_{10}$CFU/m³)	2.6 ± 0.4	3.2 ± 0.8	3.6 ± 0.7	3.0 ± 0.2

(continued overleaf)

Table 14.2 (*Continued*)

Year	Characteristic	Winter	Spring	Summer	Autumn
2006	Aerosol mass concentration ($\mu g/m^3$)	51 ± 17	75 ± 57	41 ± 15	41 ± 34
	Total protein concentration ($\mu g/m^3$)	–	1.18 ± 1.58	1.04 ± 0.39	0.45 ± 0.28
	Percentage of total protein in aerosol mass	–	1.34 ± 1.0	3.27 ± 2.57	1.37 ± 1.01
	Culturable microorganisms ($\log_{10}CFU/m^3$)	2.4 ± 0.3	–	2.9 ± 0.1	3.1 ± 0.6
2007	Aerosol mass concentration ($\mu g/m^3$)	43 ± 23	71 ± 54	46 ± 21	30 ± 18
	Total protein concentration ($\mu g/m^3$)	0.08 ± 0.04	0.65 ± 0.37	1.84 ± 0.86	0.21 ± 0.19
	Percentage of total protein in aerosol mass	0.21 ± 0.12	1.19 ± 0.70	4.65 ± 2.83	0.70 ± 0.58
	Culturable microorganisms ($\log_{10}CFU/m^3$)	2.4 ± 0.3	2.8 ± 0.7	4.0 ± 0.7	2.7 ± 0.5
2008	Aerosol mass concentration ($\mu g/m^3$)	55 ± 18	55 ± 30	50 ± 18	37 ± 21
	Total protein concentration ($\mu g/m^3$)	0.08 ± 0.06	0.71 ± 0.32	0.56 ± 0.29	0.23 ± 0.09
	Percentage of total protein in aerosol mass	0.16 ± 0.13	1.58 ± 1.11	1.20 ± 0.64	0.81 ± 0.48
	Culturable microorganisms ($\log_{10}CFU/m^3$)	2.6 ± 0.7	2.4 ± 0.9	2.9 ± 0.8	–

Second, for altitude samples of atmospheric air, the values for each month of the year averaged over 10 years of observations are given for the whole atmospheric mass from 500 to 7000 m. It is known that the concentration of aerosols in southwestern Siberia with diameter of more than 0.4 μm decreases at an altitude of 7 km by more than an order of magnitude as compared with that in the surface layer [236]. For bioaerosols, the situation is entirely different. As follows from the data presented in Figure 14.4, the vertical profile of the concentrations of total protein and viable microorganisms at altitudes from 500 to 7000 m averaged over the observation period is practically constant (the average concentration decrease does not exceed 50%). Averaging was performed as follows. Experimental data for all altitudes obtained in one probing were normalized by the mean of the determined value to exclude their variation over the year. After that, these normalized values were averaged for each altitude, and a confidence interval at 95% confidence level was constructed for the obtained value. We will return to the analysis of the observed altitude profiles below. It should also be noted that, according to [237], there is a 80-fold decrease in bacterial concentration at altitudes from 500 to 6000 m, while

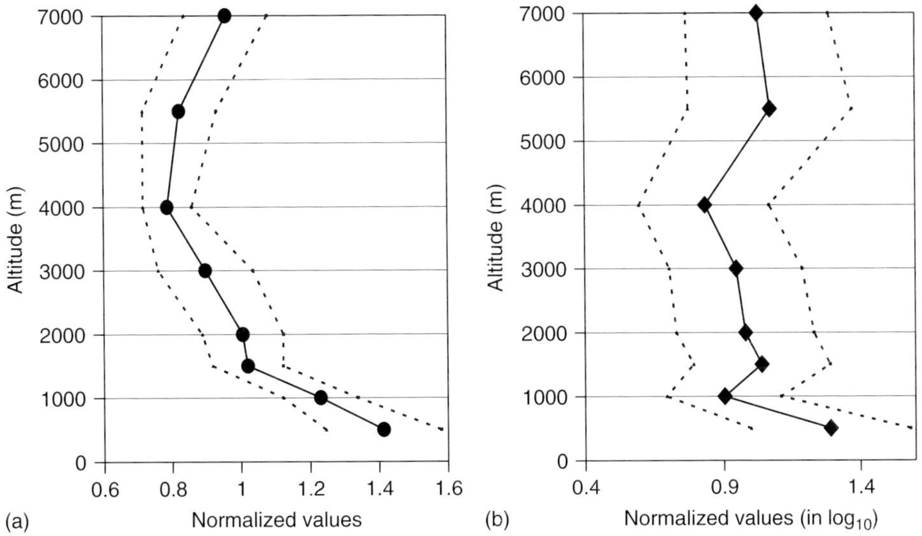

Figure 14.4 Normalized altitude profiles of concentrations of (a) total protein and (b) culturable microorganisms in southwestern Siberia.

in experiments above the North Sea weak variation of bioaerosol concentration was observed at altitudes from 500 to 3000 m, and in one experiment also at altitudes from 0 to 500 m [238].

Now let us analyze the variety of culturable microorganisms detected in atmospheric air samples. On the whole, the percentage of microorganisms in samples varies, and sometimes very sharply, both between altitudes and between measurements (both at ground level and at different altitudes). As the variety of genera of detected microorganisms is very large, it proves to be impossible to reveal the temporal variations of their concentrations in the atmosphere. That is why the detected microorganisms were combined in the following groups: cocci, rods, non-sporiferous bacteria, actinomycetes, yeast, and fungi. Figures 14.5 and 14.6 show the contributions of these groups of viable microorganisms to their total number in the samples. As can be seen from Figures 14.5 and 14.6, even for these groups it is difficult to reveal any regularities of the variation of their concentrations in the atmosphere. There are literature data on the time variation of the concentration of different microorganisms in the atmosphere [185, 216, 217, 221, 230–233, 239]. The authors of these works sometimes observed a very sharp change in both the number and the type of microorganisms in successive samples. The relative amounts of different viable microorganisms in atmospheric air samples collected at altitudes of 500–7000 m averaged over all annual samples also varies between years (Figure 14.7), though these differences are not as pronounced as for individual samples. Probably, only long-term observations will allow the collection of enough experimental data to reveal statistically significant seasonal and altitude dependences for individual microorganisms as done for their total number.

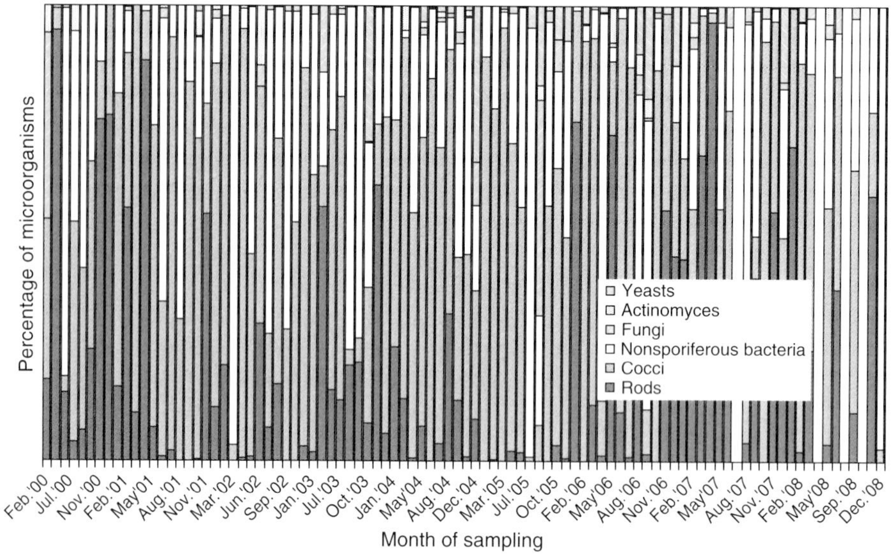

Figure 14.5 Long-term variations of types of culturable microorganisms in atmospheric air of southwestern Siberia samples at the altitudes of 500–7000 m (normalization by the total number of microorganisms detected in each sample).

An important characteristic of bioaerosol is its proportion in the full mass of the atmospheric aerosol. Data [240] and later results of on-land measurements for different seasons demonstrate that the proportion of total protein in the samples makes up from 0.01% to approximately 10% of the mass of aerosol particles determined with gravimetric method. However, according to [90], the mass proportion of bioaerosol during the year in the vicinity of Lake Baikal makes up from 10% to 80% of the full aerosol mass (it should be noted that these values correlate well with the authors' measurements in different regions near Baikal in 2008), and, according to [241], the proportion of biogenic particles can reach 95%. Other authors [91, 112, 242, 243] estimate this proportion at 3–11% by mass and even up to 25% depending on the sampling place and meteorological conditions. Estimates made taking into account the fact that the proportion of viable microorganisms in the total number of microorganisms in different samples is 0.001–40% [22, 129, 175, 196, 226, 244, 245] show that the detected viable microorganisms make a significant contribution to the observed total protein concentrations.

Previously it has been shown that the fraction of particles with aerodynamic diameters of 0.16–0.4 μm contains the maximal number of protein molecules [240, 246]. The mass proportion of total protein in all the particles is maximal in the range from 2.1 to 10 μm. These results are consistent with the data in [243] on the presence of cellulose in bioaerosol particles of different sizes and in [174, 177, 247] where it is shown that the fraction of atmospheric aerosol with particle diameters

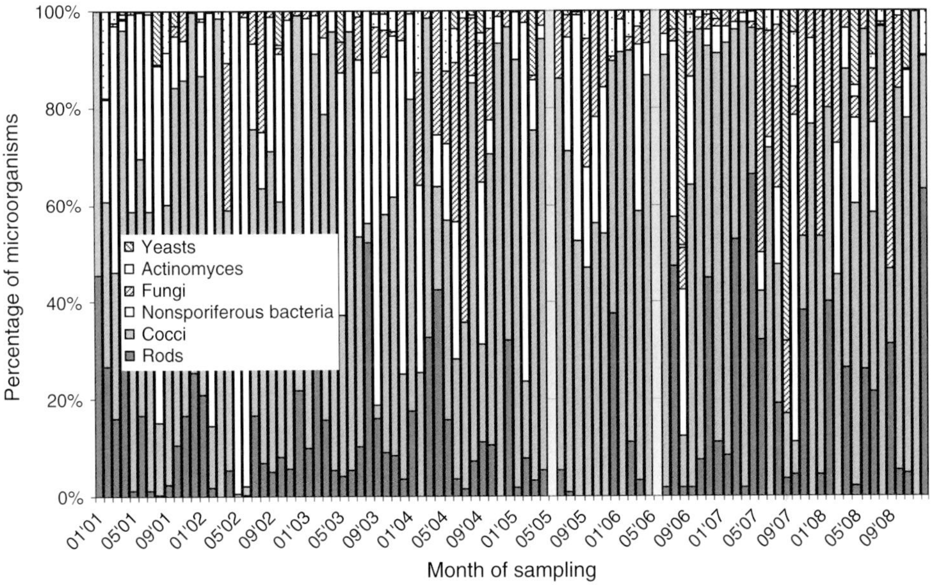

Figure 14.6 Eight-year variation of types of viable microorganisms in atmospheric air samples from southwestern Siberia at point A (normalization by the total number of microorganisms detected in all samples collected during 24 hours).

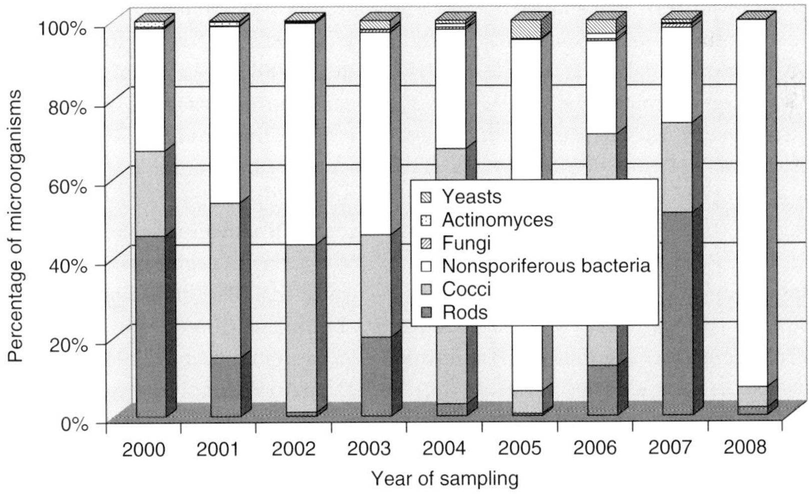

Figure 14.7 Nine-year annual variation of numbers of viable microorganisms in atmospheric air samples from southwestern Siberia at altitudes of 500–7000 m (normalization by the full number of microorganisms detected in samples collected at all studied altitudes for the whole year).

exceeding 2 μm contains the largest number of microorganisms (always containing a considerable amount of protein), and the mean diameter of particles containing microorganisms is 3.9 μm. They also agree with recently published data for OC [248] showing that its amount is maximal in particles with an aerodynamic diameter of 0.3 μm and its proportion in the total mass of particles slightly varies in the range from 0.1 to 10 μm. It is clear that mechanical destruction of non-living biological material (the remains of plants, animals, and their individual cells) into particles of submicrometer size requires considerable energy expenditure. Therefore, it is natural that particles with diameters exceeding 0.1 μm are most rich in biogenic components.

Data obtained in on-land experiments allow us to evaluate the diurnal course of the measured values. The observed differences in the concentrations of total protein and viable microorganisms in atmospheric air samples collected during 24 hours of measurements proved to be statistically uncertain. Nor were considerable differences revealed in the amounts of culturable microorganisms in samples collected at different times of the day. In other words, these values remain practically constant during the day. This somewhat contradicts data of other authors, who recorded a clear diurnal course of the concentrations of different viable microorganisms or other bioaerosols [175, 226, 233, 249, 250]. This is connected with the fact that sampling was performed at an altitude of approximately 15 m above ground level, which could slightly smooth the daily course of the observed concentrations; and our samples were sampled 6 hourly but in the above-mentioned works it was performed much more frequently. Probably, a longer integration time "blurs" the existing diurnal course of bioaerosol concentration in the surface atmospheric layer. It should be noted that the authors of [230] also did not manage to reveal the daily course of microorganism concentration in the atmosphere for two observation points.

Statistical analysis of the results of measuring the concentrations of total protein and culturable microorganisms in the atmosphere at altitudes of 500–7000 m was performed in [251]. A conclusion was drawn about the different physical nature of the statistics of the concentrations of total protein and culturable microorganisms. The concentration of culturable microorganisms is described by a discrete law, that is, a Poisson distribution, whereas the total protein concentration obeys a continuous law, that is, a modification of the log-normal distribution proposed in [252]. The correlation between the fluctuations of the total protein and culturable microorganism concentrations was estimated. The correlation coefficient of the obtained estimates of the concentration dispersions was found to be 0.02. This means that, in spite of equal conditions of pollutant spread, the sources of total protein and culturable microorganisms probably have different natures, that is, they act independently of one another.

The *wavelet analysis* of cumulative time-series data was performed to reveal quasi-periodic processes in the variation of bioaerosol concentrations in southwestern Siberia [253]. We used *Morlet wavelet* to analyze the concentration variations, as it is good for the analysis of quasi-periodic processes such as atmospheric variations resulting from the atmosphere's own fluctuations. The Morlet wavelet also provides a pictorial representation and easy interpretation of the data in terms of *Fourier analysis*. The analysis of data on the total protein concentration in aerosols obtained in the environs of Novosibirsk described in [254, 255] allowed a relation to be revealed between the variations of the mass concentration of aerosols and the total protein concentration in the surface atmospheric layer and at altitudes of 500–7000 m with its typical periodic synoptic processes. It was shown that variations of the concentrations of total protein and viable microorganisms in southwestern Siberia are mainly determined by the revealed seasonal periodic processes.

Wavelet analysis of longer time series refined the previously obtained results and showed that the concentration of viable microorganisms determined in on-land measurements has a strong variation, with a time scale of one-and-a-half years, and a weak variation, with a period of approximately 20–28 months for site B. For site A, the annual harmonics of the concentration of viable microorganisms is the strongest, and the semi-annual one is weaker. In these concentration series, one can observe strong irregular variations with time periods from 2 to 5 months during the whole period of measurement. For the total protein concentrations at site B, a strong variation with a period of one year and very weak variations with periods of 6 and 20–26 months were revealed. For site A, periods of 20–28, 16, and 12 months are distinguished. The wavelet analysis showed that the inter-year and intra-seasonal processes determining the variations of the studied concentrations at these points are most probably nonlinear. The presence of one year periodicity is explained by seasonal cycles. The inter-year variation with a time scale of about 20–27 months might be associated with stratospheric circulation, in particular, with quasi-biennial oscillation of equatorial stratospheric winds [256].

Thirty atmospheric air samples collected on fibrous filters at site A in different seasons were analyzed to see if virus-containing aerosols could be present in the atmosphere. The volume of air sampled was 10–15 m^3. All the samples were analyzed with the PCR method for the presence of viruses causing diseases with respiratory syndromes: influenza viruses of all types, including avian influenza virus type A subtype H5N1; severe acute respiratory syndrome virus; other coronaviruses; respiratory syncytial virus; adenoviruses of all serotypes; and some others. No genetic material of the above viruses was found in any of the tested samples. Taking into account the sensitivity of the PCR method used, the detection threshold of this technique is not more than one viral particle in 1 m^3 of atmospheric air. The samples were also analyzed with our own ELISA tests for the presence of the above viruses. All the tests carried out gave negative results – these viruses were not detected in atmospheric air samples. It should be noted that western Siberia is not an endemic area for airborne infections. It is located in the center of the Eurasian continent far from powerful sources of influenza virus aerosols [257], and the sampling points are rather far from the habitats of migrant birds transmitting influenza viruses [258].

Electron microscopy did not detect viruses in atmospheric air samples either. Usually, bacteria, fungal spores, their microscopic fragments, fragments of plants, and sometimes virus-like particles are found in the samples. But their identification was not performed, and the above PCR methods did not confirm the presence of viruses in the samples. The presence of viral particles in the region's atmospheric aerosol might be a rare event.

14.3.2
Spatial Variation of the Concentrations and Composition of Atmospheric Bioaerosol Components

In the previous sections, we mentioned a large number of works containing data on the variation of the concentration of different bioaerosols in the atmosphere in different time periods. Practical data on the spatial variations of the concentration and composition of atmospheric bioaerosol components are lacking in the literature. That is why we will use our own data on the variations of the concentrations of total protein and culturable microorganisms in practically simultaneous altitude measurements (sampling at altitudes of 500–7000 m took about 2 h) and simultaneous measurements in different on-land sampling points.

It was noted above that in southwestern Siberia the concentrations of total protein and viable microorganisms remain practically constant at altitudes of 500–7000 m

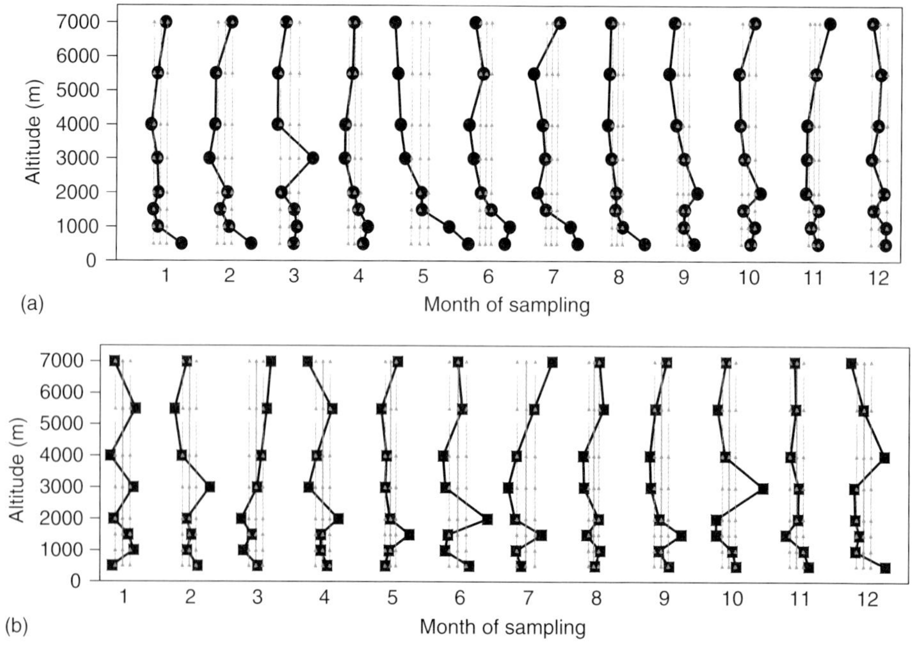

Figure 14.8 Annual variation of normalized altitude profiles of concentrations of (a) total protein and (b) viable microorganisms averaged for 1999–2008. Average values at each altitude and ±95% confidence interval of the whole data collection for this year are presented.

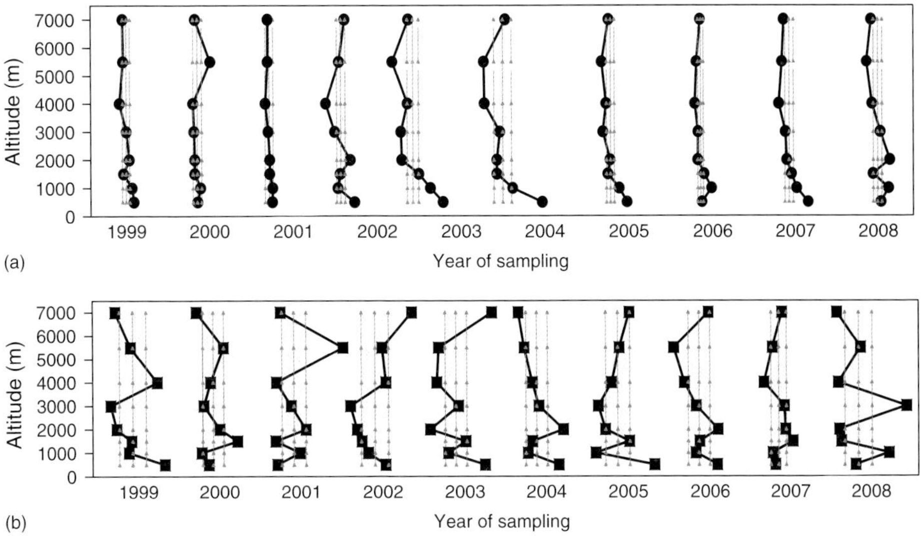

Figure 14.9 Nine-year variation of normalized altitude profiles of concentrations of (a) total protein and (b) viable microorganisms. Average values at each altitude and ±95% confidence interval of the whole data collection for this year are presented.

(Figure 14.4). It is natural to expect that local bioaerosol sources contribute to the studied profiles during the warm period of the year. However, the results of the construction of normalized altitude profiles for the concentrations of total protein and viable culturable microorganisms averaged for each of the months of 1999–2008 show that it is impossible to reveal variations over the course of a year for these altitude profiles in that region (Figure 14.8). This indicates that the contribution of local sources to the measured concentration values at altitudes of 500–7000 m is small. Figure 14.9 presents average annual altitude profiles of these concentrations calculated for 1999–2008 and normalized by the average annual values. As follows from the figures, no expressed trends in the variations of the observed altitude profiles for these years are revealed: all of them are practically constant at altitudes of 500–7000 m, with insignificant deviations from the average values, which could be caused by differences in atmospheric processes in the region.

The revealed profiles of the bioaerosol components determined in the atmosphere can be formed by very powerful remote sources such as vast forest lands, water surfaces, and soil. Aerosols from such sources rising to considerable altitudes mix and are transferred over the whole Northern Hemisphere, creating the observed profiles as particles deposit. Different mechanisms causing particles to rise in the atmosphere exist in nature [259, 260], which is why the rise of bioaerosols (with typical sizes lying within the micrometer range) to considerable altitudes

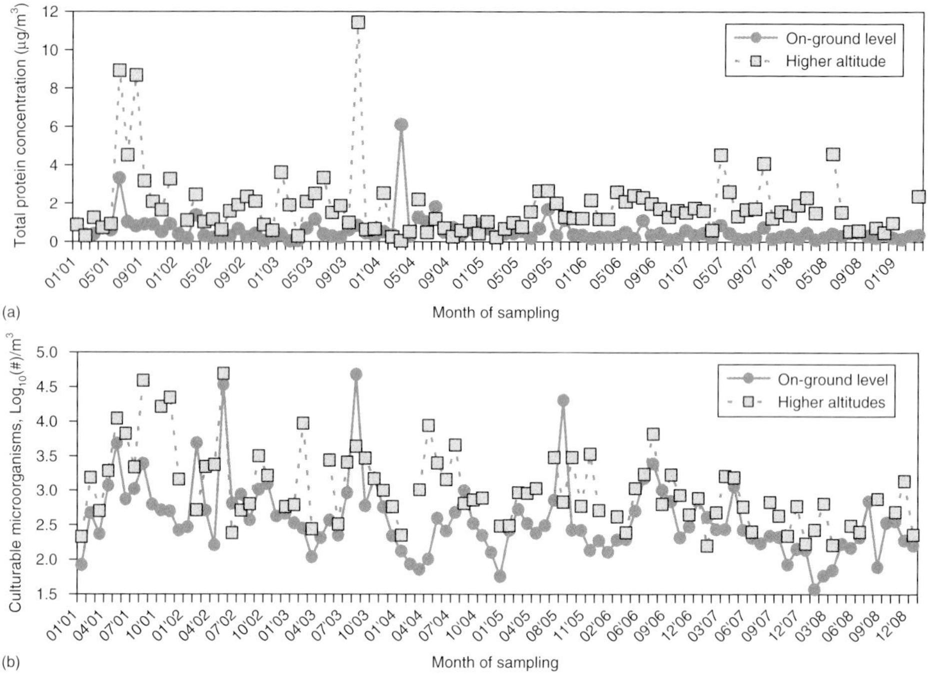

Figure 14.10 The comparison of on-land and altitude concentrations of (a) total protein and (b) viable microorganisms measured during particular months.

is theoretically possible. The fact that the mass concentration of atmospheric bioaerosols is rather large even at altitudes up to 7000 m is indicative of the remarkable role that the mechanisms of aerosol ascent to the upper atmospheric layers play in nature.

Data on culturable microorganism concentrations obtained for non-simultaneous on-land and altitude measurements in the same month are close to each other (Figure 14.10). However, the total protein concentrations determined during altitude measurements and presented in Figure 14.10 are approximately twice as high as those for on-land measurements. This indicates that, first, atmospheric bioaerosols containing culturable microorganisms and total protein originate from different sources, though microorganisms make their contribution to the observed total protein concentration. Second, local sources of bioaerosols containing total protein do not make a significant contribution to the total protein concentrations observed in the atmosphere.

It should also be noted that, in spite of the weak variation of the concentrations of culturable microorganisms in the atmosphere, their amounts can change significantly even at the neighboring altitudes at which the measurements were performed [261, 262]. The latter fact, as well as the small number of culturable

Table 14.3 The comparison of pairs of values measured simultaneously on different sites.

Compared sites	The comparison of two values	Number of pairs and correlation coefficients
Site A and site B, 2001–2008	Aerosol mass concentration 0.9*/÷ 2.0	24 pairs, +0.47
Site A and site B, 2001–2008	Total protein mass concentration 1.8*/÷ 5.5	28 pairs, +0.36
Site B and site C, 2005	Aerosol mass concentration 1.0*/÷ 1.8	15 pairs, −0.38
Site B and site C, 2005	Total protein mass concentration 2.0*/÷ 1.9	15 pairs, −0.19
Site B and site C, 2005	Ratio of total protein in aerosol mass 2.0*/÷ 1.7	15 pairs, +0.27
Site B and site C, 2006	Aerosol mass concentration 1.0*/÷ 1.6	17 pairs, +0.37
Site B and site C, 2006	Total protein mass concentration 0.6*/÷ 1.6	17 pairs, +0.19
Site B and site C, 2006	Ratio of total protein in aerosol mass 0.6*/÷ 1.9	17 pairs, +0.54
Site B and site C, 2008	Aerosol mass concentration 2.7*/÷ 1.7	15 pairs, +0.13
Site B and site C, 2008	Total protein mass concentration 2.0*/÷ 1.9	15 pairs, −0.15
Site C and site C, 2008	Ratio of total protein in aerosol mass 0.7*/÷ 2.1	15 pairs, +0.38

microorganisms of each species (genus) in samples collected at each altitude, does not allow the construction of reliable altitude profiles for individual microorganism species (genera).

Now let us proceed to the analysis of on-land data on bioaerosol concentrations determined simultaneously at different sampling points. Table 14.3 summarizes the results of the comparison of all the simultaneous measurements performed in 2001–2008. As follows from the results presented in Table 14.3, the mass aerosol concentration at site B tends to exceed that measured at sites C and A. The observed differences are statistically indistinguishable on account of the sparseness of the data.

The total protein concentration in the full aerosol mass at point A tend to be higher than those measured at sites B and C. The concentrations and proportions of total protein in the full aerosol mass behave differently in different years at the latter two sites. However, the observed differences are statistically indistinguishable, as for the mass concentrations of aerosol. The compared pairs of measurements correlate with each other, though the maximal value of the correlation coefficients does not exceed 0.54. Only in one case do the values compared in Table 14.3 have a negative correlation coefficient; in the other cases it varies from 0.13 to 0.47.

Thus, preliminary data on the spatial variation of the concentrations of atmospheric aerosols and bioaerosols in the region show that at distances up to 80 km and altitudes up to 7000 m their concentrations can be considered approximately constant though weakly related with each other. This could be explained by the absence of powerful local sources of bioaerosols in the region under study.

14.3.3
Possible Sources of Atmospheric Bioaerosols and their Transfer in the Atmosphere

The question arises why we observe such a diversity in the composition of culturable microorganisms in samples collected on the same day at different altitudes or at different times of the day at the same sampling point. We suppose that this is explained by the fact that samples contain bioaerosols (and non-biogenic aerosols) that have come to the sampling point from different sources. In the case when unique chemical, physical, or biological markers for definite sources are known, it is possible to determine these sources (even in the case when there are several sources with the same marker in the region under study) using the approaches developed in [29, 35, 263–265]. These approaches give good results for the identification of anthropogenic sources and the evaluation of their contribution to observed atmospheric air pollution.

In most cases, natural sources of aerosols have been poorly characterized and are often unknown at all. As follows from the previous data, the main sources of bioaerosols observed in southwestern Siberia are most probably located outside the region. In this case, we can use one of the models described in [266], or the model HYSPLIT-4 (which can be found at *http://www.arl.noaa.gov/ready/hysplit4.html*). These models allow the reconstruction of the most probable backward trajectory of the motion of air masses from the sampling point to the area where possible sources of bioaerosols are located. However, they take into account neither the possibility of contributions to the observed bioaerosol concentrations from sources located in different regions nor the possibility of "mixing" of the calculated backward trajectories.

The approaches developed in [267–269] for local sources do not have these disadvantages. They allow the coordinates, type, and characteristics of local sources to be determined from a small number of measurements and meteorological data. This task is very difficult for remote sources.

At the same time, approaches allowing the calculation of the probability that an aerosol particle starting from a certain point on the ground surface will reach the sampling point have been developed [270–272]. The calculations are performed with a complex of hemispherical models of transfer in a coordinate system following the relief of the ground surface. Conjugate problems are solved in a "reverse" time of 30 days. The algorithm was constructed based on the combination of the Lagrange approach with the Monte Carlo method. The reconstruction of the space–time structure of the atmospheric circulation uses the NCEP/NCAR Reanalysis data set (from the US National Centers for Environmental Prediction and the National Center for Atmospheric Research) with a time step of 30 minutes.

The probabilities that an aerosol particle starting from a certain point on the ground surface will reach the sampling point are calculated as a result of averaging the backward trajectories of the motion of air masses that touch the ground surface within a certain period of time. A feature of the calculated trajectories is that air masses gathering at the sampling point have different prehistories. While moving, they have been at different altitudes above different surface areas – often

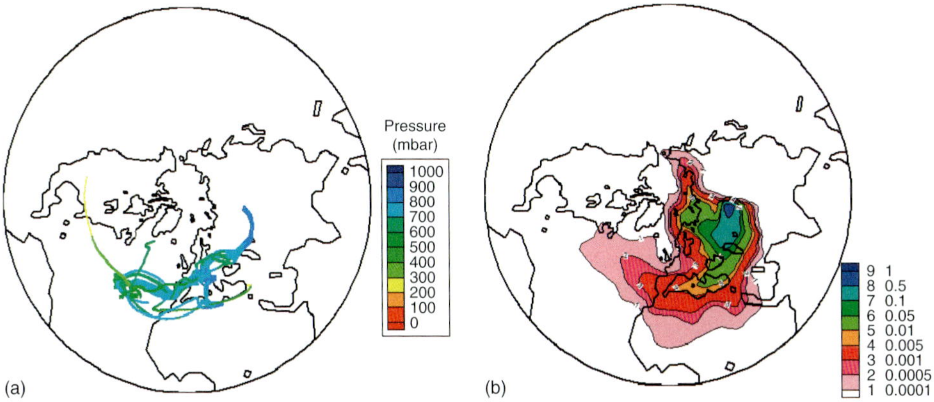

Figure 14.11 (a) Adjoint trajectories starting from the sampling point in the vicinity of Novosibirsk at an altitude of 1.5 km on 31 May 2002. They demonstrate the possible pathways to the receptor point. (b) The distribution of sensitivity function (relative units) characterizes the level of contribution of the sources situated at different points on the Earth's surface to the amount of aerosols collected at sampling points in the vicinity of Novosibirsk on 31 May 2002 at altitudes 0.5–5.5 km.

even above different continents and oceans (see Figure 14.11a) – and only on approaching the sampling point do the trajectories mix and converge to a single point. Aerosol particles coming into the atmosphere from different sources can get to the sampling point. As particles in the micrometer size range can be present in the atmosphere for a rather long time, they can reach the observation point together with air masses. Moving air masses travel from one altitude to another and can touch the surface (where, probably, they become rich in bioaerosols), and in certain places they mix intensively (these places may be different for each backward trajectory), after which such masses reach the measuring point.

As an example, the total sensitivity function for the functional describing the measurements carried out at the point with coordinates 54.3°N and 82.09°E at altitudes of approximately 0.5–5 km on 31 May 2002 is given in Figure 14.11b. It turned out that, at that time point, particles from northwest Kazakhstan were the most likely to be found. The calculations carried out for other times of measurements show that, for a greater part of the considered measurements, it is most probable to detect particles from middle Asia and northwest Kazakhstan in southwestern Siberia.

However, even such approaches that allow the probability of bioaerosol getting into the sample to be evaluated do not provide an answer to the question of what the source of detected bioaerosols was like. Soil, water reservoirs, vegetation, and animals, as well as various anthropogenic sources, are known to be bioaerosol sources [12, 13, 182, 188, 266, 273–277]. Taking into account the above estimate of the mass of bioaerosols present in the atmosphere [8–10], the major bioaerosol sources by mass have a great power. In principle, each bioaerosol source can be characterized by a single marker or a set of markers. However, as a sample usually contains bioaerosols from different sources, it is necessary to determine different bioaerosol markers in the sample. The complexity consists in the fact that many markers (for example, total protein) characterize the classes of compounds or microorganisms, but are insufficiently specific to characterize bioaerosol sources.

All the aforesaid is confirmed by studies carried out at Lake Baikal [278] to search for markers allowing aerosol originating from the surface microlayer (SML) of water to be detected in the lake atmosphere. The search for such a marker was performed by the elemental composition of atmospheric air samples and the SML of the water, their molecular composition, as well as by the presence of bacterial genetic material in the SML of the lake water. The research results demonstrated a close relation between the atmosphere and the water of the lake; however, it was impossible to determine unambiguously if the SML of the water was the source of the observed aerosol or whether atmospheric aerosol from other sources depositing onto the lake surface created the observed composition of the SML of the water [278].

On the other hand, even if we manage to detect individual markers for each bioaerosol source, their inventory in nature will take too much time before it becomes possible to identify all bioaerosols detected in samples by their sources. Probably, for remote bioaerosol sources, only identification by type will be possible in the very near future, and only local bioaerosol sources situated close to the sampling points or unique, powerful sources – such as fires on swamps near Moscow, the bioaerosols from which were found in southwestern Siberia (see data in Figure 14.10 for August 2001) – will be identified by location, emission power, and so on.

14.3.4
The Use of Snow Cover Samples to Analyze Atmospheric Bioaerosols

Snow cover is a good accumulator of all aerosol pollutants getting into it. These pollutants can get there in different ways: (i) dry deposition of atmospheric aerosols, including the deposition of discharge from sources located near sampling points; and (ii) washout of these pollutants from the atmosphere by snow. In this later case two variants are also possible: aerosols of pollutants present in the atmosphere near the source under consideration can be captured by snowflakes, and pollutants can get into the volume of snowflakes in the place of their formation. Correspondingly, the task of the study of snow cover pollutants should include two subtasks: the determination of background pollutants (fresh snow) and those created by local and remote sources of different discharges.

Snow cover samples were collected in 1999–2006 near powerful city sources of anthropogenic discharges (the complexes of the Novosibirsk Heat Power Plant) and suburban sources (Berdsk Chemical Plant (BCP) and Novosibirsk Electrode Plant). Besides, fresh snow samples were collected at site A. For the sources studied for several years, the sampling route was the same. Among the above sources, only BCP, which produced protein–vitamin concentrate, was a source of bioaerosols. A dependence of the total protein concentration on the distance from the source

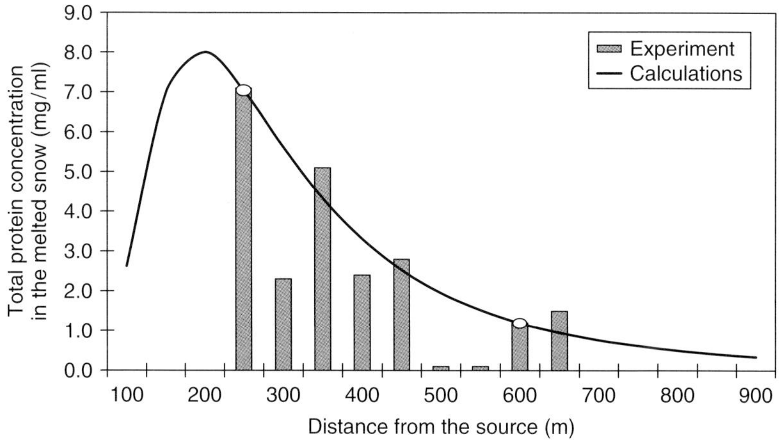

Figure 14.12 Experimental and calculated data of protein concentration in snow cover in the winter of 1999–2000 depending on the distance from the chimney of the BCP. Reference points used to evaluate the regression function are marked on the calculation curve.

was revealed for it, as shown in Figure 14.12. This figure also presents the curve approximating the experimental data using selected reference points [279] for the estimates given below. The values of total protein discharge made up from 16.2 to 26.4 kg/year in 2000–2002.

The dependence of the total protein concentration in snow cover on the distance from the source was also revealed for sources that are not sources of bioaerosols [280, 281]. Such a situation can be realized in the case when falling coarse particles from the studied source wash out rather small particles present in the atmosphere and containing the total protein through coagulation [282] (it was noted above that the largest mass proportion of total protein is in particles of submicrometer size). But in this case, no expressed dependence of the concentration of culturable microorganisms in snow cover on the distance from the source should be observed as microorganisms are concentrated mainly in 2–7 μm particles [176, 247], which less actively coagulate with coarse particles [282] from anthropogenic sources. The results of the studies carried out [280, 281] confirm the hypothesis advanced and do not allow a dependence of the concentration of culturable microorganisms in snow cover on the distance from the source to be revealed.

In snow cover samples collected in the vicinity of the studied anthropogenic sources, a wide variety of viable microorganisms was observed even for neighboring points. It is natural to expect such a situation, as snow accumulating the particles getting into it and preserving them during the whole winter performs a natural time integration of the existing changes in the variety of microorganisms present in the atmosphere. It should also be noted that the obtained data on the concentration and type of microorganisms in snow cover samples agree with those for other regions [283–286]. The variety of culturable microorganisms detected in snow cover

Table 14.4 The observed variety (percentage) of viable microorganisms in snow cover samples in the lower atmospheric layers in the winter period in southwestern Siberia.

Winters		Rods	Cocci	Nonsporiferous bacteria	Actinomyces	Fungi	Yeasts
2000–2001	Atmosphere	40.28	26.26	13.95	16.31	2.03	1.18
	Snow cover	2.93	8.51	79.32	0.03	9.20	0
2001–2002	Atmosphere	29.81	32.97	36.87	0.07	0.28	0
	Snow cover	7.28	25.75	66.61	0.08	0.21	0.07
2002–2003	Atmosphere	7.00	72.23	20.23	0	0.40	0.13
	Snow cover	3.92	4.03	91.87	0	0.11	0.07
2003–2004	Atmosphere	19.65	44.95	32.30	1.62	1.48	0
	Snow cover	0.60	10.09	88.85	0.26	0.11	0.09
2004–2005	Atmosphere	2.39	68.11	25.70	0	1.78	2.02
	Snow cover	0.11	12.80	86.68	0	0.15	0.26

generally retraces that observed during the winter period in the lower atmospheric layers in our region (Table 14.4). Certainly, there is no full coincidence. In particular, snow samples contain significantly larger numbers of fungi and non-sporiferous bacteria while atmospheric air contains greater numbers of different cocci. This non-coincidence may be caused, first, by insufficient number of samples (aircraft probing was performed once a month during four winter months, and snow was studied only on three sites in the suburbs and two in the city), and second, the contribution of local microorganism sources to the analyzed snow samples. The data accumulated up to now are insufficient for more reliable conclusions.

Snow cover pollutants, which were evaluated in [280, 281], do not allow a reliable evaluation of discharge from the studied anthropogenic sources to be performed, as the "background" values of pollutants in snow cover are not known. That is why the study of pollutants brought into the snow cover by fresh snow was carried out. The results of measurements for four studied episodes described in [287] show that fresh snow contains a significantly lower concentration of microorganisms than snow cover samples collected at the end of the winter: one gram of fresh snow contains from 0.9 to 1.8 $\log_{10}$CFU while "old" snow contains 2.1–4.8 $\log_{10}$CFU in different years near different sources [246, 280, 281]. The total protein concentration in 1 g of fresh snow was from 0.7 to 3.4 μg, which correlates with the results obtained previously (0.4–5.25 μg/g) [246, 280, 281]. The study of the morphology of cells and colonies of microorganisms detected in fresh snow samples revealed the prevalence of non-sporiferous bacteria [287]. It should be noted that, according to the data obtained during previous years, non-sporiferous bacteria also prevail in snow cover [246, 280, 281], while atmospheric aerosol contains large numbers of cocci and rods. Probably, the wider number of non-sporiferous bacteria in snow cover is observed not only due to their getting into fresh snow but also due to dry deposition of atmospheric aerosol containing microorganisms.

Microorganisms found in fresh snow were tested for the presence of plasmo-coagulation, fibrinolytic, hemolytic, and gelatinase activities. All these activities characterize the potential *pathogenicity* of microorganisms as the latter reflect the ability to reproduce in blood and other human tissues. It has been discovered that 13–62% of the studied microorganisms possess at least one of the four activities. Most of these microorganisms refer to non-sporiferous bacteria and rods; cocci detected in fresh snow samples rarely possess such activities.

The ability of microorganisms to grow at different temperatures and in media with high NaCl concentration reflects the possibility of their long-term survival in the atmosphere. A large number of microorganisms were shown to preserve growth properties at high temperatures and high NaCl concentrations. With the growth of NaCl concentration and temperature, their number, naturally, decreases [287]. The obtained results indicate that fresh snow contains many microorganisms that are stable in different environments. It should be noted that they include not only bacteria of the genus *Bacillus* (which can form endospores and, therefore, are rather stable), but also some strains of cocci, non-sporiferous bacteria, and so on.

It was shown that the parameters of ionic composition differed both between samples [287] and from those published for measurements at another observation point [288]. Differences for each observation point can be associated with the variability of the meteorological situation, and differences in the values determined for different observation points can be associated with their local environment and the proximity to Novosibirsk city. The total concentration of water-soluble salts in the tested samples, which can be evaluated by the electroconductivity value, proved to be higher for two of four episodes. Probably, the most effective washout of ions by snowflakes was realized in these episodes. All samples were shown to contain a small proportion of ammonium and a high proportion of calcium and magnesium cations [287]. The source of the former in snow may be fine aerosol particles and to a smaller extent ammonia gas; and the source of the latter, having no volatile precursors, may be coarsely dispersed aerosol articles, including biogenic compounds and microorganisms.

The concentrations of some elements found in fresh snow were compared. As expected, they proved to be significantly lower than those for snow cover samples in the area affected by an anthropogenic source (Novosibirsk Heat Station 4) [287]. High concentrations of aluminum, silicon, calcium, and magnesium were detected in fresh snow in episode 1. The combination of these elements is typical for aluminosilicates, and therefore their presence in snowfall can be explained by the effects of erosive sources. For episodes 2–4, no abnormalities of the elemental composition were revealed. The high aluminum concentration in fresh snow of episode 1 can be explained by the fact that the air masses from which the snow fell turned out to be located within an area affected by an industrial enterprise or a gas flare situated near the trajectory of motion of these air masses. Besides, as mentioned above, aluminum is the main component of the row of minerals making up part of the soil; therefore, its getting into the sample can also have an absolutely random character. In fact, the simulation of the 10-day backward trajectories of the motion of the air masses that brought the snow showed that such cities as Pavlodar, Ekibastuz, Omsk, and further Nizhnevartovsk and Surgut, were on their path and could have contributed to the observed pollution. For episodes 2–4, for which no abnormalities of the elemental composition were revealed, the backward trajectories of the motion of the air masses were distant from any powerful sources of aerosol atmospheric pollutants that were working in the period when the air masses were passing. The simulation of the 10-day backward trajectories of the motion of the air masses was performed using the program HYSPLIT-4 (see above), for the conditions described above at the sampling point with coordinates 54.94°N and 83.23°E at an altitude of 500 m.

The differences in the numbers of viable microorganisms in the fresh snow samples in the studied episodes can most probably be explained by the particular features of the actual trajectories of motion of the air masses from which the snow fell. However, the currently available data are insufficient to reveal the relation between the numbers of microorganisms in fresh snow samples and the trajectory of motion of the air masses from which it has fallen.

The studies of fresh snow demonstrate that samples differ significantly in the concentrations of biological and chemical pollutants from snow cover samples collected at the end of winter. Having studied all the important snowfall episodes in the period of snow cover formation, one can determine the pollutants that get into the snow cover with deposited atmospheric aerosol by subtracting from the observed concentrations of snow cover pollutants those values referring to fresh snow. These values can be compared with the chemical and biological composition of atmospheric aerosol integrated over the whole winter period of observations in the same region. The presented results demonstrate the importance of such studies for understanding the processes of local and global transport of bioaerosols in the atmosphere and revealing all possible sources and flows of these aerosols.

14.3.5 Potential Danger of Atmospheric Bioaerosols for Humans and Animals

Microorganisms in atmospheric aerosol represent a potential danger for humans, causing or provoking infectious diseases, allergic reactions, and worsening the state of health. For pollen and fungi, it is known which bioaerosols cause allergic reactions [1, 149, 168, 169, 174, 181, 212, 217]. Therefore, the detection of particles containing them in the atmosphere provides information on the danger of such bioaerosols for humans. However, there is no currently available quantitative method allowing the comparison of the potential danger for humans represented by culturable bacteria detected in different atmospheric air samples.

It is proposed to use four groups of tests for quantitative evaluation of the potential dangers represented by all culturable microorganisms detected in atmospheric aerosols [81]. (i) Microorganisms in atmospheric aerosols must be pathogenic or conventionally pathogenic for humans in order to pose a health risk. Therefore, the first group of tests should characterize the potential danger represented by bacteria to humans. (ii) Bioaerosols are more dangerous when they have a higher concentration of viable microorganisms and, therefore, have a larger portion of microorganisms that are pathogenic or conventionally pathogenic for humans. Consequently, the second group of tests should evaluate the number of various culturable microorganisms in an atmospheric air sample. (iii) The danger *hazardous bioaerosol* represented by a certain microorganism increases in the case where it displays increased resistance to unfavorable environmental factors causing inactivation of the microorganism. Consequently, the third group of tests should evaluate their resistance to unfavorable environmental factors. (iv) If potentially pathogenic microorganisms that are not inactivated in the environment affect humans, then those displaying drug resistance are the most dangerous. The results obtained in each of the four groups of tests can be numerically characterized by a certain integral index quantitatively reflecting the contribution of the experimentally determined characteristics of bacteria.

The analysis of the literature [81] shows that the most important indirect characteristics of bacteria determining their pathogenicity are as follows: plasmocoagulation activity (significance index of the characteristic is estimated at 7, see below), the presence of capsules in the bacterium and hemolytic activity (significance index estimated at 5), the presence of endonucleases in the microorganism (significance index estimated at 4), the presence of lecithinase, fibrinolytic, and lipase activities in the microorganism (the significance index for all of these is estimated at 3), and the presence of gelatinase activity (significance index estimated at 2). In addition, there exist a number of microorganism characteristics that also influence the potential pathogenicity of defined strains, but to a smaller extent. These include the presence of proteolytic activity with respect to other substrates (significance index estimated at 1), the presence of pigment in cells (significance index estimated at 0.5), and the presence of amylolytic activity and mobility of bacteria on the growth medium (the significance index for all of them is estimated at 0.1). All these characteristics can be determined quantitatively or qualitatively in experiments according to the methods described above and, generally speaking, are unique for each microorganism genus, species or strain. At the same time, with some assumptions, it is possible to construct a unified *integral index of potential pathogenicity*, which can be used for all bacteria detected in the samples. In this case, the value of the potential pathogenicity of a sample will be determined as the average for all the bacteria detected in the sample.

The value of the integral index of potential pathogenicity was calculated as the sum of the characteristics present in a microorganism multiplied by the significance coefficients (it was considered that, if the characteristic was present, its index equaled 1, if not, the index equaled 0) divided by the maximum possible value of this sum. In other words, the most dangerous bacteria have a value of the integral index of potential pathogenicity that equals 1, and microorganisms that do not represent a danger have a value that equals 0. Other integral indices (except for the second one, which was standardized by the maximal value of culturable microorganism concentration detected in atmospheric air samples during the year of measurement) were calculated and standardized in a similar way.

The methods to determine the concentrations and types of different bacteria in air samples are presented above. The degrees of increased resistance of bacteria to unfavorable environmental factors as well as the manifestation of pathogenicity were determined by performing an expert evaluation of the significance of a certain bacterial characteristic influencing bacterial resistance to environmental factors and evaluated on the basis of a literature data analysis [81].

This index provides a certain generalized characteristic of bacterial resistance. It is obvious that, like other integral indices, it does not reflect all the nuances of bacterial resistance in the environment. As a literature analysis shows, the detailed characteristics of bacterial resistance to unfavorable environmental effects such as temperature, relative humidity, freezing–thawing, irradiation at different wavelengths, mechanical injuries, sampling-related stresses, and so on, have only been determined for a few strains from the total number of currently known microorganisms – according to the data presented in [289], it has been estimated that not more than 10% of bacteria existing on the Earth are known to scientists. Data on bacterial resistance to unfavorable environmental effects can undoubtedly be used for more precise definition of the *integral index of bacterial resistance*, but only in cases where it has been possible to determine the presence of a definite strain in an atmospheric air sample and to prove its viability.

The most significant characteristics determining the resistance of bacteria to unfavorable environmental factors are the presence of restriction endonucleases and plasmid DNAs as well as the ability to form quiescent forms, in particular, endospores allowing a species to survive under unfavorable environmental conditions (the significance index for all of these is estimated at 2). The ability to grow at increased NaCl concentrations (significance index estimated at 1) and the presence of pigmentation of cells (significance index estimated at 0.5) are less significant for survival.

The evaluation of *drug resistance* only included the study of bacterial resistance to the effect of 15 antibiotics used in medicine. All the antibiotics were considered to have equal significance.

The work [81] presents annual variations of different indices measured for all culturable bacteria detected in atmospheric air samples collected at altitudes of 500–7000 m. It was shown that each individual characteristic of bacteria varies during the year rather randomly, and it does not seem possible to draw any conclusions on regularities of variations in each of these values. The situation differs considerably when we turn from individual characteristics of bacteria to integral indices constructed on their basis. For example, the value of the second integral index varies during the year by approximately an order of magnitude.

In contrast to the second integral index, the dependences for three other integral indices do not show such a sharp change. For example, the value of the integral index of potential pathogenicity during two years of observation changes smoothly from about 0.15 to about 0.3. Thus, the atmosphere contains twice as many bacteria that are harmful for human health in the "most dangerous periods" than in the

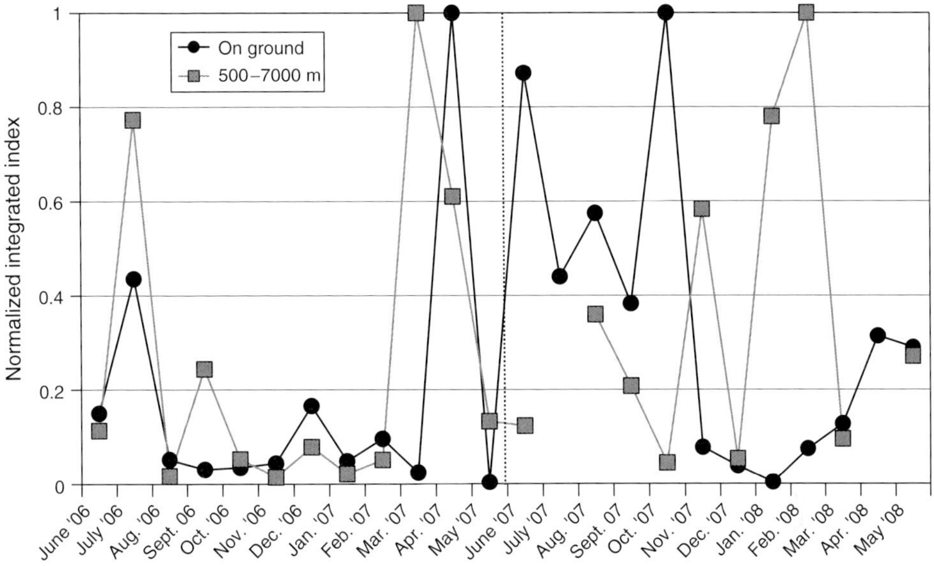

Figure 14.13 Biannual dynamics of the variation in the integral index of danger of viable bacteria detected at three points of atmospheric air sampling. Normalization to unity was performed separately for each year. This index was not performed for altitude samples in July 2007 and April 2008.

"least dangerous periods." A similar situation was observed for the integral index of bacterial resistance to environmental factors, which also varied smoothly during the observation period and made up approximately 0.25 from June to September 2006, sharply increased approximately to 0.4–0.5 from October 2006 to June 2007, and smoothly varied further in the range of 0.3.

Superposition (multiplication) of all four integral indices for each sampling allows us to evaluate the danger represented by viable bacteria that are present in atmospheric aerosol of the region at the moment of sampling, and to trace its variation during the year (Figure 14.13).

To compare the integral index of potential pathogenicity of bacteria for humans formulated in [81] with the existing approaches to risk evaluation, it should be interpreted proceeding from a simple dose–response dependence for the "dose of the active factor" and the "reaction of the organism." The value of the integral index of potential bacterial pathogenicity for humans is an analog of the probability that the organism's reaction can be caused by a single microorganism – the p value in the formula

$$p = 1 - \exp(-pD) \qquad (14.1)$$

where p is the probability of causing a reaction in the organism, and D is the active factor dose. This formula describes the dependence of the active factor dose on the organism reaction on the basis that all the microorganisms getting into the organism act independently and with equal probability p can cause a corresponding reaction in it [81, 290]. The higher the p value or the value of the integral index of potential bacterial pathogenicity for humans, the greater the probability of causing the organism reaction and, correspondingly, the higher the potential bacterial pathogenicity. For the proposed index, its values calculated based on experimental data vary from 0 to 0.68. In each of the studied air samples, the maximal value of this index for individual bacteria exceeds the value 0.36, and in 50% of cases it exceeds the value 0.48. In other words, along with potentially quite non-pathogenic bacteria, whose integral index of potential pathogenicity strictly equals 0 (note that a sample on average contains approximately only 2.5% of such bacteria), there are bacteria of potentially medium pathogenicity. Thus, practically all the bacteria detected in atmospheric air samples are to some extent potentially pathogenic for humans. No known highly pathogenic bacteria were detected in the studied atmospheric air samples.

The proposed procedure of expert evaluation does not attempt to be definitive. It is possible that not all significant indices have been included in the above groups of tests. In addition, the performed expert evaluation of the significance of various bacterial characteristics may not be very accurate. However, the proposed method allows one to compare the potential dangers represented to humans by all bacteria (including unculturable ones) detected in an atmospheric air sample with other samples, containing significantly different concentrations and types of bacteria.

Thus, the work [81] proposes an approach to the evaluation of danger represented by viable bacteria in an atmospheric aerosol for the region's population. The results obtained by the realization of this approach for the first time allow the evaluation of the dangers represented by atmospheric microorganisms for humans. At the same time, since the observations have only been performed for two years, it has not been possible to reveal the annual recurrence and inter-year differences between these indices. In addition, the dependences obtained are not statistically significant. Only the continuation of the work will facilitate the reliable determination of the dynamics of the values studied and their use for reasonable prediction of variations in the dangers represented by viable bacteria in atmospheric aerosols for the region's population.

14.4 Conclusion

The review of the methods and equipment used for atmospheric bioaerosol research presented in this chapter as well as the results of 10-year studies in southwestern Siberia using some of them allow us to formulate some conclusions and prospects of atmospheric bioaerosol research and the development of methods and equipment used.

First of all, it should be noted that, in spite of the considerable progress achieved in the development of methods for atmospheric bioaerosol sampling and analysis, this task still remains ongoing. Since bioaerosols are influenced by different factors (time, temperature, humidity, the presence of different compounds in the atmosphere, and so on), it is necessary to perform their real-time complete analysis *in situ*. However, the necessary equipment is not yet available. That is why the first task is to develop new methods and equipment for the analysis of atmospheric bioaerosols. As atmospheric bioaerosols present a mixture of bioaerosols from different sources, methods allowing the study of the composition of individual particles are required. Such techniques can be based, for example, on the latest developments of mass spectrometric methods for individual particles, but it is necessary to extend the size range of particles tested (now it makes up 3 μm by aerodynamic diameter, and many microorganisms and especially their aggregates are 10 μm and even larger) and to extend the range of detected masses to more than 9000 Da.

The next task is to study the processes of formation and transformation of bioaerosols under the influence of the above factors under laboratory and field conditions. These works will provide a better understanding of the processes occurring in bioaerosols and will allow the construction of prediction models of complex processes, which should be taken into account in models of bioaerosol transfer in the atmosphere.

Data on atmospheric bioaerosols accumulated up to now give information on the space–time variation of their concentrations and components. However, these data were obtained at the regional scale, and global observation of bioaerosols is currently not performed. Therefore, the next task is to organize global observation of bioaerosols to reveal their space–time variation and methods of transfer, and to determine their most powerful sources, in order to construct prediction models at the planetary level.

The construction of these models and the study of the processes occurring in bioaerosols under the influence of different factors require detailed information on all the bioaerosols observed in each region. That is why the tasks of obtaining a more detailed inventory of bioaerosol sources in various regions, their characterization in terms of emission power and its variation with time, and the component composition of initial particles and its dependence on aerosolization conditions (season, time of the day, atmospheric conditions, and so on) have become urgent.

The last task is mathematical modeling of bioaerosol spread from sources to receptors, taking into account their stochastic mixing and transformation in the atmosphere, and possible ascent to high altitudes.

Thus, in spite of the more than 100-year history of atmospheric bioaerosol research, it is necessary to conclude that humankind is only at the beginning of this research. Only the main tasks of atmospheric bioaerosol research have been formulated, and only the first – but nevertheless very significant – results have been obtained. Great efforts should be made over the coming years and decades.

References

1. Cox, C.S. and Wathes, C.M. (eds) (1995) *Bioaerosols Handbook*, Lewis, Boca Raton, FL.
2. Andreae, M.O. and Crutzen, P.J. (1997) *Science*, **276** (5315), 1052–1058.
3. O'Dowd, C.D. *et al.* (2002) *Nature*, **416** (6880), 497–498.
4. Kulmala, M. *et al.* (2008) *Tellus B*, **60** (3), 300–317.
5. Ray, R.C. *et al.* (2009) *J. Environ. Sci. Health*, **27C** (1), 1–35.
6. Nel, A. *et al.* (2006) *Science*, **311** (5761), 622–627.
7. Fields, B.N. and Knipe, D.M. (eds) (1986) *Fundamental Virology*, Raven, New York.
8. Jaenicke, R. (2005) *Science*, **308** (5718), 73.
9. Winiwater, W. *et al.* (2009) *Atmos. Environ.*, **43** (7), 1403–1409.
10. Griffin, R.J. *et al.* (1999) *Geophys. Res. Lett.*, **26** (17), 2721–2724.
11. Berlyand, M.E. (1985) *Predication and Regulation of Air Pollution*, Gidrometeoizdat, Leningrad (in Russian).
12. Work, K. and Warner, S. (1980) *Air Pollution: Sources and Control*, Myr, Moscow (in Russian).
13. Aller, J.Y. *et al.* (2005) *J. Aerosol Sci.*, **36** (5–6), 801–812.
14. Choi, M.E. and Chan, C.K. (2002) *Environ. Sci. Technol.*, **36** (11), 2422–2428.
15. Pöschl, U. (2002) *J. Aerosol Med.*, **15** (2), 203–212.
16. Pöschl, U. (2005) *Angew. Chem.*, **44** (46), 7520–7540.
17. Gao, J. *et al.* (2009) *Atmos. Environ.*, **43** (4), 829–836.
18. Donahue, N.M. *et al.* (2009) *Atmos. Environ.*, **43** (1), 94–106.
19. Deguillaume, L. *et al.* (2008) *Biogeoscience*, **5** (4), 1073–1084.
20. Tong, Y. and Lighthart, B. (1997) *Atmos. Environ.*, **31** (6), 897–900.
21. Lighthart, B. and Shaffer, B.T. (1997) *Aerosol Sci. Technol.*, **27** (3), 439–446.
22. Bauer, H. *et al.* (2008) *Atmos. Environ.*, **42** (22), 5542–5549.
23. Endo, M. *et al.* (2004) *Atmos. Environ.*, **38** (36), 6263–6267.
24. Elbert, W. *et al.* (2007) *Atmos. Chem. Phys.*, **7** (17), 4569–4588.
25. Putaud, J.-P. *et al.* (2004) *Atmos. Environ.*, **38** (16), 2579–2595.
26. Lee, S.J. *et al.* (2008) *Sensors Actuators, B*, **132** (2), 443–448.
27. Conte, M.H. and Weber, J.C. (2002) *Nature*, **417** (6889), 639–641.
28. Kourtchev, I. *et al.* (2008) *Chemosphere*, **73** (8), 1308–1314.
29. Lau, A.S.P. *et al.* (2007) *Atmos. Environ.*, **41** (13), 2831–2843.
30. Lee, A.K.Y. *et al.* (2006) *Atmos. Environ.*, **40** (2), 249–259.
31. Bauer, H. *et al.* (2008) *Atmos. Environ.*, **42** (3), 588–593.
32. Spencer, M.T. and Prather, K.A. (2006) *Aerosol Sci. Technol.*, **40** (8), 585–494.
33. Wijnands, L.M. *et al.* (2000) *Allergy*, **55** (9), 850–855.
34. Yttri, K.E. *et al.* (2007) *Atmos. Chem. Phys.*, **7** (16), 4267–4279.
35. Schauer, J.J. *et al.* (1996) *Atmos. Environ.*, **30** (22), 3837–3855.
36. Yee, E. *et al.* (1992) *Appl. Opt.*, **31** (15), 2900–2913.
37. Grishin, A.I. *et al.* (2008) *Int.J. Remote Control*, **29** (9), 2549–2565.
38. Kanaani, H. *et al.* (2008) *J. Aerosol Sci.*, **39** (2), 175–180.
39. Agranovski, V. and Ristovski, Z. (2005) *J. Aerosol Sci.*, **36** (5–6), 665–676.
40. Ecker, D.J. *et al.* (2006) *JALA*, **11** (6), 341–351.
41. Takegawa, N. *et al.* (2005) *Aerosol Sci. Technol.*, **39** (8), 760–770.
42. Ryzhov, V. *et al.* (2000) *Appl. Environ. Microbiol.*, **66** (9), 3828–3834.
43. Stowers, M.A. *et al.* (2000) *Rapid Commun. Mass Spectrom.*, **14** (10), 829–833.
44. Szponar, B. and Larsson, L. (2001) *Ann. Agric. Environ. Med.*, **8** (2), 111–117.
45. Kleefsman, I. *et al.* (2007) *Part. Part. Syst. Char.*, **24** (2), 85–90.
46. Vinopal, R.T. *et al.* (2002) *Anal. Chem. Acta*, **457** (1), 83–95.
47. Griest, W.H. *et al.* (2001) *Field Anal. Chem. Technol.*, **5** (4), 177–184.
48. VerBerkmoes, N.C. *et al.* (2005) *Anal. Chem.*, **77** (3), 923–932.
49. Hindson, B.J. *et al.* (2005) *Biosens. Bioelectron.*, **20** (10), 1925–1931.

50. McBride, M.T. *et al.* (2003) *Anal. Chem.*, **75** (8), 1924–1930.
51. Regan, J.F. *et al.* (2008) *Anal. Chem.*, **80** (19), 7422–7429.
52. Luoma, G.A. *et al.* (1999) *Field Anal. Chem. Technol.*, **3** (4–5), 260–273.
53. Griffiths, W.D. and DeCosemo, G.A.L. (1994) *J. Aerosol Sci.*, **25** (8), 1425–1458.
54. Henningson, E.W. and Ahlberg, M.S. (1994) *J. Aerosol Sci.*, **25** (8), 1459–1492.
55. Albrecht, A. *et al.* (2008) *Int. J. Hyg. Environ. Health*, **211** (1–2), 121–131.
56. Terzieva, S. *et al.* (1996) *Appl. Environ. Microbiol.*, **62** (7), 2264–2272.
57. Cage, B.R. *et al.* (1996) *Ann. Allergy, Asthma Immunol.*, **77** (5), 401–406.
58. Lighthart, B. and Tong, Y. (1998) *Aerobiologia*, **14** (4), 325–332.
59. Lin, W.-H. and Li, C.-S. (1998) *Aerosol Sci. Technol.*, **28** (6), 511–522.
60. Li, C.-S. (1999) *Aerosol Sci. Technol.*, **30** (3), 280–287.
61. Lin, W.-H. (1999) *Aerosol Sci. Technol.*, **30** (2), 119–126.
62. Kenny, L.C. *et al.* (1999) *Ann. Occup. Hyg.*, **43** (6), 393–404.
63. Aizenberg, V. *et al.* (2000) *Am. Ind. Hyg. Assoc. J.*, **61** (6), 855–864.
64. Mehta, S.K. *et al.* (2000) *Am. Ind. Hyg. Assoc. J.*, **61** (6), 850–854.
65. Maus, R. *et al.* (2001) *Atmos. Environ.*, **35** (1), 105–113.
66. Radosevich, J.L. *et al.* (2002) *Lett. Appl. Microbiol.*, **34** (3), 162–167.
67. An, H.R. *et al.* (2004) *Indoor Air*, **14** (6), 385–393.
68. Ho, J. *et al.* (2005) *J. Aerosol Sci.*, **36** (5–6), 557–573.
69. Tseng, C.-C. and Li, C.-S. (2005) *J. Aerosol Sci.*, **36** (5–6), 593–607.
70. Mainelis, G. *et al.* (2006) *J. Aerosol Sci.*, **37** (5), 645–657.
71. Sigaev, G.I. *et al.* (2006) *Aerosol Sci. Technol.*, **40** (5), 293–308.
72. Yao, M. and Mainelis, G. (2006) *J. Aerosol Sci.*, **37** (11), 1467–1483.
73. Dillner, A.M. *et al.* (2007) *Aerosol Sci. Technol.*, **41** (1), 75–85.
74. Engelhart, S. *et al.* (2007) *Int. J. Hyg. Environ. Health*, **210** (6), 733–739.
75. Thummes, K. *et al.* (2007) *System. Appl. Microbiol.*, **30** (8), 634–643.
76. Carvalho, E. *et al.* (2008) *Aerobiologia*, **24** (4), 191–201.
77. Dart, A. and Thornburg, J. (2008) *Atmos. Environ.*, **42** (4), 828–832.
78. Han, T. and Mainelis, G. (2008) *J. Aerosol Sci.*, **39** (12), 1066–1078.
79. Macher, J. *et al.* (2008) *J. Occup. Environ. Hyg.*, **5** (11), 724–734.
80. Madsen, A.M. and Sharma, A.K. (2008) *Ann. Occup. Hyg.*, **52** (3), 167–176.
81. Safatov, A.S. *et al.* (2008) *CLEAN – Soil, Air, Water*, **36** (7), 564–571.
82. Agranovski, I.E. *et al.* (2004) *Atmos. Ocean. Opt.*, **17** (5–6), 429–432.
83. Fuchs, N.A. (1961) *Progress of Aerosol Mechanics*, Academy of Sciences of the USSR, Moscow (in Russian).
84. Willeke, K. and Baron, P.A. (eds) (1993) *Aerosol Measurement: Principles, Techniques, and Applications*, Van Nostrand Reinhold, New York.
85. Grinshpun, S. *et al.* (1994) *J. Aerosol Sci.*, **25** (8), 1425–1458.
86. Schneider, J. *et al.* (2006) *J. Aerosol Sci.*, **37** (7), 839–857.
87. Nazarov, L.E. (1985) *Tr. Inst. Exp. Meteorol.*, **9** (124), 76–81 (in Russian).
88. Zuev, V.E. *et al.* (1992) *Atmos. Ocean. Opt.*, **5** (10), 658–663.
89. AFA Filters (1970) *Catalogue–Handbook*, Atomizdat, Moscow (in Russian).
90. Mattias-Maser, S. and Jaenicke, R. (1994) *J. Aerosol Sci.*, **25** (8), 1605–1613.
91. Mattias-Maser, S. *et al.* (2000) *Atmos. Environ.*, **34** (22), 3805–3811.
92. Leck, C. and Bigg, E.K. (2005) *Tellus*, **57B** (4), 305–316.
93. Larsen, P.A. and Rawlings, J.B. (2009) *Part. Part. Syst. Char.*, **25** (5–6), 420–4330.
94. Hobbie, J.E. *et al.* (1977) *Appl. Environ. Microbiol.*, **33** (5), 1225–1228.
95. Yu, W. *et al.* (1995) *Appl. Environ. Microbiol.*, **61** (9), 3367–3372.
96. Wittmaack, K. *et al.* (2005) *Sci. Total Environ.*, **346** (1–3), 244–255.
97. Barkay, Z. *et al.* (2005) *Microsc. Res. Tech.*, **68** (2), 107–114.
98. Carrera, M. *et al.* (2005) *Aerosol Sci. Technol.*, **39** (10), 960–965.
99. Leck, C. and Bigg, E.K. (2008) *Tellus*, **60B** (1), 118–126.

100. Pósfai, M. *et al.* (2003) *Atmos. Res.*, **66** (4), 231–240.

101. Laskin, A. *et al.* (2006) *J. Electron. Spectrosc.*, **150** (2–3), 260–274.

102. Bluhm, H. and Siegmann, H.C. (2009) *Surf. Sci.*, **603** (10–12), 1969–1978.

103. Kildesø, J. and Nielsen, B.H. (1997) *Ann. Occup. Hyg.*, **41** (2), 201–216.

104. Dyukhina, E.I. and Belenko, O.A. (2004) *Atmos. Ocean. Opt.*, **17** (5–6), 462–465.

105. Beus, A.A. *et al.* (1976) *Geochemistry of the Environment*, Nedra, Moscow (in Russian).

106. Petrenchuk, O.P. (1979) *Experimental Investigations of Atmospheric Aerosol*, Hydrometeoizdat, Leningrad (in Russian).

107. Snyder, A.P. *et al.* (2001) *Field Anal. Chem. Technol.*, **5** (4), 190–204.

108. Pashynska, V. *et al.* (2002) *J. Mass Spectrom.*, **37** (2), 1249–1257.

109. Williams, B.J. *et al.* (2006) *Aerosol Sci. Technol.*, **40** (8), 627–638.

110. Kreisberg, N.M. *et al.* (2009) *Aerosol Sci. Technol.*, **43** (1), 38–52.

111. Thrane, U. *et al.* (2001) *IFEMS Microbiol. Lett.*, **203** (2), 249–255.

112. Womiloju, T.O. *et al.* (2003) *Atmos. Environ.*, **37** (31), 4335–4344.

113. Caseiro, A. *et al.* (2007) *J. Chromatogr. A*, **1171** (1–2), 37–45.

114. Hermandez, M. (1999) *Aerosol Sci. Technol.*, **30** (2), 145–160.

115. Cheng, Y.S. (1999) *Aerosol Sci. Technol.*, **30** (2), 186–201.

116. Hill, S.C. *et al.* (1999) *Field Anal. Chem. Technol.*, **3** (4–5), 221–239.

117. Stopa, P.J. *et al.* (1999) *Field Anal. Chem. Technol.*, **3** (4–5), 283–290.

118. Brosseau, L.M. *et al.* (2000) *Aerosol Sci. Technol.*, **32** (6), 545–558.

119. Eversole, J.D. *et al.* (2001) *Field Anal. Chem. Technol.*, **5** (4), 205–212.

120. Ho, J. (2002) *Anal. Chem. Acta*, **457** (1), 125–148.

121. Yacoub-George, E. *et al.* (2002) *Anal. Chem. Acta*, **457** (1), 3–12.

122. Squirrell, D.J. *et al.* (2002) *Anal. Chem. Acta*, **457** (1), 109–114.

123. Pan, Y.L. *et al.* (2003) *Aerosol Sci. Technol.*, **37** (8), 628–639.

124. Kunnil, J. *et al.* (2005) *Aerosol Sci. Technol.*, **39** (9), 842–848.

125. Green, B.J. *et al.* (2006) *Anal. Biochem.*, **354** (2), 151–153.

126. Previte, M.J. *et al.* (2007) *Anal. Chem.*, **79** (18), 7042–7052.

127. Aslan, K. *et al.* (2008) *Anal. Chem.*, **80** (11), 4125–4132.

128. Clarke, R.G. and Pinder, A.C. (1998) *J. Appl. Microbiol.*, **84** (4), 577–584.

129. Lange, J.L. *et al.* (1997) *Appl. Environ. Microbiol.*, **63** (4), 1557–1563.

130. Day, J.P. *et al.* (2002) *Appl. Environ. Microbiol.*, **68** (1), 37–45.

131. Prigione, V. *et al.* (2004) *Appl. Environ. Microbiol.*, **70** (3), 1360–1365.

132. Vanhee, L.M.E. *et al.* (2008) *J. Microbiol. Methods*, **72** (1), 12–19.

133. Sengupta, A. *et al.* (2007) *J. Colloid Interface Sci.*, **309** (1), 36–43.

134. Grow, A.E. *et al.* (2003) *J. Microbiol. Methods*, **53** (2), 221–233.

135. Orsini, F. *et al.* (2000) *J. Microbiol. Methods*, **42** (1), 17–27.

136. Gottardini, E. *et al.* (2007) *Aerobiologia*, **23** (3), 211–219.

137. Rouxhet, P.G. *et al.* (1994) *Colloid Surf. B: Biointerfaces*, **2** (1–3), 347–369.

138. Zhu, Y.-J. *et al.* (2001) *Environ. Sci. Technol.*, **35** (15), 3113–3121.

139. Pastuszka, J.S. *et al.* (2005) *Aerobiologia*, **21** (3–4), 181–192.

140. Srogi, K. (2008) *Anal. Lett.*, **41** (5), 677–724.

141. Osán, J. *et al.* (2000) *Microchim. Acta*, **132** (2–4), 349–355.

142. Pushkin, S.G. and Mikhailov, V.L. (1989) Comparator neutron activation analysis, in *Atmospheric Aerosol Research*, Nauka, Siberian Branch, Novosibirsk (in Russian).

143. Hancock, J.R. and D'Agostino, P.A. (2002) *Anal. Chem. Acta*, **457** (1), 71–82.

144. Simoneit, B.R.T. (2005) *Mass Spectrosc. Rev.*, **24** (5), 719–765.

145. De, B.K. *et al.* (2002) *Emerging Infect. Dis.*, **8** (10), 1060–1065.

146. Dunbar, S.A. *et al.* (2003) *J. Microbiol. Methods*, **53** (2), 245–252.

147. Petrenko, V.A. and Vodyanoy, V.J. (2003) *Microbiol. Methods*, **53** (2), 253–262.

148. Straub, T.M. *et al.* (2005) *J. Microbiol. Methods*, **62** (3), 303–316.

149. Dillon, H.K. *et al.* (2007) *J. Occup. Environ. Hyg.*, **4** (7), 509–513.

150. Wilson, K.H. *et al.* (2002) *Appl. Environ. Microbiol.*, **68** (5), 2535–2541.

151. Belgrader, P. *et al.* (2003) *Anal. Chem.*, **75** (14), 3446–3450.

152. Greene, E.A. and Voordouw, G.A.J. (2003) *Microbiol. Methods*, **53** (2), 165–174.

153. An, H.R. *et al.* (2006) *Atmos. Environ.*, **40** (40), 7924–7939.

154. Peccia, J. and Hernandez, M. (2006) *Atmos. Environ.*, **40** (21), 3941–3961.

155. Brodie, E.L. *et al.* (2007) *Proc. Natl. Acad. Sci.*, **104** (1), 299–304.

156. Chen, P.-S. *et al.* (2009) *Aerosol Sci. Technol.*, **43** (4), 290–297.

157. Burge, H.A. (2002) *J. Allergy Clin. Immunol.*, **110** (4), 544–552.

158. Georgakopoulos, D.G. *et al.* (2008) *Biogeosci. Disc.*, **5** (2), 1469–2008.

159. You, W.W. *et al.* (1997) *Ann. Biochem.*, **244** (2), 277–282.

160. State Committee of USSR for Hydrometeorology (1991) The Detection of Polycyclic Aromatic Hydrocarbons (The Method of High Performance Liquid Chromatography). Guidance for Atmospheric Pollution Control, RD 52.04.186-89, Moscow: Ministry of Health of USSR, pp. 647–657 (in Russian).

161. Kutsenogii, K.P. (ed.) (2006) *Aerosols of Siberia*, Publishing House of the Siberian Branch of the Russian Academy of Sciences, Novosibirsk (in Russian).

162. Yudelevich, I.G. *et al.* (1980) *Chemical–Spectral Analysis of High-Purity Substances*, Nauka, Novosibirsk (in Russian).

163. Miquel, P. (1883) *Les Organismes Vivants de L'Atmosphere*, Gauthier-Villars, Paris.

164. Fuzzi, S. *et al.* (1998) *Atmos. Environ.*, **31** (2), 287–290.

165. Möhler, O. *et al.* (2007) *Biogeosci.*, **4** (6), 1059–1071.

166. Langmann, B. *et al.* (2009) *Atmos. Environ.*, **43** (1), 107–116.

167. Nicas, M. *et al.* (2005) *J. Occup. Environ. Hyg.*, **2** (3), 143–154.

168. O'Gorman, C.M. and Fuller, H.T. (2008) *Atmos. Environ.*, **42** (18), 4355–4368.

169. Douwes, J. *et al.* (2003) *Ann. Occup. Hyg.*, **47** (3), 187–200.

170. Gloster, J. and Alexandersen, S. (2004) *Atmos. Environ.*, **38** (3), 503–505.

171. Alexandersen, S. *et al.* (2003) *J. Compar. Pathol.*, **129** (1), 1–36.

172. Hammond, G.W. *et al.* (1989) *Rev. Infect. Dis.*, **11** (3), 494–497.

173. Pillai, S.D. *et al.* (1996) *Appl. Environ. Microbiol.*, **62** (1), 296–299.

174. Lin, W.-H. and Li, C.-S. (1996) *Aerosol Sci. Technol.*, **25** (2), 93–100.

175. Tong, Y. and Lighthart, B. (1999) *Aerosol Sci. Technol.*, **30** (2), 246–254.

176. Lin, W.-H. and Li, C.-S. (2000) *Aerosol Sci. Technol.*, **32** (4), 359–368.

177. Tong, Y. and Lighthart, B. (2000) *Aerosol Sci. Technol.*, **32** (5), 393–403.

178. Fischer, G. *et al.* (2008) *Int. J. Hyg. Environ. Health*, **211** (1–2), 132–142.

179. Griffin, W.D. (2008) *Aerobiologia*, **24** (1), 19–25.

180. Kummer, V. and Thiel, W.R. (2008) *Int. J. Hyg. Environ. Health*, **211** (3–4), 299–307.

181. Fierer, N. *et al.* (2008) *Appl. Environ. Microbiol.*, **74** (1), 200–207.

182. Mims, S.A. and Mims, F.M. (2004) *Atmos. Environ.*, **38** (5), 651–655.

183. Kellog, C.A. and Griffin, D.W. (2006) *Trends Ecol. Evol.*, 21 (11), 638–644.

184. Keer, J.T. and Birch, L. (2003) *J. Microbiol. Methods*, **53** (2), 175–183.

185. Pyrri, I. and Kapsanski-Gotsi, E. (2008) *Aerobiologia*, **23** (1), 3–15.

186. Chi, M.-C. and Li, C.S. (2006) *Aerosol Sci. Technol.*, **40** (12), 1071–1079.

187. Kogure, K. *et al.* (1979) *Can. J. Microbiol.*, **25** (3), 415–420.

188. Rodriguez, G.G. *et al.* (1992) *Appl. Environ. Microbiol.*, **58** (6), 1801–1808.

189. Gerhard, P. *et al.* (eds) (1981) *Manual of Methods for General Bacteriology*, vol. 1, American Society for Microbiology, Washington, DC, p. 3.

190. Ngundi, M.M. *et al.* (2006) *Expert Rev. Proteom.*, **3** (5), 511–524.

191. Stärk, K.D.C. *et al.* (1998) *Appl. Environ. Microbiol.*, **64** (2), 543–548.

192. Makino, S.-I. and Cheun, H.-I. (2002) *J. Microbiol. Methods*, **53** (2), 141–147.

193. Wu, Z. *et al.* (2002) *J. Environ. Monit.*, **4** (3), 377–382.
194. McCartney, H.A. *et al.* (2003) *Pest Manage. Sci.*, **59** (2), 129–142.
195. Maron, P.-A. *et al.* (2005) *Atmos. Environ.*, **39** (20), 3686–36951.
196. Paez-Rubio, T. *et al.* (2005) *Appl. Environ. Microbiol.*, **71** (2), 804–810.
197. Després, V.R. *et al.* (2007) *Biogeosciences*, **4** (6), 1127–1141.
198. Chang, C.-W. *et al.* (1995) *Am. Ind. Hyg. Assoc. J.*, **56** (10), 979–986.
199. Griffiths, W.D. *et al.* (2001) *Aerobiologia*, **17** (2), 109–119.
200. Miller, G. (1976) *Experiments in Molecular Genetics*, Mir, Moscow (in Russian).
201. Saggie, E. (1983) *The Methods of Soil Microbiology*, Mir, Moscow (in Russian).
202. Melnik, V.A. (2000) *Classifier of Fungi of Russia*, Nauka, St Petersburg (in Russian).
203. Hibbett, D.S. (2007) *Mycol. Res.*, **111** (5), 509–547.
204. Starr, M.P. *et al.* (eds) (1981) *The Prokaryotes. A Handbook on Habitats, Isolation, and Identification of Bacteria*, Springer, Berlin.
205. Lebedeva, M.N. (1973) *A Guide for Practical Studies in Medical Microbiology*, Medicine, Moscow (in Russian).
206. Weisburg, W.G. *et al.* (1991) *J. Bacteriol.*, **173** (2), 697–703.
207. Ashmarin, I.P. and Vorobyov, A.A. (eds) (1962) *Statistical Methods in Microbiological Studies*, Medgiz, Leningrad (in Russian).
208. Puchkova, L.I. *et al.* (2002) *Appl. Biochem. Microbiol. (Russia)*, **38** (1), 15–19.
209. Laassri, M. *et al.* (2003) *J. Virol. Methods*, **112** (1–2), 67–78.
210. Lapa, S. *et al.* (2002) *J. Clin. Microbiol.*, **40** (3), 753–757.
211. Pyankov, O.V. *et al.* (2007) *Environ. Microbiol.*, **9** (4), 992–1000.
212. Millington, W.M. and Corden, J.M. (2005) *Aerobiologia*, **21** (2), 105–113.
213. Corden, J.M. *et al.* (2003) *Aerobiologia*, **19** (3–4), 191–199.
214. Spieksma, F.T.M. *et al.* (2003) *Aerobiologia*, **19** (3–4), 171–184.
215. Bianchi, M.M. and Olabuenga, S.E. (2006) *Aerobiologia*, **22** (4), 247–257.
216. Katial, R.K. *et al.* (1997) *Int. J. Biometeorol.*, **41** (1), 17–22.
217. Adhikari, A. *et al.* (2004) *Sci. Total Environ.*, **326** (1–3), 123–141.
218. Kushner, D.J. (ed.) (1978) *Microbial Life in Extreme Environments*, Academic Press, New York.
219. Gregory, P.H. (1961) *Microbiology of the Atmosphere*, Leonard Hill, London.
220. Proctor, B.E. (1935) *J. Bacteriol.*, **30** (4), 363–375.
221. Kelly, C.D. and Pady, S.M. (1954) *Can. J. Bot.*, **32** (5), 591–600.
222. Fulton, J.D. (1966) *Appl. Microbiol.*, **14** (3), 237–240.
223. Imshenetsky, A.A. *et al.* (1978) *Appl. Environ. Microbiol.*, **35** (1), 1–5.
224. Griffin, D.W. (2004) *Aerobiologia*, **20** (2), 135–140.
225. Wainwright, M. *et al.* (2003) *FEMS Microbiol. Lett.*, **218** (2), 161–165.
226. Lighthart, B. (1997) *FEMS Microbiol. Ecol.*, **23** (4), 263–274.
227. Hryhorczuk, D. *et al.* (2001) *Ann. Agric. Environ. Med.*, **8** (2), 177–185.
228. Harrison, R.M. *et al.* (2005) *Int. J. Biometeorol.*, **49** (3), 167–178.
229. Kellogg, C.A. and Griffin, D.W. (2006) *Trends Ecol. Evol.*, 21 (11), 638–644.
230. Di Giorgio, C. *et al.* (1996) *Atmos. Environ.*, **30** (1), 155–160.
231. Lighthart, B. (2000) *Aerobiologia*, **16** (1), 7–16.
232. Fang, Z. *et al.* (2005) *Sci. Total Environ.*, **350** (1–3), 47–58.
233. Hasnain, S.N. *et al.* (2005) *Aerobiology*, **21** (2), 139–145.
234. Angelosante Bruno, A. *et al.* (2007) *Aerobiology*, **23** (3), 221–228.
235. Oliveira, M. *et al.* (2009) *Int. J. Biometeorol.*, **53** (1), 61–73.
236. Panchenko, M.V. and Pol'kin, V.V. (2001) *Atmos. Ocean. Opt.*, **14** (6–7), 478–488.
237. Slavianski, V.M., Pimenov, E.V. *et al.* (2009) *Introduction to Aerobiology*, Vyatka State University Publications, Kirov, p. 222 (in Russian).
238. Gruber, S. *et al.* (1998) *J. Aerosol. Sci.*, **29** (Suppl. 1), S771–S772.
239. Lighthart, B. and Shaffer, B.T. (1995) *Aerobiologia*, **11** (1), 19–25.
240. Safatov, A.S. *et al.* (2003) *Atmos. Ocean. Opt.*, **16** (5–6), 491–495.

241. Artaxo, P. *et al.* (1990) *J. Geophys. Res.*, **95D** (10), 16971–16985.

242. Jones, A.M. and Harrison, R.M. (2004) *Sci. Total Environ.*, **326** (1–3), 151–180.

243. Paxbaum, H. and Tenze-Kunit, M. (2003) *Atmos. Environ.*, **37** (28), 3693–3699.

244. Hysek, J. *et al.* (1991) *Grana*, **29** (5), 450–453.

245. Wilson, M. and Lindow, S.E. (1992) *Appl. Environ. Microbiol.*, **59** (12), 3908–3913.

246. Andreeva, I.S. *et al.* (2002) *Chem. Sustainable Dev.*, **10** (5), 523–537.

247. Che, F. *et al.* (1992) *Aerobiology*, **8** (2), 297–300.

248. Maenhaut, W. *et al.* (2008) Final Report (Phase I). Formation Mechanisms, Marker Compounds and Source Appointment for BIOgenic Atmospheric Aerosols "BIOSOL", section 7.5, pp. 62–64. (*http://www.belspo.be/belspo/ssd/science/FinalReports/Reports/SDAT02A_en.pdf*). (Accessed 29 March 2010).

249. Lighthart, B. and Shaffer, B.T. (1994) *Atmos. Environ.*, **28** (7), 1267–1274.

250. Krejci, R. *et al.* (2005) *Atmos. Chem. Phys.*, **5** (6), 1527–1543.

251. Borodulin, A.I. *et al.* (2005) *J. Aerosol Sci.*, **36** (5–6), 785–800.

252. Borodulin, A.I. *et al.* (1992) *Statistical Description of the Process of Aerosol Turbulent Diffusion in Atmosphere. Method and Application*, Novosibirsk University, Novosibirsk (in Russian).

253. Anderson, T.W. (1958) *An Introduction to Multivariate Statistical Analysis*, John Wiley and Sons, Inc., New York.

254. Borodulin, A.I. *et al.* (2003) *Dokl. Biol. Sci.*, **392** (1–6), 422–424.

255. Borodulin, A.I. *et al.* (2004) *Dokl. Earth Sci.*, **399** (8), 1125–1127.

256. Holton, J.R. (1972) *An Introduction to Dynamic Meteorology*, Academic Press, New York.

257. Chen, P.S. *et al.* (2008) European Aerosol Conference 2008, Thessaloniki, Abstract T02A098P.

258. Liu, J. *et al.* (2005) *Science*, **309** (5738), 1206.

259. Beresnev, S.A. and Gryazin, V.I. (2007) *Atmos. Ocean. Opt.*, **20** (6), 492–498.

260. Chernyak, V. and Beresnev, S. (1993) *J. Aerosol Sci.*, **24** (7), 857–866.

261. Belan, B.D. *et al.* (2000) *Dokl. Biochem.*, **374** (1–6), 196–199.

262. Andreeva, I.S. *et al.* (2001) *Dokl. Biol. Sci.*, **381** (1–6), 530–534.

263. Viana, M. *et al.* (2008) *J. Aerosol Sci.*, **39** (10), 827–849.

264. Kim, E. and Hopke, P.K. (2007) *J. Air Waste Manage. Assoc.*, **57** (7), 811–819.

265. Hopke, P.K. (2008) *J. Toxicol. Environ. Health*, **71A** (9–10), 555–563.

266. Fraile, R. *et al.* (2006) *Aerobiologia*, **22** (1), 34–45.

267. Desyatkov, B.M. *et al.* (2001) *Atmos. Ocean. Opt.*, **14** (6–7), 557–560.

268. Borodulin, A.I. *et al.* (2003) *Atmos. Ocean. Opt.*, **16** (8), 705–708.

269. Desyatkov, B.M. *et al.* (2003) *Atmos. Ocean. Opt.*, **16** (8), 702–704.

270. Penenko, V.V. and Tsvetova, E.A. (1999) *Atmos. Ocean. Opt.*, **12** (6), 462–468.

271. Penenko, V. *et al.* (2002) *Future Gener. Comput. Syst.*, **18** (5), 661–671.

272. Penenko, V. and Tsvetova, E. (2009) *J. Comput. Appl. Math.*, **226** (2), 319–330.

273. Awad, A.H.A. (2005) *Aerobiologia*, **21** (1), 53–61.

274. Abdel Hameed, A.A. and Khodr, M.I. (2001) *J. Environ. Monit.*, **3** (2), 206–209.

275. Green, C.F. *et al.* (2006) *J. Occup. Environ. Hyg.*, **3** (1), 9–15.

276. Gelencsér, A. *et al.* (2007) *J. Geophys. Res.*, **112**, D23S04.

277. Lindow, S.E. and Brandl, M.T. (2003) *Appl. Environ. Microbiol.*, **69** (4), 1875–1883.

278. Sergeev, A.N. *et al.* (2009) *Atmos. Ocean. Opt.*, **22** (4), 467–477.

279. Raputa, V.F. *et al.* (1997) *Meteorol. Hydrol.* (2), 33–41.

280. Andreeva, I.S. *et al.* (2001) *Atmos. Ocean. Opt.*, **14** (6–7), 497–500.

281. Andreeva, I.S. *et al.* (2002) *Atmos. Ocean. Opt.*, **15** (5–6), 425–428.

282. Fuchs, N.A. (1964) *The Mechanics of Aerosols*, Pergamon, New York.

283. Koroleva, G.P. *et al.* (1998) *Chem. Sustainable Dev.*, **6** (3), 327–337 (in Russian).

284. Kul'ko, A.B. and Marfenina, O.E. (1998) *Microbiology (Russia)*, **67** (4), 470–472.
285. Gruber, S. and Jaenicke, R. (2000) *J. Aerosol Sci.*, **31** (Suppl. 1), S737–S738.
286. Amato, P. *et al.* (2007) *FEMS Microbiol. Ecol.*, **59** (2), 255–264.
287. Buryak, G.A. *et al.* (2007) *Atmos. Ocean. Opt.*, **20** (10), 842–846.
288. Shinkarenko, M.P. and Smolyakov, B.S. (2004) *Chem. Sustainable Dev.*, **12** (5), 631–640.
289. Oren, A. (2004) *Phil. Trans. R. Soc. Lond., B*, **359** (1444), 623–638.
290. Peto, S. (1953) *Biometrics*, **9** (9), 320–335.

Index

Aerosols – Science and Technology. Edited by Igor Agranovski

ISBN: 978-3-527-32660-0